T0338550

Advance Praise

"This book provides a comprehensive overview of SRv6, including its background, fundamentals, benefits, and applications. I firmly believe that the publication of this book will further promote the large-scale deployment and development of IPv6."

— Hequan Wu, Academician of Chinese Academy of Engineering

"The book *SRv6 Network Programming: Ushering in a New Era of IP Networks*, written by Huawei's Zhenbin Li and his team, provides a complete description of SRv6 innovations and standards, and systematically summarizes their experience in SRv6 R&D. It gives a holistic view into SRv6. I sincerely hope that the publication of this book can positively promote the development of IPv6 core technologies in China and make a strong impact on the world."

— Xing Li, Professor of Department of Electronic Engineering, Tsinghua University

"This book takes a much-needed step-by-step approach to explaining the SR technologies that apply to IPv6 networks. In doing so, it clarifies and builds on the standardization work of the IETF's SPRING working group that has successfully brought together all of the large networking equipment vendors with a number of future-looking network operators to 'make the Internet work better.' The future of the Internet is notoriously hard to predict, but this book will assist readers to embrace its potential."

— Adrian Farrel, Former IETF Routing Area Director

"In the past 20+ years, MPLS played an important role in the development of IP networks. As the development of 5G and cloud progresses, SRv6 is winning much attention in the IP industry as the promising technology. In the process, IP experts from Huawei make great contributions to the innovation and standardization work in IETF. The book written by the team details the principles of SRv6 and corresponding applications for 5G and cloud according to years of extensive experience. I believe that it can be of much help to readers to master SRv6 and will facilitate the development of SRv6 technology and industry."

— Loa Andersson, MPLS WG Chair of IETF

"In this, Zhenbin Li and his team walk the reader through the history leading to SRv6 and introduce the basic concepts as well as more advanced ideas for the technology. Contributions for this book are written by technologists who have been actively involved in the development and standardization of SRv6 in the IETF from the beginning and have personally helped to evolve not only the base technology but also many more advanced features that are critical for successful service creation to meet the demands placed upon today's networks. The reader will gain much insight into how SRv6 has and continues to evolve, and in addition will read about many new concepts that are actively being worked on in the standards bodies."

— James N. Guichard, SPRING WG Chair of IETF

"This book is a comprehensive introduction and tutorial to SRv6. It comprehensively covers all aspects from background to SRv6 principles and basics, and to services built upon SRv6, such as traffic engineering, reactions to network failures, VPN, and multicast. It concludes with a preview of future evolution directions such as 5G and SRv6 header compression. In addition to the technology, it includes personal reflections on the background and reasons for the technical choices."

— Bruno Decraene, Orange expert and senior network architect at Orange/SPRING WG Co-chair of IETF

"SRv6 is maturing, and the family of standards defining all aspects of the solution keeps growing — greatly supported by the authors' enduring efforts in the IETF. Our lab and other organizations have successfully completed multi-vendor interoperability tests in the past three years, with a growing number of participants confirming that the technology becomes adopted. The authors systematically and comprehensively describe the evolution and application of SRv6 as well as the benefits that it brings to the industry. I truly believe that this book provides excellent value in guiding SRv6 application and deployment."

— Carsten Rossenhovel, Managing Director of EANTC (European Advanced Networking Test Center)

SRv6 Network Programming

Data Communication Series

Cloud Data Center Network Architectures and Technologies
Lei Zhang and Le Chen

Campus Network Architectures and Technologies
Ningguo Shen, Bin Yu, Mingxiang Huang, and Hailin Xu

Enterprise Wireless Local Area Network Architectures and Technologies
Rihai Wu, Xun Yang, Xia Zhou, and Yibo Wang

Software-Defined Wide Area Network Architectures and Technologies
Cheng Sheng, Jie Bai, and Qi Sun

SRv6 Network Programming
Ushering in a New Era of IP Networks
Zhenbin Li, Zhibo Hu, and Cheng Li

For more information on this series, please visit: https://www.routledge.com/Data-Communication-Series/book-series/DCSHW

SRv6 Network Programming

Ushering in a New Era of IP Networks

Zhenbin Li, Zhibo Hu, and Cheng Li

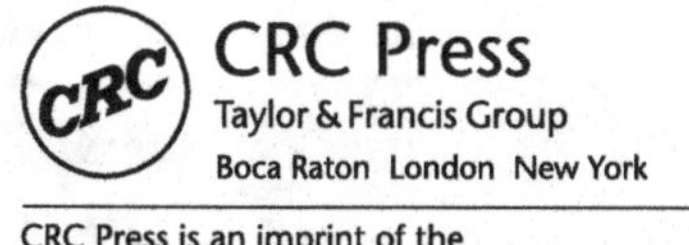

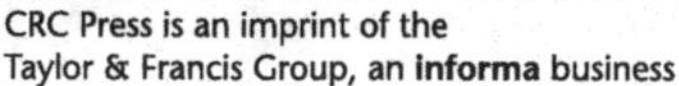

First edition published 2021
by CRC Press
6000 Broken Sound Parkway NW, Suite 300, Boca Raton, FL 33487-2742

and by CRC Press
2 Park Square, Milton Park, Abingdon, Oxon, OX14 4RN

Translated by Yanqing Zhao, Wei Li

CRC Press is an imprint of Taylor & Francis Group, LLC

Library of Congress Cataloging-in-Publication Data
Names: Li, Zhenbin (Telecommunications engineer), author. | Hu, Zhibo, author. |
Li, Cheng (Telecommunications engineer), author.
Title: SRv6 network programming: ushering in a new era of IP networks / Zhenbin Li, Zhibo Hu, Cheng Li.
Description: First edition. | Boca Raton: CRC Press, 2021. | Series: Data communication series | Includes bibliographical references. | Summary: "SRv6 Network Programming, beginning with the challenges for Internet Protocol version 6 (IPv6) network development, describes the background, design roadmap, and implementation of Segment Routing over IPv6 (SRv6), as well as the application of this technology in traditional and emerging services. With rich, clear, practical, and easy-to-understand content, the volume is intended for network planning engineers, technical support engineers and network administrators who need a grasp of the most cutting-edge IP network technology. It is also intended for communications network researchers in scientific research institutions and universities"— Provided by publisher.
Identifiers: LCCN 2020056776 (print) | LCCN 2020056777 (ebook) | ISBN 9781032016245 (hbk) |
ISBN 9781003179399 (ebk)
Subjects: LCSH: Self-routing (Computer network management)—Technological innovations. |
TCP/IP (Computer network protocol)
Classification: LCC TK5105.54873.S45 L53 2021 (print) | LCC TK5105.54873.S45 (ebook) |
DDC 004.6/2—dc23
LC record available at https://lccn.loc.gov/2020056776
LC ebook record available at https://lccn.loc.gov/2020056777

ISBN: 978-1-032-01624-5 (hbk)
ISBN: 978-1-032-01635-1 (pbk)
ISBN: 978-1-003-17939-9 (ebk)

Typeset in Minion
by codeMantra

Contents

Foreword I

DEFINING AND INTRODUCING A novel technology not only requires the ability to address and solve use cases or critical problems, but must also come at the right moment. Innovation always requires the right timing: you need to come with the right answer at the right moment. Timing is critical for the success of any innovation. There are so many examples of brilliant and innovative technologies that came too early or too late and that disappeared without leaving any trace.

It's now been more than four decades since the Internet was invented. The IP underlying technology is a powerful, flexible, and scalable connectionless protocol on top of which we have multiple transport protocols leveraged by application protocols that allowed the web to become what we have today. IP demonstrated a very strong scalability, flexibility, and efficiency. A simple datagram service based on a connectionless model that leverages a shortest-path tree routing paradigm is the foundation of all communications today.

IPv6 was invented more than 25 years ago. It was the time when we thought that the only major problem to solve was the exhaustion of IPv4 addresses, and extending the address space to 128 bits would have solved the problem for a long time. In theory, everything worked well, but in practice, the industry addressed and did overcome the problem with a workaround: NAT and later Carrier-Grade NAT. So, for many years, IPv6 remained on the side.

A few years later (around the end of the 20th century), MPLS was invented and introduced an interesting and powerful mechanism: separation of control and data planes. Thanks to MPLS, we've been able to override routing decisions based on policies and labels. This was the time we implemented Traffic Engineering, Fast Reroute, and different flavors of VPNs.

The separation of control and data planes opened a brand-new era. From there on, the ability to implement specific policies for different services and data flows became more and more a necessity for most network operators. The era of SDN started. However, the way policies were implemented using MPLS required state in the network and we arrived at a scalability boundary.

Segment Routing (RFC8402) finally brought a powerful way to introduce source routing into modern networks running either MPLS or IPv6 data plane. Why suddenly has source routing become so popular? Because of timing! Now is the time when you want to control data flows without incurring heavy scale state machinery (that would not work anyway). Source routing brings the state in the packet and leaves the control plane of the network very light and very easy to manage.

Now the times are mature for the introduction of source routing but revisited with a modern technology that copes with both MPLS and IPv6 data planes. In addition, Segment Routing applied to IPv6 data plane (SRv6) allows the leveraging of the two main strong points of IPv6 architecture. First, SRv6 leverages the IPv6 128 bit address space that allows the definition of a node with enough bits so that an augmented semantic can be defined (e.g., a node identity and a node location). Second, IPv6 defines the concept of Extension Headers (EH) which can be added to the packet for different purposes. One of these purposes is source routing, and especially with the Segment Routing Header (SRH, RFC8754), you can now specify the path the packet should follow.

The innovation SRv6 brought to the industry has multiple aspects: it allows the implementation of a straightforward, highly scalable, backward compatible mechanism for source routing. This allows the implementation of large-scale policies addressing the variety of use cases network operators have to address these days (5G, IoT, Connected Objects, Virtualization, etc.). Another advantage is the ability to express not only a path (that the packet should take) but also the process a packet should go through during its journey. A typical example is the Service Chain use case where a packet has to traverse service application elements before reaching its destination.

SRv6 introduces the ability to combine forwarding decisions with application/processing decisions, and this is called SRv6 Network Programming. Finally, and this is the timing factor, we are able to leverage a mechanism that was invented many years ago (source routing, IPv6) and combine it with modern capabilities, allowing to address current use cases and requirements and taking into account the scale and performance dimensions.

SR and SRv6 began in 2012 and 2013, and at that time, we were a very small team. Rapidly, the industry understood the potential, and large collaboration between vendors and operators took place. The standardization of SR and SRv6 took place in IETF, and now we have a large community participating in the SR and SRv6 effort.

Zhenbin Li and the Huawei team he leads are part of this community, and their contribution to the evolution and standardization of SR technologies became substantial over the last years. Huawei SRv6 team has been instrumental in the way SRv6 has been defined and standardized in IETF, adopted by the industry and still extended to cope with increasing and ever-emerging new requirements. Also, at the time of this writing, Zhenbin Li is also a member of IAB.

This book gives an exhaustive view of the SRv6 technology including the network programming capability, and I'm sure it will help the reader to understand SRv6 technology, the use cases it applies to, and eventually, to appreciate SRv6 technology as much as I have appreciated participating in its invention.

Stefano Previdi
SR-MPLS and SRv6 Pioneer and Original Contributor

Foreword II

THE INTERNET PROTOCOL VERSION 6 (IPv6) was designed by the IPv6 Task Force within the IETF under the co-chairmanship of Steve Deering and Robert Hinden and a small group of 40 engineers. Steve is also the designer of Multicast. The last draft standard of IPv6 (RFC 2460) was released in December 1998, winning against IPv7, IPv8, and IPv9 proposals. This release was basically the last effort of the IPv6 Task Force as it had to close down its working group at the February 2–5, 1999, meeting in Grenoble. In this meeting, I proposed the formation of the IPv6 Forum to promote IPv6 to industry, ISPs/ MNOs, governments, academia, and research ecosystem to start large-scale pilots and initial deployments. There was a nice coincidence with the creation of 3GPP in December 1998 within ETSI in Sophia Antipolis. It was quite clear to the core IPv6 team that 3G is the first innovative driver that needs plenty of IP addresses. I contacted Karl-Heinz Rosenbruck, Director General of ETSI and at that time he was the chairman of 3GPP and proposed to him to adopt IPv6 for 3G instead of WAP. After 6 months of discussions with the various 3GPP WGs, 3GPP announced in May 10, 2000, the adoption of IPv6 for 3G. However, the 3G MNOs did not have the capacity building and the skills to adopt IPv6 but preferred to use IPv4 and especially the Network Address Translation (NAT) for 3G. This was in itself a great achievement to get the wireless world to adopt the Internet for the telecom world as they were at that time not that enthusiastic about the open Internet. The client/ server model became the norm and the wireless Internet especially with 4G boomed in our hands and the rest is history. The Internet has reached 5 billion users with a kind of economy class service.

The public IPv4 address space managed by IANA (http://www.iana.org) was completely depleted back in February 1, 2011. This creates by itself a critical challenge when adding new IoT networks and enabling machine learning services on the Internet. Without publicly routable IP addressing, the Internet of Things, and anything that's part of Machine Learning services on the Internet, would be greatly reduced in its capabilities and then limited in its potential success. Most discussions of IP over everything have been based on the illusionary assumption that the IP address space is an unlimited resource or it's even taken for granted that IP is like oxygen produced for free by nature.

The introduction of IPv6 provides enhanced features that were not tightly designed or scalable in IPv4 like IP mobility, end-to-end (e2e) connectivity, ad hoc services, etc. IPv6 will be addressing the extreme scenarios where IP becomes a commodity service. This new address platform will enable lower cost network deployment of large-scale sensor networks, RFID, IP in the car, to any imaginable scenario where networking adds value to commodity.

IPv6 deployment is now in full swing with some countries, such as Belgium, achieving over 50% penetration. India has taken the lead by having over 350 million IPv6 users. China has over 200 million IPv6 users while the US has over 100 million users using IPv6 without the users even knowing it.

There are many inflections happening this decade to influence the design of the first tangible IoT, 5G, and Smart Cities. It will take a combination of IoT, SDN-NFV, Cloud Computing, Edge Computing, Big IoT Data, and 5G, to sift through to realize the paradigm shift from current research-based work to advanced IoT, 5G, and Smart Cities.

However, the move to NAT has basically killed the end-to-end model that IPv6 restored so that applications and devices can peer directly with each other. The IPv6 deployment has continued its growth and reached today over 1 billion IPv6 users. These users can be qualified as business class beneficiaries. The restoration of the end-to-end model has created new opportunities for designing new end-to-end protocols such as IPv6-based Segment (SRv6) which uses the prime and clean state design of IPv6 of end-to-end services. SRv6 is the first big attempt to usher in the era of IPv6 innovation, and boasts the following two key characteristics:

- SRv6 is an IPv6-native network technology designed to simplify the transport network. In this era where IPv4, IPv6, MPLS, and SR-MPLS coexist, SRv6 will achieve optimal simplification at the transport layer under the minimalism and centralization guiding principles. The real benefits of SRv6 include the support of VPN, FRR, TE, network slicing, and BIERv6, enabling application-driven path programming, providing differentiated SLA assurance for users, and helping carriers transform from "offering bandwidth" to "offering services."

- SRv6 is a rare project that the IETF has focused on over the past decade. It provides a programmable network architecture for 5G and cloud transport, while also heralding the era of IPv6+ network system innovation. Datacom industry players, such as Huawei, Cisco, China Telecom, China Mobile, China Unicom, and SoftBank, have invested a tremendous amount of effort into jointly promoting SRv6 maturity, and the SRv6 transport solution is being applied to an increasing number of networks.

Huawei is the first in the industry to launch SRv6-capable data communication products. In the initial phase of SRv6's history, the company actively participated in innovation and standardization, and achieved a plethora of significant milestones. To add to that, Zhenbin Li was elected as an IETF Internet Architecture Board (IAB) member in early 2019. This book is a summary of the SRv6 research conducted by him and his team, as well as their corresponding experiences. The book systematically interprets the innovative technologies and standards of SRv6, especially with regard to SRv6 implementation practices, which is not only inspiring and helpful but also beneficial to the continuous promotion of IPv6 innovation.

Latif Ladid
Founder and President, IPv6 Forum; Member of 3GPP PCG (MRP), Co-chair, IEEE FNI 5G, World Forum/5G Summits. Chair, ETSI IPv6 Integration ISG.

Preface

SEGMENT ROUTING OVER IPv6 (SRv6) is an emerging IP technology. The development of 5G and cloud services creates many new requirements on network service deployment and automated O&M. SRv6 provides comprehensive network programming capabilities to better meet the requirements of new network services and is compatible with IPv6 to simplify network service deployment.

IP transport networks are built around connections. 5G changes the attributes of connections, and cloud changes their scope. These changes bring big opportunities for SRv6 development. The development of 5G services poses higher requirements on network connections, such as stronger Service Level Agreement (SLA) guarantee and deterministic latency. As such, packets need to carry more information. These requirements can be well met through SRv6 extensions. The development of cloud services makes service processing locations more flexible. Some cloud services (such as telco cloud) further break down the boundary between physical and virtual network devices, integrating services and transport networks. All of this changes the scope of network connections. The unified programming capability of SRv6 in service and transport, as well as the native IP attribute, enables the rapid establishment of connections and meets the requirements for flexible adjustment of the connection scope. As mentioned in this book, SRv6 network programming has ushered in a new network era, redefining the development of IP technologies significantly.

Technical experts from Huawei Data Communication Product Line have been conducting long-term and in-depth research and development in IP and SRv6 fields, and contributing to numerous SRv6 standards in the IETF. In addition, many experts have participated in SRv6 network deployments, accumulating extensive experience in the field. We compiled this book based on comprehensive research, development, and network

O&M experience to provide a complete overview of SRv6 in the hope of better understanding its fundamentals and the new network technologies that derive from it. We also hope it will spark participation in the research, application, and deployment of these new technologies, promoting the development of communications networks.

OVERVIEW

This book begins with the challenges services face regarding IP technology development. It describes the background and mission of SRv6 and presents a comprehensive overview of the technical principles, service applications, planning and design, network deployment, and industry development of next-generation IP networks. This book consists of 13 chapters and is divided into four parts:

Part I: Introduction

Chapter 1 focuses on the development of IP technologies and reveals why SRv6 is developing so quickly.

Part II: SRv6 1.0

Chapters 2 through 8 introduce SRv6 1.0, covering basic SRv6 capabilities and demonstrating how SRv6 supports existing services in a simple and efficient manner.

Part III: SRv6 2.0

Chapters 9 through 12 introduce SRv6 2.0, covering new SRv6-based network technologies targeted at 5G and cloud services and demonstrating service model innovation brought by SRv6 network programming.

Part IV: Summary and Future Developments

Chapter 13 summarizes the development of the SRv6 industry and forecasts developmental trends from SRv6 to IPv6+, which stretches people's imagination about future networks.

CHAPTER 1: SRv6 BACKGROUND

This chapter comprehensively describes the development of IP technologies, introduces SRv6 based on historical experience and requirements for

IP technology development, and summarizes the value and significance of SRv6 from a macro perspective.

CHAPTER 2: SRv6 FUNDAMENTALS

This chapter describes SRv6 fundamentals to demonstrate how SRv6 is used to implement network programming, and summarizes the advantages of SRv6 network programming and its mission from a micro perspective.

CHAPTER 3: BASIC PROTOCOLS FOR SRv6

This chapter describes the fundamentals of and extensions to Intermediate System to Intermediate System (IS-IS) and Open Shortest Path First version 3 (OSPFv3) for SRv6. Protocols such as Resource Reservation Protocol-Traffic Engineering (RSVP-TE) and Label Distribution Protocol (LDP) are not required on an SRv6 network, thereby simplifying the network control plane.

CHAPTER 4: SRv6 TE

This chapter describes the fundamentals and extensions of SRv6 Traffic Engineering (TE), which is a basic feature of SRv6. SRv6 Policy is the main mechanism for implementing SRv6 traffic engineering and draws on the source routing mechanism of Segment Routing to encapsulate an ordered list of instructions on the headend, guiding packets through the network.

CHAPTER 5: SRv6 VPN

This chapter describes the principles and protocol extensions of Virtual Private Network (VPN), a basic SRv6 feature. SRv6 supports existing Layer 2 VPN (L2VPN), Layer 3 VPN (L3VPN), and Ethernet VPN (EVPN) services, which can be deployed as long as edge nodes are upgraded to support SRv6, shortening the VPN service provisioning time.

CHAPTER 6: SRv6 RELIABILITY

This chapter describes the fundamentals and protocol extensions of SRv6 reliability technologies, including FRR, TI-LFA, midpoint protection, egress protection, and microloop avoidance. These technologies ensure E2E local protection switching within 50 ms on an SRv6 network.

CHAPTER 7: SRv6 NETWORK EVOLUTION

This chapter describes the challenges and technical solutions of SRv6 network evolution, that is, how to evolve an existing IP/MPLS network to an SRv6 network. SRv6 supports incremental deployment, which protects existing investments and allows for new services.

CHAPTER 8: SRv6 NETWORK DEPLOYMENT

This chapter describes SRv6 network deployment, including SRv6 application scenarios and how to design and configure features such as TE, VPN, and reliability. Network deployment practices show that SRv6 is advantageous over traditional solutions in terms of inter-AS, scalability, protocol simplification, and incremental deployment.

CHAPTER 9: SRv6 OAM AND ON-PATH NETWORK TELEMETRY

This chapter describes the fundamentals and protocol extensions of SRv6 OAM and on-path network telemetry. SRv6 OAM and on-path network telemetry provide a solid basis for SRv6 network quality guarantee, allowing for extensive use of SRv6 on carrier networks.

CHAPTER 10: SRv6 FOR 5G

This chapter describes new SRv6 technologies applicable to 5G services, including VPN+ network slicing, Deterministic Networking (DetNet), and SRv6-to-mobile core solutions and relevant protocol extensions.

CHAPTER 11: SRv6 FOR CLOUD SERVICES

This chapter describes the application of SRv6 in cloud services, including the concept, challenges, and SRv6-based solution for telco cloud as well as the fundamentals and protocol extensions of SRv6 for SFC and SRv6 for SD-WAN.

CHAPTER 12: SRv6 MULTICAST/BIERv6

This chapter describes SRv6-based multicast technologies, focusing on the fundamentals and protocol extensions of Bit Index Explicit Replication (BIER) and Bit Index Explicit Replication IPv6 Encapsulation (BIERv6). Working with SRv6, BIERv6 allows both unicast and multicast services to be transmitted on a unified IPv6 data plane based on an explicit forwarding path programmed on the ingress.

CHAPTER 13: SRv6 INDUSTRY AND FUTURE

This chapter summarizes the development of the SRv6 industry and provides forecasts for its future, focusing on SRv6 extension header compression, Application-aware IPv6 Networking (APN6), and the three phases that may exist during its development from SRv6 to IPv6+.

As SRv6 involves a significant number of IPv6 basics, this book expands on them in Appendix A to help better understand SRv6. In addition, this book describes IS-IS and OSPFv3 TLVs in Appendices B and C, respectively, as SRv6 is implemented based on these IS-IS and OSPFv3 extensions. This will help to demonstrate how SRv6 is implemented as well as the relationships between IS-IS and SRv6, and between OSPFv3 and SRv6.

In the Postface "SRv6 Path," Zhenbin Li draws on his personal experience to summarize the development history of SRv6 and Huawei's participation in promoting SRv6 innovation and standardization. In addition, "Stories behind SRv6 Design" is provided at the end of each chapter, touching on the experience and philosophy behind the protocol design, interpreting its technical nature, or describing the experience and lessons learned during the development of technologies and standards. Through this content, it is hoped that readers can obtain a greater understanding of SRv6 and the reasoning behind its design principles. Some of the content constitutes the author's opinion and should be used for reference only.

The editorial board for this book gathered the technical experts from Huawei's data communication research team, standards and patents team, protocol development team, solution team, technical documentation team, and translation team. Team members include the developers and promoters of SRv6 standards, R&D staff on SRv6 design and implementation, and solution experts who helped customers successfully complete SRv6 network design and deployment. Their achievements and experience are systematically summed up in this book. Information department staff, including Lanjun Luo, edited the text and worked on the figures. The publication of this book is the collaborative result of team effort and collective wisdom. Heartfelt thanks go to every member of the editorial board. Though the process is hard, we really enjoyed the teamwork in which everyone learned so much from each other.

Teams

TECHNICAL COMMITTEE

Chair

Kewen Hu (Kevin Hu), President of Huawei Data Communication Product Line

Vice Chair

Zhipeng Zhao, Vice President of Huawei Data Communication Product Line

Members

Shaowei Liu, Director of Huawei Data Communication R&D Mgmt Dept

Chenxi Wang, Director of Huawei Data Communication Strategy & Business Development Dept

Su Feng, Director of Huawei Carrier IP Marketing & Solution Sales Dept

Jinzhu Chen, President of Service Router Domain, Huawei Data Communication Product Line

Meng Zuo, President of Core Router Domain, Huawei Data Communication Product Line

Liang Sun, Director of Huawei Data Communication Marketing Execution Dept

Zhiqiang Du, Director of Service Router PDU, Huawei Data Communication Product Line

Jianbo Wang, Director of Core Service Product Dept, Huawei Data Communication Product Line

Xiao Qian, Director of Huawei Data Communication Research Dept

Jianbing Wang, Director of Huawei Data Communication Architecture & Design Dept

Zhaokun Ding, Director of Huawei Data Communication Protocol Development Dept

Minwei Jin, Director of IP Technology Research Dept, NW, Huawei Data Communication Product Line

Dawei Fan, Director of Huawei Data Communication Standard & Patent Dept

Jianping Sun, Director of Huawei Data Communication Solutions Dept

Wenjun Meng, Director of Huawei Information Digitalization and Experience Assurance (IDEA) Dept, DC

EDITORIAL BOARD

Editor-in-Chief: Zhenbin Li (Robin Li)

Deputy Editors-in-Chief: Zhibo Hu, Cheng Li

Members: Lanjun Luo, Ting Liao, Shunwan Zhuang, Haibo Wang, Huizhi Wen, Yaqun Xiao, Guoyi Chen, Tianran Zhou, Jie Dong, Xuesong Geng, Shuping Peng, Lei Li, Jingrong Xie, Jianwei Mao

Technical Reviewers: Rui Gu, Gang Yan

Translators: Yanqing Zhao, Wei Li, Xue Zhou, Yufu Yang, Zhaodi Zhang, Huan Liu, Yadong Deng, Chen Jiang, Ruijuan Li, Junjie Guan, Jun Peng, Samuel Luke Winfield-D'Arcy, George Fahy, Rene Okech, Evan Reeves

TECHNICAL REVIEWERS

Rui Gu: Chief Solution Architect of Huawei Data Communication Product Line. Mr. Gu joined Huawei in 2007 and now leads the Data Communication Solutions Design Dept. He has extensive experience working in the Versatile Routing Platform (VRP) department and has conducted in-depth research on IP/MPLS protocols. He once led the R&D team of Huawei's next-generation backbone routers and also has a wealth of experience in implementing end-to-end data communication products and solutions. He worked in Europe from 2012 to 2017, led the development and innovation of data communication solutions across Europe, and took the lead in the solution design of numerous data communication projects for leading European carriers.

Gang Yan: Chief IGP expert of Huawei Data Communication Product Line. Mr. Yan joined Huawei in 2000 and has been working in the VRP department. From 2000 to 2010, he was responsible for the architecture design of the VRP IGP sub-system and completed the design and delivery of fast convergence, Loop-Free Alternate (LFA) Fast Reroute (FRR)/multi-source FRR, and Non-Stop Routing (NSR). He has extensive experience in protocol design, and since 2011, he has been responsible for improving the O&M capabilities of VRP protocols, building the VRP compatibility management system, and building and delivering the YANG model baseline. He is currently leading the team to innovate IP technologies, including SRv6 slicing, E2E 50 ms protection, and Bit Index Explicit Replication (BIER)-related protocols.

Acknowledgments

While promoting the innovation of SRv6 standards, numerous people have contributed heavily throughout the process from the initial concept to actual products and solutions, including strategic decision-making, technical research, standards promotion, product development, commercial deployment, and industry ecosystem construction. We have received extensive support and help from both inside and outside Huawei. We would like to take this opportunity to express our heartfelt thanks to Kewen Hu, Shaowei Liu, Chenxi Wang, Jinzhu Chen, Meng Zuo, Suning Ye, Xiao Qian, Jianbing Wang, Zhaokun Ding, Minwei Jin, Dawei Fan, Jianping Sun, Mingzhen Xie, Jiandong Zhang, Nierui Yang, Yue Liu, Zhiqiang Fan, Yuan Zhang, Xinbing Tang, Shucheng Liu, Juhua Xu, Xinjun Chen, Lei Bao, Naiwen Wei, Xiaofei Wang, Shuying Liu, Wei Hu, Banghua Chen, Su Feng, Jianwu Hao, Dahe Zhao, Jianming Cao, Yahao Zhang, Minhu Zhang, Yi Zeng, Jiandong Jin, Weiwei Jin, Xing Huang, Xiaohui Li, Ming Yang, Wenjiang Shi, Xiaoxing Xu, Yiguang Cao, Qingjun Li, Gang Zhao, Xiaoqi Gao, Jialing Li, Peng Zheng, Peng Wu, Xiaoliang Wang, Fu Miao, Fuyou Miao, Min Liu, Xia Chen, Yunan Gu, Jie Hou, Yuezhong Song, Ling Xu, Bing Liu, Ning Zong, Qin Wu, Liang Xia, Bo Wu, Zitao Wang, Yuefeng Qiu, Huaru Yang, Xipeng Xiao, Dhruv Dhody, Shucheng Liu, Hongjie Yang, Shuanglong Chen, Ruizhao Hu, Xin Yan, Hong Shen, Wenxia Dong, Guanjun Zhou, Yuanyi Sun, Leyan Wang, Shuhui Wang, Xiaohui Tong, Mingyan Xi, Xiaoling Wang, Yonghua Mao, Lu Huang, Kaichun Wang, Chen Li, Jiangshan Chen, Guoyou Sun, Xiuren Dou, Ruoyu Li, Cheng Yang, Ying Liu, Hongkun Li, Mengling Xu, Taixu Tian, Jun Gong, Yang Xia, Fenghua Zhao, Pingan Yang, Yongkang Zhang, Guangying Zheng, Sheng Fang, Chuang Chen, Ka Zhang, Yu Jiang, Hanlin Li, Ren Tan, Yunxiang Zheng, Yanmiao Wang, Weidong Li, Zhidong Yin, Yuanbin Yin, Zhong Chen, Chun Liu, Xinzong Zeng, Mingliang Yin,

Fengqing Yu, Weiguo Hao, Shuguang Pan, Haotao Pan, Wei Li, and other leaders and colleagues from Huawei. We sincerely thank Hui Tian, Feng Zhao, Yunqing Chen, Huiling Zhao, Chongfeng Xie, Fan Shi, Bo Lei, Qiong Sun, Aijun Wang, Yongqing Zhu, Huanan Chen, Xiaodong Duan, Weiqiang Cheng, Fengwei Qin, Zhenqiang Li, Liang Geng, Peng Liu, Xiongyan Tang, Chang Cao, Ran Pang, Xiaohu Xu, Ying Liu, Zhonghui Li, Suogang Li, Tao Huang, Jiang Liu, Bin Yang, Fengyu Zhang, Dong Liu, Dujuan Gu, and other technical experts in China's IP field who have innovated in technology and promoted the formulation of standards for a long time now. We also sincerely thank Adrian Farrel, Loa Andersson, James Guichard, Bruno Decraene, Carsten Rossenhovel, Clarence Filsfils, Keyur Patel, Pablo Camarillo Garvia, Daniel Voyer, Satoru Matsushima, Alvaro Retana, Martin Vigoureux, Deborah Brungard, Susan Hares, John Scudder, Joel Halpern, Matthew Bocci, Stephane Litkowski, Lou Berger, Christian Hopps, Acee Lindem, Gunter Van de Velde, Daniel King, Nicolai Leymann, Tarek Saad, Sam Aldrin, Stewart Bryant, Andy Malis, Julien Meuric, Mike McBride, Jeff Tantsura, Chris Bowers, Zhaohui Zhang, Jeffrey Haas, Reshad Rahman, Tony Przygienda, Zafar Ali, Darren Dukes, Francois Clad, Kamran Raza, Ketan Talaulikar, Daniel Bernier, Daisuke Koshiro, Gaurav Dawra, Jaehwan Jin, Jongyoon Shin, Ahmed Shawky, Linda Dunbar, Donald Eastlake, Yingzhen Qu, Haoyu Song, Parviz Yegani, Huaimo Chen, Young Lee, Iftekhar Hussain, Himanshu Shah, Wim Henderickx, Ali Sajassi, Patrice Brissette, Jorge Rabadan, and other IP experts in the industry who helped us with work on SRv6. Finally, we would like to express our special thanks to Stefano Previdi and Latif Ladid, both of whom wrote Forewords for this book.

We hope that this book can, as comprehensively as possible, present basic SRv6 technologies and emerging technologies targeted at 5G and cloud services to help readers comprehensively understand the fundamentals of SRv6, its benefits to the industry, and its far-reaching impact on communications networks. As SRv6 is an emerging technology and our capabilities are limited, errors and omissions are inevitable. We will be grateful for any information that helps to rectify these. Please send any comments and feedback by email to lizhenbin@huawei.com.

Authors

Zhenbin Li is the Chief Protocol Expert of Huawei and member of the IETF IAB, responsible for IP protocol research and standards promotion at Huawei. Zhenbin Li joined Huawei in 2000. For more than a decade, he had been responsible for the architecture, design, and development of Huawei's IP operating system—Versatile Routing Platform (VRP)—and MPLS sub-system as an architect and system engineer. From 2015 to 2017, Zhenbin Li worked as an SDN architect and was responsible for the research, architecture design, and development of network controllers. Since 2009, Zhenbin Li has been actively participating in the innovation and standardization work of the IETF, and in the past 6 years, he has been continuously promoting the innovation and standardization of BGP, PCEP, and NETCONF/YANG protocols for SDN transition. His research currently focuses on Segment Routing over IPv6 (SRv6), 5G transport, telemetry, and network intelligence, and he has led and participated in more than 100 IETF RFCs/drafts as well as applied for more than 110 patents. In 2019, he was elected a member of the IETF IAB to undertake Internet architecture management from 2019 to 2021.

Zhibo Hu is a Senior Huawei Expert in SR and IGP, responsible for SR and IGP planning and innovation. Currently, Zhibo Hu is mainly engaged in the research of SR-MPLS/SRv6 protocols and 5G network slicing technologies. Since 2017, Zhibo Hu has been actively participating in the innovation and standardization work of the IETF and leading in standards related to SRv6 reliability, SRv6 YANG, 5G network slicing, and IGP, and is committed to leveraging SRv6 innovation to support network evolution toward 5G and cloudification.

Cheng Li is a Huawei Senior Pre-research Engineer and IP standards representative, responsible for Huawei's SRv6 research and standardization, involving SRv6 extension header compression, SRv6 OAM/path segments, SFC, and security. Cheng Li started to participate in IETF meetings from 2018. Up to now, Cheng Li has submitted more than 30 individual drafts, more than ten of which have been promoted to working group drafts.

I

Introduction

CHAPTER 1

SRv6 Background

IN THIS CHAPTER, WE expand on the history of Internet technology development, from the competition between Asynchronous Transfer Mode (ATM) and IP to the emergence of Multiprotocol Label Switching (MPLS), the dawning of the All IP 1.0 era and the consequent challenges, as well as the rise of Software Defined Networking (SDN). We then sum up by introducing Segment Routing over IPv6 (SRv6), which is a key technology to enable the All IP 2.0 era.

1.1 OVERVIEW OF INTERNET DEVELOPMENT

Humans develop in tandem, as opposed to individually, through communication and collaboration. Over our extensive history, various communications modes have emerged, ranging from beacon fires to emails, and from messenger birds to quantum entanglement. The scope of communications has also expanded from people within a certain vicinity to people in distant localities, across a whole country, around the globe, and even in outer space. That said, we as a people have never stopped our pursuit of communication technologies, which in turn have boosted human prosperity. It is, therefore, safe to say that we are no longer content with just human-to-human communication, or put differently, we now aim at achieving the connectivity of everything anytime and anywhere using the Internet. It is within this context that the development of the Internet has profoundly impacted the path of human society.

After years of development, the Internet has become almost as essential as water and electricity. Although the Internet has made life more

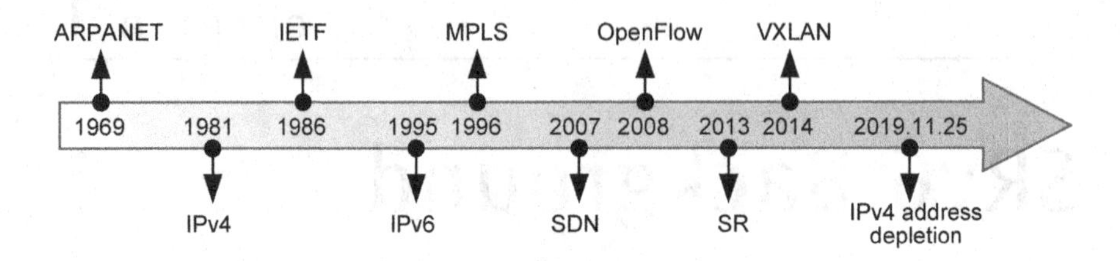

FIGURE 1.1 Internet development milestones.

convenient through the information age, few people know the ins and outs of its technological development history. With this in mind, we briefly summarize the history of Internet technology development in Figure 1.1.

In 1969, Advanced Research Projects Agency Network (ARPANET) — the major predecessor of the Internet — came into existence.

In 1981, IPv4[1] was defined.

In 1986, the Internet Engineering Task Force (IETF), dedicated to formulating Internet standards, was founded.

In 1995, IPv6,[2] the next generation of IPv4, was standardized.

In 1996, MPLS[3] was proposed.

In 2007, SDN[4] debuted.

In 2008, the OpenFlow[5] protocol was introduced.

In 2013, SR[6] was proposed, including Segment Routing over MPLS (SR-MPLS)[7] and SRv6.[8]

In 2014, Virtual eXtensible Local Area Network (VXLAN)[9] was released.

On November 25, 2019, IPv4 addresses were exhausted.[10]

1.2 START OF ALL IP 1.0: A COMPLETE VICTORY FOR IP

1.2.1 Competition between ATM and IP

In the initial stage of network development, multiple types of networks, such as X.25, Frame Relay (FR), ATM, and IP, coexisted to meet different service requirements. These networks could not interwork with each other, and also competed, with mainly ATM and IP networks taking center stage.

ATM is a transmission mode that uses fixed-length cell switching. It establishes paths in connection-oriented mode and can provide better Quality of Service (QoS) capabilities than IP. Its design philosophy involves centering on networks and providing reliable transmission, and its design concepts reflect the reliability and manageability requirements of telecommunications networks. This is the reason why ATM was widely deployed on early telecommunications networks.

The design concepts of IP differ greatly from those of ATM. To be more precise, IP is a connectionless communication mechanism that provides the best-effort forwarding capability, and the packet length is not fixed. On top of that, IP networks mainly rely on the transport-layer protocols (e.g., TCP) to ensure transmission reliability, and the requirement for the network layer involves ease of use. To add on to this, the design concept of IP networks embodies the "terminal-centric and best-effort" notion of the computer network. We can therefore say that IP is widely used on computer networks because it meets the corresponding service requirements.

The competition between ATM and IP networks can essentially be represented as a competition between telecommunications and computer networks. In other words, telecommunications practitioners sought to use ATM for network interconnection to protect network investments. On the flip side, computer practitioners aimed at using ATM as only a link-layer technology to provide QoS guarantee for IP networks, while setting aside the task of establishing network connections for IP.

Computer networks subsequently evolved toward broadband, intelligence, and integration, with mainly burst services. Despite this, the QoS requirements that traffic places on computer networks are not as high as those on telecommunications networks, and the length of packets is not fixed. As such, the advantages of ATM — fixed-length cell switching and good QoS capabilities — cannot be brought into full play on computer networks. Not only that, the QoS capabilities of ATM are based on connection-oriented control with a certain packet header overhead. Therefore, ATM is inefficient in carrying computer network traffic, and it yields high transmission and switching costs.

To sum up, as network scale expanded and network services increased in number, ATM networks became more complex than IP networks, while also bearing higher management costs. Within the context of costs versus benefits, ATM networks exited the arena as they were gradually replaced by IP networks.

1.2.2 MPLS: The Key to All IP 1.0

Although with relation to the development of computer networks, the IP network is more fitting than the ATM network, a certain level of QoS guarantee is still required. To compensate for the IP network's insufficient

QoS capabilities, numerous technologies integrating IP and ATM, such as Local Area Network Emulation (LANE), IP over ATM (IPoA),[11] and tag switching,[12] have been proposed. However, these technologies only addressed part of the issue, until 1996 when MPLS technology was proposed[3] to provide a better solution to this issue.

MPLS is considered as a Layer 2.5 technology that runs between Layer 2 and Layer 3. It supports multiple network-layer protocols, such as IPv4 and IPv6, and is compatible with multiple link-layer technologies, such as ATM and Ethernet. Some of its other highlights include the fact that it incorporates ATM's Virtual Channel Identifier (VCI) and Virtual Path Identifier (VPI) switching concepts, combines the flexibility of IP routing and simplicity of label switching, and adds connection-oriented attributes to connectionless IP networks. By establishing virtual connections, MPLS provides better QoS capabilities for IP networks.

However, this is not the only reason why it was initially proposed. Point in case being that MPLS also forwards data based on the switching of fixed-length 32-bit labels, and therefore it features a higher forwarding efficiency than IP, which forwards data based on the Longest Prefix Match (LPM). That said, as hardware capabilities have and continue to improve, MPLS no longer features distinct advantages in forwarding efficiency. Nevertheless, MPLS provides a good QoS guarantee for IP through connection-oriented label forwarding and also supports Traffic Engineering (TE), Virtual Private Network (VPN), and Fast Reroute (FRR).[13] These advantages play a key role in the continuous expansion of IP networks, while also catapulting the IP transformation of telecom networks.

In general, the success of MPLS depends mainly on its three important features: TE, VPN, and FRR.

- TE: Based on Resource Reservation Protocol-Traffic Engineering (RSVP-TE),[14] MPLS labels can be allocated and distributed along the MPLS TE path, and TE features (such as resource guarantee and explicit path forwarding) can be implemented. This overcomes IP networks' lack of support for TE.
- VPN: MPLS labels can be used to identify VPNs[15] for isolation of VPN services. As one of the major application scenarios of MPLS, VPN is a key technology for enterprise interconnection and multi-service transport as well as an important revenue source for carriers.

- FRR: The IP network cannot provide complete FRR protection, which in turn means that it is unable to meet the high-reliability requirements of carrier-grade services. MPLS improves the FRR capabilities of IP networks and supports 50 ms carrier-grade protection switching in most failure scenarios.

Because IP networks are cost-effective and MPLS provides good TE, VPN, and FRR capabilities, IP/MPLS networks gradually replaced dedicated networks, such as ATM, FR, and X.25. Ultimately, MPLS was applied to various networks, including IP backbone, metro, and mobile transport, to support multiservice transport and implement the Internet's All IP transformation. In this book, we refer to the IP/MPLS multiservice transport era as the All IP 1.0 era.

1.3 CHALLENGES FACING ALL IP 1.0: IP/MPLS DILEMMA

Although IP/MPLS drove networks into the All IP 1.0 era, the IPv4 and MPLS combination has also set forth numerous challenges, which are becoming more prominent as network scale expands and cloud services develop, and are thereby hindering the further development of networks.

1.3.1 MPLS Dilemma

From one perspective, MPLS plays an important role in All IP transport, while from another perspective, it complicates inter-domain network interconnection by causing isolated network islands.

To put it more precisely, consider the fact that on the one hand, MPLS is deployed in different network domains, such as IP backbone, metro, and mobile transport networks, forming independent MPLS domains and creating new network boundaries. However, many services require E2E deployment, and this means that services need to be deployed across multiple MPLS domains, which in turn results in complex inter-domain MPLS solutions. In that regard, multiple inter-Autonomous System (AS) solutions, such as Option A, Option B, and Option C,[15,16] have been proposed for inter-AS MPLS VPN, and each one involves relatively complex service deployment.

On the other hand, as the Internet and cloud computing develop, more and more cloud data centers are built. To meet the requirements of multi-tenant networking, multiple overlay technologies were proposed, among which VXLAN is a typical example. At the same time, quite a few attempts were made to provide VPN services by introducing MPLS to data centers. However, these attempts all wound up in failure due to multiple factors, including numerous network boundaries, complex management, and insufficient scalability.

In Figure 1.2, traffic from end users to the cloud data center needs to travel multiple network domains. It passes through the MPLS-based Fixed Mobile Convergence (FMC) transport network. The traffic then enters the MPLS-based IP backbone network through the native IP network, accesses the data center's IP network at the edge, and reaches the VXLAN gateway. From there, it travels along the VXLAN tunnel, arrives at the Top of Rack (TOR) switch at the egress of the VXLAN tunnel, and finally accesses the Virtual Network Function (VNF) device. We can therefore envision how complex the service access process is due to an excessive number of network domains.

Other major factors hindering the development of MPLS are scalability and extensibility, which involves two aspects: scalability of the label space and extensibility of encapsulation.

In the MPLS label space, as shown in Figure 1.3, there are 20 bits for the MPLS label, which equates to a 2^{20} label space.

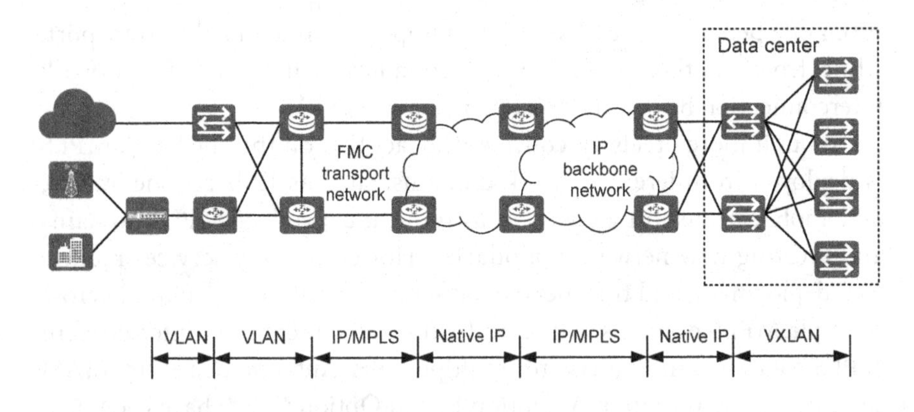

FIGURE 1.2 Isolated MPLS network islands.

0 … 19	20 … 22	23	24 … 31
Label	EXP	S	TTL

FIGURE 1.3 MPLS label encapsulation format.

As the network scale expands, the label space is no longer sufficient. Moreover, due to the limitation of the RSVP-TE protocol, the control plane of MPLS networks also faces challenges such as complexity and a lack of scalability. These make it difficult for MPLS to satisfy the requirements of network development.

On the other hand, the encoding of fields of MPLS encapsulation is fixed. Although the MPLS label stack provides certain extensibility, as the new network services develop and require more flexible encapsulation in the forwarding plane (e.g., carrying metadata[17] in Service Function Chaining (SFC) or In-situ Operations, Administration, and Maintenance (IOAM)[18] packets), the extensibility of MPLS encapsulation faces more challenges.

1.3.2 IPv4 Dilemma

One of the biggest problems regarding IPv4 is its insufficient address resources. Since the 1980s, IPv4 addresses were consumed at an unexpectedly fast pace. Put differently, the Internet Assigned Numbers Authority (IANA) announced that the last five IPv4 address blocks were allocated on February 3, 2011. At 15:35 (UTC+1) on November 25, 2019, the final/22 IPv4 address block was allocated in Europe, thereby signifying the depletion of global IPv4 public addresses. Although technologies such as Network Address Translation (NAT) help alleviate this issue by reusing private network address blocks, this is by no means the ultimate solution.

As shown in Figure 1.4, NAT not only requires extra network configurations but also needs the maintenance of network state mappings, which further complicates network deployment. On top of that, NAT does not support source tracing of IPv4 addresses because the actual addresses are hidden, and this generates management risks.

IPv4 also faces another dilemma: the insufficient extensibility of packet headers results in inadequate programmability. Given this, it is difficult for IPv4 networks to support many new services that require more extensions of the header, such as source routing, SFC, and IOAM. Although IPv4 defines the Options field for extension, it is rarely implemented and used. This means that the insufficient extensibility of the IPv4 header will constrain IPv4 development to a certain extent. Taking this into account, the Internet Architecture Board (IAB) stated in 2016 that formulating standards about new features based on IPv4 will not be considered in the future.

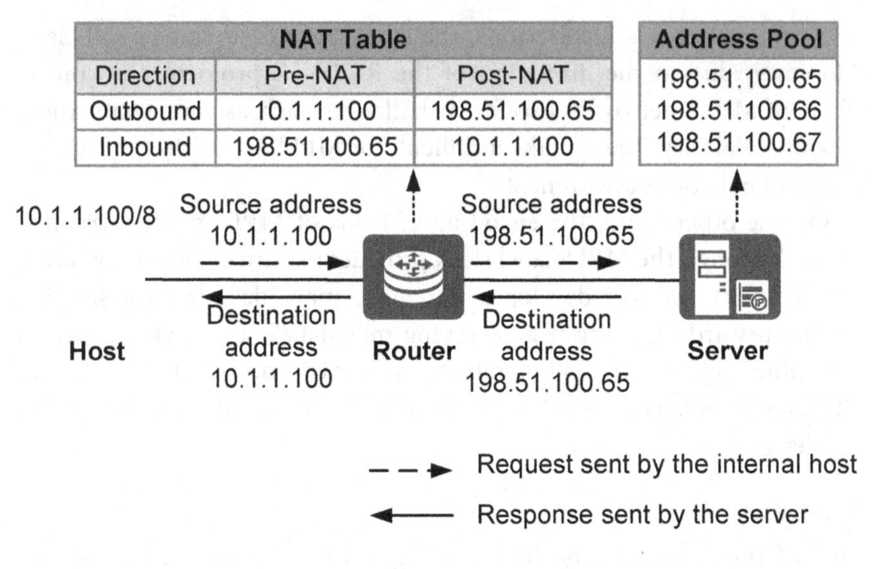

FIGURE 1.4 NAT.

To find a solution against IPv4 address exhaustion and poor extensibility, the industry designed the next-generation upgrade solution for IPv4 — IPv6.[2]

1.3.3 Challenges for IPv6

As the next-generation IP protocol, IPv6 aims to solve two major problems of IPv4: limited address space and insufficient extensibility.[19] To this end, IPv6 offers certain improvements to IPv4.

One such improvement is the expanded address space, or stated differently, the increase from 32 bits for IPv4 addresses to 128 bits for IPv6 addresses. The magnitude of this expansion can be compared to allocating an IPv6 address to every grain of sand on the Earth, effectively solving the problem of insufficient IPv4 addresses.

The extension header mechanism counts as another noteworthy enhancement. Requirement For Comments (RFC) 8200[19] defines the following IPv6 extension headers, which ideally should appear in a packet in the following order (Appendix A expands on the details of IPv6):

1. IPv6 header
2. Hop-by-Hop Options header

3. Destination Options header
4. Routing header
5. Fragment header
6. Authentication header
7. Encapsulating Security Payload header
8. Destination Options header
9. Upper-Layer header

Figure 1.5 shows the encapsulation format of a common IPv6 extension header that carries a TCP message.

As far as IPv6 is concerned, extension headers provide good extensibility and programmability. For example, the Hop-by-Hop Options header can be used to implement hop-by-hop IPv6 data processing, and the Routing header to implement source routing.

Over 20 years have passed since IPv6 was proposed, yet IPv6 development is still quite slow. Only within the last few years have technology development and policies propelled IPv6 deployment. In retrospect, the slow IPv6 development can mainly be attributed to the following factors:

1. Incompatibility with IPv4 and high costs for network upgrades: Although IPv6 offers an address space of 128 bits compared to 32 bits in IPv4, it is incompatible with IPv4, meaning that hosts using IPv6 addresses cannot directly communicate with those using IPv4 addresses. As such, a transition solution is required, resulting in high network upgrade costs.

2. Insufficient service driving force and low network upgrade benefits: IPv6 advocates have been promoting the 128-bit address space,

IPv6 Header Next Header = Routing	Routing Header Next Header = Fragment	Fragment Header Next Header = TCP	Fragment of TCP Header + Data

FIGURE 1.5 Encapsulation format of an IPv6 extension header carrying a TCP message.

which is viewed as a solution to IPv4 address exhaustion. However, technologies such as NAT can also solve this problem, and NAT does indeed serve as the main remedy. The specific process involves leveraging private network addresses and address translation technologies to temporarily alleviate the problem which may hinder network service development.

A key advantage of NAT is the fact that it can be deployed at a lower cost than that needed to upgrade IPv4 networks to IPv6. Considering the fact that existing services run well on IPv4 networks, there is no point in upgrading them, especially as this does not yield new revenue and, on the contrary, would lead to higher costs. This is the major reason why carriers are unwilling to upgrade.

In light of this, the key to solving slow IPv6 deployment lies in finding more attractive services supported by IPv6, as opposed to IPv4. This way, business benefits can drive carriers to upgrade their networks to IPv6.

1.4 OPPORTUNITIES FOR ALL IP 1.0: SDN AND NETWORK PROGRAMMING

In addition to the insufficient extensibility and programmability of IPv4 and MPLS data planes, the All IP 1.0 era also faces the following challenges[20]:

- Lack of a global network view and traffic visualization capability. As a consequence of this, it is difficult to make optimal decisions from a global perspective of the network, or quickly respond to TE requirements.
- Lack of a unified abstract model in the data plane prevents the control plane from supporting new functions by programming with data-plane Application Programming Interfaces (APIs).
- Lack of automation tools and a long service rollout period.
- The data and control planes of a device are tightly coupled and bundled for sale. As such, these two planes rely on each other as far as evolution is concerned, and the control plane on a device provided by one vendor cannot control the data plane on a device provided by another.

The traditional network devices in the All IP 1.0 era are similar to the IBM mainframes of the 1960s. To put it more exactly, the network device hardware, Operating System (OS), and network applications are tightly coupled and rely on each other. Because of this, if we aim to conduct innovation or evolution on one part, we must also upgrade the other parts correspondingly, and this architecture impedes network innovation.[20]

In contrast, personal computers have a vastly different development model, which involves them using a universal processor, based on which they implement software-defined functions. Therefore, the computer has a more flexible programming capability, resulting in explosive growth in software applications. Furthermore, the open-source model of computer software breeds a large amount of open-source software, accelerates the software development process, and promotes the rapid development of the entire computer industry. We can look at the Linux open-source OS as the best example of this.

Drawing on universal hardware, software-defined functions, and the open-source model in the computer field, Professor Nick McKeown's team proposed a new network architecture — SDN.[4]

In the SDN architecture, the control plane of a network is separated from its data plane: The data plane becomes more generalized, similar to the universal hardware of a computer. It no longer needs to specifically implement the control logics of various network protocols; instead, it only needs to execute the operation instructions received from the control plane. The control logic of a network device is defined by the SDN controller and applications, and as such, network functions are defined by software. With the emergence of open-source SDN controllers and open-source SDN open interfaces, the network architecture includes three elements: universal hardware, software-defined support, and open-source model. Figure 1.6 depicts the evolution from the traditional network architecture to the SDN architecture.

SDN has the following three characteristics[19]:

1. Open network programmability: SDN provides a new network abstraction model with a complete set of universal APIs for users, who can then program on the controller to configure, control, and manage networks.

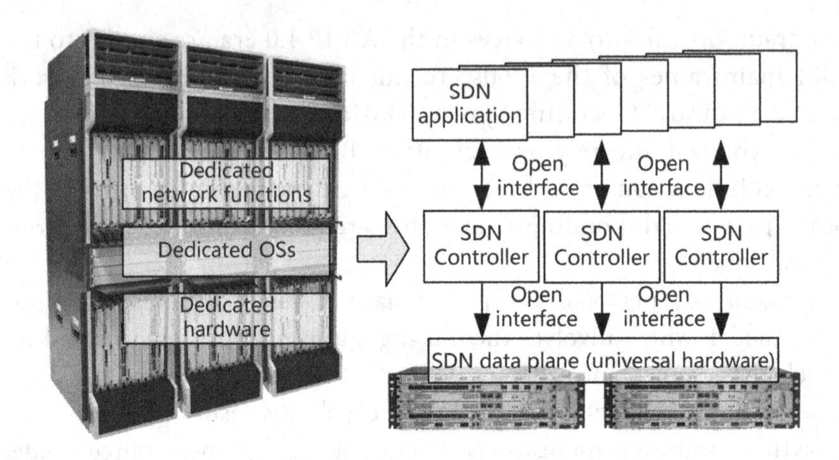

FIGURE 1.6 Evolution from the traditional network architecture to the SDN architecture.

2. Separation of the control and data planes: Separation refers to the decoupled control and data planes, which can evolve independently and communicate using a set of open APIs.

3. Logical centralized control: This refers to the centralized control of distributed network states. Logical centralized control architecture is the basis for SDN, propelling automated network control into the realm of possibilities.

A network that has the preceding three characteristics can be called an SDN network. Among the three characteristics, separation of the control and data planes creates the necessary conditions for logical centralized control, which in turn provides the architectural basis for open network programmability. Note that open network programmability is the core characteristics of SDN.

That being said, it is worth noting that SDN is only a network architecture, and multiple technologies have been proposed to implement it, such as OpenFlow, Protocol Oblivious Forwarding (POF),[21] Programming Protocol-independent Packet Processors (P4),[22] and SR.[7]

1.4.1 OpenFlow

On March 14, 2008, Professor Nick McKeown and others proposed OpenFlow, which is a protocol used between the SDN control plane and data plane.

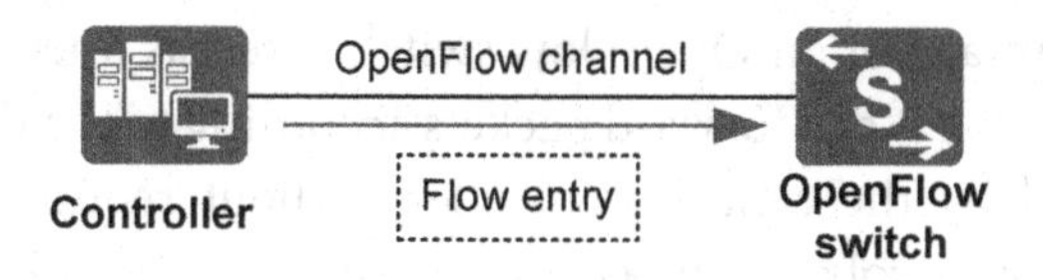

FIGURE 1.7 OpenFlow 1.0 architecture.

In the OpenFlow protocol architecture shown in Figure 1.7, an OpenFlow channel is established between an OpenFlow switch and OpenFlow controller to exchange information. The controller can deliver flow table entries to the OpenFlow switch using the OpenFlow protocol. Each flow table entry defines a type of flow and corresponding forwarding actions. That is, if the matching for a specific type of flow succeeds, corresponding actions will be executed to process and forward packets.

Essentially, the forwarding behavior on a device can be abstracted into "matching and action." For example, Layer 2 switches and Layer 3 routers forward packets by searching for a destination Media Access Control (MAC) address and destination IP address, respectively. The design principle of OpenFlow involves abstracting matching and forwarding actions into specific operations and using the controller to deliver flow table entries to switches to guide packet forwarding. In short, OpenFlow is an SDN technology that abstracts and generalizes network processing rules and supports centralized programming.

In OpenFlow, a packet can be forwarded to a specific interface or discarded by matching protocol fields such as Ethernet, IPv4, or IPv6. The controller can program a matching+action rule via a flow table entry to implement network programming. For example, the controller delivers a flow table entry to switch A, instructing the switch to forward a packet whose destination IP address is 192.168.1.20 to outbound interface 1.

OpenFlow's advantage lies in the flexible programming of forwarding rules, but that's not to say it does not have apparent problems.

1. The flow table entries are stored in the expensive Ternary Content Addressable Memory (TCAM), meaning that only 1k–10k entries can generally be supported in OpenFlow switches. The limited OpenFlow flow table specifications result in OpenFlow switches offering insufficient scalability. Currently, OpenFlow is mainly deployed in data centers for simple data switching and cannot be deployed at a location requiring numerous flow table entries.

2. The advantage of an OpenFlow switch (used as a Layer 3 switch or router) is that it can forward packets according to the flow table entry generated by the central controller without requiring distributed routing protocols, such as an Interior Gateway Protocol (IGP). However, no carrier would like to abandon distributed routing protocols on their live networks, causing the distributed routing protocols like IGPs to continue running on OpenFlow switches. In this case, the basic shortest path forwarding can be performed by the IGP while OpenFlow can only be used to optimize traffic steering, so the benefits of OpenFlow are limited comparing to the cost brought by it. Taking this into account, the OpenFlow switch does not simplify the protocols, but on the contrary, it introduces additional complexity brought by OpenFlow.

3. OpenFlow supports packet forwarding by adding corresponding flow table entries based on the existing forwarding logic. It cannot program the forwarding logic of a switch. For this reason, when new features are added to OpenFlow, the protocol stack implementation on the controller and switch has to be updated. Sometimes, even the switch's chip and other hardware need to be redesigned. Therefore, the cost of supporting new features is significant in OpenFlow.

4. OpenFlow does not have sufficient capability to support stateful network processing in the data plane that covers a full range of L4–L7 services. Therefore, the limited expressivity impacts the forwarding plane programmability. The network states must be maintained on the controller and synchronized up with the OpenFlow Switch. Overdependence on the controller significantly burdens it, as well as brings problems in extensibility and performance.

Due to the preceding limitations, OpenFlow has not been widely deployed.

1.4.2 POF

Huawei proposed POF to provide programmable forwarding logic for switches, which is missing from OpenFlow.

Similar to OpenFlow, the architecture of POF includes two parts: the control plane (POF controller) and the data plane (POF Forwarding Element (FE)).

POF makes the forwarding plane totally protocol-oblivious. The POF FE has no need to understand the packet format. All an FE needs to do is, under the instruction of its controller, extract and assemble the search keys (located by one or more<offset, length>tuples) from the packet header, conduct table lookups, and then execute the associated instructions. As a result, the FE can easily support new protocols and forwarding requirements in the future.[21]

In summary, POF is an SDN technology that comprehensively abstracts network processing into a more interoperable, protocol-oblivious process, while providing the capabilities of programming forwarding rules and forwarding logic. Figure 1.8 shows the architecture of the POF hardware and software switches.

Because POF supports protocol-oblivious forwarding, it can be deployed on any network, including non-Ethernet networks such as Named Data Network (NDN) and Content-Centric Network (CCN). In addition, the POF FE supports adding metadata to packets so that stateful processing can be supported.

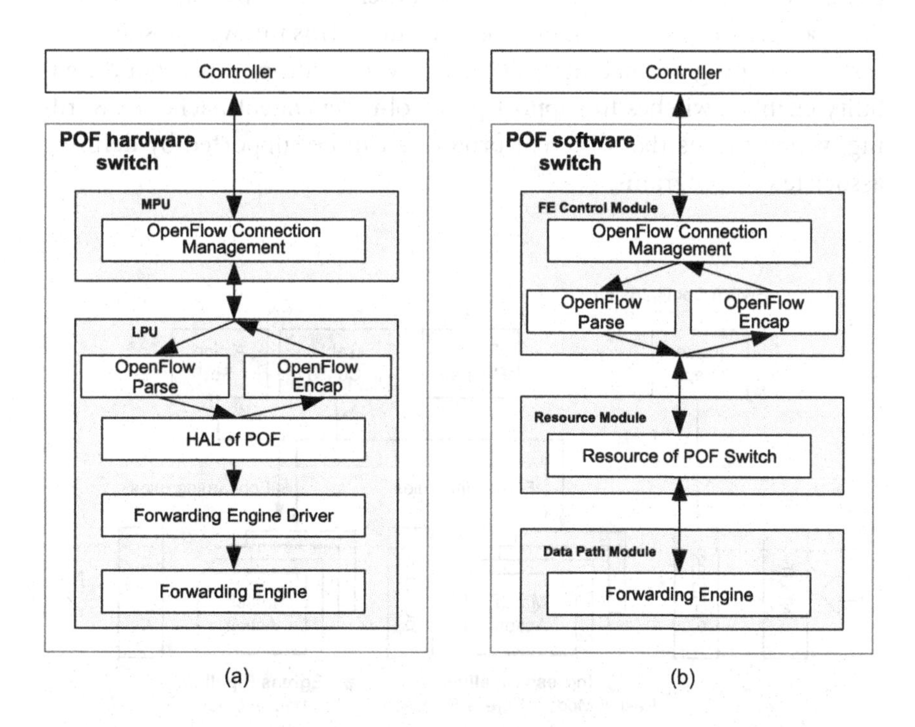

FIGURE 1.8 POF hardware switch (a) and POF software switch (b).

However, POF is more complex than OpenFlow. Furthermore, a Flow Instruction Set[21] needs to be defined to implement complex instruction-based scheduling, which affects forwarding performance to some extent. Therefore, POF has not made much progress in commercial use.

1.4.3 P4

In order to solve the insufficient programmability of OpenFlow, Professors Nick McKeown (Stanford) and Jennifer Rexford (Princeton), among others, proposed P4.[21] P4 is a high-level language for programming protocol-independent packet processors. P4 can be used to configure switches to express how to parse, process, and forward packets.

P4 supports programmable packet processing on a device by defining a P4 program containing the following components: headers, parsers, tables, actions, and control programs.[22] A P4 program can be compiled and run on the common abstract forwarding model-based devices. Figure 1.9 shows the common abstract forwarding model.

With P4 you can configure and program how a switch processes packets even after the switch is deployed, thereby eliminating the need to purchase new devices to support new features. This innovation solves the insufficient programmability of OpenFlow. In addition, P4 programmability enables switches to support protocol-independent packet forwarding, which means that different protocols can be supported by defining associated P4 programs.

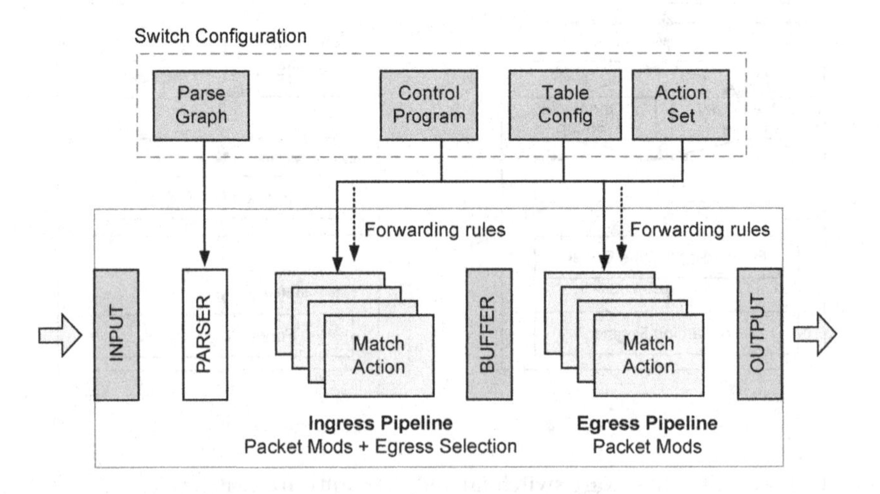

FIGURE 1.9 Common abstract forwarding model.

Although P4 has certain technical advantages, it has not made much progress in commercial deployment. One reason is that networks are evolving nowhere fast. The standardization of new network features is not solely up to one device vendor. It is usually discussed for several years among carriers, device vendors, and other relevant parties in the industry. With such a long process, the benefit of supporting new features in a short time provided by P4 is no longer significant. Another reason is that fully centralized SDN has insufficient reliability and response speed but high requirements on the controller. However, carriers do not need to completely reconstruct their networks by overturning the distributed routing protocol architecture. Instead, the better solution is to provide the global traffic optimization based on existing distributed routing protocols. So far, P4 has not yet been put into large-scale commercial use. On June 11, 2019, Barefoot, a P4 startup, was ultimately acquired by Intel.

1.4.4 SR

OpenFlow, POF, and P4 were originally designed to provide programmability for networks. But is revolutionary innovation necessary for network programmability? The answer is no.

In 2013, SR was proposed, which is a transitional extension based on the existing network and provides network programmability. SR is a source routing paradigm which can program the forwarding path of packets by allowing the ingress of the path to insert forwarding instructions into packets. The core idea of SR is to combine different segments into a path and insert segment information into packets at the ingress of the path to guide packet forwarding. A segment is an instruction, which is executed on a node for packet forwarding or processing. It can refer to a specific interface or the shortest path to a node through which a packet is forwarded. Such a segment is identified by a Segment Identifier (SID). Currently, there are two data planes for SR: MPLS and IPv6. When SR is applied to the MPLS data plane, it is called SR-MPLS, and the SID is encoded as an MPLS label. When SR is applied to the IPv6 data plane, it is called SRv6, and the SID is encoded as an IPv6 address.

The design of SR is easily comparable to many real-life examples, for instance, traveling by train or plane. The following uses travel as an example to further explain SR.

If flying from Haikou to London requires a stop in Guangzhou and in Beijing, we need to buy three tickets: Haikou to Guangzhou, Guangzhou to Beijing, and Beijing to London.

With three tickets in hand, we know we will fly from Haikou to Guangzhou on flight HU7009 according to the first ticket; when arriving in Guangzhou, we take flight HU7808 to Beijing, according to the second ticket; when arriving in Beijing, we fly to London by flight CA937, according to the last ticket. Ultimately, we fly to London segment by segment, using the three tickets.

The packet forwarding on an SR network is similar. As shown in Figure 1.10, a packet enters the SR network from node A. From the destination address, node A knows that the packet needs to pass through nodes B and C before reaching node D. Therefore, node A inserts the SIDs of nodes B, C, and D into the packet header to indicate how to steer the packet. The packet will be sent to node B, node C, and finally node D according to the SID information in the packet header.

Compared with RSVP-TE MPLS, SR-MPLS has the following advantages:

- Simplifies the control plane. In SR-MPLS, label distribution protocols such as RSVP-TE are not needed any more. Instead, only IGP and Border Gateway Protocol (BGP) extensions are required in the SR-MPLS control plane, thereby reducing the number of control plane protocols.
- Simplifies network states. When RSVP-TE is used on an MPLS network, nodes need to maintain per-flow states. In contrast, on an SR-MPLS network, only the ingress node needs to maintain per-flow states, while the transit and egress nodes do not.

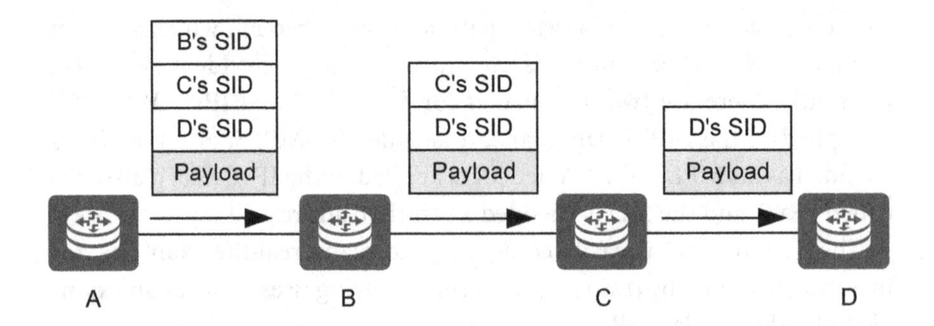

FIGURE 1.10 Packet forwarding process with SR.

SR-MPLS provides network programming capabilities by reusing the existing MPLS forwarding mechanism. Therefore, it can support smooth upgrades from existing MPLS networks to SR-MPLS networks. In this way, SR-MPLS, as the transitional innovation, is much easier to be adopted by the industry. In addition, SR retains the distributed intelligence of the network while introducing the global traffic optimization of the SDN controller. This makes implementation more practical, and it will go further. So far, SR has already become the de facto SDN standard.

Although SR-MPLS based on MPLS data plane can provide good programmability, it cannot satisfy services that need to carry metadata, such as SFC and IOAM, as MPLS encapsulation has relatively poor extensibility. Compared with SR-MPLS, SRv6, which is based on IPv6 data plane, not only inherits all the advantages of SR-MPLS, but also provides better extensibility.

1.5 KEY TO ALL IP 2.0: SRv6

As mentioned before, the key to speeding up IPv6 deployment lies in finding the services supported by IPv6, as opposed to IPv4. Driven by business benefits, carriers will upgrade their networks to IPv6. But what drives the need for IPv6 over IPv4? The answer lies in the network programmability of IPv6, which allows new services to be deployed quickly and easily to create revenue for carriers.

Although IPv4 also provides the programmable Options field, this field is not often used. IPv6, however, takes the extensibility of packet headers into account from the very beginning. A number of extension headers, including Hop-by-Hop Options, Destination Options, and Routing headers,[19] were designed to support further extension.

However, after 20-plus years, IPv6 extension headers are still rarely applied to their full potential. With the rise of new services such as 5G and cloud, and the development of network programming technologies, services require the forwarding plane of a network to provide stronger programming capabilities and a simpler converged network solution. This is where SRv6 comes into play.

SRv6 is an SR network paradigm based on IPv6 data plane by making use of a new IPv6 Routing Header, called Segment Routing Header (SRH), allowing the ingress to insert forwarding instructions to guide data packet forwarding. As shown in Figure 1.11, SRv6 combines the advantages of

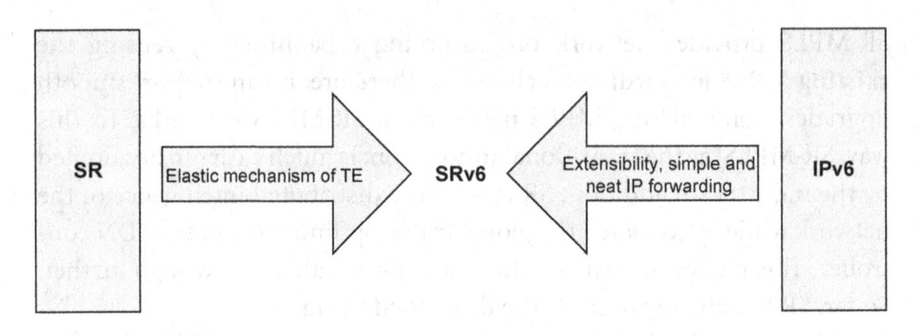

FIGURE 1.11 SR+IPv6=SRv6.

SR-MPLS's programmability and IPv6 headers' extensibility, giving IPv6 an edge.

As of now, SRv6 has been commercially deployed by multiple carriers around the world just after 2 years since the draft *SRv6 Network Programming*[23] was submitted to the IETF. Such rapid development is uncommon among IP technologies. During these 2-plus years when we promoted SRv6 innovation and standardization, we have communicated extensively with industry experts and made a number of reflections on the experience and lessons learned in the development of Internet technologies. This has given us a further understanding of the value and significance of SRv6.

To sum up, the MPLS-based All IP 1.0 era has achieved great success but has also brought some problems and challenges:

1. Isolated IP transport network islands: Although MPLS unified the technologies for transport networks, the IP backbone, metro, and mobile transport networks are separated and need to be interconnected using complex technologies such as inter-AS VPN, making E2E service deployment difficult.

2. Limited programming space in IPv4 and MPLS encapsulation: Many new services require more forwarding information to be added to packets. However, the IETF announced that it has stopped formulating further standards for IPv4. In addition, the format of the MPLS Label field is fixed and lacks extensibility. These reasons make it difficult for IPv4 and MPLS to meet the requirements of new services for network programming.

3. Decoupling of applications and transport networks: This makes it difficult to optimize networks and improve the value of networks. Many carriers find themselves stuck as a provider of pipes and cannot benefit from value-added applications. Moreover, the lack of application information means that carriers can only implement network adjustment and optimization in a coarse-granularity way, resulting in wasting resources. Attempts have been made over the years to apply network technologies to user terminals, but all have failed. An example of such attempts is ATM-to-desktop. Other attempts have been made to deploy MPLS closer to hosts and applications, for example, deploying MPLS for the cloud. But the fact is, deploying MPLS in data centers is very difficult while VXLAN becomes the de facto standard of data centers.

SRv6 technology is the answer to these problems.

1. SRv6 is compatible with IPv6 forwarding and can implement interconnection of different network domains easily through IPv6 reachability. Unlike MPLS, SRv6 does not require additional signaling or networkwide upgrades.
2. SRv6, based on SRHs, supports encapsulation of more information into packets, meeting diversified requirements of new services.
3. SRv6's affinity to IPv6 enables it to seamlessly integrate IP transport networks with IPv6-capable applications and provide more potential value-added services for carriers through application-aware networks.

The development of IPv6 over the past 20-plus years proves that the demand for address space alone cannot promote the large-scale deployment of IPv6. But the rapid development of SRv6 indicates that IPv6 development can be boosted by requirements on new services. As shown in Figure 1.12, along with the development of services such as 5G, cloud, and Internet of Things (IoT), the increasing number of network devices require more addresses and network programmability. SRv6 can better meet the requirements of these services, promote the development of network services, and drive networks into a new All IP era, that is, an

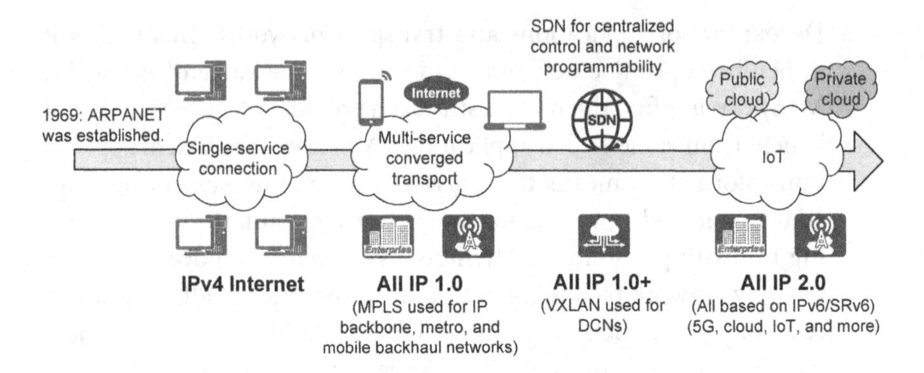

FIGURE 1.12 IP technology development generations.

intelligent IP era where all things are connected based on IPv6. In this book, we refer to this era as the All IP 2.0 era.

1.6 STORIES BEHIND SRv6 DESIGN

I. SRv6 and SDN

Since its proposal in 2007, SDN has exerted a lasting impact on industries. The concept of SDN is also made more applicable over different scenarios. From the originally revolutionary OpenFlow and POF that require complete separation of forwarding and control, SDN has undergone gradual evolution amid a heated debate in the industry. The most important driving force behind this is the IETF. From 2013 to 2016, an important task of the IETF was the standardization of southbound protocols of SDN controllers, such as BGP, Path Computation Element Communication Protocol (PCEP), and Network Configuration Protocol (NETCONF)/Yet Another Next Generation (YANG). After more than 4 years of effort, the IETF completed the main work of SDN transition in the control plane and started to work on SRv6 Network Programming in 2017. Compared with SR-MPLS, SRv6 has stronger network programming capabilities. Programming, once implemented based on an OpenFlow- or POF-capable forwarding plane, is now implemented through SRv6, which provides better compatibility. This reflects the SDN transition in the forwarding plane. In other words, the SDN transition is still ongoing, and the focus has shifted from the control plane to the forwarding plane.

II. Rethinking the Value and Significance of SR

While exploring this technology and discussing it with others, I was often struck with some excellent thoughts, which were of great help for me to understand the essence of technologies. Section 1.5 of this chapter summarizes the problems faced by MPLS and the value and significance of SRv6. Later I discussed with experts in the industry and acquired more enlightening viewpoints which are shared in the following for your reference:

1. MPLS is also essentially an extension of IP functions. However, due to the limitations in the past, it was implemented by using the Shim layer, which requires networkwide upgrades to support the extended functions. After more than 20 years of software and hardware development, many of these limitations have been removed. SRv6 uses a new method to integrate IP and MPLS functions (SRv6 SIDs reflect both IP-like and MPLS-like identifiers), which better complies with the trend of technology development.

2. SRv6 enables SDN on carrier IP networks. VXLAN is an important foundation for the development of SDN on data center networks. However, no VXLAN-like technology was available to boost the development of SDN on carrier IP networks. SRv6 solves this problem.

   During the development of SDN, there was a bias toward the construction of SDN controller capabilities, but the impact of network infrastructure was largely ignored. Carrier IP networks are a lot more complex than data center networks, and this has much to do with MPLS as a basic transport technology. In turn, this has resulted in complex functions and difficult deployment of the SDN controller. The native IP attribute of SRv6 greatly simplifies basic transport technologies. Furthermore, SRv6 can be deployed from the edge to the core, in a staggered manner, from SRv6 VPN to loose TE, strict TE, and more. This also enables SDN to be gradually rolled out on carrier IP networks in a simple-to-complex manner.

3. 5G changes the attributes of connections, and cloud changes the scope of connections. These changes bring big opportunities for SRv6 development.

Network connections are the core of IP transport networks, and the development of 5G services poses more requirements on network connections, such as stronger Service Level Agreement (SLA) assurance and deterministic latency. 5G also changes or, more precisely, enhances the attributes of connections, requiring more information to be carried in packets. In addition, with the development of cloud services, service processing locations have become more flexible. Some cloud services (such as telco cloud) further break the boundary between physical and virtual network devices, integrating services and transport networks. All these have changed the scope of network connections. SRv6 enables the programming of services and transport networks based on a single data plane. Also, thanks to its native IP attribute, SRv6 allows rapid setup of connections and satisfies requirements of flexible adjustment of the connection scope.

III. Development of IP Generations

We used to believe that IP was developed in an incremental and compatible manner. That is, IP does not have clear definitions of generations such as 2G, 3G, 4G, and 5G like wireless. However, when we look back upon the development of IP over the past few decades, we find some intergenerational characteristics.

First is the rise and decline of network protocols. In the 1990s, the competition between ATM and IP led to the decline of telecommunications. IP transport networks unified networks by replacing independent networks such as ATM, FR, and Time Division Multiplexing (TDM). With the rise of SR, traditional MPLS signaling Label Distribution Protocol (LDP) and RSVP-TE are declining, and the entire MPLS will gradually decline as its data plane is replaced by IPv6 extensions due to the increasing popularity of SRv6.

Second, IP has the habit of expanding application scenarios. IP was first applied to the Internet. Later, IP-based MPLS was applied to IP backbone, metro, and mobile transport networks. With the success of SDN, IP is widely used in data centers to replace the traditional Layer 2 networking. SRv6 is in the process of continuing this trend, as it can well meet the requirements of new scenarios such as 5G and cloud. To give another analogy of this trend, the development of ring roads in Beijing is similar to the development of IP. As the city keeps expanding, new ring roads are built around the city, going from two to three, and even to six rings. Construction of each ring road requires the development and improvement of a new solution, along with the reconstruction of areas that have already been encircled within the ring roads. Just like SRv6 is used for new 5G and cloud scenarios, MPLS will be replaced by SRv6 on existing IP transport networks.

These significant changes in network development act as a reference for defining IP generations. Summarizing the history of network development and defining IP generations help us better seize opportunities for the future.

REFERENCES

[1] Postel J. Internet Protocol[EB/OL]. (2013-03-02)[2020-03-25]. RFC 791.

[2] Deering S, Hinden R. Internet Protocol Version 6 (IPv6) Specification[EB/OL]. (2013-03-02)[2020-03-25]. RFC 2460.

[3] Rosen E, Viswanathan A, Callon R. Multiprotocol Label Switching Architecture[EB/OL]. (2020-01-21)[2020-03-25]. RFC 3031.

[4] Casado M, Freedman M J, Pettit J, Luo J, Mckeown N, Shenker S. Ethane: Taking Control of the Enterprise[EB/OL]. (2007-08-31)[2020-03-25]. ACM SIGCOMM Computer Communication Review, 2007.

[5] Mckeown N, Anderson T, Balakrishnan H, Parulkar G, Peterson L, Rexford J, Shenker S, Turner J. OpenFlow: Enabling Innovation in Campus Networks[EB/OL]. (2008-03-31)[2020-03-25]. ACM SIGCOMM Computer Communication Review, 2008.

[6] Filsfils C, Previdi S, Insberg L, Decraene B, Litkowski S, Shakir R. Segment Routing Architecture[EB/OL]. (2018-12-19)[2020-03-25]. RFC 8402.

[7] Bashandy A, Filsfils C, Previdi S, Decraene B, Litkowski S, Shakir R. Segment Routing with MPLS Data Plane[EB/OL]. (2019-12-06)[2020-03-25]. draft-ietf-spring-segment-routing-mpls-22.

[8] Filsfils C, Dukes D, Previdi S, Leddy J, Matsushima S, Voyer D. IPv6 Segment Routing Header (SRH)[EB/OL]. (2020-03-14)[2020-03-25]. RFC 8754.

[9] Mahalingam M, Dutt D, Duda K, Agarwal P, Kreeger L, Sridhar T, Bursell M, Wright C. Virtual eXtensible Local Area Network (VXLAN): A Framework for Overlaying Virtualized Layer 2 Networks over Layer 3 Networks[EB/OL]. (2020-01-21)[2020-03-25]. RFC 7348.

[10] Réseaux IP Européens Network Coordination Centre. The RIPE NCC has run out of IPv4 Addresses[EB/OL]. (2019-11-25)[2020-03-25].

[11] Huang S, Liu J. *Computer Network Tutorial Problem Solving and Experiment Guide[M]*. Beijing: Tsinghua University Press, 2006.

[12] Rekhter Y, Davie B, Katz D, Rosen E, Swallow G. Cisco Systems' Tag Switching Architecture - Overview[EB/OL]. (2013-03-02)[2020-03-25]. RFC 2105.

[13] Pan P, Swallow G, Atlas A. Fast Reroute Extensions to RSVP-TE for LSP Tunnels[EB/OL]. (2020-01-21)[2020-03-25]. RFC 4090.
[14] Awduche D, Berger L, Gan D, Li T, Srinivasan V, Swallow G. RSVP-TE: Extensions to RSVP for LSP Tunnels[EB/OL]. (2020-01-21)[2020-03-25]. RFC 3209.
[15] Rosen E, Rekhter Y. BGP/MPLS IP Virtual Private Networks (VPNs)[EB/OL]. (2020-01-21)[2020-03-25]. RFC 4364.
[16] Leymann N, Decraene B, Filsfils C, Konstantynowicz M, Steinberg D. Seamless MPLS Architecture[EB/OL]. (2015-10-14)[2020-03-25]. draft-ietf-mpls-seamless-mpls-07.
[17] Halpern J, Pignataro C. Service Function Chaining (SFC) Architecture[EB/OL]. (2020-01-21)[2020-03-25]. RFC 7665.
[18] Brockners F, Bhandari S, Pignataro C, Gredler H, Leddy J, Youell S, Mizrahi T, Mozes D, Lapukhov P, Chang R, Bernier D. Data Fields for In-situ OAM[EB/OL]. (2020-03-09)[2020-03-25]. draft-ietf-ippm-ioam-data-09.
[19] Deering S, Hinden R. Internet Protocol Version 6 (IPv6) Specification[EB/OL]. (2020-02-04)[2020-03-25]. RFC 8200.
[20] Yang Z, Li C. *Network Reconstruction: SDN Architecture and Implementation[M]*. Beijing: Electronic Industry Press, 2017.
[21] Song H. Protocol-Oblivious Forwarding: Unleash the Power of SDN through a Future-Proof Forwarding Plane[EB/OL]. (2013-08-16)[2020-03-25]. *Proceedings of the Second ACM SIGCOMM Workshop on Hot Topics in Software Defined networking*, 2013.
[22] Bosshart P, Daly D, Gibb G, Izzard M, Mckeown N, Rexford J, Schlesinger C, Talayco A, Vahdat A, Varghese G. P4: Programming Protocol-Independent Packet Processors[EB/OL]. (2014-07-28)[2020-03-25]. ACM SIGCOMM Computer Communication Review, 2014.
[23] Filsfils C, Camarillo P, Leddy J, Voyer D, Matsushima S, Li Z. SRv6 Network Programming[EB/OL]. (2019-12-05)[2020-03-25]. draft-ietf-spring-srv6-network-programming-05.

II

SRv6 1.0

CHAPTER 2

SRv6 Fundamentals

THIS CHAPTER DESCRIBES SRv6 fundamentals, including the basic concepts, SRv6 extension header, instruction sets, packet forwarding processes, and technical advantages of SRv6. Inheriting the advantages of both IPv6 and source routing technology, SRv6 is more suitable for cross-domain deployment than MPLS as it supports incremental evolution and ensures better scalability, extensibility, and programmability. These advantages mean that SRv6 is strategically important for future technical evolution.

2.1 SRv6 OVERVIEW

SRv6 provides outstanding network programming capabilities.[1]

In Chapter 1, we introduced the development of network programming and the advantages of SRv6 in this regard. Network programming stems from computer programming, through which we can convert our intent into a series of instructions that computers can understand and execute to meet our requirements. Similarly, if service intent can be translated into a series of device-executable forwarding instructions on a network, network programming can be achieved, ultimately meeting service customization requirements. Figure 2.1 illustrates the implementation of network programming as compared to computer programming.

SRv6 was introduced to translate network functions into instructions and encapsulate the instructions into 128-bit IPv6 addresses. As such, service requirements can be translated into an ordered list of instructions, which can then be executed by network devices along the service

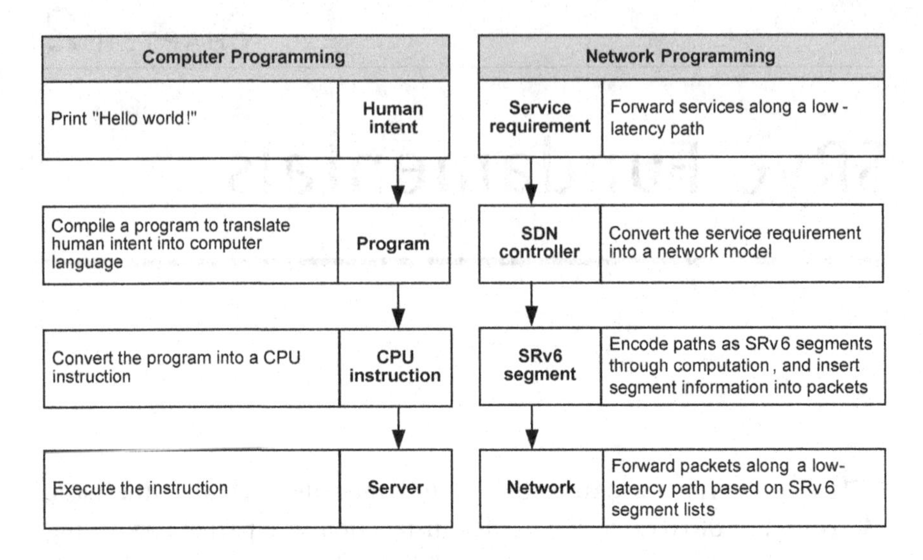

FIGURE 2.1 Computer programming and network programming.

forwarding path, thereby achieving flexible orchestration and on-demand customization of services on an SRv6 network.

2.2 NETWORK INSTRUCTIONS: SRv6 SEGMENTS

A computer instruction typically consists of an opcode and an operand. The former determines the operation to be performed; the latter determines the data, memory address, or both to be used in the computation. Similarly, network instructions, which are called SRv6 segments, need to be defined for SRv6 network programming, and they are identified using 128-bit SIDs.[2] Each SRv6 SID usually consists of three fields, as shown in Figure 2.2.

The three fields are described as follows:

1. The Locator field identifies the location of a network node and is used for other nodes to route and forward packets to this identified node. A locator has two important characteristics: routable and aggregatable. "Routable" means that the corresponding locator route can be advertised by a node to other nodes on the network through

Locator	Function	(Optional) Arguments

FIGURE 2.2 SRv6 SID.

an IGP so that those nodes can forward packets to the node through the locator route. "Aggregatable" means that the locator route can be aggregated. The two characteristics help tackle problems such as high network complexity and large network scale.

The locator length is variable, so SRv6 SIDs can be used on small to large networks.

The following is an example of locator configuration.

```
<HUAWEI> system-view
[~HUAWEI] segment-routing ipv6
[~HUAWEI-segment-routing-ipv6] locator test1 ipv6-prefix 2001:db8:100:: 64
```

After the preceding configurations are complete, a locator with the 64-bit prefix 2001:db8:100:: is generated to guide packet forwarding to the node generating the corresponding instruction, thereby making the instruction addressable.

2. The Function field specifies the forwarding behavior to be performed and is similar to the opcode in a computer instruction. In SRv6 network programming, forwarding behaviors are expressed using different functions. You can think of SIDs as similar to computer instructions, in that each type of SID identifies corresponding forwarding behaviors, such as forwarding packets to a specified link or searching a specified table for packet forwarding.

   The following is an example of End.X function configuration.

```
[~HUAWEI-segment-routing-ipv6-locator] opcode ::1 end-x interface
GigabitEthernet3/0/0 next-hop 2001:db8:200::1
```

The preceding example defines an End.X function with an opcode of ::1. If the Arguments field is not specified, the locator prefix 2001:db8:100:: and function opcode ::1 form the SRv6 SID 2001:db8:100::1, which is a manually configured End.X SID. An End.X SID identifies one or a group of Layer 3 network adjacencies connected to a network node. The forwarding behavior bound to the SID is to update the next SID into the IPv6 Destination Address (DA) field and forward packets to the corresponding neighboring node through the interface specified by the End.X SID. In the preceding example, the End.X function instructs the local node to forward packets from GigabitEthernet 3/0/0 to the next hop at 2001:db8:200::1.

3. The Arguments (Args) field is optional. It is used to define parameters for instruction execution and can contain flow, service, and any other related information. For example, the Arg.FE2 parameter can be specified for an End.DT2M SID to exclude a specific or a group of outbound interfaces when the corresponding Layer 2 forwarding table is searched for multicast replication in Ethernet Segment Identifier (ESI) filtering and Ethernet Virtual Private Network (EVPN) Ethernet Tree (E-Tree) scenarios.

Much like computers that use limited instruction sets to implement various computing functions, SRv6 also defines instruction sets for forwarding purposes.

Instructions are essential to network programming. Any E2E service connection requirement can be expressed using an ordered set of instructions. In current SR implementation, the source node encapsulates ordered segment lists into packets to instruct specified nodes to execute corresponding instructions, thereby achieving network programmability. With the increase of SRv6 application scenarios, SRv6 instruction sets will evolve continuously.

2.3 NETWORK NODES: SRv6 NODES

An SRv6 network may contain nodes of the following roles[3]:

- SRv6 source node: a source node that encapsulates packets with SRv6 headers
- Transit node: an IPv6 node that forwards SRv6 packets but does not perform SRv6 processing
- SRv6 segment endpoint node (endpoint node for short): a node that receives and processes SRv6 packets whose IPv6 DA is a local SID of the node

A node plays a role based on the task it takes in SRv6 packet forwarding, and it may play two or more roles. For example, it can be the source node on one SRv6 path and a transit or endpoint node on another SRv6 path.

2.3.1 SRv6 Source Node

An SRv6 source node steers a packet using an SRv6 segment list. If the SRv6 segment list contains only one SID, and no Type Length Value (TLV) or other information needs to be added to the packet, the DA field of the packet is set to the SID, without requiring SRH encapsulation.

An SRv6 source node can be either an SRv6-capable host where IPv6 packets originate or an edge device in an SRv6 domain.

2.3.2 Transit Node

A transit node is an IPv6 node that does not participate in SRv6 processing on the SRv6 packet forwarding path; that is, the transit node just forwards IPv6 packets. After receiving an SRv6 packet, a transit node parses the IPv6 DA field in the packet. If the value of this field is neither a locally configured SRv6 SID nor a local interface address, the transit node considers the SRv6 packet as an ordinary IPv6 packet. As such, it searches the corresponding IPv6 routing table according to the longest match rule for packet processing and forwarding. In this process, processing the DA as an SRv6 SID or processing SRHs is not required.

A transit node can be either an ordinary IPv6 node or an SRv6-capable node.

2.3.3 Endpoint Node

An endpoint node is a node that receives a packet destined for itself (a packet of which the IPv6 DA is a local SID). Unlike transit nodes, endpoint nodes have to process both SRv6 SIDs and SRHs.

To sum up, an SRv6 source node encapsulates packets with SRv6 headers, a transit node processes and forwards the packets as common IPv6 packets, and an endpoint node processes both SRv6 SIDs and SRHs in the packets, as shown in Figure 2.3.

Note: As recommended by RFC 3849, to avoid address conflicts with live-network addresses, IPv6 addresses with the prefix 2001:DB8::/32 are usually used as examples in books. However, 2001:DB8::/32 is too long to fit neatly into figures. Therefore, short addresses such as A1::1 and 1::1 are often used in this book. Unless otherwise specified, all IPv6 addresses in this book are examples only.

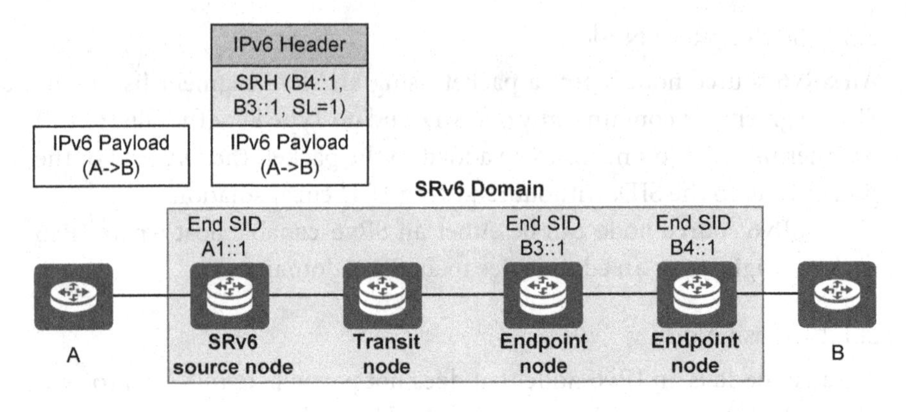

FIGURE 2.3 SRv6 nodes.

2.4 NETWORK PROGRAM: SRv6 EXTENSION HEADER

2.4.1 SRv6 Extension Header Design

To implement SRv6, the original IPv6 header is extended to define a new type of routing header called SRH,[3] which carries segment lists and other related information to explicitly specify an SRv6 path.

Figure 2.4 shows the format of the SRH.

Table 2.1 describes the fields in an SRH.

The SRH stores a SID list used for implementing network services, which is similar to an ordered list of computer instructions. The segments represented by Segment List [0] through Segment List [n] are similar to

Version	Traffic Class	Flow Label		
Payload Length		Next Header = 43	Hop Limit	IPv6 header
Source Address				
Destination Address				
Next Header	Hdr Ext Len	Routing Type = 4	Segments Left	SRH
Last Entry	Flags	Tag		
Segment List [0] (128-bit IPv6 address)				
...				
Segment List [n] (128-bit IPv6 address)				
Optional TLV objects (variable)				
IPv6 Payload				

FIGURE 2.4 IPv6 SRH format.

TABLE 2.1 Fields in an SRH

Field	Length	Description
Next Header	8 bits	Type of the header following the SRH. Common header types are as follows: • 4: IPv4 encapsulation • 41: IPv6 encapsulation • 43: IPv6 routing header (IPv6-Route) • 58: Internet Control Message Protocol version 6 (ICMPv6) • 59: no Next Header for IPv6
Hdr Ext Len	8 bits	Length of an SRH, excluding the first 8 bytes.
Routing Type	8 bits	Type of the extension routing header. SRHs have a routing type value of 4.
Segments Left (SL)	8 bits	Number of remaining segments.
Last Entry	8 bits	Index of the last element in a segment list.
Flags	8 bits	Reserved for special processing, such as Operations, Administration, and Maintenance (OAM).
Tag	16 bit	Whether a packet is part of a group of packets, such as packets sharing the same set of properties.
Segment List [n]	128 bits × number of segments	The nth segment in a segment list, expressed using a 128-bit IPv6 address.
Optional TLV	Variable	Optional TLVs, such as Padding TLVs and Hash-based Message Authentication Code (HMAC) TLVs.

the instructions of a computer program, and Segment List [n] indicates the first instruction that needs to be executed. The SL field, which is similar to the Program Counter (PC) of a computer program, points to the instruction that is being executed and can be initially set to n (the index of the last segment in a segment list). Each time an instruction is executed, the SL value is decremented by 1 to point to the next instruction to be executed. From this, we can see that the SRv6 forwarding process can be easily simulated using a computer program.

To make it easier to explain the SRv6 data forwarding process in this book, a simplified version of the SRH shown in Figure 2.5 (a) is used and can be further simplified as shown in Figure 2.5 (b).

2.4.2 SRH TLVs

As shown in Figure 2.4, SRHs use optional TLVs to offer a higher level of network programmability.[3] The TLVs provide better extensibility and

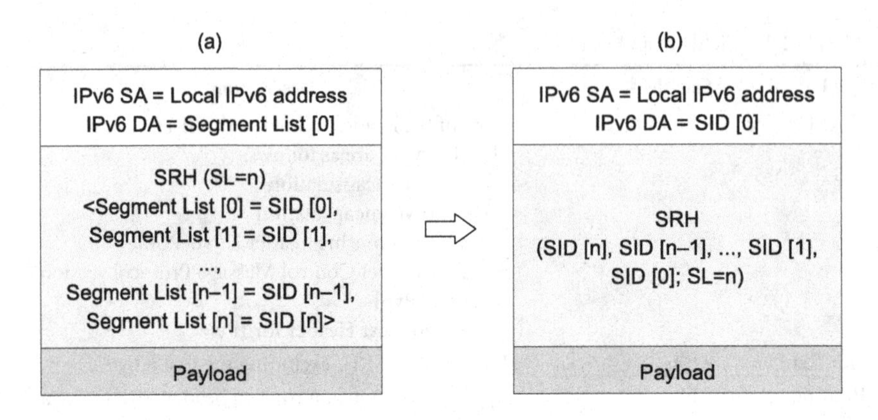

FIGURE 2.5 Abstract SRH.

can carry variable-length data, including encryption, authentication, and performance monitoring information. Additionally, as we described in previous sections, the SIDs in a segment list define instructions to be executed by nodes. This means that each SID can be executed only by the node advertising it, and other nodes only perform basic table lookup and forwarding operations. In contrast, the information carried in TLVs can be processed by any node advertising a SID in the corresponding segment list. To improve forwarding efficiency, a local TLV processing policy that determines, for example, whether to ignore all TLVs or process only certain types of TLVs according to interface configurations, needs to be defined.

SRHs currently support two types of TLVs: Padding TLVs and HMAC TLVs.[3,4]

2.4.2.1 Padding TLV

Two types of Padding TLVs (Pad1 TLV and PadN TLV) are defined for SRHs. If the length of subsequent TLVs is not a multiple of 8 bytes, Padding TLVs can be used for alignment, thereby ensuring that the overall SRH length is a multiple of 8 bytes. Because Padding TLVs are not instructive, they are ignored by nodes during SRH processing.

The Pad1 TLV defines only the Type field, which is used for padding a single byte. If two or more bytes are required, this TLV cannot be used. Figure 2.6 shows the format of the Pad1 TLV.

The PadN TLV is used for padding multiple bytes. Figure 2.7 shows the format of the PadN TLV.

Table 2.2 describes the fields in the PadN TLV.

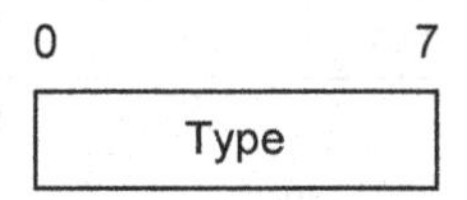

FIGURE 2.6 Pad1 TLV.

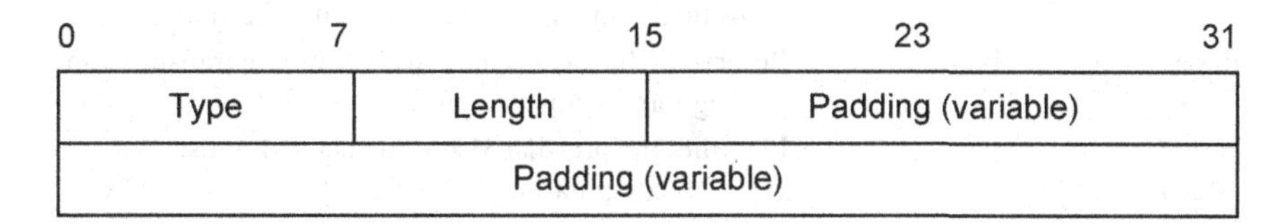

FIGURE 2.7 PadN TLV.

TABLE 2.2 Fields in the PadN TLV

Field	Length	Description
Type	8 bits	TLV type.
Length	8 bits	Length of the Padding field.
Padding	Variable	Padding bits, which are used when the packet length is not a multiple of 8 bytes. As the padding bits are not instructive, they must be set to 0s before transmission and ignored upon receipt.

2.4.2.2 HMAC TLV

The HMAC TLV is used to prevent key information in an SRH from being tampered with. It is optional and requires 8-byte alignment. Figure 2.8 shows the format of the HMAC TLV.

Table 2.3 describes the fields in the HMAC TLV.

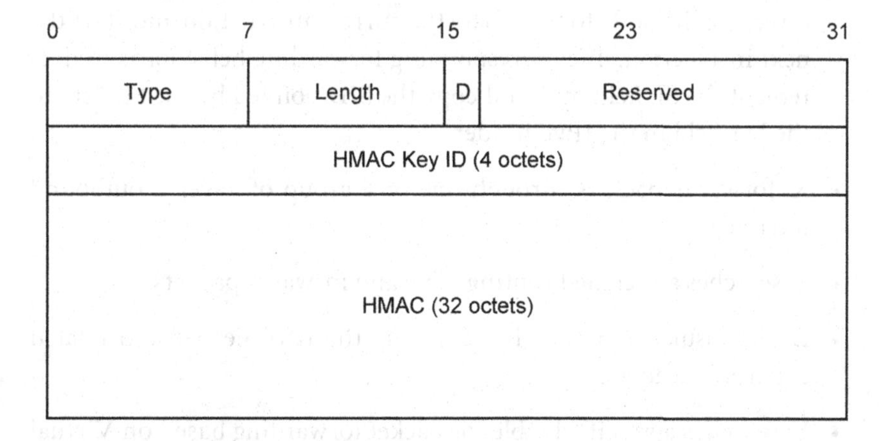

FIGURE 2.8 HMAC TLV.

TABLE 2.3 Fields in the HMAC TLV

Field	Length	Description
Type	8 bits	TLV type.
Length	8 bits	Length of the HMAC field.
D	1 bit	When this bit is set to 1, DA verification is disabled. This field is mainly used in reduced SRH mode.
Reserved	15 bits	Reserved. This field is filled with 0s before transmission and ignored upon receipt.
HMAC Key ID	32 bits	Identifies the pre-shared key and algorithm used for HMAC computation. If this field is set to 0, the TLV does not contain the HMAC field.
HMAC	Variable (up to 256 bits)	HMAC computation result.

2.4.3 SRv6 Instruction Set: Endpoint Node Behaviors

After describing SRv6 instructions and SRHs in previous sections, let's move onto the details of these instructions. The IETF draft *SRv6 Network Programming*[1] defines multiple behaviors or instructions. Each SID is bound to an instruction to specify the action to be taken during SID processing. An SRH can encapsulate an ordered list of SIDs, providing packet-related services including forwarding, encapsulation, and decapsulation.

SRv6 instructions define the atomic functions of networks. Before introducing SRv6 instructions, let's learn the naming rules of these instructions.

- End: the most basic instruction executed by an endpoint node, directing the node to terminate the current instruction and start the next instruction. The corresponding forwarding behavior is to decrement the SL field by 1 and copy the SID pointed by the SL field to the DA field in the IPv6 header.
- X: forwards packets through one or a group of Layer 3 outbound interfaces.
- T: searches a specified routing table and forwards packets.
- D: decapsulates packets by removing the IPv6 header and related extension headers.
- V: searches a specified table for packet forwarding based on Virtual Local Area Network (VLAN) information.

- U: searches a specified table for packet forwarding based on unicast MAC address information.
- M: searches a Layer 2 forwarding table for multicast forwarding.
- B6: binding to an SRv6 Policy.
- BM: binding to an SR-MPLS Policy.

Table 2.4 describes the functions of common SRv6 instructions, each of which integrates one or more atomic functions previously listed.

TABLE 2.4 Functions of Common SRv6 Instructions

Instruction	Function Description	Application Scenario
End	Terminates the current instruction and executes the next one.	Used for packet forwarding through a specified node. An End SID is similar to an SR-MPLS node SID.
End.X	Forwards a packet through a specified outbound interface.	Used for packet forwarding through a specified outbound interface. An End.X SID is similar to an SR-MPLS adjacency SID.
End.T	Searches a specified IPv6 routing table for packet forwarding.	Used in scenarios where multiple routing tables exist.
End.DX6	Decapsulates a packet and forwards it over a specified IPv6 Layer 3 adjacency.	Used in L3VPNv6 scenarios where packets are forwarded to a Customer Edge (CE) through a specified IPv6 adjacency.
End.DX4	Decapsulates a packet and forwards it over a specified IPv4 Layer 3 adjacency.	Used in L3VPNv4 scenarios where packets are forwarded to a CE through a specified IPv4 adjacency.
End.DT6	Decapsulates a packet and searches a specified IPv6 routing table for packet forwarding.	Used in L3VPNv6 scenarios.
End.DT4	Decapsulates a packet and searches a specified IPv4 routing table for packet forwarding.	Used in L3VPNv4 scenarios.
End.DT46	Decapsulates a packet and searches a specified IPv4 or IPv6 routing table for packet forwarding.	Used in L3VPNv4/L3VPNv6 scenarios.

(*Continued*)

TABLE 2.4 (*Continued*) Functions of Common SRv6 Instructions

Instruction	Function Description	Application Scenario
End.DX2	Decapsulates a packet and forwards it through a specified Layer 2 outbound interface.	Used in EVPN VPWS scenarios.
End.DX2V	Decapsulates a packet and searches a specified Layer 2 table for packet forwarding based on inner VLAN information.	Used in EVPN Virtual Private LAN Service (VPLS) scenarios.
End.DT2U	Decapsulates a packet, learns the inner source MAC address in a specified Layer 2 table, and searches the table for packet forwarding based on the inner destination MAC address.	Used in EVPN VPLS unicast scenarios.
End.DT2M	Decapsulates a packet, learns the inner source MAC address in a specified Layer 2 table, and forwards the packet to Layer 2 outbound interfaces excluding the specified one.	Used in EVPN VPLS multicast scenarios.
End.B6.Insert	Inserts an SRH and applies a specified SRv6 Policy.	Used in scenarios such as traffic steering into an SRv6 Policy in Insert mode, tunnel stitching, and Software Defined Wide Area Network (SD-WAN) route selection.
End.B6.Insert. Red	Inserts a reduced SRH and applies a specified SRv6 Policy.	Used in scenarios such as traffic steering into an SRv6 Policy in Insert or Reduced mode, tunnel stitching, and SD-WAN route selection.
End.B6.Encaps	Encapsulates an outer IPv6 header and SRH, and applies a specified SRv6 Policy.	Used in scenarios such as traffic steering into an SRv6 Policy in Encaps mode, tunnel stitching, and SD-WAN route selection.
End.B6.Encaps. Red	Encapsulates an outer IPv6 header and reduced SRH, and applies a specified SRv6 Policy.	Used in scenarios such as traffic steering into an SRv6 Policy in Encaps or Reduced mode, tunnel stitching, and SD-WAN route selection.
End.BM	Inserts an MPLS label stack and applies a specified SR-MPLS Policy.	Used in SRv6 and SR-MPLS interworking scenarios where traffic needs to be steered into an SR-MPLS Policy.

2.4.3.1 End SID

End is the most basic SRv6 instruction. A SID bound to the End instruction is called an End SID, which identifies a node. An End SID instructs a network node to forward a packet to the node that advertises the SID. After receiving the packet, the node performs the operations defined by the SID for packet processing.

The End SID instruction includes the following operations:

1. Decrements the SL value by 1.
2. Obtains the next SID from the SRH based on the SL value.
3. Updates the DA field in the IPv6 header with the next SID.
4. Searches the routing table for packet forwarding.

Other parameters, such as Hop Limit, are processed according to the normal forwarding process.

The pseudocode of an End SID is as follows:

```
S01. When an SRH is processed {
S02. If (Segments Left == 0) {
S03.     Send an ICMP Parameter Problem message to the Source Address,
        code 4 (SR Upper-layer Header Error), pointer set to the
        offset of the upper-layer header, interrupt packet processing
        and discard the packet
S04. }
S05. If (IPv6 Hop Limit <= 1) {
S06.     Send an ICMP Time Exceeded message to the Source Address,
        code 0 (Hop limit exceeded in transit), interrupt packet
        processing and discard the packet
S07. }
S08. max_LE = (Hdr Ext Len / 2) - 1
S09. If ((Last Entry > max_LE) or (Segments Left > Last Entry+1)) {
S10.     Send an ICMP Parameter Problem to the Source Address, code
        0 (Erroneous header field encountered), pointer set to the
        Segments Left field, interrupt packet processing and discard
        the packet
S11. }
S12. Decrement Hop Limit by 1
S13. Decrement Segments Left by 1
S14. Update IPv6 DA with Segment List[Segments Left]
S15. Resubmit the packet to the egress IPv6 FIB lookup and transmission
      to the new destination
S16. }
```

2.4.3.2 End.X SID

End.X (short for Layer-3 cross-connect) supports packet forwarding to a Layer 3 adjacency over a specified link. End.X SIDs can be used in scenarios including Topology Independent Loop Free Alternate (TI-LFA) and strict explicit path-based TE.

In essence, End.X SIDs are developed based on End SIDs. End.X can be disassembled into End+X, where X indicates cross-connect (meaning that a packet needs to be directly forwarded to a specified Layer 3 adjacency). Therefore, each End.X SID needs to be bound to one or a group of Layer 3 adjacencies.

The End.X SID instruction includes the following operations:

1. Decrements the SL value by 1.
2. Obtains the next SID from the SRH based on the SL value.
3. Updates the DA field in the IPv6 header with the next SID.
4. Forwards the IPv6 packet to the Layer 3 adjacency bound to the End.X SID.

The pseudocode of an End.X SID is based on that of an End SID, but with line S15 being modified as follows:

```
S15. Resubmit the packet to the IPv6 module for transmission to the new
   IPv6 destination via a member of specific L3 adjacencies
```

2.4.3.3 End.T SID

End.T (short for specific IPv6 table lookup) supports packet forwarding by searching the specified IPv6 routing table. End.T SIDs can be used in common IPv6 routing and VPN scenarios.

End.T SIDs are also developed based on End SIDs. End.T can be disassembled into End+T, where T indicates table lookup for packet forwarding. Therefore, each End.T SID needs to be bound to an IPv6 routing table.

The End.T SID instruction includes the following operations:

1. Decrements the SL value by 1.
2. Obtains the next SID from the SRH based on the SL value.
3. Updates the DA field in the IPv6 header with the next SID.
4. Searches the specified routing table for IPv6 packet forwarding.

The pseudocode of an End.T SID is based on that of an End SID, but with line S15 being modified as follows:

```
S15.1. Set the packet's associated FIB table to the specific IPv6 FIB
S15.2. Resubmit the packet to the egress IPv6 FIB lookup and transmission
       to the new IPv6 destination
```

2.4.3.4 End.DX6 SID

End.DX6 (short for decapsulation and IPv6 cross-connect) supports packet decapsulation and forwarding to a specified IPv6 Layer 3 adjacency. End.DX6 SIDs are mainly used in L3VPNv6 scenarios as per-CE VPN SIDs.

End.DX6 can be disassembled into End+D+X6, where D indicates decapsulation, and X6 indicates IPv6 cross-connect (meaning that a packet needs to be directly forwarded to a specified IPv6 Layer 3 adjacency). Therefore, each End.DX6 SID needs to be bound to one or a group of IPv6 Layer 3 adjacencies.

The End.DX6 SID instruction includes the following operations:

1. Removes the outer IPv6 header and SRH.
2. Forwards the inner IPv6 packet to the IPv6 Layer 3 adjacency bound to the End.DX6 SID.

The pseudocode of an End.DX6 SID is as follows:

```
S01. If (Upper-Layer Header type != 41) {
S02. Send an ICMP Parameter Problem message to the Source Address,
       code 4 (SR Upper-layer Header Error), pointer set to the offset
       of the upper-layer header, interrupt packet processing and discard
       the packet
S03. }
S04. Remove the outer IPv6 Header with all its extension headers
S05. Forward the exposed IPv6 packet to a member of
     specific L3 adjacencies
```

2.4.3.5 End.DX4 SID

End.DX4 (short for decapsulation and IPv4 cross-connect) supports packet decapsulation and forwarding to a specified IPv4 Layer 3 adjacency. End.DX4 SIDs are mainly used in L3VPNv4 scenarios as per-CE VPN SIDs.

End.DX4 can be disassembled into End+D+X4, where D indicates decapsulation, and X4 indicates IPv4 cross-connect (meaning that a packet needs to be directly forwarded to a specified IPv4 Layer 3 adjacency).

Therefore, each End.DX4 SID needs to be bound to one or a group of IPv4 Layer 3 adjacencies.

The End.DX4 SID instruction includes the following operations:

1. Removes the outer IPv6 header and SRH.
2. Forwards the inner IPv4 packet to the IPv4 Layer 3 adjacency bound to the End.DX4 SID.

The pseudocode of an End.DX4 SID is as follows:

```
S01. If (Upper-Layer Header type != 4) {
S02. Send an ICMP Parameter Problem message to the Source Address,
       code 4 (SR Upper-layer Header Error), pointer set to the
     offset of the upper-layer header, interrupt packet processing and
       discard the packet
S03. }
S04. Remove the outer IPv6 Header with all its extension headers
S05. Forward the exposed IPv4 packet to a member of specific
    L3 adjacencies
```

2.4.3.6 End.DT6 SID

End.DT6 (short for decapsulation and specific IPv6 table lookup) supports packet decapsulation and lookup in a specified IPv6 routing table for packet forwarding. End.DT6 SIDs are mainly used in L3VPNv6 scenarios as per-VPN SIDs.

End.DT6 can be disassembled into End+D+T6, where D indicates decapsulation and T6 indicates IPv6 table lookup for forwarding. Therefore, each End.DT6 SID needs to be bound to an IPv6 routing table, which can be either the IPv6 routing table of a VPN instance or a public IPv6 routing table.

The End.DT6 SID instruction includes the following operations:

1. Removes the outer IPv6 header and SRH.
2. Searches the IPv6 routing table bound to the End.DT6 SID to forward the inner IPv6 packet.

The pseudocode of an End.DT6 SID is as follows:

```
S01. If (Upper-Layer Header type != 41) {
S02. Send an ICMP Parameter Problem message to the Source Address,
       code 4 (SR Upper-layer Header Error), pointer set to the
       offset of the upper-layer header, interrupt packet processing
```

```
        and discard the packet
S03. }
S04. Remove the outer IPv6 Header with all its extension headers
S05. Set the packet's associated FIB table to the specific IPv6 FIB
S06. Resubmit the packet to the egress IPv6 FIB lookup and transmission
      to the new IPv6 destination
```

2.4.3.7 End.DT4 SID

End.DT4 (short for decapsulation and specific IPv4 table lookup) supports packet decapsulation and lookup in a specified IPv4 routing table for packet forwarding. End.DT4 SIDs are mainly used in L3VPNv4 scenarios as per-VPN SIDs.

End.DT4 can be disassembled into End+D+T4, where D indicates decapsulation and T4 indicates IPv4 table lookup for forwarding. Therefore, each End.DT4 SID needs to be bound to an IPv4 routing table, which can be either the IPv4 routing table of a VPN instance or a public IPv4 routing table.

The End.DT4 SID instruction includes the following operations:

1. Removes the outer IPv6 header and SRH.
2. Searches the IPv4 routing table bound to the End.DT4 SID to forward the inner IPv4 packet.

The pseudocode of an End.DT4 SID is as follows:

```
S01. If (Upper-layer Header type == 4) {
S02. Remove the outer IPv6 Header with all its extension headers
S03. Set the packet's associated FIB table to the specific IPv4 FIB
S04. Resubmit the packet to the egress IPv4 FIB lookup and transmission
       to the new IPv4 destination
S05. } Else if (Upper-layer Header type == 41) {
S06. Remove the outer IPv6 Header with all its extension headers
S07. Set the packet's associated FIB table to the specific IPv6 FIB
S08. Resubmit the packet to the egress IPv6 FIB lookup and transmission
       to the new IPv6 destination
S09. } Else {
S10. Send an ICMP Parameter Problem message to the Source Address,
       code 4 (SR Upper-layer Header Error), pointer set to the offset
       of the upper-layer header, interrupt packet processing and discard
     the packet
S11. }
```

2.4.3.8 End.DT46 SID

End.DT46 (short for decapsulation and specific IP table lookup) supports packet decapsulation and lookup in a specified IPv4 or IPv6 routing table for packet forwarding. End.DT46 SIDs are mainly used in L3VPN scenarios as per-VPN SIDs.

End.DT46 can be disassembled into End+D+T46, where D indicates decapsulation and T46 indicates IPv4 or IPv6 table lookup for forwarding. Therefore, each End.DT46 SID needs to be bound to an IPv4 or IPv6 routing table, which can be either the IPv4 or IPv6 routing table of a VPN instance or a public IPv4 or IPv6 routing table.

The End.DT46 SID instruction includes the following operations:

1. Removes the outer IPv6 header and SRH.
2. According to the Layer 3 protocol type of the inner IP packet, searches the IPv4 or IPv6 routing table bound to the End.DT46 SID to forward the inner IP packet.

The pseudocode of an End.DT46 SID is as follows:

```
S01. If (Upper-layer Header type == 4) {
S02. Remove the outer IPv6 Header with all its extension headers
S03. Set the packet's associated FIB table to the specific IPv4 FIB
S04. Resubmit the packet to the egress IPv4 FIB lookup and transmission
       to the new IPv4 destination
S05. } Else if (Upper-layer Header type == 41) {
S06. Remove the outer IPv6 Header with all its extension headers
S07. Set the packet's associated FIB table to the specific IPv6 FIB
S08. Resubmit the packet to the egress IPv6 FIB lookup and transmission
       to the new IPv6 destination
S09. } Else {
S10. Send an ICMP Parameter Problem message to the Source Address,
       code 4 (SR Upper-layer Header Error), pointer set to the offset
       of the upper-layer header, interrupt packet processing and discard
     the packet
S11. }
```

2.4.3.9 End.DX2 SID

End.DX2 (short for decapsulation and L2 cross-connect) supports packet decapsulation and forwarding through a specified Layer 2 outbound interface. End.DX2 SIDs are mainly used in Layer 2 Virtual Private Network (L2VPN) and EVPN Virtual Private Wire Service (VPWS)[5] scenarios.

End.DX2 can be disassembled into End+D+X2, where D indicates decapsulation and X2 indicates cross-connect (meaning that a packet needs to be directly forwarded through a specified Layer 2 interface). Therefore, each End.DX2 SID needs to be bound to a Layer 2 outbound interface.

The End.DX2 SID instruction includes the following operations:

1. Removes the outer IPv6 header and SRH.
2. Forwards the inner Ethernet frame to the Layer 2 outbound interface bound to the End.DX2 SID.

The pseudocode of an End.DX2 SID is as follows:

```
S01. If (Upper-Layer Header type != 143) {
S02. Send an ICMP Parameter Problem message to the Source Address,
       code 4 (SR Upper-layer Header Error), pointer set to the offset
       of the upper-layer header, interrupt packet processing and
       discard the packet
S03. }
S04. Remove the outer IPv6 Header with all its extension headers and
     forward the Ethernet frame to the specific outgoing L2 interface
```

2.4.3.10 End.DX2V SID

End.DX2V (short for decapsulation and VLAN L2 table lookup) supports packet decapsulation and lookup in a specified Layer 2 table based on the inner VLAN information of the packet to be forwarded. End.DX2V SIDs are mainly used in EVPN flexible cross-connect scenarios.[5]

End.DX2V can be disassembled into End+D+X2V, where D indicates decapsulation and X2V indicates VLAN Layer 2 table lookup. Therefore, each End.DX2V SID needs to be bound to a Layer 2 table.

The End.DX2V SID instruction includes the following operations:

1. Removes the outer IPv6 header and SRH.
2. Searches the Layer 2 table bound to the End.DX2V SID based on the VLAN information in the inner Ethernet frame.

The pseudocode of an End.DX2V SID is based on that of an End.DX2 SID, but with line S04 being modified as follows:

```
S04. Remove the outer IPv6 Header with all its extension headers, lookup
   the exposed inner VLANs in the specific L2 table, and forward via
   the matched table entry.
```

2.4.3.11 End.DT2U SID

End.DT2U (short for decapsulation and unicast MAC L2 table lookup) supports packet decapsulation, inner source MAC address learning and saving to a specified Layer 2 table, and lookup in the table for packet forwarding based on the inner destination MAC address. End.DT2U SIDs are mainly used in EVPN bridging unicast scenarios.[5]

End.DT2U can be disassembled into End+D+T2U, where D indicates decapsulation and T2U indicates Layer 2 table lookup for unicast forwarding. Therefore, each End.DT2U SID needs to be bound to a Layer 2 table.

The End.DT2U SID instruction includes the following operations:

1. Removes the outer IPv6 header and SRH.
2. Learns the source MAC address of the inner Ethernet frame.
3. Saves the source MAC address to the Layer 2 table bound to the End.DT2U SID.
4. Searches the Layer 2 table bound to the End.DT2U SID based on the destination MAC address of the inner Ethernet frame.
5. Forwards the corresponding packet accordingly.

The pseudocode of an End.DT2U SID is as follows:

```
S01. If (Upper-Layer Header type != 143) {
S02. Send an ICMP Parameter Problem message to the Source Address,
       code 4 (SR Upper-layer Header Error), pointer set to the
       offset of the upper-layer header, interrupt packet processing
       and discard the packet
S03. }
S04. Remove the IPv6 header and all its extension headers
S05. Learn the exposed inner MAC Source Address in the specific
    L2 table (T)
S06. Lookup the exposed inner MAC Destination Address in table T
S07. If (matched entry in T) {
S08. Forward via the matched table T entry
S09. } Else {
S10. Forward via all outgoing L2 interfaces entries in table T
S11. }
```

2.4.3.12 End.DT2M SID

End.DT2M (short for decapsulation and L2 table flooding) supports packet decapsulation, inner source MAC address learning and saving to a specified Layer 2 table, and packet forwarding to Layer 2 outbound interfaces excluding the specified one. End.DT2M SIDs are mainly used in EVPN bridging Broadcast & Unknown-unicast & Multicast (BUM)[5] and EVPN E-Tree[6] scenarios.

End.DT2M can be disassembled into End+D+T2M, where D indicates decapsulation and T2M indicates Layer 2 table lookup for multicast forwarding. Therefore, each End.DT2M SID needs to be bound to a Layer 2 table and may carry EVPN ESI filtering or EVPN E-Tree parameters in order to exclude specified outbound interfaces during flooding.

The End.DT2M SID instruction includes the following operations:

1. Removes the outer IPv6 header and SRH.

2. Learns the source MAC address of the inner Ethernet frame.

3. Saves the source MAC address to the Layer 2 table bound to the End. DT2M SID.

4. Forwards the inner Ethernet frame through Layer 2 outbound interfaces excluding the interface specified by the parameters carried in the End.DT2M SID.

The pseudocode of an End.DT2M SID is as follows:

```
S01. If (Upper-Layer Header type != 143) {
S02. Send an ICMP Parameter Problem message to the Source Address,
       code 4 (SR Upper-layer Header Error), pointer set to the offset
       of the upper-layer header, interrupt packet processing and
       discard the packet
S03. }
S04. Remove the IPv6 header and all its extension headers
S05. Learn the exposed inner MAC Source Address in the specific L2 table
S06. Forward via all outgoing L2 interfaces excluding the one specified
     in Arg.FE2
```

2.4.3.13 End.B6.Insert SID

End.B6.Insert (short for endpoint bound to an SRv6 Policy with Insert) supports the application of specified SRv6 Policies[7] to packets. End. B6.Insert SIDs instantiate binding SIDs in SRv6 and are used in scenarios where TE can be flexibly implemented across SRv6 domains.

End.B6.Insert can be disassembled into End+B6+Insert, where B6 indicates the application of an SRv6 Policy and Insert indicates the insertion of an SRH after the IPv6 header. Therefore, each End.B6.Insert SID needs to be bound to an SRv6 Policy.

The End.B6.Insert SID instruction includes the following operations:

1. Inserts an SRH (including segment lists) after the IPv6 header.

2. Sets the DA to the first SID of a specified SRv6 Policy.

3. Searches the IPv6 routing table.

4. Forwards the new IPv6 packet accordingly.

The pseudocode of an End.B6.Insert SID is as follows:

```
S01. When an SRH is processed {
S02. If (Segments Left == 0) {
S03.     Send an ICMP Parameter Problem message to the Source Address,
        code 4 (SR Upper-layer Header Error), pointer set to the
        offset of the upper-layer header, interrupt packet processing
        and discard the packet
S04. }
S04. If (IPv6 Hop Limit <= 1) {
S05.     Send an ICMP Time Exceeded message to the Source Address,
        code 0 (Hop limit exceeded in transit), interrupt packet
        processing and discard the packet
S06. }
S07. max_LE = (Hdr Ext Len / 2) - 1
S08. If ((Last Entry > max_LE) or (Segments Left > (Last Entry+1))){
S09.     Send an ICMP Parameter Problem to the Source Address, code 0
        (Erroneous header field encountered), pointer set to the
        Segments Left field, interrupt packet processing and discard
        the packet
S11. }
S12. Decrement Hop Limit by 1
S13. Insert a new SRH after the IPv6 Header and the SRH
      contains the list of segments of the specific SRv6 Policy
S14. Set the IPv6 DA to the first segment of the SRv6 Policy
S15. Resubmit the packet to the egress IPv6 FIB lookup and transmission
      to the new IPv6 destination
S16. }
```

In addition, the End.B6.Insert.Red instruction is an optimization of the End.B6.Insert instruction, except that the former requires a reduced SRH (i.e., the SRH excluding the first SID of the involved SRv6 Policy) to be added.

2.4.3.14 End.B6.Encaps SID

End.B6.Encaps (short for endpoint bound to an SRv6 Policy with Encaps) supports the application of specified SRv6 Policies[7] to packets. End.B6.Encaps SIDs also instantiate binding SIDs in SRv6 and are used in scenarios where TE can be flexibly implemented across SRv6 domains.

End.B6.Encaps can be disassembled into End+B6+Encaps, where B6 indicates the application of an SRv6 Policy and Encaps indicates the encapsulation of an outer IPv6 header and SRH. Therefore, each End.B6.Encaps SID needs to be bound to an SRv6 Policy.

The End.B6.Encaps SID instruction includes the following operations:

1. Decrements the SL value of the inner SRH by 1.
2. Updates the DA field of the inner IPv6 header with the SID to which the SL field is pointing.

3. Encapsulates an IPv6 header and SRH (including segment lists).
4. Sets the source address to the address of the current node and the DA to the first SID of the involved SRv6 Policy.
5. Sets other fields in the outer IPv6 header.
6. Searches the IPv6 routing table.
7. Forwards the new IPv6 packet accordingly.

The pseudocode of an End.B6.Encaps SID is as follows:

```
S01. When an SRH is processed {
S02. If (Segments Left == 0) {
S03.     Send an ICMP Parameter Problem message to the Source Address,
        code 4 (SR Upper-layer Header Error), pointer set to the offset
        of the upper-layer header, interrupt packet processing and
        discard the packet
S04. }
S05. If (IPv6 Hop Limit <= 1) {
S06.    Send an ICMP Time Exceeded message to the Source Address, code
        0 (Hop limit exceeded in transit), interrupt packet processing
        and discard the packet
S07. }
S08. max_LE = (Hdr Ext Len / 2) - 1
S09. If ((Last Entry > max_LE) or (Segments Left > (Last Entry+1)) {
S10.     Send an ICMP Parameter Problem to the Source Address, code 0
        (Erroneous header field encountered), pointer set to the
        Segments Left field, interrupt packet processing and discard
        the packet
S11. }
S12. Decrement Hop Limit by 1
S13. Decrement Segments Left by 1
S14. Update the inner IPv6 DA with inner Segment List[Segments Left]
S15. Push a new IPv6 header with its own SRH containing the list of
     segments of the SRv6 Policy
S16. Set the outer IPv6 SA to itself
S17. Set the outer IPv6 DA to the first SID of the SRv6 Policy
S18. Set the outer Payload Length, Traffic Class, Flow Label and
     Next Header fields
S19. Resubmit the packet to the egress IPv6 FIB lookup and transmission
     to the new IPv6 destination
S20. }
```

In addition, the End.B6.Encaps.Red instruction is an optimization of the End.B6.Encaps instruction, with the difference lying in the fact that the former requires a reduced SRH to be added.

2.4.3.15 End.BM SID

End.BM (short for endpoint bound to an SR-MPLS Policy) supports the application of a specified SR-MPLS Policy[7] to packets. End.BM SIDs instantiate SR-MPLS binding SIDs in SRv6 and are used in scenarios where TE can be flexibly implemented across MPLS domains.

End.BM can be disassembled into End+BM, where BM indicates the application of an SR-MPLS Policy. Therefore, each End.BM SID needs to be bound to an SR-MPLS Policy.

The End.BM SID instruction includes the following operations:

1. Decrements the SL value of the inner SRH by 1.
2. Inserts an MPLS label stack contained in an SR-MPLS Policy before the IPv6 header.
3. Searches the corresponding MPLS label forwarding table.
4. Forwards the new MPLS packet accordingly.

The pseudocode of an End.BM SID is as follows:

```
S01. When an SRH is processed {
S02. If (Segments Left == 0) {
S03.     Send an ICMP Parameter Problem message to the Source Address
        code 4 (SR Upper-layer Header Error), pointer set to the
        offset of the upper-layer header, interrupt packet processing
        and discard the packet
S04. }
S05. If (IPv6 Hop Limit <= 1) {
S06.     Send an ICMP Time Exceeded message to the Source Address, code
        0 (Hop limit exceeded in transit), interrupt packet processing
        and discard the packet
S07. }
S08. max_LE = (Hdr Ext Len / 2) - 1
S09. If ((Last Entry > max_LE) or (Segments Left > (Last Entry+1)) {
S10.     Send an ICMP Parameter Problem to the Source Address,
        code 0 (Erroneous header field encountered),
        pointer set to the segments left field,
        interrupt packet processing and discard the packet
S11. }
S12. Decrement Hop Limit by 1
S13. Decrement Segments Left by 1
S14. Push the MPLS label stack for SR-MPLS Policy
S15. Submit the packet to the MPLS engine for transmission to the
      topmost label
S16. }
```

2.4.4 SRv6 Instruction Set: Source Node Behaviors

As we previously described, an SRv6 source node steers packets into an SRv6 Policy and, if possible, encapsulates SRHs into the packets. Table 2.5 describes the behaviors of an SRv6 source node.

TABLE 2.5 Source Node Behaviors

Behavior	Function Description
H.Insert	Inserts an SRH into a received IP packet and searches the corresponding routing table for packet forwarding.
H.Insert.Red	Inserts a reduced SRH into a received IP packet and searches the corresponding routing table for packet forwarding.
H.Encaps	Encapsulates an outer IPv6 header and SRH for a received IP packet, and searches the corresponding routing table for packet forwarding.
H.Encaps.Red	Encapsulates an outer IPv6 header and reduced SRH for a received IP packet, and searches the corresponding routing table for packet forwarding.
H.Encaps.L2	Encapsulates an outer IPv6 header and SRH for a received Layer 2 packet, and searches the corresponding routing table for packet forwarding.
H.Encaps.L2.Red	Encapsulates an outer IPv6 header and reduced SRH for a received Layer 2 packet, and searches the corresponding routing table for packet forwarding.

2.4.4.1 H.Insert

H.Insert (short for headend with insertion of an SRv6 Policy) is a behavior used to steer IP packets into an SRv6 Policy so that they are forwarded over a new path. This behavior is usually used in TI-LFA scenarios and must be bound to an SRv6 Policy.

This behavior includes the following operations:

1. Inserts an SRH after the IPv6 header.
2. Sets the DA to the first SID of the involved SRv6 Policy.
3. Searches the IPv6 routing table.
4. Forwards the new IPv6 packet accordingly.

The pseudocode of this behavior is as follows:

```
S01. insert the SRH containing the list of segments of SRv6 Policy
S02. update the IPv6 DA to the first segment of SRv6 Policy
S03. forward along the shortest path to the new IPv6 destination
```

The H.Insert.Red behavior is an optimization of the H.Insert behavior. Their difference is that the former requires a reduced SRH to be added.

2.4.4.2 H.Encaps

H.Encaps (short for headend with encapsulation in an SRv6 Policy) is another behavior used to steer IP packets into an SRv6 Policy so that they are forwarded over a new path. As for its application, it is usually used in L3VPNv4 or L3VPNv6 scenarios and must be bound to an SRv6 Policy.

This behavior includes the following operations:

1. Encapsulates an IPv6 header and SRH before the original IP header.
2. Sets the source address to the address of the current node and the DA to the first SID of the involved SRv6 Policy.
3. Sets other fields in the outer IPv6 header.
4. Searches the IPv6 routing table.
5. Forwards the new IPv6 packet accordingly.

The pseudocode of this behavior is as follows:

```
S01. push an IPv6 header with its own SRH containing the list of
    segments of SRv6 Policy
S02. set outer IPv6 SA to itself and outer IPv6 DA to the first segment
    of SRv6 Policy
S03. set outer payload length, traffic class and flow label
S04. update the Next-Header value
S05. decrement inner Hop Limit or TTL
S06. forward along the shortest path to the new IPv6 destination
```

The H.Encaps.Red behavior is an optimization of the H.Encaps behavior, with the distinction being that the former requires a reduced SRH to be added.

2.4.4.3 H.Encaps.L2

H.Encaps.L2 (short for headend with encapsulation of L2 frames) is a behavior used to steer Layer 2 frames into an SRv6 Policy so that they are forwarded over a new path. As such, H.Encaps.L2 must be bound to an SRv6 Policy. Because this behavior applies to Layer 2 frames, the last SID of the involved SRv6 Policy must be of the End.DX2, End.DX2V, End. DT2U, or End.DT2M type.

This behavior includes the following operations:

1. Encapsulates a new IPv6 header and SRH for a Layer 2 frame.
2. Searches the IPv6 routing table.
3. Forwards the frame through a new SRv6 Policy accordingly.

The pseudocode of this behavior is similar to that of the H.Encaps behavior.

Compared side by side, the H.Encaps.L2.Red behavior is an optimization of the H.Encaps.L2 behavior, and it requires a reduced SRH to be added.

2.4.5 SRv6 Instruction Set: Flavors

In this section, we describe three optional behaviors called flavors, which are defined to enhance End series behaviors, thereby meeting diverse service requirements. Table 2.6 lists the flavors and their functions.

2.4.5.1 PSP

The PSP flavor is used to remove the SRH on the penultimate endpoint node and needs to be used with End, End.X, and End.T behaviors as an additional behavior.

This flavor includes the following operations:

1. After the End behavior is performed, checks whether the updated SL value is 0.
2. If the value is 0, removes the SRH.

The pseudocode of this flavor is as follows:

```
S14.1. If (updated SL == 0) {
S14.2. Pop the SRH
S14.3. }
```

TABLE 2.6 Flavors

Flavor	Function Description
Penultimate Segment Pop of the SRH (PSP)	Removes the SRH on the penultimate endpoint node.
Ultimate Segment Pop of the SRH (USP)	Removes the SRH on the ultimate endpoint node.
Ultimate Segment Decapsulation (USD)	Decapsulates the outer IPv6 header on the ultimate endpoint node.

The SRH can be removed on the penultimate endpoint node by using a PSP-flavored SID to relieve the pressure on the egress.

2.4.5.2 USP

The USP flavor is used to remove the SRH on the ultimate endpoint node and needs to be used with End, End.X, and End.T behaviors as an additional behavior.

This flavor includes the following operations:

1. Before the End behavior is performed, checks whether the current SL value is 0.
2. If the value is 0, removes the SRH.

The pseudocode of this flavor is as follows:

```
S02. If (Segments Left == 0) {
S03. Pop the SRH
S04. }
```

2.4.5.3 USD

The USD flavor is used to decapsulate the outer IPv6 header on the ultimate endpoint node and needs to be used with End, End.X, and End.T behaviors as an additional behavior.

This flavor includes the following operations:

1. Before the corresponding endpoint behavior is performed, checks whether the current SL value is 0.
2. If the value is 0, removes the outer IPv6 header with all its extension headers and processes the next packet header.

The pseudocode of this flavor is as follows:

```
S02. If (Segments Left == 0) {
S03. Skip the SRH processing and proceed to the next header
S04. }
```

If the upper-layer protocol header is an IPv4 or IPv6 header, the node decapsulates the outer IPv6 header and executes the table lookup and forwarding instructions corresponding to the End, End.X, or End.T behavior on the inner packet header.

The pseudocode of the USD flavor corresponding to the End behavior is as follows:

```
S01. If (Upper-layer Header type == 41 || 4) {
S02. Remove the outer IPv6 Header with all its extension headers
S03. Resubmit the packet to the egress IP FIB lookup and transmission
       to the new destination
S04. } Else {
S05. Send an ICMP Parameter Problem message to the Source Address,
       code 4 (SR Upper-layer Header Error), pointer set to the offset
       of the upper-layer header, interrupt packet processing and
       discard the packet
S06. }
```

The pseudocode of the USD flavor corresponding to the End.T behavior is as follows:

```
S01. If (Upper-layer Header type == 41 || 4) {
S02. Remove the outer IPv6 Header with all its extension headers
S03. Set the packet's associated FIB table to the specific IP FIB
S04. Resubmit the packet to the egress IP FIB lookup and transmission
       to the new destination
S05. } Else {
S06. Send an ICMP Parameter Problem message to the Source Address,
       code 4 (SR Upper-layer Header Error), pointer set to the
       offset of the upper-layer header, interrupt packet processing
       and discard the packet
S07. }
```

The pseudocode of the USD flavor corresponding to the End.X behavior is as follows:

```
S01. If (Upper-layer Header type == 41 || 4) {
S02. Remove the outer IPv6 Header with all its extension headers
S03. Forward the exposed IP packet to a member of specific
       L3 adjacencies
S04. } Else {
S05. Send an ICMP Parameter Problem message to the Source Address,
       code 4 (SR Upper-layer Header Error), pointer set to the offset
       of the upper-layer header, interrupt packet processing and
       discard the packet
S06. }
```

The preceding three flavors define additional SRH removal and IPv6 header decapsulation functions corresponding to the End, End.T, and End.X behaviors, and can be combined as needed. For example, an End SID carrying the PSP and USP flavors indicates that both PSP and USP capabilities are supported. This means that the pseudocode of both PSP and USP is inserted into the pseudocode of the End behavior. In this way, when the SID is encapsulated as the penultimate SID of an SRH, PSP-defined operations are performed; if it is encapsulated as the last SID, USP-defined operations are performed.

2.5 NETWORK PROGRAM EXECUTION: SRv6 PACKET FORWARDING

2.5.1 Local SID Table

Each SRv6 node maintains a Local SID Table containing all SRv6 SIDs generated on the node. This table provides the following functions:

- Stores locally generated SIDs, such as End.X SIDs.
- Specifies instructions bound to the SIDs.
- Stores forwarding information related to the instructions, such as VPN instance, outbound interface, and next hop information.

In this section, we list some common Local SID Tables.

Local SID Table containing End SIDs:

```
<HUAWEI> display segment-routing ipv6 local-sid end forwarding

               My Local-SID End Forwarding Table
               ---------------------------------

SID         : 2001:DB8:10::1:0:0/128              FuncType : End
Flavor      : --
LocatorName : as1                               LocatorID: 1

SID         : 2001:DB8:10::1:0:1/128              FuncType : End
Flavor      : PSP
LocatorName : as1                               LocatorID: 1

Total SID(s): 2
```

Local SID Table containing End.X SIDs:

```
<HUAWEI> display segment-routing ipv6 local-sid end-x forwarding

            My Local-SID End.X Forwarding Table
            -----------------------------------

SID        : 2001:DB8::101:0:1/128               FuncType : End.X
Flavor      : --
LocatorName: as2                               LocatorID: 1
NextHop      :                 Interface :        ExitIndex:
FE80::3A00:10FF:FE03:1         GE2/0/0              0x0000000a

SID        : 2001:DB8::101:0:2/128               FuncType : End.X
Flavor      : PSP
LocatorName: as2                               LocatorID: 1
NextHop      :                 Interface :        ExitIndex:
FE80::3A00:10FF:FE03:0         GE1/0/0              0x00000009

Total SID(s): 2
```

Local SID Table containing End.DT4 SIDs:

```
<HUAWEI> display segment-routing ipv6 local-sid end-dt4 forwarding

             My Local-SID End.DT4 Forwarding Table
             -------------------------------------

SID        : 2001:DB8:1234::40/128                FuncType : End.DT4
VPN Name    : vpn1                               VPN ID    : 67
LocatorName: locator_1_locator_1_locator_1_3        LocatorID: 7

SID        : 2001:DB8:1234::41/128                FuncType : End.DT4
VPN Name    : vpn2                               VPN ID    : 68
LocatorName: locator_1_locator_1_locator_1_3        LocatorID: 7

Total SID(s): 2
```

Local SID Table containing End.DT6 SIDs:

```
<HUAWEI> display segment-routing ipv6 local-sid end-dt6 forwarding

             My Local-SID End.DT6 Forwarding Table
             -------------------------------------

SID        : 2001:DB8:12::4/128               FuncType : End.DT6
VPN Name    : 1                              VPN ID    : 3
LocatorName: l1                               LocatorID: 3

Total SID(s): 1
```

Local SID Table containing End.DX4 SIDs:

```
<HUAWEI> display segment-routing ipv6 local-sid end-dx4 forwarding

              My Local-SID End.DX4 Forwarding Table
              -------------------------------------

SID        : 2001:DB8:3::13/128                       FuncType : End.DX4
VPN Name   : test2                                  VPN ID   : 13
LocatorName: 2                                      LocatorID: 11
NextHop    :                 Interface :             ExitIndex:
 3::3                         Vbdif11                 0x0000003f

Total SID(s): 1
```

Local SID Table containing End.DX6 SIDs:

```
<HUAWEI> display segment-routing ipv6 local-sid end-dx6 forwarding

              My Local-SID End.DX6 Forwarding Table
              -------------------------------------

SID        : 2001:DB8:1::13/128                       FuncType : End.DX6
VPN Name   : test1                                   VPN ID    : 3
LocatorName: 1                                        LocatorID: 1
NextHop    :               Interface :                 ExitIndex:
 1::3                       Vbdif1                     0x0000002c

Total SID(s): 1
```

Local SID Table containing End.DT2U SIDs:

```
<HUAWEI> display segment-routing ipv6 local-sid end-dt2u forwarding

                My Local-SID End.DT2U Forwarding Table
                ---------------------------------------

SID             : 2001:DB8:1::8/128             FuncType : End.DT2U
Bridge-domain ID: 10
LocatorName     : 2                           LocatorID : 1

Total SID(s): 1
```

Local SID Table containing End.DT2M SIDs:

```
<HUAWEI> display segment-routing ipv6 local-sid end-dt2m forwarding

                My Local-SID End.DT2M Forwarding Table
                ---------------------------------------

SID             : 2001:DB8:1::3/128            FuncType : End.DT2M
Bridge-domain ID: 20
LocatorName     : 2                          LocatorID: 1

Total SID(s): 1
```

Local SID Table containing End.DX2 SIDs:

```
<HUAWEI> display segment-routing ipv6 local-sid end-dx2 forwarding

              My Local-SID End.DX2 Forwarding Table
              -------------------------------------

SID         : 2001:DB8:2::1/128                    FuncType : End.DX2
EVPL ID     : 1
LocatorName: l1                                   LocatorID: 1

Total SID(s): 1
```

2.5.2 Packet Forwarding Process

In this section, we use an example to describe the SRv6 packet forwarding process.

Assume that an IPv4 packet needs to be forwarded from host H1 to host H2, and host H1 first sends the packet to node A for processing, as shown in Figure 2.9. Note that only nodes A, B, D, and F support SRv6.

Network programming is implemented on the SRv6 source node A for sending the packet to node F over B–C and D–E links, and then to host H2 from node F. The specifics of how the packet is processed when being forwarded from node A to node F are as follows:

2.5.2.1 Step 1: Processing on SRv6 Source Node A

As shown in Figure 2.10, node A encapsulates SRv6 path information into an SRH and specifies End.X SIDs for the B–C and D–E links. Additionally, node A needs to encapsulate the End.DT4 SID A6::100 advertised by node F, which corresponds to an IPv4 VPN instance of node F. Be aware that the SIDs are encapsulated in reverse order, and that because three SIDs exist in total, the SL value in the encapsulated packet is 2. The SL field points to the segment list to be processed, which is Segment List.[2] Given this, node A copies the SID A2::23 to the DA field in the outer IPv6 header, searches the corresponding IPv6 routing table according to the longest match rule, and then forwards the packet to node B.

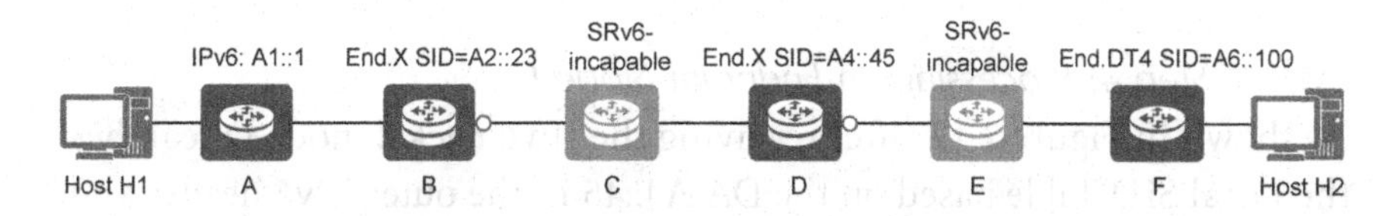

FIGURE 2.9 SRv6 packet forwarding process.

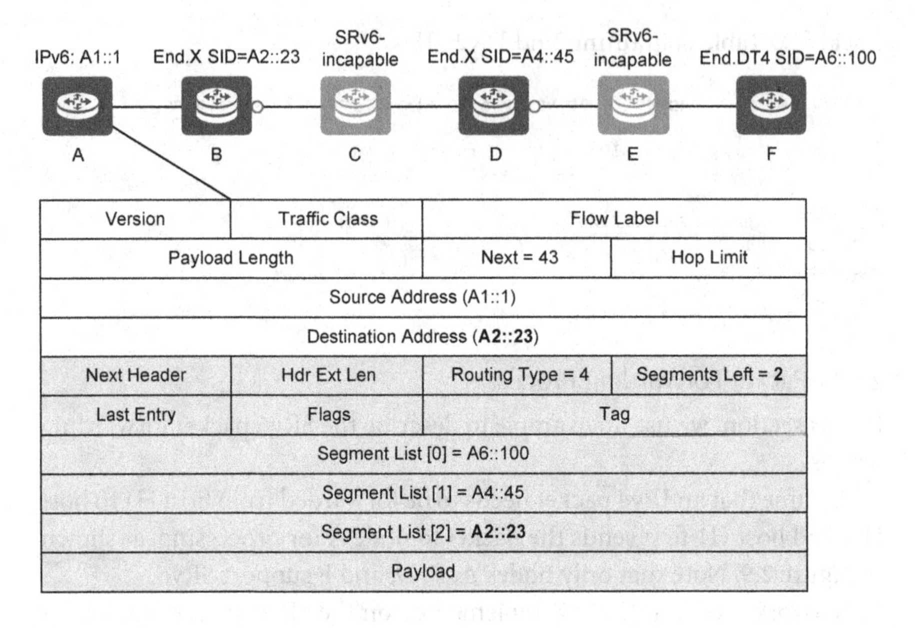

FIGURE 2.10 Processing on SRv6 source node A.

2.5.2.2 Step 2: Processing on Endpoint Node B

As shown in Figure 2.11, after receiving the IPv6 packet, node B searches the Local SID Table based on the DA A2::23 in the outer IPv6 header and finds a matching End.X SID. According to the instructions defined by the End.X SID, node B decrements the SL value by 1, updates the DA field in the outer IPv6 header with the SID specified by the SL field, and then sends the packet over the link bound to the End.X SID.

2.5.2.3 Step 3: Processing on Transit Node C

Node C is a transit node that can process only IPv6 headers, not SRHs. For this reason, node C processes the received packet as it does to a common IPv6 packet. To put it more precisely, it searches the corresponding IPv6 routing table according to the longest match rule and then forwards the packet to node D represented by the current DA.

2.5.2.4 Step 4: Processing on Endpoint Node D

As shown in Figure 2.12, after receiving the IPv6 packet, node D searches the Local SID Table based on the DA A4::45 in the outer IPv6 header and finds a matching End.X SID. According to the instructions defined by the

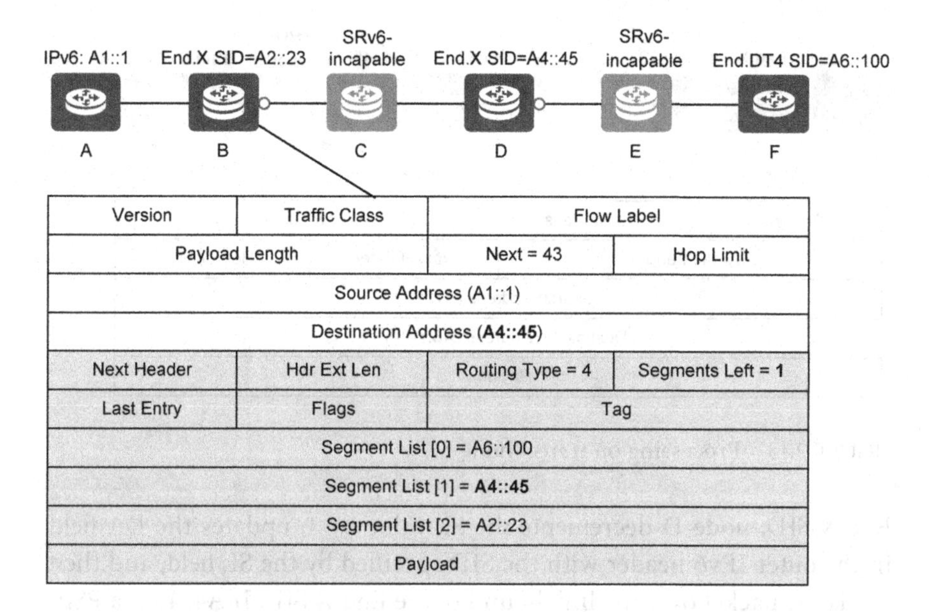

FIGURE 2.11 Processing on endpoint node B.

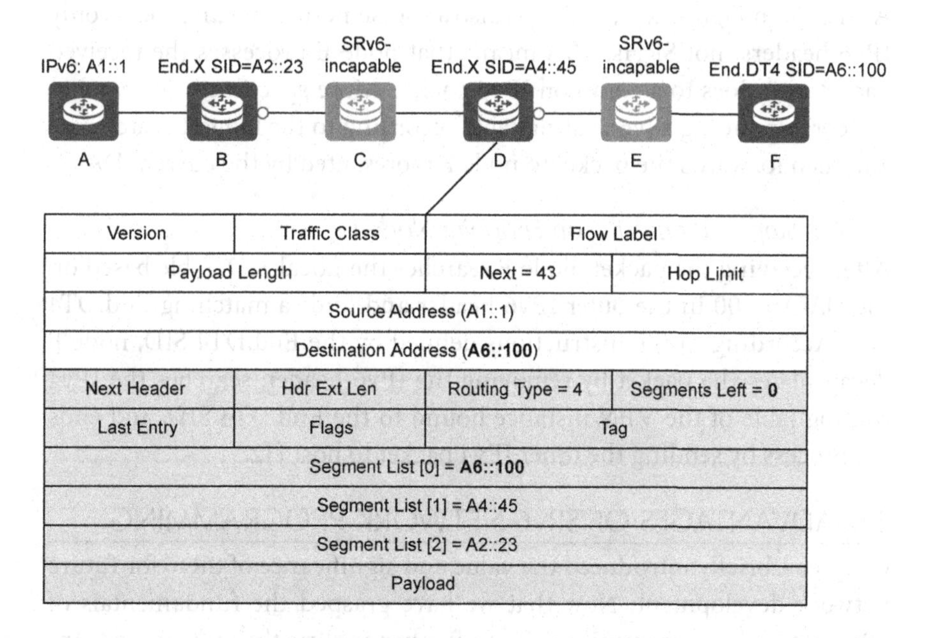

FIGURE 2.12 Processing on endpoint node D.

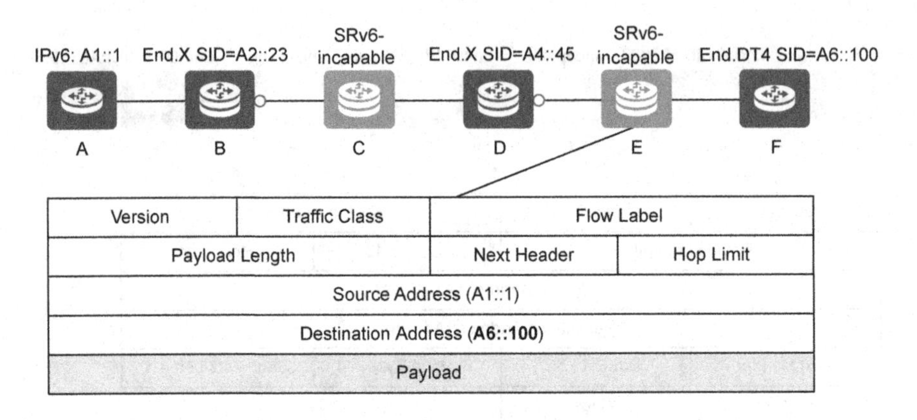

FIGURE 2.13 Processing on transit node E.

End.X SID, node D decrements the SL value by 1, updates the DA field in the outer IPv6 header with the SID specified by the SL field, and then sends the packet over the link bound to the End.X SID. If A4::45 is a PSP-flavored SID, the SRH can be removed as instructed by this flavor, turning the packet into a common IPv6 packet.

2.5.2.5 Step 5: Processing on Transit Node E

As shown in Figure 2.13, node E is also a transit node that can process only IPv6 headers, not SRHs. This means that node E processes the received packet as it does to a common IPv6 packet. More specifically, it searches the corresponding IPv6 routing table according to the longest match rule and then forwards the packet to node F represented by the current DA.

2.5.2.6 Step 6: Processing on Endpoint Node F

After receiving the packet, node F searches the Local SID Table based on the DA A6::100 in the outer IPv6 header and finds a matching End.DT4 SID. According to the instructions defined by the End.DT4 SID, node F decapsulates the packet by removing the IPv6 header, searches the IPv4 routing table of the VPN instance bound to the End.DT4 SID, and ends the process by sending the inner IPv4 packet to host H2.

2.6 ADVANTAGES OF SRv6 NETWORK PROGRAMMING

Chapter 1 briefly introduced the value and significance of SRv6 for future network development. Now that we have grasped the fundamentals of SRv6 network programming, we can further analyze the technical advantages of SRv6 over MPLS and SR-MPLS.

2.6.1 Superior Backward Compatibility and Smooth Evolution

Carriers' IP networks are facing a wide array of challenges, including various device types and inter-vendor device incompatibility.

Compatibility is imperative to network deployment, and issues pertaining to this have arisen during IPv6 development. Due to the exhaustion of 32-bit IPv4 addresses, 128-bit IPv6 addresses are being introduced. The issue, however, is that 128-bit IPv6 and 32-bit IPv4 addresses are incompatible, which means that IPv6 can be supported only through means of networkwide upgrades, making network deployment difficult. SRv6 overcomes this issue as it is compatible with the routing and forwarding design of IPv6, allowing IPv6 to smoothly evolve to SRv6.[8] On top of that, using different functions, SRv6 not only supports the traffic engineering capability provided by traditional MPLS, but also defines a wide range of forwarding behaviors, making networks programmable.

MPLS has already been widely deployed on IP transport networks, where typically, MPLS labels must be bound to routable IP addresses. This means that all devices along an MPLS path have to maintain the mappings between labels and Forwarding Equivalence Classes (FECs).

As shown in Figure 2.14, SR-MPLS is implemented based on the MPLS forwarding plane. A transport network can evolve from MPLS to SR-MPLS in either of the following modes:

- MPLS/SR-MPLS dual stack

  In this mode, SR-MPLS can be deployed only after the entire network is upgraded.

- MPLS and SR-MPLS interworking

  In this mode, the Segment Routing Mapping Server (SRMS) function needs to be deployed for each border node between the MPLS and SR-MPLS domains to stitch together MPLS and SR paths.

Regardless of the evolution mode, the existing network needs to undergo significant changes in the initial phase, and this evolution will take time. On the flip side, SRv6 allows the existing network to be upgraded on demand. Specifically, in order to support specific services based on SRv6, only the related devices need to be upgraded to support SRv6, whereas other devices only need to support common IPv6 route forwarding. As SRv6 features easy incremental deployment, customers can enjoy

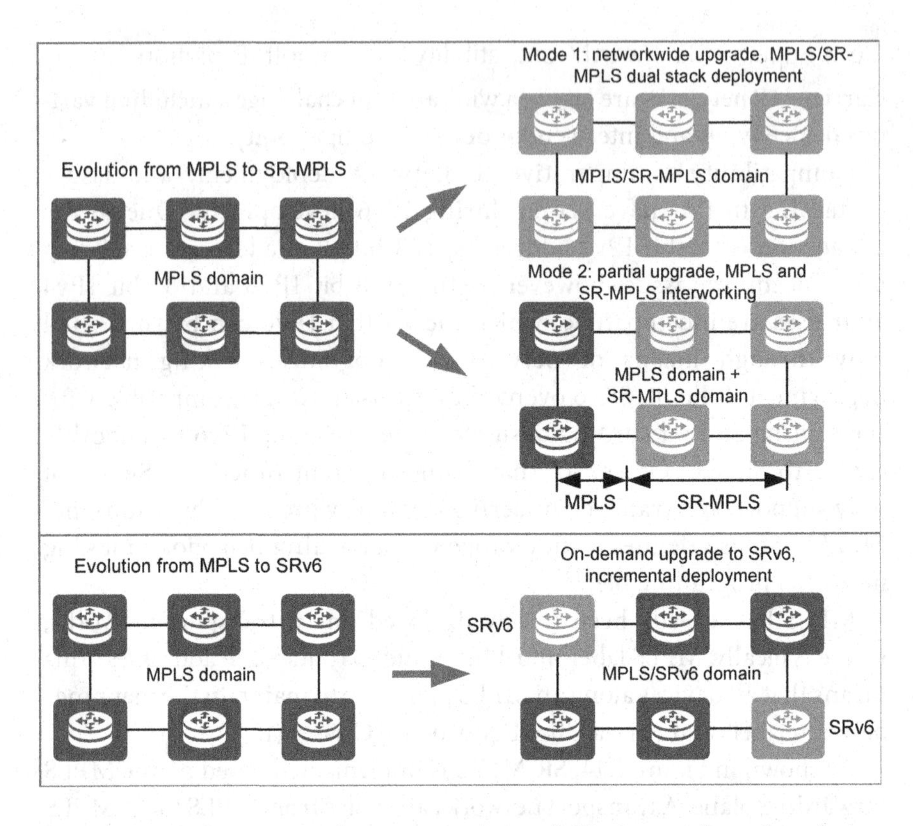

FIGURE 2.14 Comparison between on-demand SRv6 upgrade and evolution from MPLS to SR-MPLS.

maximized Return on Investment (ROI). More details on SRv6 network evolution are provided in Chapter 7.

2.6.2 High Scalability and Simple Deployment in Cross-Domain Scenarios

Traditionally, it has been difficult for carriers to deploy services across MPLS domains. This is by virtue of the fact that to achieve cross-domain deployment, carriers have to use traditional inter-AS MPLS VPN technologies, which are complex and consequently slow down service provisioning.

It is therefore not surprising that MPLS deployment is complex in cross-domain scenarios. If SR-MPLS is used in such a scenario, SIDs need to be imported from one domain to another in order to establish an E2E SR path. In SR-MPLS, Segment Routing Global Blocks (SRGBs) and node SIDs must be planned networkwide, and SID conflicts must be avoided during SID import. This further complicates network planning.

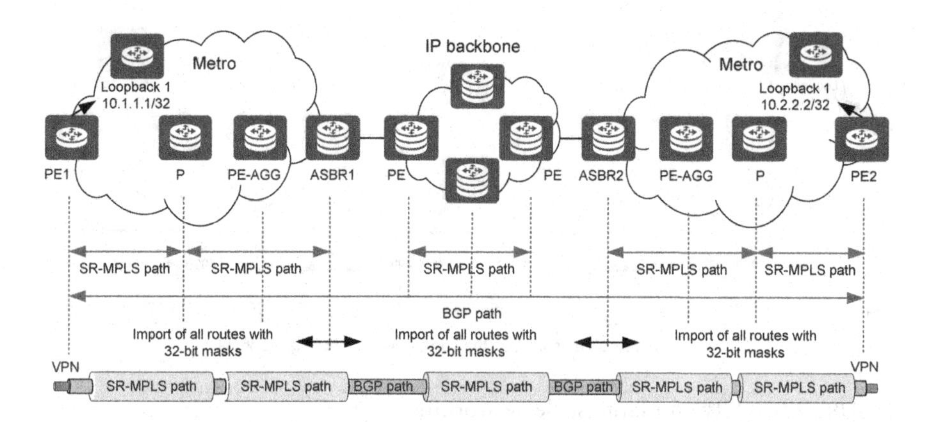

FIGURE 2.15 SR-MPLS-based large-scale networking.

It is also worth noting that scalability is limited when MPLS is used in cross-domain scenarios. Figure 2.15 provides an example where seamless MPLS or SR-MPLS is used to achieve cross-domain networking. Because MPLS labels do not contain reachability information, they must be bound to routable IP addresses and the binding must be propagated across domains. On a large-scale network, numerous MPLS entries need to be generated on border nodes, placing a heavy burden on the control and forwarding planes and reducing network scalability.

One highlight of SRv6 is that it makes cross-domain deployment easier, as it supports native IPv6 and can work purely based on IPv6 reachability. For this reason, services such as SRv6 L3VPN services (Chapter 5 delves into more details) can be deployed across domains only by importing IPv6 routes from one domain to another through BGP4+ (BGP IPv6). This significantly simplifies service deployment.

In comparison with MPLS, SRv6 provides higher scalability in cross-domain scenarios. Because the native IPv6 feature enables SRv6 to work based on aggregated routes, to achieve large-scale cross-domain networking, operators only need to configure border nodes to import a limited number of aggregated routes, as shown in Figure 2.16. This not only reduces the requirements for network device capabilities but also improves network scalability.

2.6.3 Networking Programming for Building Intelligent Networks

Leveraging programmable SRHs, SRv6 offers more powerful network programming capabilities than SR-MPLS. Generally, SRHs provide a three-dimensional programming space, as shown in Figure 2.17.

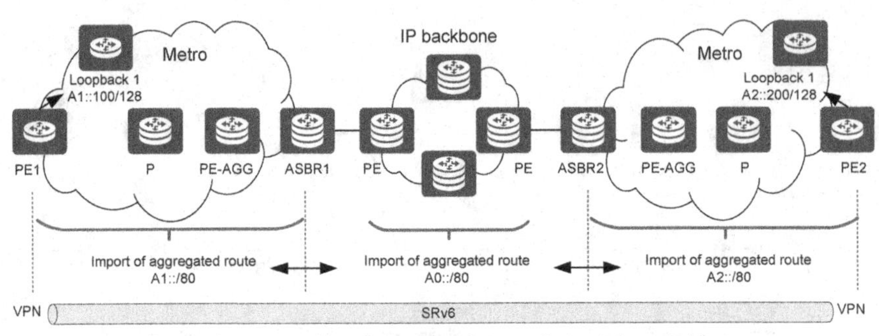

FIGURE 2.16 SRv6 for large-scale networking.

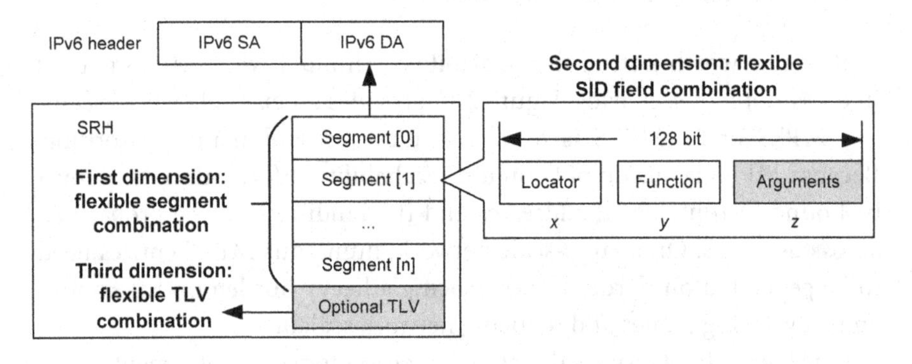

FIGURE 2.17 Three-dimensional programming space provided by SRHs.

First dimension: flexible segment combination. In SRv6, multiple segments are sequentially combined to represent a specific SRv6 path. This is similar to the application of an MPLS label stack.

Second dimension: flexible combination of fields in the 128-bit SRv6 SID. It is known that each MPLS label contains four fixed-length fields: 20-bit label field, 8-bit Time to Live (TTL) field, 3-bit Traffic Class field, and 1-bit S field. In contrast, each SRv6 SID has 128 bits that can be flexibly divided into fields with variable lengths. This further demonstrates the programmability of SRv6.

Third dimension: flexible combination of optional TLVs following segment lists. During packet transmission on a network, irregular information can be encapsulated in the forwarding plane by flexibly combining TLVs.

The three-dimensional programming space allows SRv6 to deliver more powerful network programming capabilities, better meeting different network path requirements, such as network slicing, deterministic latency, and IOAM requirements. Taking a step further, SRv6 integrates SDN-powered global network management and control capabilities to implement flexible programming, thereby facilitating fast deployment of new services and helping build truly intelligent networks.

2.6.4 All Things Connected through an E2E Network

On networks today, services usually need to be deployed across ASs. For example, in the Data Center Interconnect (DCI) scenario shown in Figure 2.18, the IP backbone network uses MPLS or SR-MPLS, whereas data center networks use VXLAN. In this case, gateways need to be deployed to implement mapping between VXLAN and MPLS, which in turn complicates service deployment, without yielding any corresponding benefits.

SRv6 inherits three important features (TE, VPN, and FRR) of MPLS and can therefore replace MPLS/SR-MPLS on IP backbone networks. In a similar fashion to VXLAN, SRv6 can work by simply depending on IP reachability, making SRv6 deployment on data center networks possible. Furthermore, because host applications support IPv6 with which SRv6 is compatible, SRv6 may be directly used for host applications in the future.

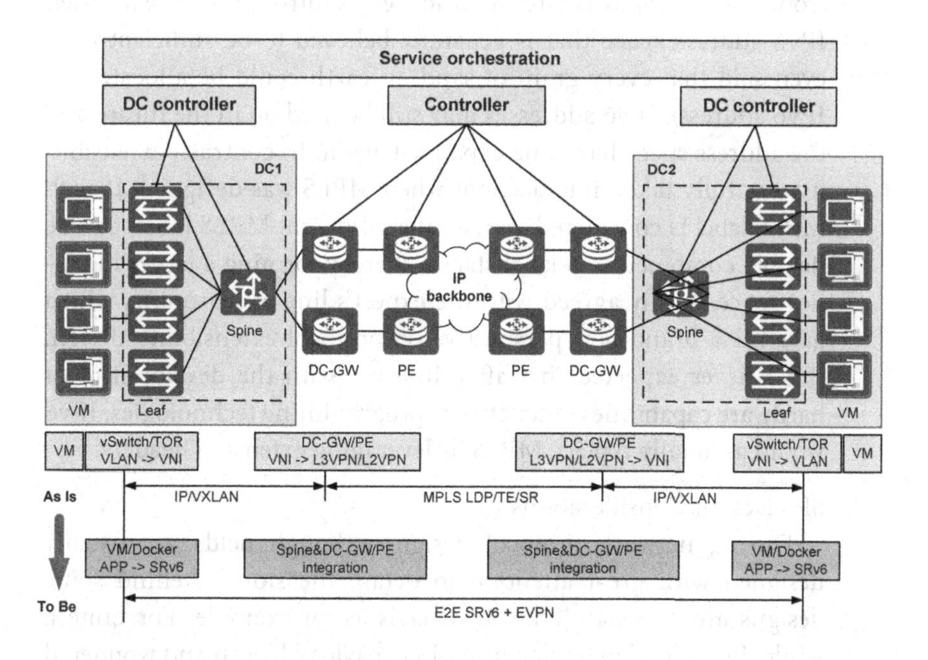

FIGURE 2.18 SRv6 E2E network.

Another highlight of SRv6 is that it defines overlay services and underlay transport as SIDs with different behaviors as well as combining the two through network programming. This effectively avoids protocol interworking problems resulting from service-transport separation, and also more flexibly supports a wide range of function requirements.

SRv6 unleashes the value of IPv6 scalability and helps build intelligent and simplified E2E programmable networks, thereby implementing unified service forwarding and connection of all things through an E2E network. This is precisely what technologies such as ATM, IPv4, MPLS, and SR-MPLS cannot achieve over the course of network technology development, and more importantly, it is the principal vision of SRv6.

2.7 STORIES BEHIND SRv6 DESIGN

I. Extensibility of MPLS and IPv6

At the end of 2008, when I just started participating in the IP innovation work, I consulted a well-versed expert in the MPLS field who told me that extensibility is crucial to protocol design. In his opinion, IPv6 is less extensible than MPLS. Although the extension from 32-bit IPv4 to 128-bit IPv6 addresses introduces the expanded IPv6 address space that is generally believed to be sufficient (it is even said that every grain of sand on earth could be allocated an IPv6 address), IPv6 addresses may still be used up in the future and the address space has to be expanded again. In contrast, extensibility was fully taken into account when MPLS was designed. If each MPLS label is considered an identifier, different MPLS labels can be flexibly combined into label stacks, thereby forming a larger identifier space. I fully agreed with the expert's line of reasoning, which gave me a brand-new perspective on protocol extensibility design. But I never expected that after 10 years, with the development of hardware capabilities and network programming technologies, IPv6 would gradually replace MPLS by leveraging extension headers.

II. SID Design Considerations

During network protocol design, although fields are usually designed with great attention to detail, the stories behind some designs are unusual. Take ATM cells as an example. For quite a while, I was confused by their 48-byte payload length and wondered why it was not a power of two. A pioneer who participated in the

ATM design surprised me by revealing that this was the result of a compromise between one party who insisted on a payload length of 32 bytes and the other party who insisted on 64 bytes, with the average of the two finally being settled on.

SRv6 SIDs are designed cleverly. First, all SRv6 SIDs consist of 128 bits, making them compatible with the IPv6 address length. This enables SRv6 SIDs to function as IPv6 addresses and allows us to reuse existing IPv6 implementation. Second, we can consider the Locator and Function fields in SRv6 SIDs to be a combination of routing and MPLS technologies, with the latter known as a 2.5-layer technology. Similarly, in my opinion, SRv6 is a 3.5-layer technology that combines the advantages of both routing and MPLS technologies. Finally, the meaning and length of each field in an SRv6 SID can be customized, whereas MPLS labels are encapsulated using fixed fields. In fact, this is also a reflection of the POF concept, but implemented in a more practical manner.

III. SRH Design Considerations

There are many debates regarding SRH design, and they mainly include:

1. Why are segments in the SRv6 SRH not popped after being processed by nodes?

   People who are familiar with SR-MPLS naturally raise this question, which can be answered in several ways. First, because the earliest IPv6 RH design was not closely related to MPLS, the segment removal option did not exist at the time. Second, in contrast to MPLS labels that are independently placed on the top of packets and therefore can be directly removed, SRv6 segments are placed in the SRH and associated with information (such as security encryption and verification information) in other extension headers. Third, SRv6 segments can be used for path backtracking based on the path information kept in the SRv6 header. In addition, some innovative solutions regarding SRv6 segments have been taken into account. For example, the remaining segments in the SRH may be reused to develop new functions, such as SRv6 Light IOAM.[9]

2. An SRv6 SRH may be of a considerable length. How can it be shortened to lower the requirements on network devices?

In Chapter 13, we introduce research on SRv6 extension header compression. In fact, SRv6 designers discussed this problem early on. But back then, it was not the right time to propose an SRv6 compression or optimization solution because it would weaken the rationality behind the SRv6 solution and lead to divergence in the industry. On top of that, SRv6 is a future-oriented technology, meaning it will take time for the industry to fully accept the corresponding standards. Considering the continuous development of network hardware capabilities, the challenges on hardware may be greatly reduced when SRv6 becomes more mature.

If we give the SRv6 solution full marks, SRv6 compression or optimization solutions can be seen as solutions that pass muster, and there will be diverse options. I did not initially view multiple choices of solutions as a problem. However, during the subsequent promotion of relevant standards and industries, I gradually realized that the existence of different solutions to the same problem would inevitably decentralize and deplete industry forces, and this is not conducive to industry development.

3. Why do SRHs need to introduce TLVs?

   TLVs are usually used to flexibly define variable-length fields or field combinations for IP control protocols. Introducing TLVs into SRv6 SRHs poses higher challenges to hardware forwarding, while boosting network programming capabilities. Also worth noting is that the new network services supported by SRv6 have some irregular parameters, which cannot be carried by segments or subsegments. In this case, TLVs can be used to carry them. With that in mind, the possible applications of SRH TLVs are as follows:

   - HMAC TLV (described in this chapter): used for security purposes
   - SRv6 IOAM (described in Chapter 9): uses TLVs to carry IOAM parameters.
   - SRv6 DetNet (described in Chapter 10): uses TLVs to carry flow IDs and sequence numbers in order to achieve redundancy protection.

According to IPv6 standards, some functional options can be placed in the Destination Options header before the RH to represent the functions to be processed by the destinations carried in the RH. This design also leads to certain problems, for example:

- Functions are carried through different extension headers and therefore need to be parsed separately, reducing forwarding performance. In contrast, SRv6 uses SRHs to uniformly carry the destinations and related functions, and this is more reasonable.
- In scenarios such as IOAM, if the IOAM information of an SRv6 path is recorded hop by hop in the Destination Options header, the header length increases, causing SRHs to move downward. This affects packet parsing and reduces forwarding performance.

To sum up, as a future-oriented design, SRv6 SRHs bring challenges but also boost hardware development while gaining more industry support. Considering the pace at which hardware capabilities are developing, we can certainly achieve the design objectives of SRv6 SRHs.

REFERENCES

[1] Filsfils C, Camarillo P, Leddy J, Voyer D, Matsushima S, Li Z. SRv6 Network Programming[EB/OL]. (2019-12-05)[2020-03-25]. draft-ietf-spring-srv6-network-programming-05.

[2] Filsfils C, Previdi S, Ginsberg L, Decraene B, Litkowski S, Shakir R. Segment Routing Architecture. (2018-12-19)[2020-03-25]. RFC 8402.

[3] Filsfils C, Dukes D, Previdi S, Leddy J, Matsushima S, Voyer D. IPv6 Segment Routing Header (SRH)[EB/OL]. (2020-03-14)[2020-03-25]. RFC 8754.

[4] Krawczyk H, Bellare M, Canetti R. HMAC: Keyed-Hashing for Message Authentication[EB/OL]. (2020-01-21)[2020-03-25]. RFC 2104.

[5] Sajassi A, Aggarwal R, Bitar N, Isaac A, Uttaro J, Drake J, Henderickx W. BGP MPLS-Based Ethernet VPN[EB/OL]. (2020-01-21)[2020-03-25]. RFC 7432.

[6] Sajassi A, Salam S, Drake J, Uttaro J, Boutros S, Rabadan J. Ethernet-Tree (E-Tree) Support in Ethernet VPN (EVPN) and Provider Backbone Bridging EVPN (PBB-EVPN)[EB/OL]. (2018-01-31)[2020-03-25]. RFC 8317.

[7] Filsfils C, Sivabalan S, Voyer D, Bogdanov A, Mattes P. Segment Routing Policy Architecture[EB/OL]. (2019-12-15)[2020-03-25]. draft-ietf-spring-segment-routing-policy-06.
[8] Tian H, Zhao F, Xie C, Li T, Ma J, Peng S, Li Z, Xiao Y. SRv6 Deployment Consideration[EB/OL]. (2019-11-04)[2020-03-25]. draft-tian-spring-srv6-deployment-consideration-00.
[9] Li C, Cheng W, Previdi S, et al. A Light Weight IOAM for SRv6 Network Programming[EB/OL]. (2019-06-27)[2020-03-25]. draft-li-spring-light-weight-srv6-ioam-01.

CHAPTER 3

Basic Protocols for SRv6

To implement SRv6, we need to extend existing link-state routing protocols, of which there are currently two types available for IPv6 networks: Intermediate System to Intermediate System (IS-IS)[1,2] and Open Shortest Path First version 3 (OSPFv3).[3] Put differently, we can extend these protocols to carry SRv6 information, which in turn makes it possible to implement SRv6 control plane functions while also avoiding the need to maintain certain control plane protocols, such as RSVP-TE and LDP. From this perspective, SRv6 simplifies the network control plane. This chapter describes the IS-IS and OSPFv3 extensions required by the SRv6 control plane.

3.1 IS-IS EXTENSIONS

3.1.1 IS-IS SRv6 Fundamentals

A link-state routing protocol computes the shortest path to a specified address using Dijkstra's Shortest Path First (SPF) algorithm through the following process: Adjacent nodes establish neighbor relationships by exchanging Hello packets, and flood their local link-state information on the entire network to form consistent Link-State Databases (LSDBs). Each node runs the SPF algorithm based on the LSDB to compute routes.

To support SRv6, IS-IS needs to advertise two types of SRv6 information: locator information and SID information. The former is used for routing to a specific node, whereas the latter completely describes the functions of SIDs, for example, behaviors bound to SIDs.

On an SRv6 network, a locator generally needs to be unique in the SRv6 domain since it provides the routing function. That being said, in some special scenarios (such as anycast protection), multiple devices may be configured with the same locator. Figure 3.1 shows the advertisement of IS-IS SRv6 TLVs. IS-IS uses two TLVs with different functions to advertise a locator's routing information.[4] These two TLVs are as follows:

- SRv6 Locator TLV: contains the locator's prefix and mask and is used to advertise the locator information. With this TLV, the other SRv6-capable nodes on the network can learn the locator route. The SRv6 Locator TLV carries not only information used to guide routing, but also the SRv6 SIDs that do not need to be associated with IS-IS neighbors, for example, End SIDs.
- IPv6 Prefix Reachability TLV: contains the same prefix and mask as those of the SRv6 Locator TLV. The IPv6 Prefix Reachability TLV was already supported before SRv6 and can also be processed by common IPv6 nodes (SRv6-incapable nodes). On that account, a common IPv6 node can also use this TLV to generate a locator route, which guides packet forwarding to the node that advertises the locator. This renders it possible to deploy common IPv6 nodes together with SRv6-capable nodes on the same network.

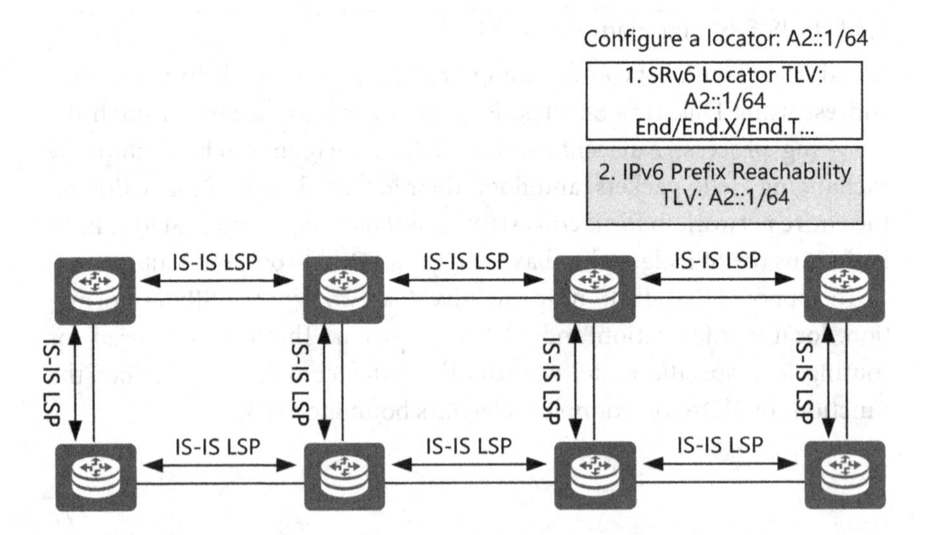

FIGURE 3.1 Advertisement of the IS-IS SRv6 TLVs.

IS-IS goes a step further and provides SRv6 with a function for flooding both SRv6 SID information and SID-specific SRv6 endpoint behavior information through the various SID sub-TLVs of IS-IS, so that the path or service orchestration unit can perform network programming. The specifics of this implementation are expanded on in the next section. Table 3.1 lists the endpoint behaviors advertised by IS-IS.

The following three flavors are defined for the End, End.X, and End.T behaviors: PSP, USP, and USD. We can use different combinations of these flavors to extend the functions of the behaviors. Put differently, End, End.X, and End.T can be used with different flavors to implement different behaviors. Taking the End behavior as an example, different IDs represent different behaviors:

- 1: End
- 2: End + PSP
- 3: End + USP
- 4: End + PSP&USP
- 28: End + USD
- 29: End + PSP&USD
- 30: End + USP&USD
- 31: End + PSP&USP&USD

TABLE 3.1 Endpoint Behaviors Advertised by IS-IS

Endpoint Behavior/ Behavior ID	Whether Carried by the SRv6 End SID Sub-TLV	Whether Carried by the SRv6 End.X SID Sub-TLV	Whether Carried by the SRv6 LAN End.X SID Sub-TLV
End (PSP, USP, and USD)/1–4 and 28–31	Yes	No	No
End.X (PSP, USP, and USD)/5–8 and 32–35	No	Yes	Yes
End.T (PSP, USP, and USD)/9–12 and 36–39	Yes	No	No
End.DX6/16	No	Yes	Yes
End.DX4/17	No	Yes	Yes
End.DT6/18	Yes	No	No
End.DT4/19	Yes	No	No
End.DT64/20	Yes	No	No

3.1.2 IS-IS Extensions for SRv6

Table 3.2 describes the IS-IS extensions for SRv6.[4,5]

3.1.2.1 SRv6 Capabilities Sub-TLV

In SRv6, segment list information is stored in the SRH. SRv6-capable nodes must be able to process the SRH, and each node needs to advertise its own SRH processing capabilities using the SRv6 Capabilities sub-TLV. Figure 3.2 shows its format.

Table 3.3 describes the fields in the SRv6 Capabilities sub-TLV.

TABLE 3.2 IS-IS TLV Extensions for SRv6

Name	Function	Carried In
SRv6 Capabilities sub-TLV	Advertises SRv6 capabilities.	IS-IS Router Capability TLV
Node MSD sub-TLV	Advertises Maximum SID Depths (MSDs) that a device can accept.	IS-IS Router Capability TLV
SRv6 Locator TLV	Advertises an SRv6 locator and its associated SIDs. (This TLV is the only top-level TLV introduced by SRv6.)	IS-IS Link State PDU (LSP)
SRv6 End SID sub-TLV	Advertises SRv6 SIDs associated with endpoint behaviors.	SRv6 Locator TLV
SRv6 End.X SID sub-TLV	Advertises an SRv6 SID associated with a Point-to-Point (P2P) adjacency.	IS-IS NBR TLV
SRv6 LAN End.X SID sub-TLV	Advertises an SRv6 SID associated with a Local Area Network (LAN) adjacency.	IS-IS NBR TLV
SRv6 SID Structure sub-sub-TLV	Advertises the formats of SRv6 SIDs.	SRv6 End SID sub-TLV, SRv6 End.X SID sub-TLV, and SRv6 LAN End.X SID sub-TLV

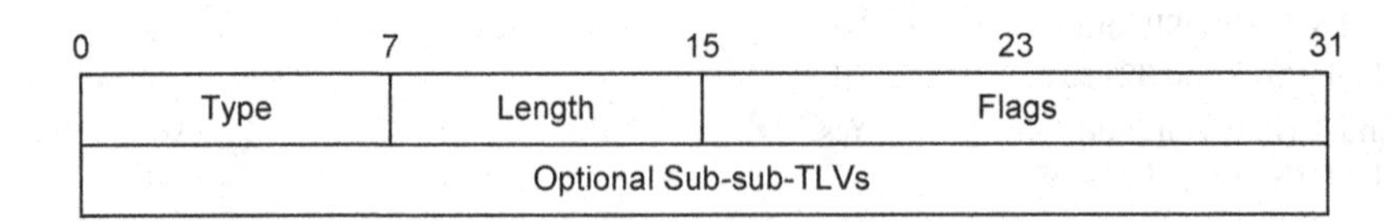

FIGURE 3.2 SRv6 Capabilities sub-TLV.

TABLE 3.3 Fields in the SRv6 Capabilities Sub-TLV

Field	Length	Description
Type	8 bits	Type of the sub-TLV.
Length	8 bits	Length of the sub-TLV.
Flags	16 bits	The second bit in this field is used as the OAM flag. If this field is set, the node supports the SRH O-bit.
Optional Sub-sub-TLVs	Variable	Optional sub-sub-TLVs.

3.1.2.2 Node MSD Sub-TLV

The Node MSD sub-TLV is used to advertise Maximum SID Depths that a node can process. Figure 3.3 shows its format.

Table 3.4 describes the fields in the Node MSD sub-TLV.

3.1.2.3 SRv6 Locator TLV

The SRv6 Locator TLV is used to advertise an SRv6 locator's routing information and the SRv6 SIDs that do not need to be associated with IS-IS neighbors, for example, End SIDs. Figure 3.4 shows its format.

Table 3.5 describes fields in the SRv6 Locator TLV.

As we previously covered, SRv6 SIDs are inherently equipped with the routing capability, which is implemented by using the SRv6 Locator TLV. To break this process down, after receiving the SRv6 Locator TLV, a network node generates a locator route accordingly, which all SIDs allocated from the locator can match based on the LPM rule.

A locator can also be advertised by using an IPv6 Prefix Reachability TLV (236 or 237), and such advertisement is required if the Algorithm field value in the locator is 0 or 1. This ensures that SRv6-incapable devices can deliver a forwarding entry for SRv6 traffic associated with the

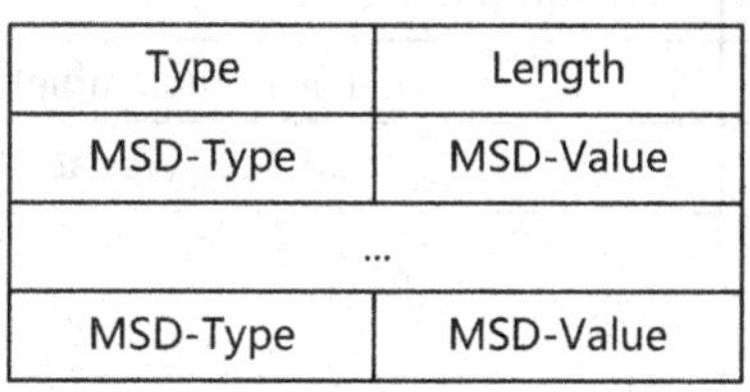

0 — 7	— 15
Type	Length
MSD-Type	MSD-Value
...	
MSD-Type	MSD-Value

FIGURE 3.3 Node MSD sub-TLV.

TABLE 3.4 Fields in the Node MSD Sub-TLV

Field	Length	Description
Type	8 bits	Type of the sub-TLV.
Length	8 bits	Length of the sub-TLV.
MSD-Type	8 bits	MSD type: • Maximum Segments Left MSD Type: specifies the maximum value of the Segments Left (SL) field in the SRH of a received packet before the SRv6 endpoint function associated with a SID is applied. • Maximum End Pop MSD Type: specifies the maximum number of SIDs in the SRH when a node applies the PSP or USP flavor to the SRH. The value 0 indicates that the advertising node cannot apply either flavor to the SRH. • Maximum H.Insert MSD Type: specifies the maximum number of SIDs that can be inserted when the H.Insert operation is performed to insert SRH information. The value 0 indicates that the advertising node cannot perform the H.Insert operation. • Maximum H.Encaps MSD Type: specifies the maximum number of SIDs that can be encapsulated when the H.Encaps operation is performed. • Maximum End D MSD Type: specifies the maximum number of SIDs before decapsulation associated with End.D (such as End.DX6 and End.DT6) functions is performed.
MSD-Value	8 bits	Value of the Maximum SID Depth.

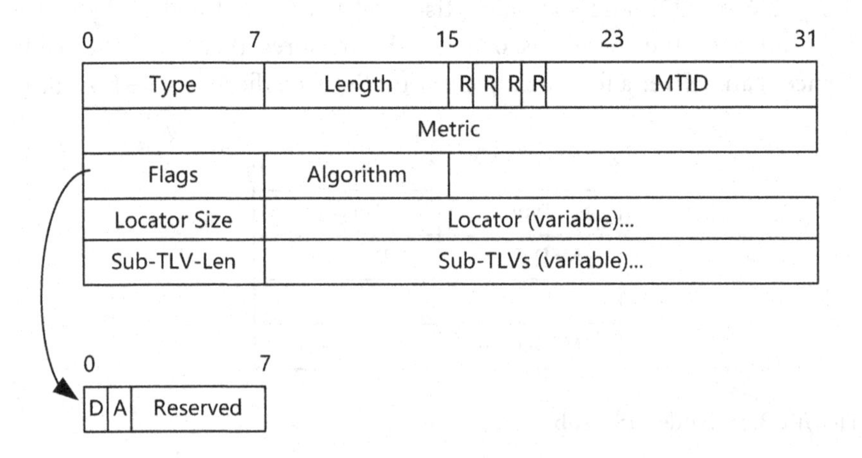

FIGURE 3.4 SRv6 Locator TLV.

TABLE 3.5 Fields in the SRv6 Locator TLV

Field	Length	Description
Type	8 bits	Type of the TLV.
Length	8 bits	Length of the TLV.
MTID	12 bits	Multi-Topology Identifier (MTID).
Metric	32 bits	Metric value associated with the locator.
Flags	8 bits	Flags: • D-flag: When a SID is leaked from a Level-2 area to a Level-1 area, the D-flag must be set. SIDs with the D-flag set cannot be leaked from the Level-1 area to the Level-2 area, thereby preventing routing loops. • A-flag: Anycast flag. This flag must be set if the locator type is set to anycast.
Algorithm	8 bits	Algorithm ID: • 0: SPF • 1: Strict SPF • 128–255: Used in Flexible Algorithm (Flex-Algo) scenarios
Locator Size	8 bits	Length of the locator.
Locator (variable)	Variable	SRv6 locator to be advertised.
Sub-TLV-Len	8 bits	Number of octets occupied by sub-TLVs.
Sub-TLVs (variable)	Variable	Included sub-TLVs, for example, SRv6 End SID sub-TLV.

Algorithm field value of 0 or 1. If a device receives both an IPv6 Prefix Reachability TLV and an SRv6 Locator TLV, the former takes precedence for installation.

3.1.2.4 SRv6 End SID Sub-TLV

The SRv6 End SID sub-TLV is used to advertise SRv6 SIDs belonging to a node, such as End SIDs, as well as associated endpoint behaviors. Figure 3.5 shows its format.

Note: cont is short for continued, indicating subsequent SIDs.

Table 3.6 describes the fields in the SRv6 End SID sub-TLV.

The SRv6 End SID sub-TLV is a sub-TLV of the SRv6 Locator TLV. Given this, if the SIDs to be advertised by IS-IS are not associated with any IS-IS neighbor, they are advertised in the SRv6 End SID sub-TLV.

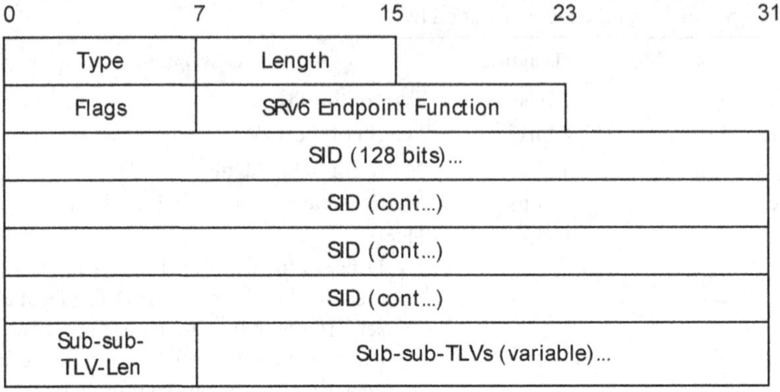

0 7 15 23 31

Type | Length

Flags | SRv6 Endpoint Function

SID (128 bits)...

SID (cont...)

SID (cont...)

SID (cont...)

Sub-sub-TLV-Len | Sub-sub-TLVs (variable)...

FIGURE 3.5 SRv6 End SID sub-TLV.

TABLE 3.6 Fields in the SRv6 End SID Sub-TLV

Field	Length	Description
Type	8 bits	Type of the sub-TLV.
Length	8 bits	Length of the sub-TLV.
Flags	8 bits	Flags of the sub-TLV.
SRv6 Endpoint Function	16 bits	Codepoint of an SRv6 endpoint behavior. Table 3.1 dives into the details on the supported values.
SID	128 bits	SRv6 SID to be advertised.
Sub-sub-TLV-Len	8 bits	Sub-sub-TLV length.
Sub-sub-TLVs (variable)	Variable	Included sub-sub-TLVs, for example, SRv6 SID Structure sub-sub-TLV.

3.1.2.5 SRv6 End.X SID Sub-TLV

The SRv6 End.X SID sub-TLV is used to advertise an SRv6 End.X SID associated with a P2P adjacency and also used as a sub-TLV of an IS-IS neighbor TLV. Figure 3.6 shows its format.

Table 3.7 describes fields in the SRv6 End.X SID sub-TLV.

3.1.2.6 SRv6 LAN End.X SID Sub-TLV

The SRv6 LAN End.X SID sub-TLV is used to advertise an SRv6 End.X SID associated with a LAN adjacency and also used as a sub-TLV of an IS-IS neighbor TLV.

As shown in Figure 3.7, each IS-IS node on the broadcast network advertises only the adjacency pointing to the pseudo-node created by the Designated Intermediate System (DIS).[1] Therefore, node A advertises

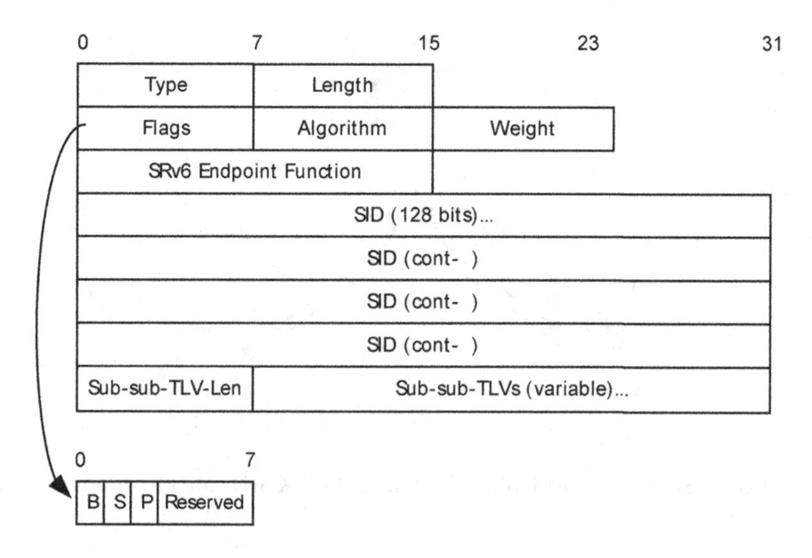

FIGURE 3.6 SRv6 End.X SID sub-TLV.

TABLE 3.7 Fields in the SRv6 End.X SID Sub-TLV

Field	Length	Description
Type	8 bits	Type of the sub-TLV.
Length	8 bits	Length of the sub-TLV.
Flags	8 bits	Flags of the sub-TLV: • B-flag: Backup flag. If this flag is set, the End.X SID has a backup path. • S-flag: Set flag. If this flag is set, the End.X SID is associated with a group of adjacencies. • P-flag: Persistent flag. If this flag is set, the End.X SID is permanently allocated. Do note that the End.X SID remains unchanged even if neighbor relationship flapping, a protocol restart, or a device restart occurs.
Algorithm	8 bits	Algorithm ID: • 0: SPF • 1: Strict SPF • 128–255: used in Flex-Algo scenarios
Weight	8 bits	Weight of the End.X SID used for load balancing.
SRv6 Endpoint Function	16 bits	Codepoint of an SRv6 endpoint behavior. Table 3.1 dives into the details on the supported values.
SID	128 bits	SRv6 SID to be advertised.
Sub-sub-TLV-Len	8 bits	Sub-sub-TLV length.
Sub-sub-TLVs	Variable	Included sub-sub-TLVs, for example, SRv6 SID Structure sub-sub-TLV.

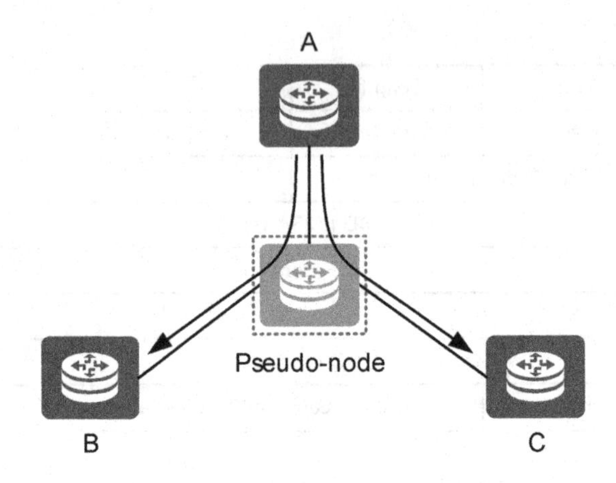

FIGURE 3.7 Scenario in which the SRv6 LAN End.X SID sub-TLV is advertised.

only the adjacency pointing to the pseudo-node. Even so, End.X SIDs need to be differentiated for different IS-IS neighbors, and in this case, the system IDs of neighbors need to be added to distinguish the End.X SIDs of neighbors B and C.

Figure 3.8 shows the format of the SRv6 LAN End.X SID sub-TLV.

If we look at the SRv6 LAN End.X SID sub-TLV side by side with the SRv6 End.X SID sub-TLV, we will notice that the former only has one additional System ID field, which is described in Table 3.8.

0	7	15	23	31
Type	Length	System ID (6 octets)		
Flags	Algorithm	Weight		
SRv6 Endpoint Function				
SID (128 bits)...				
SID (cont-)				
SID (cont-)				
SID (cont-)				
Sub-sub-TLV-Len	Sub-sub-TLVs (variable)...			

FIGURE 3.8 SRv6 LAN End.X SID sub-TLV.

TABLE 3.8 System ID Field in the SRv6 LAN End.X SID Sub-TLV

Field	Length	Description
System ID	6 octets	System ID of an IS-IS neighbor.

3.1.2.7 SRv6 SID Structure Sub-Sub-TLV

The SRv6 SID Structure sub-sub-TLV is used to advertise the lengths of different fields in an SRv6 SID. Figure 3.9 shows the SRv6 SID structure.

Figure 3.10 shows the SRv6 SID Structure sub-sub-TLV format.

Table 3.9 describes the fields in the SRv6 SID Structure sub-sub-TLV.

The SRv6 SID Structure sub-sub-TLV can appear only once in its parent sub-TLV. If it appears multiple times, its parent sub-TLV must be ignored by the receiver.

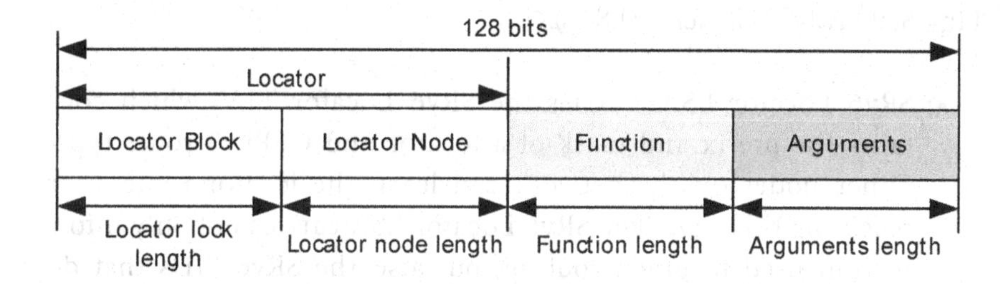

FIGURE 3.9 SRv6 SID Structure.

0	7	15	23 31
Type	Length		
LB Length	LN Length	Fun. Length	Arg. Length

FIGURE 3.10 SRv6 SID Structure sub-sub-TLV.

TABLE 3.9 Fields in the SRv6 SID Structure Sub-sub-TLV

Field	Length	Description
Type	8 bits	Type of the sub-sub-TLV. The value is 1.
Length	8 bits	Length of the sub-sub-TLV.
LB Length	8 bits	Length of the SRv6 SID Locator Block (LB), which is the address block used to allocate SIDs.
LN Length	8 bits	Length of the SRv6 SID Locator Node (LN) ID.
Fun.Length	8 bits	Length of the Function field in the SID.
Arg. Length	8 bits	Length of the Arguments field in the SID.

3.2 OSPFv3 EXTENSIONS

3.2.1 OSPFv3 SRv6 Fundamentals

OSPFv3 extensions for SRv6 share certain commonalities with IS-IS extensions for SRv6. For instance, they both provide the following functions: advertising locator information and advertising SID information. Locator information is used for routing to a specific node, whereas SID information completely describes the functions of SIDs, for example, behaviors bound to SIDs.

As we elaborated on previously, a locator generally needs to be unique in an SRv6 domain since it provides the routing function on an SRv6 network. However, in some special scenarios (such as anycast protection), multiple devices may be configured with the same locator. To advertise a locator's routing information, OSPFv3 needs to advertise two types of Link State Advertisements (LSAs).[6]

- SRv6 Locator LSA: carries the SRv6 Locator TLV, which contains the prefix and mask of a locator and OSPFv3 route type. Other nodes on the network can learn the locator route after receiving this LSA. The SRv6 Locator LSA carries not only information used to guide routing, but also the SRv6 SIDs that do not need to be associated with OSPFv3 neighbors, for example, End SIDs.

- Prefix LSA: Locator prefixes can be advertised through various types of Prefix LSAs, such as Inter-Area Prefix LSAs, AS-External LSAs, Not-So-Stubby Area (NSSA) LSAs, Intra-Area Prefix LSAs, and their equivalent extended LSAs. The Prefix LSAs were already supported before SRv6. Common IPv6 nodes (SRv6-incapable nodes) can also learn them and generate locator routes to guide packet forwarding to the nodes that advertise the locator routes. As such, common IPv6 nodes can be deployed together with SRv6-capable nodes on the same network.

OSPFv3 also provides SRv6 with a function for flooding SRv6 SID information and endpoint behavior information through various OSPFv3 SID sub-TLVs, so that the path or service orchestration unit can perform network programming. On a side note, OSPFv3 advertises the same SRv6 endpoint behavior information as IS-IS. For details, see Table 3.1.

3.2.2 OSPFv3 Extensions for SRv6

Table 3.10 describes the OSPFv3 TLV extensions for SRv6.

3.2.2.1 SRv6 Capabilities TLV

In SRv6, segment list information is stored in the SRH. SRv6-capable nodes must be able to process the SRH, and each node needs to advertise its own SRH processing capabilities using the SRv6 Capabilities sub-TLV. Figure 3.11 shows its format.

Table 3.11 describes fields in the SRv6 Capabilities TLV.

TABLE 3.10 OSPFv3 TLV Extensions for SRv6

Name	Function	Carried In
SRv6 Capabilities TLV	Advertises OSPFv3 SRv6 capabilities.	OSPFv3 Router Information LSA[7]
SR Algorithm TLV	Advertises OSPFv3 SRv6 algorithms.	OSPFv3 Router Information LSA[7]
Node MSD TLV	Advertises Maximum SID Depths that an OSPFv3 device can accept.	OSPFv3 Router Information LSA[8]
SRv6 Locator LSA	Advertises OSPFv3 SRv6 locator information.	OSPFv3 packet
SRv6 Locator TLV	Advertises an OSPFv3 SRv6 locator as well as the End SIDs and route types associated with it.	OSPFv3 SRv6 Locator LSA
SRv6 End SID sub-TLV	Advertises OSPFv3 SRv6 SIDs.	SRv6 Locator TLV
SRv6 End.X SID sub-TLV	Advertises SRv6 SIDs associated with OSPFv3 adjacencies over P2P or Point-to-Multipoint (P2MP) links, as well as the SRv6 SID associated with the adjacency pointing to the Designated Router (DR) over broadcast or Non-Broadcast Multiple Access (NBMA) links.	OSPFv3 E-Router-LSA Router-Link TLV[9]
SRv6 LAN End.X SID sub-TLV	Advertises SRv6 SIDs associated with OSPFv3 adjacencies pointing to the Backup Designated Router (BDR) or DR Others on broadcast or NBMA[3] links.	OSPFv3 E-Router-LSA Router-Link TLV[9]
Link MSD sub-TLV	Advertises Maximum SID Depths that an OSPFv3 link can process.	OSPFv3 E-Router-LSA Router-Link TLV[9]
SRv6 SID Structure sub-sub-TLV	Advertises the formats of SRv6 SIDs.	SRv6 End SID sub-TLV, SRv6 End.X SID sub-TLV, and SRv6 LAN End.X SID sub-TLV

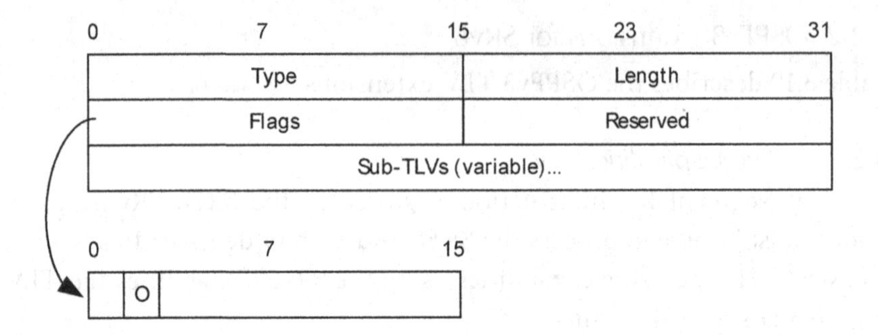

FIGURE 3.11 OSPFv3 SRv6 Capabilities TLV.

TABLE 3.11 Fields in the OSPFv3 SRv6 Capabilities Sub-TLV

Field	Length	Description
Type	16 bits	Type of the sub-TLV.
Length	16 bits	Length of the sub-TLV.
Flags	16 bits	The second bit in this field is used as the OAM flag. If this field is set, the node supports the SRH O-bit.
Sub-TLVs	Variable	Optional sub-sub-TLVs.

3.2.2.2 SR Algorithm TLV

This TLV reuses the format of the SR Algorithm TLV defined in OSPFv2. In OSPFv3, the SR Algorithm TLV is carried in an OSPFv3 Router Information LSA to advertise the algorithms used by OSPFv3 SRv6. Figure 3.12 shows its format.

Table 3.12 describes the fields in the SR Algorithm TLV.

3.2.2.3 Node MSD TLV

This TLV reuses the format of the Node MSD TLV defined in OSPFv2. In OSPFv3, the Node MSD TLV is carried in an OSPFv3 Router Information LSA to advertise specific values for the different types of MSDs that an OSPFv3 device can accept. Figure 3.13 shows its format.

Table 3.13 describes the fields in the Node MSD TLV.

0 7 15 23 31

Type | Length

Algorithm 1 | Algorithm... | Algorithm n

FIGURE 3.12 SR Algorithm TLV.

TABLE 3.12 Fields in the SR Algorithm TLV

Field	Length	Description
Type	16 bits	Type of the TLV.
Length	16 bits	Length of the TLV.
Algorithm	8 bits	Value indicating an SR algorithm type: • 0: SPF • 1: Strict SPF • 128–255: used in Flex-Algo scenarios

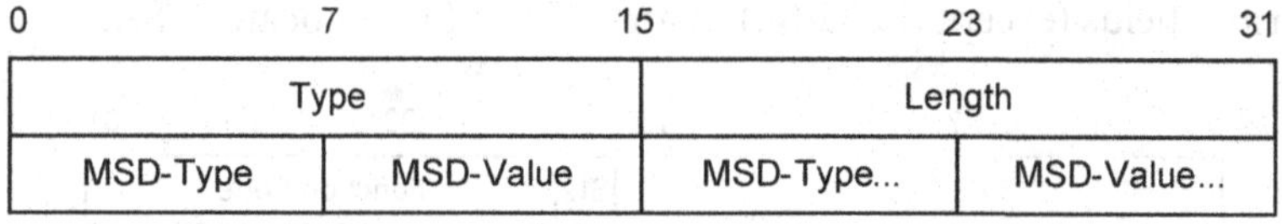

0 7	15	23	31
Type		Length	
MSD-Type	MSD-Value	MSD-Type...	MSD-Value...

FIGURE 3.13 Node MSD TLV.

TABLE 3.13 Fields in the Node MSD TLV

Field	Length	Description
Type	16 bits	Type of the TLV.
Length	16 bits	Length of the TLV.
MSD-Type	8 bits	MSD type: • Maximum Segments Left MSD Type: specifies the maximum value of the SL field in the SRH of a received packet before the SRv6 endpoint function associated with a SID is applied. • Maximum End Pop MSD Type: specifies the maximum number of SIDs in the SRH when a node applies the PSP or USP flavor to the SRH. The value 0 indicates that the advertising node cannot apply either flavor to the SRH. • Maximum H.Insert MSD Type: specifies the maximum number of SIDs that can be inserted when the H.Insert operation is performed to insert SRH information. The value 0 indicates that the advertising node cannot perform the H.Insert operation. • Maximum H.Encaps MSD Type: specifies the maximum number of SIDs that can be encapsulated when the H.Encaps operation is performed. • Maximum End D MSD Type: specifies the maximum number of SIDs before decapsulation associated with End.D (such as End.DX6 and End.DT6) functions is performed.
MSD-Value	8 bits	Value of the Maximum SID Depth.

3.2.2.4 OSPFv3 SRv6 Locator LSA

An OSPFv3 SRv6 Locator LSA is used to advertise SRv6 locator information. If an SRv6-capable node receives an SRv6 Locator LSA and supports the locator-associated algorithms, the node needs to install a routing entry in the forwarding table for the locator. Figure 3.14 shows the format of an OSPFv3 SRv6 Locator LSA.

OSPFv3 SRv6 Locator LSAs reuse the format of LSAs defined in the base OSPFv3. The U-bit in the format must be set to 1 to ensure that OSPFv3 SRv6 Locator LSAs can be flooded by devices that do not support them. Table 3.14 describes fields (except the U-bit) in an OSPFv3 SRv6 Locator LSA.

0–15			16–31
LS Age	U	S12	Function Code
Link State ID			
Advertising Router			
LS sequence number			
LS checksum			Length
TLVs (variable)...			

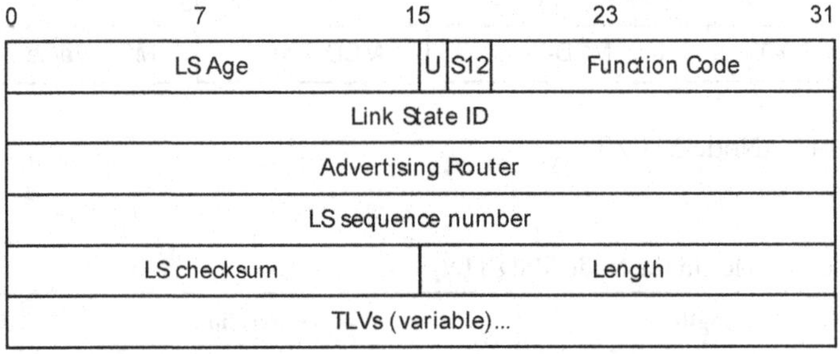

FIGURE 3.14 OSPFv3 SRv6 Locator LSA.

TABLE 3.14 Fields (Except the U-bit) in the OSPFv3 SRv6 Locator LSA

Field	Length	Description
LS Age	16 bits	Time elapsed since the LSA is generated, in seconds. The value of this field keeps increasing regardless of whether the LSA is transmitted over a link or saved in an LSDB.
S12	2 bits	S1 and S2 bits, which are used to control the LSA flooding scope.
Function Code	13 bits	Identifier of the LSA type.
Link State ID	32 bits	Originating router's identifier for the LSA. This field, together with the LS Type field (including the U-bit, S1 and S2 bits, and Function Code field), uniquely identifies the LSA in an OSPFv3 area.
Advertising Router	32 bits	Router ID of the device that generates the LSA.
LS sequence number	32 bits	Sequence number of the LSA. Neighbors can use this field to identify the latest LSA.
LS checksum	16 bits	Checksum of all fields in the LSA except the LS Age field.
Length	16 bits	Total length of the LSA, including the LSA header.
TLVs	Variable	TLVs that can be contained, for example, SRv6 Locator TLV.

TABLE 3.15 Function Code in the SRv6 Locator LSA

Function Code	Recommended Value
SRv6 Locator LSA	42

Table 3.15 describes the Function Code in the LSA.

Locators can also be advertised through various types of Prefix LSAs, such as Inter-Area Prefix LSAs, AS-External LSAs, NSSA LSAs, Intra-Area Prefix LSAs, and their equivalent extended LSAs.[3] Such advertisement is required if the Algorithm field value in the locator is 0 to ensure that SRv6-incapable devices can deliver forwarding entries for SRv6 traffic associated with the Algorithm field value of 0. If a device receives both a Prefix LSA and an SRv6 Locator LSA, the former takes precedence for installation.

3.2.2.5 SRv6 Locator TLV

The SRv6 Locator TLV is a top-level TLV in an SRv6 Locator LSA and is used to advertise an SRv6 locator as well as the associated End SIDs. Figure 3.15 shows its format.

Table 3.16 describes fields in the SRv6 Locator TLV.

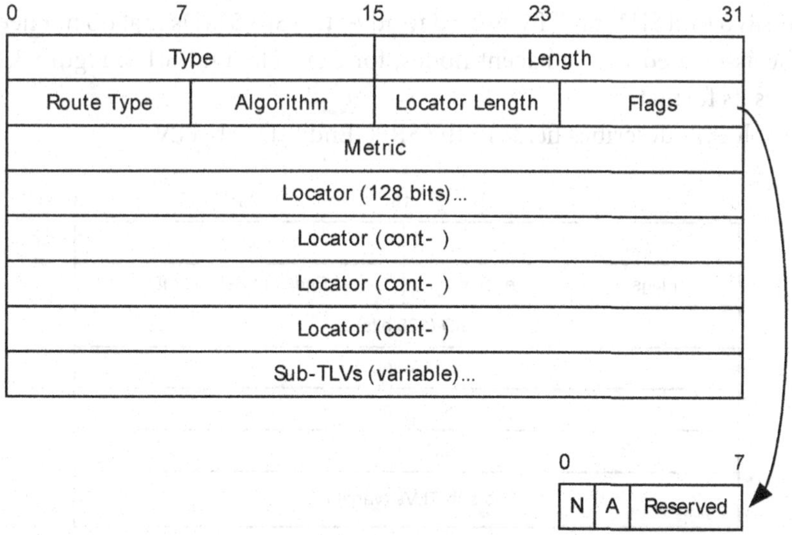

0 7 15 23 31
Type
Length
Route Type
Algorithm
Locator Length
Flags
Metric
Locator (128 bits)...
Locator (cont-)
Locator (cont-)
Locator (cont-)
Sub-TLVs (variable)...
0 7
N
A
Reserved

FIGURE 3.15 SRv6 Locator TLV.

TABLE 3.16 Fields in the SRv6 Locator TLV

Field	Length	Description
Type	16 bits	Type of the TLV.
Length	16 bits	Length of the TLV.
Route Type	8 bits	Route type: • 1: Intra-Area • 2: Inter-Area • 3: AS External • 4: NSSA External
Algorithm	8 bits	Algorithm ID: • 0: SPF • 1: Strict SPF • 128–255: used in Flex-Algo scenarios
Locator Length	8 bits	Length of the locator.
Flags	8 bits	Flags: • N-flag: Node flag. If this flag is set, the locator uniquely identifies a node on the network. • A-flag: Anycast flag. If this flag is set, the locator is configured as an anycast locator. Put differently, the locator is configured on multiple anycast nodes.
Metric	32 bits	Metric value associated with the locator.
Locator	128 bits	SRv6 locator to be advertised.
Sub-TLVs	Variable	Included sub-TLVs, for example, SRv6 End SID sub-TLV.

3.2.2.6 SRv6 End SID Sub-TLV

The SRv6 End SID sub-TLV is used to advertise SRv6 SIDs that do not need to be associated with adjacent nodes, for example, End SIDs. Figure 3.16 shows its format.

Table 3.17 describes fields in the SRv6 End SID sub-TLV.

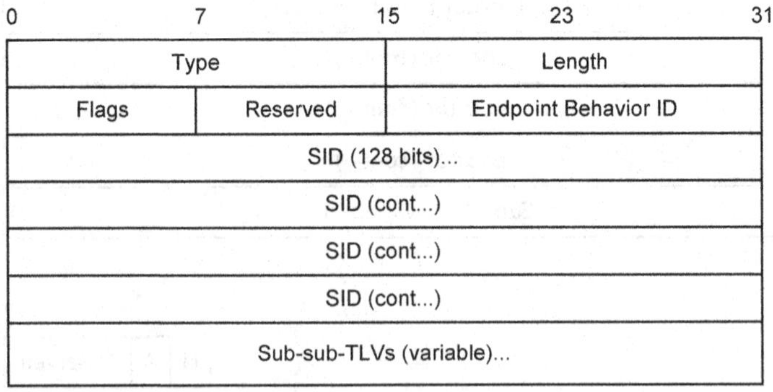

FIGURE 3.16 SRv6 End SID sub-TLV.

TABLE 3.17 Fields in the SRv6 End SID Sub-TLV

Field	Length	Description
Type	16 bits	Type of the sub-TLV.
Length	16 bits	Length of the sub-TLV.
Flags	8 bits	Flags of the sub-TLV.
Endpoint Behavior ID	16 bits	Codepoint of an SRv6 endpoint behavior.
SID	128 bits	SRv6 SID to be advertised.
Sub-sub-TLVs	Variable	Included sub-sub-TLVs, for example, SRv6 SID Structure sub-sub-TLV.

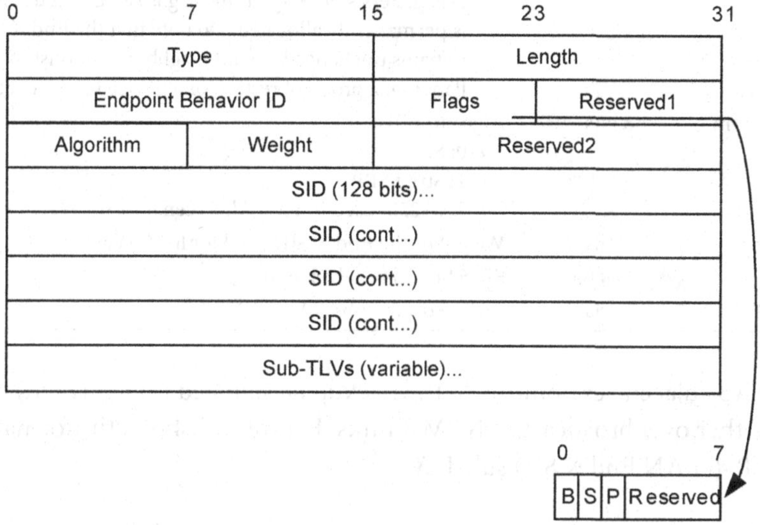

FIGURE 3.17 SRv6 End.X SID sub-TLV.

3.2.2.7 *SRv6 End.X SID Sub-TLV*

The OSPFv3 SRv6 End.X SID sub-TLV, which is carried in the OSPFv3 Router-Link TLV,[3,7] is used to advertise SRv6 End.X SIDs associated with OSPFv3 adjacencies over P2P or P2MP links, as well as an SRv6 End.X SID associated with the adjacency pointing to the DR over broadcast or NBMA links.[3] Figure 3.17 shows the format of the SRv6 End.X SID sub-TLV.

Table 3.18 describes fields in the SRv6 End.X SID sub-TLV.

3.2.2.8 *SRv6 LAN End.X SID Sub-TLV*

The OSPFv3 SRv6 LAN End.X SID sub-TLV, which is carried in the OSPFv3 Router-Link TLV, is used to advertise SRv6 End.X SIDs associated with

TABLE 3.18 Fields in the SRv6 End.X SID Sub-TLV

Field	Length	Description
Type	16 bits	Type of the sub-TLV.
Length	16 bits	Length of the sub-TLV.
Endpoint Behavior ID	16 bits	Codepoint of an SRv6 endpoint behavior.
Flags	8 bits	Flags of the sub-TLV: • B-flag: Backup flag. If this flag is set, the End.X SID has a backup path. • S-flag: Set flag. If this flag is set, the End.X SID is associated with a group of adjacencies. • P-flag: Persistent flag. If this flag is set, the End.X SID is permanently allocated. Do note that the End.X SID remains unchanged even if neighbor relationship flapping, a protocol restart, or a device restart occurs.
Algorithm	8 bits	Algorithm ID: • 0: SPF • 1: Strict SPF • 128–255: used in Flex-Algo scenarios
Weight	8 bits	Weight of the End.X SID used for load balancing.
SID	128 bits	SRv6 SID to be advertised.
Sub-TLVs	Variable	Included sub-sub-TLVs.

OSPFv3 adjacencies pointing to the Backup Designated Router (BDR) or a DR Other over broadcast or NBMA links. Figure 3.18 shows the format of the SRv6 LAN End.X SID sub-TLV.

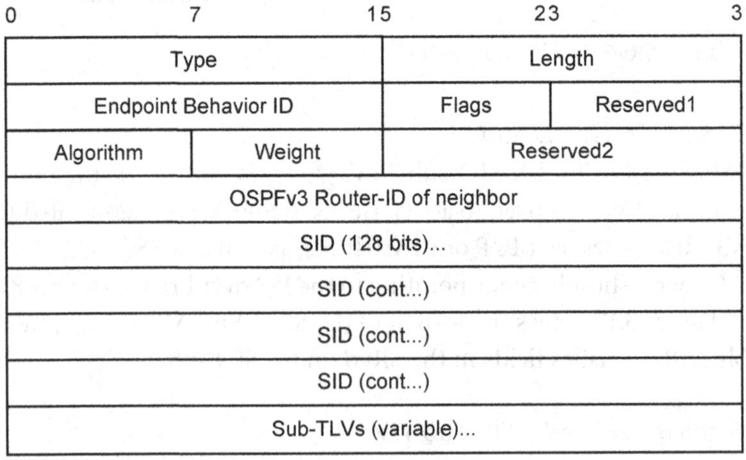

FIGURE 3.18 SRv6 LAN End.X SID sub-TLV.

TABLE 3.19 OSPFv3 Router-ID of Neighbor Field in the SRv6 LAN End.X SID Sub-TLV

Field	Length	Description
OSPFv3 Router-ID of neighbor	32 bits	Router ID of an OSPFv3 neighbor.

The only difference between the SRv6 End.X SID sub-TLV and SRv6 LAN End.X SID sub-TLV is that the latter contains one additional field: OSPFv3 Router-ID of neighbor, as shown in Table 3.19.

3.2.2.9 Link MSD Sub-TLV

The Link MSD sub-TLV in OSPFv3 shares the same format as that in OSPFv2. In OSPFv3, the Link MSD sub-TLV is carried in the OSPFv3 E-Router-LSA Router-Link TLV to advertise specific values for the different types of MSDs that an OSPFv3 link can accept. Figure 3.19 shows its format.

The descriptions of the fields in the Link MSD sub-TLV are the same as those in the Node MSD TLV and therefore are not provided here.

3.2.2.10 SRv6 SID Structure Sub-Sub-TLV

The SRv6 SID structure sub-sub-TLV is used to advertise the lengths of different fields in an SRv6 SID. Figure 3.20 shows its format.

Table 3.20 describes the fields in the SRv6 SID Structure sub-sub-TLV.

The SRv6 SID Structure sub-sub-TLV can appear only once in its parent sub-TLV. If it appears multiple times, its parent sub-TLV must be ignored by the receiver.

Appendix C dives into more details on the mappings between OSPFv3 LSAs and TLVs.

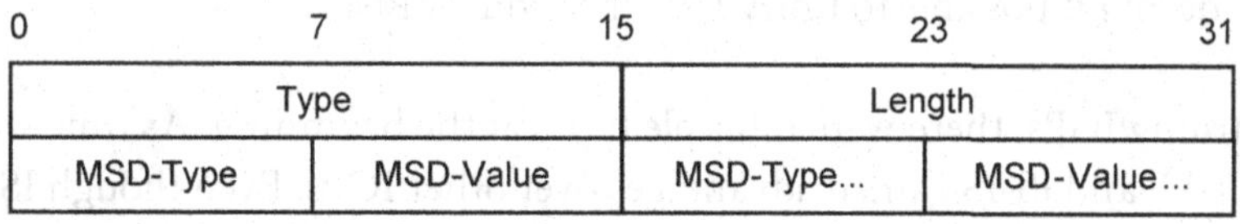

0 7	15	23	31
Type		Length	
MSD-Type	MSD-Value	MSD-Type...	MSD-Value...

FIGURE 3.19 Link MSD sub-TLV.

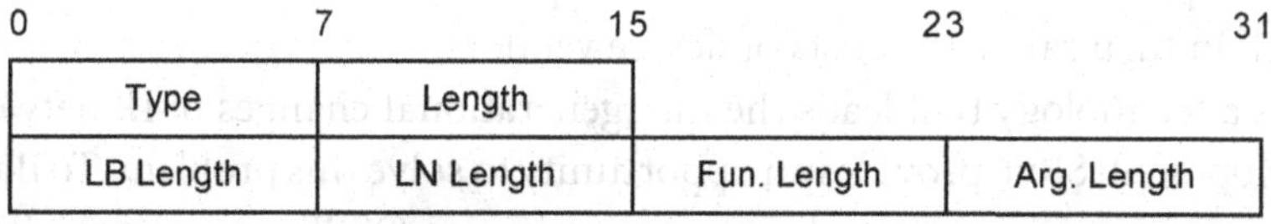

0 7	15	23	31
Type	Length		
LB Length	LN Length	Fun. Length	Arg. Length

FIGURE 3.20 SRv6 SID Structure sub-sub-TLV.

TABLE 3.20 Fields in the SRv6 SID Structure Sub-sub-TLV

Field	Length	Description
Type	8 bits	Type of the sub-sub-TLV. The value is 1.
Length	8 bits	Length of the sub-sub-TLV.
LB Length	8 bits	Length of the SRv6 SID Locator Block, which is the address block used to allocate SIDs.
LN Length	8 bits	Length of the SRv6 SID Locator Node ID.
Fun.Length	8 bits	Length of the Function field in the SID.
Arg. Length	8 bits	Length of the Arguments field in the SID.

3.3 STORIES BEHIND SRv6 DESIGN

SDN put forth the idea of simplifying protocols. Although SDN did not achieve the goal of unifying protocols with OpenFlow, it was still igniting transformation across the industry. We are witnessing a trend where simplification and unification accompany the development of IP protocols.

- With the development of SR, the traditional MPLS signaling protocols LDP and RSVP-TE are being replaced by IGPs.
- Along the same lines, as SRv6 develops, the MPLS data plane functions are being implemented by IPv6 extensions.
- Likewise, as the EVPN develops, various L2VPN technologies are unified, including VPLS and VPWS, BGP-based L2VPN, LDP-based L2VPN, and LDP-based L2VPN integrated with BGP auto-discovery. EVPN also supports Layer 3 Virtual Private Network (L3VPN), rendering it possible to unify L2VPN and L3VPN.

Regarding IGPs, there were multiple types at the beginning. As time went on, OSPF and IS-IS gained advantages over other IGPs. Even though IS-IS is deployed on a larger scale, OSPF is still deployed on many live networks due to historical reasons. In view of this, if the IGPs cannot be unified, the two protocols need to be extended separately for the same functions, which in turn raises the costs of device vendors.

As a technology that leads the intergenerational changes of IP network development, SRv6 provides an opportunity to solve this problem. To illustrate, some carriers whose legacy networks use OSPF now choose IS-IS when deploying SRv6 networks. Given that OSPFv3 (supporting IPv6) and OSPFv2 (supporting IPv4) are incompatible when networks transition from IPv4 to IPv6, a new protocol has to be introduced regardless of whether OSPFv3 or IS-IS is deployed. Against this backdrop, carriers prefer IS-IS for their target networks, so IGPs gradually converge to IS-IS. From a different vantage point, wireless networks provide numerous examples of how to solve historical issues through the intergenerational changes of network development. However, IP networks have always been developed in a compatible manner, which not only makes IP network service deployment complex but also increases network costs. In light of this, as IPv6 and SRv6 develop, we should take the chance to reduce alternative solutions and consolidate power to accelerate the development of the industry and technologies. This can be accomplished through gaining wider consensus in the industry and standards work.

REFERENCES

[1] Callon R. Use of OSI IS-IS for Routing in TCP/IP and Dual Environments [EB/OL]. (2013-03-02)[2020-03-25]. RFC 1195.

[2] Hopps C. Routing IPv6 with IS-IS[EB/OL]. (2015-10-14)[2020-03-25]. RFC 5308.

[3] Coltun R, Ferguson D, Moy J, Lindem A. OSPF for IPv6[EB/OL]. (2020-01-21)[2020-03-25]. RFC 5340.

[4] Psenak P, Filsfils C, Bashandy A, Decraene B, Hu Z. IS-IS Extension to Support Segment Routing over IPv6 Dataplane[EB/OL]. (2019-10-04)[2020-03-25]. draft-ietf-lsr-isis-srv6-extensions-03.

[5] Iana. IS-IS TLV Codepoints[EB/OL]. (2020-02-24)[2020-03-25].

[6] Li Z, Hu Z, Cheng D, Talaulikar K, Psenak P. OSPFv3 Extensions for SRv6 [EB/OL]. (2020-02-12)[2020-03-25]. draft-ietf-lsr-ospfv3-srv6-extensions-00.

[7] Lindem A, Shen N, Vasseur JP, Aggarwal R, Shaffer S. Extensions to OSPF for Advertising Optional Router Capabilities[EB/OL]. (2016-02-01)[2020-03-25]. RFC 7770.

[8] Lindem A, Roy A, Goethals D, Reddy Vallem V, Baker F. OSPFv3 Link State Advertisement (LSA) Extensibility[EB/OL]. (2018-12-19)[2020-03-25]. RFC 8362.

[9] Tantsura J, Chunduri U, Aldrin S, Psenak P. Signaling Maximum SID Depth (MSD) Using OSPF[EB/OL]. (2018-12-12)[2020-03-25]. RFC 8476.

CHAPTER 4

SRv6 TE

AS ONE OF THE MOST VITAL NETWORK SERVICES, TE is widely used in various network scenarios. It is also one of the three major features that SRv6 has inherited from MPLS. This chapter describes SRv6 TE in detail, including the SR-TE (SRv6 TE) architecture, SRv6 Policy and extensions to BGP, Border Gateway Protocol - Link State (BGP-LS), and PCEP to support SRv6 Policies. Compared with SR-MPLS and MPLS, SRv6 TE can implement inter-domain or even E2E traffic engineering more flexibly and easily by explicitly specifying forwarding paths based on IPv6 routing reachability.

4.1 SR-TE ARCHITECTURE

A conventional routing protocol selects the shortest path to route its traffic, regardless of factors such as bandwidth and latency. Traffic is not switched to another path, even if the shortest path is congested. This is not a problem with low traffic on a network. However, with the ever-increasing Internet applications, exclusive use of the SPF algorithm is likely to result in severe routing problems. In response to network evolution, TE technologies emerge.

TE focuses on optimizing the overall network performance. Its primary objectives are to provide efficient and reliable network services, optimize network resource utilization, and optimize network traffic forwarding paths. TE has two primary concerns. One relates to traffic, that is, how to guarantee the network SLA. The other relates to resources, that is, how to optimize their utilization.

A traditional way to implement TE is based on MPLS, called MPLS TE. By using an overlay model, MPLS TE can easily establish a dedicated virtual path over a physical network topology, and then map data traffic to the virtual path. MPLS TE can accurately control the path through which traffic passes in order to route traffic away from congested nodes. This addresses uneven load balancing among multiple paths and makes full use of existing bandwidth resources. MPLS TE can reserve resources during the establishment of LSPs to guarantee SLA. In addition, to ensure the reliability of network service provisioning, MPLS TE introduces path backup and FRR mechanisms, which allow traffic to be switched promptly in the case of a link or node failure.

MPLS TE uses RSVP-TE as its signaling protocol.[1] RSVP-TE needs to maintain the per-flow path states on each node along a path. As the network scale increases, so does the workload for RSVP-TE to maintain the path states, consuming excessive device resources. As a result, it is difficult to deploy RSVP-TE on large-scale networks.

This is where SR comes in, overcoming the disadvantages of RSVP-TE. SR draws on source routing and carries a sequential list of instructions (called a Segment List) in the packet header to guide packet forwarding on a network. These instructions are targeted at nodes and links instead of data flows. As such, network devices only need to maintain limited node and link states. Transit nodes only need to forward packets based on the SR path information that is explicitly carried in the packets, and do not need to maintain the per-flow path states. This solves the poor scalability issue of RSVP-TE. Because of this, the SR-TE solution, which combines SR and TE, is future-oriented, more flexible and powerful, and well accepted in the industry.

4.1.1 Traditional MPLS TE Architecture

On a traditional MPLS network, each MPLS node needs to maintain a TE architecture in order to support TE. Figure 4.1 shows a typical MPLS TE architecture.

The MPLS TE architecture consists of the following components:

1. Information advertisement

   In TE, a network device needs to obtain not only network topology information but also network load information. To obtain both types of information, MPLS TE introduces the information

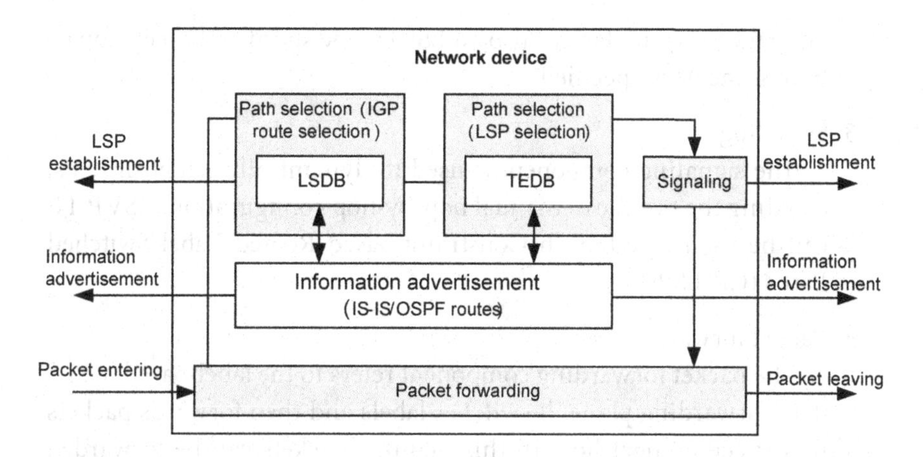

FIGURE 4.1 Typical MPLS TE architecture.

advertisement component. Specifically, MPLS TE extends the existing IGPs to advertise TE link information, for example, introducing new TLVs[2] to IS-IS and new LSAs[3] to OSPF. The TE link information includes the maximum link bandwidth, maximum reservable bandwidth, currently reserved bandwidth, and link colors.

Through IGP TE extensions, network devices can maintain TE link and topology attributes of the network to form a Traffic Engineering Database (TEDB). Based on the TEDB, paths that satisfy various constraints can be computed.

2. Path selection

The path selection component uses the Constrained Shortest Path First (CSPF) algorithm and TEDB to compute paths that meet specified constraints. As an improved variant of the SPF algorithm, CSPF takes certain constraints (such as bandwidth requirements, maximum number of hops, and management policy requirements) into consideration when computing the shortest path on a network. CSPF provides an online computation method, which can promptly respond to network changes and provide a suitable path. Alternatively, developers can develop an offline computation tool, which can be used to compute paths for all LSPs based on the networkwide topology, link attributes, and LSP requirements and then configure the computed paths to trigger LSP establishment. Such LSPs can be either strict or loose. An LSP can be specified to pass or not pass a certain device; also an LSP can be specified hop by hop or have some hops specified.

3. Signaling

   The signaling component is used to dynamically establish LSPs, avoiding the trouble of manual hop-by-hop configuration. RSVP-TE can be used to establish Constraint-based Routed Label Switched Paths (CR-LSPs).

4. Packet forwarding

   The packet forwarding component refers to the label-based MPLS data forwarding plane. It switches labels and then forwards packets to a specified next hop. In this manner, packets can be forwarded along preestablished LSPs. As LSPs support forwarding along specified paths, MPLS TE overcomes the disadvantages of IGP-based forwarding only along the shortest path.

4.1.2 Centralized SR-TE Architecture

SR-TE can also use a functional architecture similar to that used by MPLS TE as shown in Figure 4.2. Like MPLS, SR floods SR information through IGP extensions. The ingress is responsible for establishing the SR path that meets specific constraints, without the need of RSVP-TE for

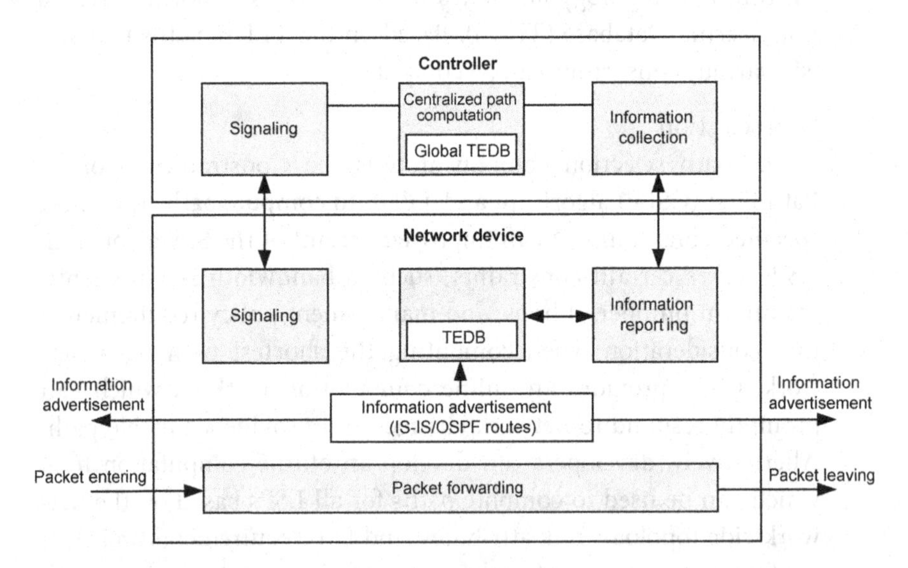

FIGURE 4.2 Centralized SR-TE architecture.

establishing LSPs. Therefore, the information advertisement and signaling components of SR-TE are integrated to some extent, but they are logically independent of each other. The information advertisement component provides the SR information flooding function, whereas the signaling component establishes the SR path that meets specific constraints.

MPLS TE uses the traditional distributed architecture, in which the ingress computes paths based on relevant constraints and uses RSVP-TE to establish CR-LSPs. As services raise more and more SLA requirements on networks and the scale of networks is gradually expanding, several problems regarding distributed path computation occur:

- Although distributed TE uses the SPF or CSPF algorithm to compute paths automatically, it does not compute paths from a global network view. Instead, each node makes its own path decision autonomously, making it impossible to obtain the globally optimal path.
- E2E inter-area path computation also has a similar problem. As IGP topology information is flooded only within an IGP area, network nodes can only detect the intra-area topology and cannot compute E2E inter-area paths. Although BGP-LS can be used to flood topology information across areas, other problems may occur. For example, path computation on a large-scale network requires more Central Processing Unit (CPU) resources.

With the emergence of SDN, controller-based centralized control has begun to show its advantages. As the brain of a centrally controlled network, the controller can collect network topology information and TE information globally on the network, compute paths in a centralized manner, and then deliver the computation results to network devices. In addition, the controller can collect E2E inter-domain topology information and compute paths centrally to obtain the optimal E2E inter-domain path. As such, controller-based centralized path computation provides traffic optimization that is hard to implement with distributed TE.

SR-TE not only solves the scalability problem of traditional RSVP-TE but also meets the requirements of the SDN architecture, helping to further the development of TE technologies and the application of SDN. Figure 4.2 shows the centralized SR-TE architecture.

The controller of the SR-TE architecture consists of the following components:

1. Information collection

   The controller can use protocol extensions such as BGP-LS to collect network topology, TE, and SR information in order to establish a global TEDB.

2. Centralized path computation

   This component can respond to TE path computation requests from network devices and compute optimal paths that meet relevant constraints based on global network information.

3. Signaling

   The signaling component receives a path computation request from the signaling component of a network device and sends the path computation result to the network device. Typically, the controller and network devices exchange signaling messages through PCEP or BGP SRv6 Policy extensions.

A network device consists of the following components:

1. Information advertisement

   SR-TE not only needs to obtain network topology and TE information from the network as MPLS TE does, but also needs to obtain SR information from the network. All of this can be done through IGP extensions.

2. Information reporting

   Network devices use protocol extensions such as BGP-LS to report information, such as network topology, TE, and SR information.

3. Signaling

   The signaling component of a network device sends a TE path computation request to the controller and receives the result from the controller. PCEP and BGP extensions are currently used as the main signaling protocols for this purpose.

4. Packet forwarding

   The packet forwarding component forwards packets based on the source routing mechanism of SR. As the SR instructions (segments) for packet forwarding are explicitly specified in a packet, the packet forwarding components of the devices along the specified path work in succession to forward the packet.

As an implementation of SR, SRv6 TE uses the same architecture as SR-TE.

4.2 BGP-LS FOR SRv6

To compute TE paths, a controller needs information regarding network-wide topology and its TE and SR attributes. A simple method is to add the controller to the IGP topology as a common routing device so that the controller can collect network topology information by listening for IGP flooding. This method requires the controller to establish an IGP neighbor relationship with at least one routing device in each IGP area and is feasible for a small-scale network with only one or a small number of IGP areas. If this method is applied to a network with a large number of IGP areas, the controller needs to maintain a large number of IGP neighbor relationships, posing challenges to scalability. In addition, an IGP does not support multi-hop connections. If an IGP is used to collect network topology information, single-hop IGP neighbor relationships need to be established and tunnels have to be established between the controller and network nodes. This increases the complexity of controller deployment.

To address this issue, BGP-LS is introduced for a controller to collect network topology information efficiently. BGP-LS has inherited many advantages of BGP, such as reliable transmission over TCP, good scalability with Route Reflectors (RRs), and flexible control using policies. Therefore, BGP-LS has become the main protocol used by the SDN controller to collect network topology information.

4.2.1 BGP-LS Overview

The typical application scenario and architecture of BGP-LS are shown in Figure 4.3, in which a consumer is equivalent to a controller[4] and BGP-LS peer relationships are established to carry topology information to the controller. Only one routing device in each IGP area needs to run BGP-LS and directly establish a BGP-LS peer relationship with a consumer. A routing device can also establish a BGP-LS peer relationship with a centralized BGP speaker, which then establishes a BGP-LS peer relationship with the consumer. After a centralized BGP speaker is deployed and the BGP route reflection mechanism is adopted, the number of external connections of

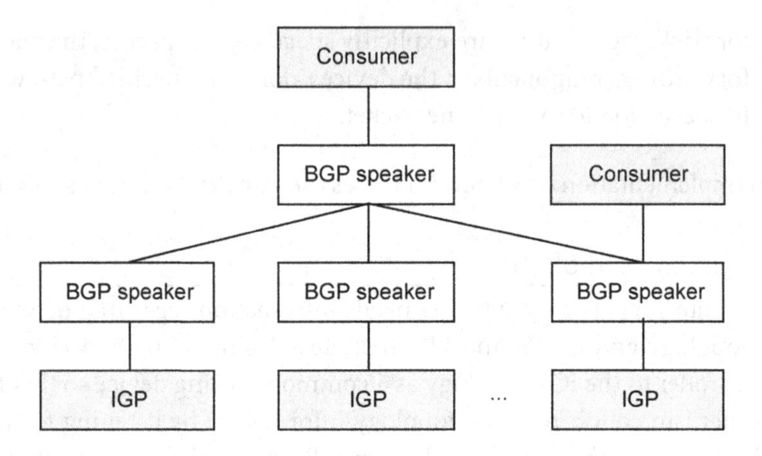

FIGURE 4.3 Typical architecture of BGP-LS.

the consumer can be reduced. As shown in Figure 4.3, the consumer needs to establish only one BGP-LS connection to collect networkwide topology information, thereby simplifying network deployment and Operations and Maintenance (O&M).

On the basis of BGP, BGP-LS introduces a series of new Network Layer Reachability Information (NLRI) attributes to carry information about links, nodes, and IPv4/IPv6 prefixes. Such new NLRIs are called Link-State NLRIs. BGP-LS includes the MP_REACH_NLRI or MP_UNREACH_NLRI attribute in BGP Update messages to carry Link-State NLRIs.

Before supporting SR, BGP-LS defines four types of Link-State NLRIs:

- Node NLRI
- Link NLRI
- IPv4 Topology Prefix NLRI
- IPv6 Topology Prefix NLRI

In addition, BGP-LS defines attributes corresponding to the preceding NLRIs to carry link, node, and IPv4/IPv6 prefix parameters. BGP-LS attributes are defined as TLVs and carried with corresponding Link-State NLRIs in BGP-LS messages. All BGP-LS attributes are optional non-transitive BGP attributes, including the following:

- Node Attribute
- Link Attribute
- Prefix Attribute

Table 4.1 describes the summary of Link-State NLRIs and their associated attributes.[4]

4.2.2 BGP-LS Extensions for SRv6

To support SR, the IETF has defined a series of extensions for the Node, Link, and Prefix Attributes. One such example is the BGP-LS extensions[5] for supporting SR-MPLS. Since the focus of this book is SRv6, in the following section, we describe the extensions added to BGP-LS for collecting SRv6 information.

First, BGP-LS defines SRv6 SID NLRI, which is used to advertise the network-layer reachability information of SRv6 SIDs. The NLRI contains the following information:

- Protocol-ID
- Local Node Descriptors
- SRv6 SID Descriptors: describes an SRv6 SID. Currently, this field is defined as the SRv6 SID Information TLV.

Second, BGP-LS defines SRv6 SID NLRI attributes to carry SRv6 SID reachability information. Currently, the following three related attributes (in the TLV format) are defined:

- SRv6 Endpoint Function: carries the endpoint function information (such as End.DT4) associated with a SID NLRI.
- SRv6 BGP Peer Node SID: carries the BGP peer information associated with a SID NLRI.
- SRv6 SID Structure: describes the length of each part of an SRv6 SID.

Last but not least, BGP-LS extends the existing attributes by defining a series of attribute TLVs, which include the following:

TABLE 4.1 Summary of Link-State NLRIs and Their Associated Attributes

NLRI Type	Description	BGP-LS Attribute

Node	A Node NLRI consists of the following three parts:
	• Protocol-ID: identifies a protocol, such as IS-IS, OSPFv2, OSPFv3, or BGP.
	• Identifier: uniquely identifies a protocol instance when IS-IS or OSPF multi-instance is running.
	• Local Node Descriptor: consists of a series of Node Descriptor sub-TLVs. Currently, the following sub-TLVs are defined:
	  • Autonomous System
	  • BGP-LS Identifier
	  • OSPF Area-ID
	  • IGP Router-ID
		The Node Attribute consists of a series of TLVs. The Node Attribute, along with the Node NLRI, describes a node. The Node Attribute contains the following TLVs:
		• Multi-Topology Identifier: carries one or more multi-topology IDs.
		• Node Flag Bits: carries a bit mask describing the node's attributes, for example, whether the node is an Area Border Router (ABR).
		• Opaque Node Attribute: carries some optional Node Attribute TLVs.
		• Node Name
		• IS-IS Area Identifier
		• IPv4 Router-ID of Local Node
		• IPv6 Router-ID of Local Node

Link	A Link NLRI consists of the following parts:
	• Protocol-ID: identifies a protocol, such as IS-IS, OSPFv2, OSPFv3, or BGP.
	• Identifier: uniquely identifies a protocol instance when IS-IS or OSPF multi-instance is running.
	• Local Node Descriptors
	• Remote Node Descriptors
	• Link Descriptors: uniquely identifies a link among multiple parallel links between two devices. Currently, the following TLVs are defined as link descriptors:
	  • Link Local/Remote Identifiers
	  • IPv4 Interface Address
		• The Link Attribute is used to describe the attributes and parameters of a link. The Link Attribute, along with the Link NLRI, describes a link. Currently, the following Link Attribute TLVs are defined:
		  • IPv4 Router-ID of Local Node
		  • IPv4 Router-ID of Remote Node
		  • IPv6 Router-ID of Local Node
		  • IPv6 Router-ID of Remote Node
		  • Administrative Group (Color): administrative group allocated to a link. It is also called the color attribute of a link.

(*Continued*)

TABLE 4.1 (*Continued*) Summary of Link-State NLRIs and Their Associated Attributes

NLRI Type	Description	BGP-LS Attribute
	• IPv4 Neighbor Address • IPv6 Interface Address • IPv6 Neighbor Address • Multi-Topology Identifier The local and remote Node Descriptors are associated with Node NLRIs.	• Maximum Link Bandwidth • Maximum Reservable Link Bandwidth • Unreserved Bandwidth

IPv4/IPv6 Prefix	An IPv4/IPv6 Prefix NLRI consists of the following parts:
	• Protocol-ID: identifies a protocol, such as IS-IS, OSPFv2, OSPFv3, or BGP.
	• Identifier: uniquely identifies a protocol instance when IS-IS or OSPF multi-instance is running.
	• Local Node Descriptors
	• Prefix Descriptors: includes the following sub-TLVs:
	  • Multi-Topology Identifier
	  • OSPF Route Type
	  • IP Reachability Information
	The preceding information is used to uniquely identify a route prefix.
		The Prefix Attribute, along with the IPv4/IPv6 Prefix NLRI, describes an IPv4/IPv6 prefix. The Prefix Attribute also consists of a series of TLVs. Currently, the following TLVs are defined:
		• IGP Flags: carries IS-IS or OSPF flags and bits originally allocated to the prefix.
		• IGP Route Tag: carries the IS-IS or OSPF route tag information.
		• IGP Extended Route Tag: carries the extended route tag information of IS-IS or OSPF.
		• Prefix Metric
		• OSPF Forwarding Address
		• Opaque Prefix Attribute

- Node Attribute extensions, with the following newly defined TLVs:
    - SRv6 Capabilities: advertises the SRv6 capabilities supported by a node.
    - SRv6 Node MSD Types: advertises an SRv6-capable node's capabilities of processing SRv6 SID depths of different types.
- Link Attribute extensions, with the following newly defined TLVs:
    - SRv6 End.X SID: advertises the End.X SIDs associated with a P2P or P2MP link.
    - SRv6 LAN End.X SID: advertises End.X SIDs associated with broadcast links.
- Prefix Attribute extension, with the following newly defined TLV:
    - SRv6 Locator: advertises the SRv6 locator of a node.

In the following section, we expand on the definitions and functions of the related NLRI and attribute TLVs.

4.2.2.1 SRv6 SID NLRI

An SRv6 SID NLRI describes the network-layer reachability information of an SRv6 SID. The SRv6 SID NLRI format is shown in Figure 4.4.

Table 4.2 describes the fields in the SRv6 SID NLRI.

4.2.2.2 SRv6 SID Information TLV

SRv6 SID Information TLV is a sub-TLV of the SRv6 SID NLRI and is used to advertise SID information. Figure 4.5 shows the SRv6 SID Information TLV format.

Table 4.3 describes the fields in the SRv6 SID Information TLV.

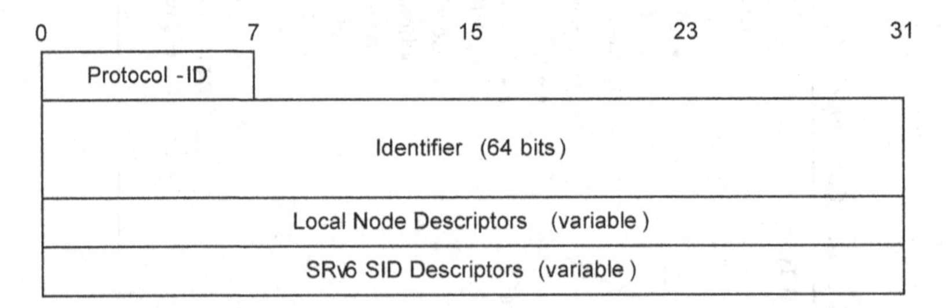

FIGURE 4.4 SRv6 SID NLRI.

TABLE 4.2 Fields in the SRv6 SID NLRI

Field	Length	Description
Protocol-ID	8 bits	Protocol ID, which specifies the protocol through which BGP-LS learns SRv6 SID information from a node.
Identifier	64 bits	Identifier of a node.
Local Node Descriptors	Variable	Description of the local node.
SRv6 SID Descriptors	Variable	SRv6 SID information. For details, see the description in SRv6 SID Information TLV.

0 7 15 23 31
Type | Length
SID (16 octets) ...
SID (cont-)
SID (cont-)
SID (cont-)

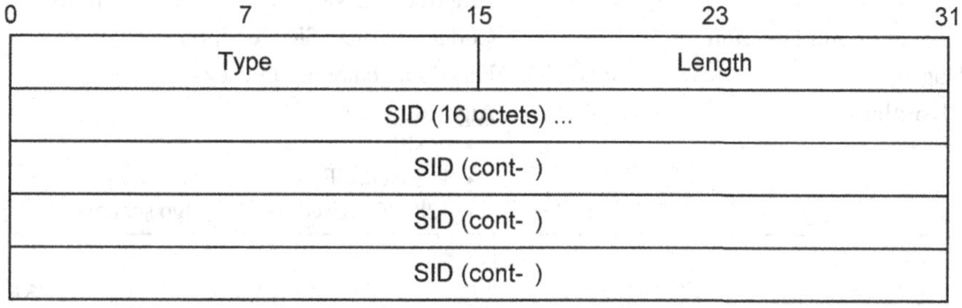

FIGURE 4.5 SRv6 SID Information TLV.

TABLE 4.3 Fields in the SRv6 SID Information TLV

Field	Length	Description
Type	16 bits	Type of the TLV.
Length	16 bits	Length of the Value field in the TLV, in octets.
SID	16 octets	Specific SRv6 SID value.

4.2.2.3 SRv6 Endpoint Function TLV

The SRv6 Endpoint Function TLV is an SRv6 SID NLRI attribute used to carry the instructions associated with each type of SID. Each SRv6 SID is associated with a network function instruction, and we dig into more details on this in Section 2.2. Figure 4.6 shows the SRv6 Endpoint Function TLV format.

Table 4.4 describes the fields in the SRv6 Endpoint Function TLV.

4.2.2.4 SRv6 BGP Peer Node SID TLV

The SRv6 BGP Peer Node SID TLV is an SRv6 SID NLRI attribute used to carry the BGP peer information related to the SID NLRI. This TLV must be used together with the SRv6 End.X SID associated with the BGP Peer

0	7	15	23	31
Type		Length		
SRv6 Endpoint Function		Flags	Algorithm	

FIGURE 4.6 SRv6 Endpoint Function TLV.

TABLE 4.4 Fields in the SRv6 Endpoint Function TLV

Field	Length	Description
Type	16 bits	Type of the TLV.
Length	16 bits	Length of the Value field in the TLV, in octets.
SRv6 Endpoint Function	16 bits	Codepoint of an SRv6 endpoint behavior.[6]
Flags	8 bits	Flags, which are not in use currently.
Algorithm	8 bits	Algorithm ID: • 0: SPF • 1: Strict SPF • 128–255: Used in Flex-Algo scenarios

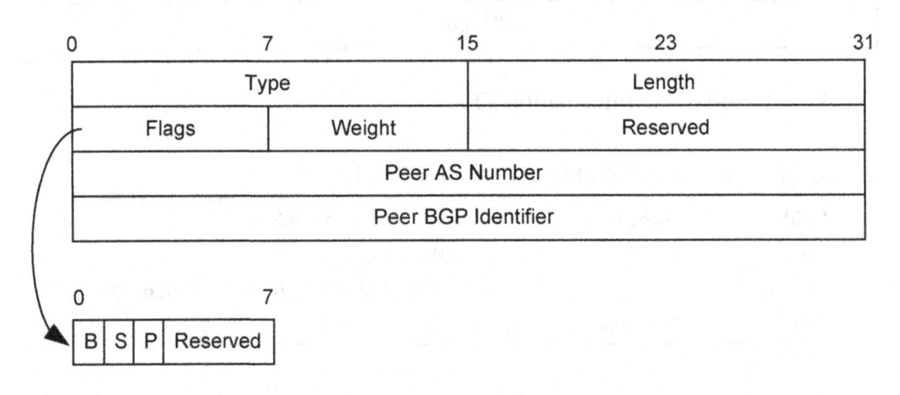

FIGURE 4.7 SRv6 BGP Peer Node SID TLV.

Node or BGP Peer Set function. Figure 4.7 shows the SRv6 BGP Peer Node SID TLV format.

Table 4.5 describes the fields in the SRv6 BGP Peer Node SID TLV.

4.2.2.5 SRv6 SID Structure TLV

The SRv6 SID Structure TLV is an SRv6 SID NLRI attribute that describes the length of each part of an SRv6 SID. As we discussed in Section 2.2, an SRv6 SID generally consists of three parts: Locator, Function, and Arguments. This TLV describes the number of bits that each part occupies in a 128-bit SID. Figure 4.8 shows the SRv6 SID Structure TLV format.

Table 4.6 describes the fields in the SRv6 SID Structure TLV.

TABLE 4.5 Fields in the SRv6 BGP Peer Node SID TLV

Field	Length	Description
Type	16 bits	Type of the TLV.
Length	16 bits	Length of the Value field in the TLV, in octets.
Flags	8 bits	Flags: • B-flag: Backup flag. If this flag is set, the End.X SID has a backup path. • S-flag: Set flag. If this flag is set, the End.X SID is associated with a group of adjacencies. • P-flag: Persistent flag. If this flag is set, the End.X SID is permanently allocated. Do note that the End.X SID remains unchanged even if peer relationship flapping, a protocol restart, or a device restart occurs.
Weight	8 bits	Weight of the End.X SID used for load balancing.
Peer AS Number	32 bits	AS number of the BGP peer.
Peer BGP Identifier	32 bits	Router ID of the BGP peer.

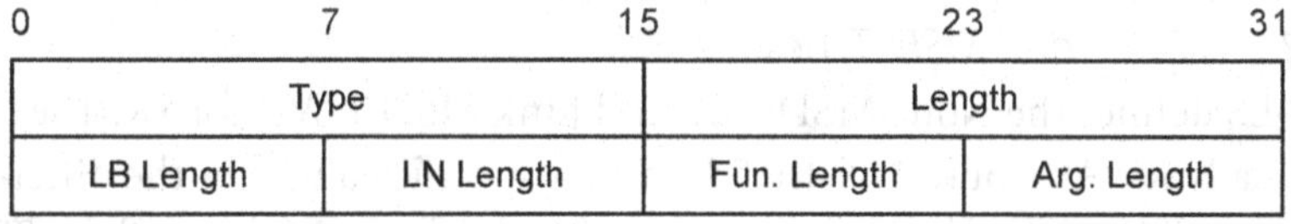

0	7	15	23	31
Type		Length		
LB Length	LN Length	Fun. Length	Arg. Length	

FIGURE 4.8 SRv6 SID Structure TLV.

TABLE 4.6 Fields in the SRv6 SID Structure TLV

Field	Length	Description
Type	16 bits	Type of the TLV.
Length	16 bits	Length of the Value field in the TLV, in octets.
LB Length	8 bits	Length of the SRv6 SID Locator Block, in bits. The Block is the address block used to allocate SIDs.
LN Length	8 bits	Length of the SRv6 SID Locator Node ID, in bits.
Fun.Length	8 bits	Length of the Function field in the SID, in bits.
Arg. Length	8 bits	Length of the Arguments field in the SID, in bits.

4.2.2.6 SRv6 Capabilities TLV

The SRv6 Capabilities TLV is an SRv6 Node Attribute used to advertise the SRv6 capabilities supported by a network node, and advertised together with a Node NLRI. Each SRv6-capable node must include this TLV in a BGP-LS attribute. The TLV corresponds to the IS-IS SRv6 Capabilities sub-TLV[7] and OSPFv3 SRv6 Capabilities TLV.[8] Figure 4.9 shows the SRv6 Capabilities TLV format.

Table 4.7 describes fields in the SRv6 Capabilities TLV.

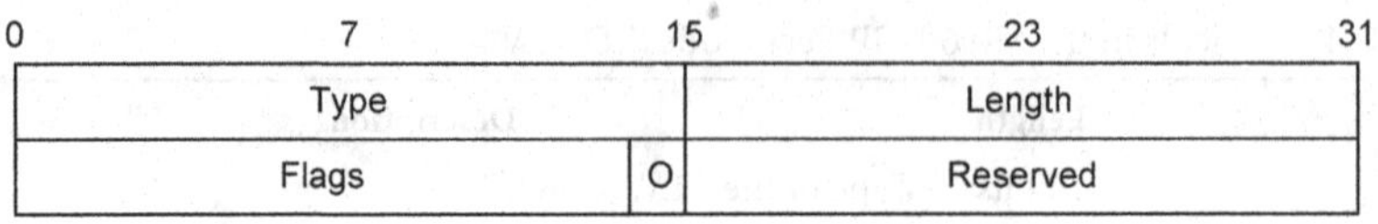

0	7	15	23	31
Type			Length	
Flags		O	Reserved	

FIGURE 4.9 SRv6 Capabilities TLV.

TABLE 4.7 Fields in the SRv6 Capabilities TLV

Field	Length	Description
Type	16 bits	Type of the TLV.
Length	16 bits	Length of the Value field in the TLV, in octets.
Flags	16 bits	Flags: • O-flag: OAM capability flag. If this flag is set, the routing device has the OAM capability corresponding to the O-bit defined in the SRH.
Reserved	16 bits	Reserved field.

4.2.2.7 SRv6 Node MSD Types

BGP-LS defines the Node MSD TLV and Link MSD TLV[9] for SR. The former is a Node Attribute TLV that describes specific values for the different types of MSDs that a device can process. The latter is a Link Attribute TLV that describes specific values for the different types of MSDs that a link can process.

The Node MSD TLV and Link MSD TLV use the same format, as shown in Figure 4.10.

Table 4.8 describes the fields in the Node or Link MSD TLV.

4.2.2.8 SRv6 End.X SID TLV

The SRv6 End.X SID TLV is an SRv6 Link Attribute TLV used to advertise SRv6 End.X SIDs. An SRv6 End.X SID is associated with a P2P or P2MP link, or with an IS-IS or OSPF adjacency. This TLV can also be used to advertise the Layer 2 member links of a Layer 3 bundle interface, which is a logical interface bundled from multiple physical interfaces.

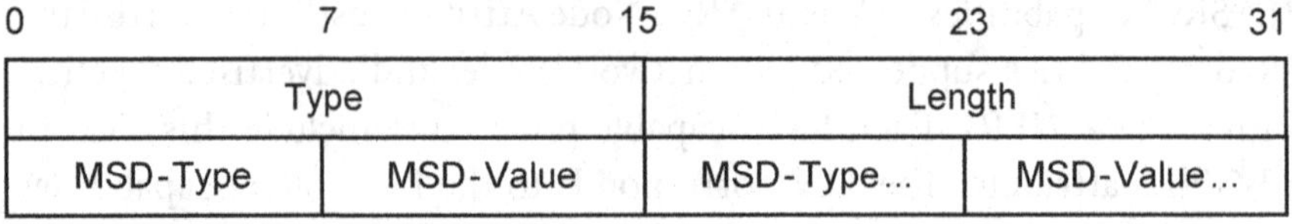

0	7	15	23	31
Type		Length		
MSD-Type	MSD-Value	MSD-Type...	MSD-Value...	

FIGURE 4.10 Node or Link MSD TLV.

TABLE 4.8 Fields in the Node or Link MSD TLV

Field	Length	Description
Type	16 bits	Type of the TLV.
Length	16 bits	Length of the Value field in the TLV, in octets.
MSD-Type	8 bits	MSD type. For SRv6, BGP-LS reuses the preceding Node and Link MSD TLVs but adds multiple new MSD types to advertise an SRv6-capable node's capabilities of processing Maximum SID Depths.[5] The new MSD types include the following: • Maximum Segments Left: specifies the maximum value of the SL field in the SRH of a received packet before the SRv6 endpoint function associated with a SID is applied. • Maximum End Pop: specifies the maximum number of SIDs in the SRH when a node applies the PSP or USP flavor to the SRH. The value 0 indicates that the advertising node cannot apply either flavor to the SRH. • Maximum H.Insert: specifies the maximum number of SIDs that can be inserted when the H.Insert operation is performed to insert SRH information. The value 0 indicates that the advertising node cannot perform the H.Insert operation. • Maximum H.Encaps: specifies the maximum number of SIDs that can be encapsulated when the H.Encaps operation is performed. • Maximum End D: specifies the maximum number of SIDs before decapsulation associated with End.D (such as End.DX6 and End.DT6) functions is performed.
MSD-Value	8 bits	Value of the Maximum SID Depth.

In addition, it can be used to advertise the End.X SIDs of the BGP Egress Peer Engineering (EPE) peers of a node running BGP. Figure 4.11 shows the SRv6 End.X SID TLV format.

Table 4.9 describes fields in the SRv6 End.X SID TLV.

4.2.2.9 SRv6 LAN End.X SID TLV

The SRv6 LAN End.X SID TLV is also an SRv6 Link Attribute TLV and is used to advertise SRv6 End.X SIDs associated with the adjacencies on a broadcast network. Figure 4.12 shows the SRv6 LAN End.X SID TLV format.

Table 4.10 describes the new field in the SRv6 LAN End.X SID TLV. The other fields are the same as those in the SRv6 End.X SID TLV (see Table 4.9 for more details).

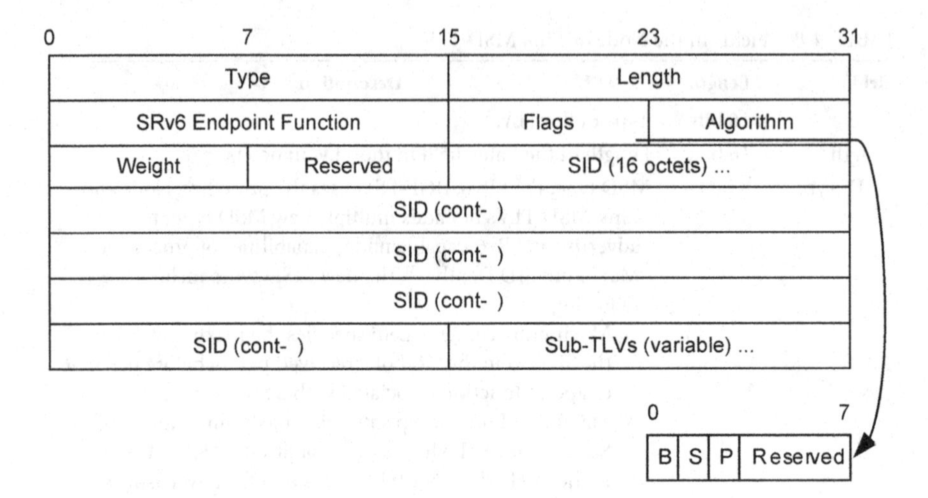

FIGURE 4.11 SRv6 End.X SID TLV.

TABLE 4.9 Fields in the SRv6 End.X SID TLV

Field	Length	Description
Type	16 bits	Type of the TLV.
Length	16 bits	Length of the Value field in the TLV, in octets.
SRv6 Endpoint Function	16 bits	Codepoint of an SRv6 endpoint behavior.
Flags	8 bits	Flags: • B-flag: Backup flag. If this flag is set, the End.X SID has a backup path. • S-flag: Set flag. If this flag is set, the End.X SID is associated with a group of adjacencies. • P-flag: Persistent flag. If this flag is set, the End.X SID is permanently allocated. Do note that the End.X SID remains unchanged even if peer relationship flapping, a protocol restart, or a device restart occurs.
Algorithm	8 bits	Algorithm ID: • 0: SPF • 1: Strict SPF • 128–255: Used in Flex-Algo scenarios
Weight	8 bits	Weight of the End.X SID used for load balancing.
SID	16 octets	SRv6 SID to be advertised.
Sub-TLVs	Variable	Sub-TLVs that can be extended. These sub-TLVs are not defined currently.

0	7	15	23	31
Type		Length		
SRv6 Endpoint Function		Flags	Algorithm	
Weight	Reserved			
IS-IS System-ID (6 octets) or OSPFv3 Router-ID (4 octets) of the neighbor				
SID (16 octets) ...				
SID (cont-)				
SID (cont-)				
Sub-TLVs (variable) ...				

FIGURE 4.12 SRv6 LAN End.X SID TLV.

TABLE 4.10 New Field in the SRv6 LAN End.X SID TLV

Field	Length	Description
IS-IS System-ID or OSPFv3 Router-ID of the neighbor	6 octets or 4 octets	Specifies the broadcast network neighbor to which the End.X SID is allocated. The value is a 6-octet system ID or 4-octet router ID.

4.2.2.10 SRv6 Locator TLV

The SRv6 Locator TLV is an SRv6 Prefix Attribute TLV. BGP-LS uses this TLV, along with the SRv6 Prefix NLRI, to collect locator information from SRv6-capable nodes. As we previously mentioned, SRv6 SIDs inherently feature the routing capability, which is implemented by IGPs advertising SRv6 locators. Figure 4.13 shows the SRv6 Locator TLV format.

Table 4.11 describes the fields in the SRv6 Locator TLV.

0	7	15	23	31
Type		Length		
Flags	Algorithm	Reserved		
Metric				
Sub-TLVs (variable) ...				

FIGURE 4.13 SRv6 Locator TLV.

TABLE 4.11 Fields in the SRv6 Locator TLV

Field	Length	Description
Type	16 bits	Type of the TLV.
Length	16 bits	Length of the Value field in the TLV, in octets.
Flags	8 bits	Flags: • D-flag: This flag must be set when a SID is leaked from a Level-2 area to a Level-1 area. SIDs with the D-flag set cannot be leaked from the Level-1 area to the Level-2 area, thereby preventing routing loops. • A-flag: Anycast flag. This flag must be set if the locator type is set to anycast.
Algorithm	8 bits	Algorithm ID: • 0: SPF • 1: Strict SPF • 128–255: Used in Flex-Algo scenarios
Metric	32 bits	Metric.
Sub-TLVs	Variable	Included sub-TLVs, for example, SRv6 End SID sub-TLV.

We have now covered BGP-LS extensions for SRv6. Based on these extensions, a controller can collect networkwide SRv6 information in addition to networkwide topology and TE information. The controller can then make use of the information to compute networkwide TE paths.

4.3 PCEP FOR SRv6

4.3.1 PCE Overview

The Path Computation Element (PCE) architecture was first proposed to solve the path computation problem on large-scale networks with multiple areas. We can use this architecture to compute inter-domain paths for TE, and it consists of the following three parts:

- PCE: a component that can compute paths meeting relevant constraints based on network topology information. The PCE can be deployed on a routing device or independent server, and in most cases, it is integrated with a controller.
- Path Computation Client (PCC): a client application that requests the PCE to perform path computation. The PCC, which is usually deployed on a routing device, sends a path computation request to the PCE and receives the corresponding result from it.

- PCEP: a protocol that runs over TCP using the port number 4189 and is used for communication between a PCC and PCE. A PCEP message consists of a common header and a collection of either mandatory or optional objects for carrying TE path computation requests and path computation results.

The PCE and centralized SR-TE architectures share certain similarities. In fact, we can say that the latter is derived from the former. Traditional PCEs adopt centralized path computation, and the PCEP of the initial version defines RSVP-TE-oriented stateless PCE protocol extensions and interaction processes.[10] Stateless PCEs independently compute paths based on the TEDB, without needing to maintain any of the previously computed paths. That said, stateless PCEs do not have information about the currently active TE paths and therefore cannot perform any type of re-optimization on the active paths. To add on to that, traditional PCEs have poor scalability when working with RSVP-TE, and that is why they have not been widely deployed.

With the development of SDN and emergence of SR, PCEP now includes stateful PCE[11,12] and relevant extensions for SR, which effectively support centralized SR-TE functions.

- Stateful PCE: A stateful PCE maintains a TEDB that is strictly synchronized with a network. The TEDB includes not only the IGP link states, TE information, and TE path information, but also reserved resources used by TE paths on the network. Comparatively speaking, stateful PCEs can compute better network paths than stateless PCEs by performing reliable state synchronization with networks.

- PCE extensions for SR: SR has advantages over RSVP-TE in terms of scalability due to the fact that it provides the same capability of explicitly specifying paths, but without needing to maintain per-flow states on transit nodes. On the other hand, as SR does not maintain states on transit nodes, it lacks the capability of computing paths based on the bandwidth usage on the ingress. That is where PCE-based SR comes into play by being able to address this issue. To put it more precisely, as a PCE stores the topology, TE, path, and reserved resource information of the entire network, it can compute paths based on networkwide resources, ultimately optimizing resources on the entire network.

4.3.2 Stateful PCE

Stateful PCE is a significant extension of PCE. A stateful PCE not only maintains topology and TE information used for path computation, but also retains TE path information. As such, the stateful PCE can optimize networkwide resources during path computation based on existing TE paths, rather than simply computing a path that meets relevant constraints.

Stateful PCEs work in either passive or active mode:

- Passive stateful PCE: A PCE performs path computation and optimization based on the LSP states learned from a PCC, which it is synchronized with, and does not update the LSP states proactively.
- Active stateful PCE: A PCE can proactively deliver path suggestions to a network to update paths. For example, a PCC can delegate its control over LSPs to an active stateful PCE through the LSP delegation mechanism. In this way, the PCE can update the LSP parameters of the PCC.

For ease of understanding, the following uses the SRv6 Policy as an example to describe how a stateful PCE computes paths, as shown in Figure 4.14.

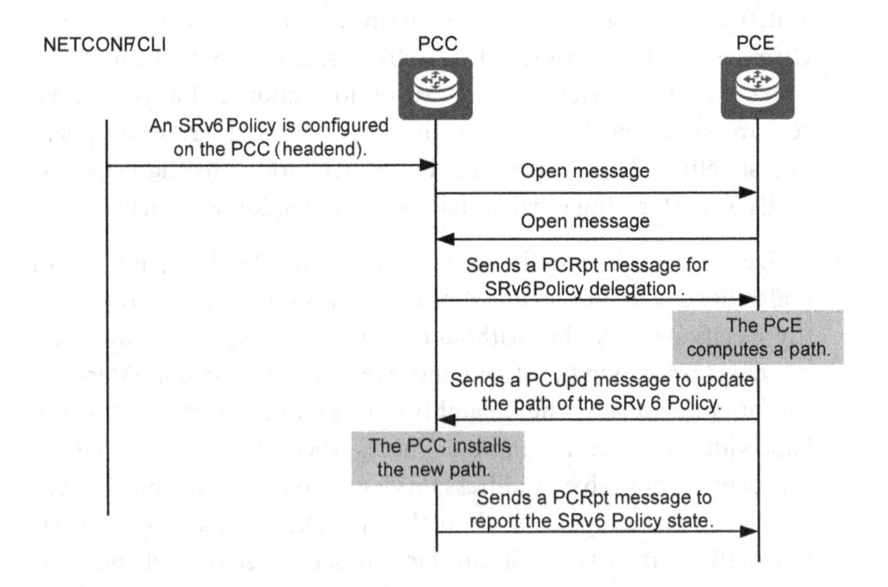

FIGURE 4.14 Process in which a stateful PCE computes paths.

1. An SRv6 Policy is configured on the PCC (headend) using NETCONF or the Command-Line Interface (CLI). More details on SRv6 Policy are provided in the following sections of this chapter.

2. The PCC and PCE send Open messages to each other in order to set up a PCEP session and negotiate supported capabilities. To support SRv6, PCEP is extended to support the SRv6 PCE Capability sub-TLV, which is used to negotiate SRv6 capabilities.

3. To delegate the SRv6 Policy and request the PCE to compute a path, the PCC (headend) sends a Path Computation LSP State Report (PCRpt) message with the Delegate flag set to 1 to the PCE. The PCRpt message carries various objects that specify a collection of the SRv6 Policy's constraints and attributes.

4. Upon receipt, the PCE computes an SRv6 path based on the Delegate flag, path constraints, and path attributes carried in the message.

5. After path computation completes, the PCE sends a Path Computation LSP Update Request (PCUpd) message, in which the Explicit Route Object (ERO) carries the path information, to the PCC (headend). To support SRv6, SRv6-ERO subobjects are added to PCEP to carry SRv6 path information.

6. The PCC (headend) receives the path information from the PCE and installs the target path.

7. After path installation completes, the PCC (headend) sends a PCRpt message to the PCE to report the SRv6 Policy state information. The message uses the Reported Route Object (RRO) to carry information about the actual forwarding path of the PCC and uses the ERO to carry information about the path computed by the PCE. To support SRv6, SRv6-RRO subobjects are added to PCEP to carry information about actual SRv6 paths, and SRv6-ERO subobjects to carry information about the SRv6 paths computed by the PCE.

In addition to the PCEP messages mentioned in the preceding process, other types of PCEP messages also exist. Table 4.12 lists different types of PCEP messages and their functions.

TABLE 4.12 Types of PCEP Messages and Their Functions

Message Type	Function
Open	Used to establish PCEP sessions and describe PCEP capabilities of the PCC/PCE.[10]
Keepalive	Sent by a PCC or PCE in order to keep the PCEP session in the active state.[10]
PCReq	Sent by a PCC to a PCE to request path computation.[10]
PCRep	Sent by a PCE to a PCC in response to a path computation request from the PCC.[10]
PCRpt	Sent by a PCC to a PCE to report the current state of an LSP.[11]
PCUpd	Sent by a PCE to a PCC to update LSP attributes.[11]
PCInitiate	Sent by a PCE to a PCC to trigger the initiation of an LSP.[12]
PCErr	Used to notify a PCEP peer that a request does not comply with the PCEP specifications or meets a protocol error condition.[10]
Close	Used to close a PCEP session.[10]

Each message may contain one or more objects describing specific functions, such as the OPEN and LSP objects. Table 4.13 lists objects in stateful PCE mode.[10–12]

TABLE 4.13 PCEP Objects in Stateful PCE Mode

Type	Name	Function	Carried In
Object	OPEN	Used to establish a PCEP session and negotiate various service capabilities.	Open and PCErr messages
Object	Stateful PCE Request Parameters (SRP)	Used to associate the update requests sent by a PCE with the error and state reports sent by a PCC.	PCRpt, PCUpd, PCErr, and PCInitiate messages
Object	LSP	Used to carry information that identifies the target LSP.	PCRpt, PCUpd, PCInitiate, PCReq, and PCRep messages
Object	ERO	Used to carry information about the SRv6 path computed by a PCE.	PCUpd, PCInitiate, PCRpt, and PCRep messages
Object	RRO	Used to carry information about the actual SRv6 path on a PCC.	PCRpt and PCReq messages
Object	ERROR	Used to carry error information.	PCErr message
Object	CLOSE	Used to close a PCEP session.	Close message

4.3.3 PCEP Extensions for SRv6

In the preceding sections, we went over the PCE architecture and provided a basic understanding of stateful PCEs. Next, we will describe the PCEP extensions for SRv6.

These extensions include three parts: the new type PATH-SETUP-TYPE for SRv6, the SRv6 PCE Capability sub-TLV used to advertise SRv6 capabilities, and the SRv6 ERO and SRv6 RRO subobjects used to carry SRv6 SIDs of a specific SRv6 path. Table 4.14 expands on the details.[13]

4.3.3.1 SRv6 PATH-SETUP-TYPE

When a PCEP session is established between a PCC and PCE, they send Open messages to notify each other of their capabilities, such as the type of path that can be established. The capability of establishing paths is described in the PATH-SETUP-TYPE-CAPABILITIES TLV, and its format is shown in Figure 4.15.[14]

Table 4.15 describes the fields in the PATH-SETUP-TYPE-CAPABILITIES TLV.

To support SRv6, PCEP introduces a new path setup type for identifying SRv6 paths, and its type value is 2. PCEP goes a step further by also defining the SRv6 PCE Capability sub-TLV for describing SRv6 capabilities in detail.[13]

During capability advertisement, if the PATH-SETUP-TYPE-CAPABILITIES TLV carries a PST with the value of 2, SRv6 path setup

TABLE 4.14 PCEP TLV Extensions for SRv6

Type	Name	Function	Carried In
Type	SRv6 PATH-SETUP-TYPE	Used to indicate the TE path created using SRv6.	PATH-SETUP-TYPE and PATH-SETUP-TYPE-CAPABILITIES TLVs
Sub-TLV	SRv6 PCE Capability sub-TLV	Used to advertise SRv6 capabilities.	PATH-SETUP-TYPE-CAPABILITIES TLV
Subobject	SRv6-ERO subobject	Used to carry information about the SRv6 path computed by a PCE.	ERO object
	SRv6-RRO subobject	Used to carry information about the actual SRv6 path on a PCC.	RRO object

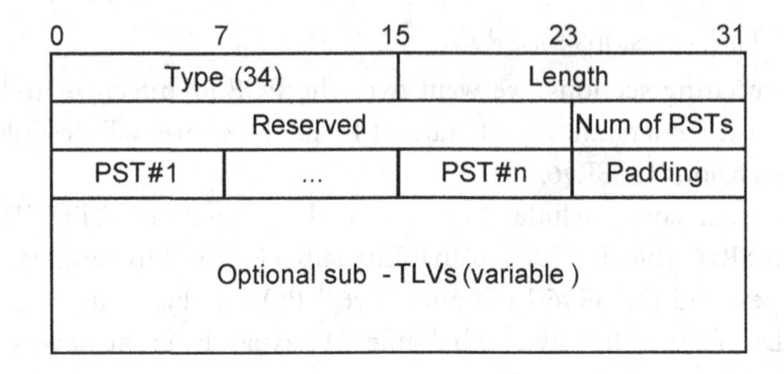

FIGURE 4.15 PATH-SETUP-TYPE-CAPABILITIES TLV.

TABLE 4.15 Fields in the SRv6 PATH-SETUP-TYPE-CAPABILITIES TLV

Field	Length	Description
Type	16 bits	Type of the TLV.
Length	16 bits	Length of the TLV.
Reserved	24 bits	Reserved field.
Num of PSTs	8 bits	Total number of Path Setup Types (PSTs) carried in the TLV.
PST #n	8 bits	Supported PST: • 0: RSVP-TE • 1: Segment Routing • 2: SRv6
Padding	Variable	Used for 4-octet alignment.
Optional sub-TLVs	Variable	Optional sub-TLVs.

is supported. In this case, the PATH-SETUP-TYPE-CAPABILITIES TLV must carry an SRv6 PCE Capability sub-TLV to describe the SRv6 capability in detail.

It is also worth noting that when path computation is requested, the SRP object needs to carry the PATH-SETUP-TYPE TLV to describe a path type.[11] The PATH-SETUP-TYPE TLV also carries a PST with the value of 2 to support SRv6 path computation requests.

4.3.3.2 SRv6 PCE Capability Sub-TLV

The SRv6 PCE Capability sub-TLV describes SRv6 capabilities in detail. Figure 4.16 shows its format.

Table 4.16 describes the fields in the SRv6 PCE Capability sub-TLV.

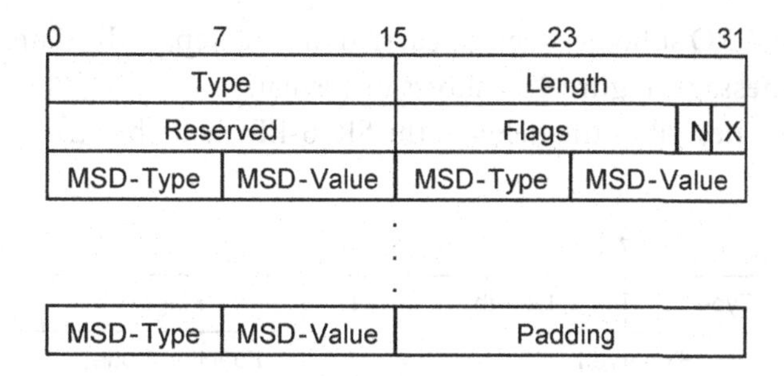

FIGURE 4.16 SRv6 PCE Capability sub-TLV.

TABLE 4.16 Fields in the SRv6 PCE Capability Sub-TLV

Field	Length	Description
Type	16 bits	Type of the sub-TLV.
Length	16 bits	Length of the sub-TLV.
Reserved	16 bits	Reserved field.
Flags	16 bits	Flags: • N-flag: If a PCC sets this flag to 1, the PCC can resolve a Node or Adjacency Identifier (NAI) into an SRv6 SID. • X-flag: If a PCC sets this flag to 1, the PCC does not pose any restriction on the MSD.
MSD-Type	8 bits	Type of the MSD: • Maximum Segments Left MSD Type: specifies the maximum value of the SL field in the SRH of a received packet that the PCC can process. • Maximum End Pop MSD Type: specifies the maximum number of SIDs supported when the POP operation is performed. • Maximum H.Encaps MSD Type: specifies the maximum number of SIDs supported when the H.Encaps operation is performed. • Maximum End D MSD Type: specifies the maximum number of SIDs supported when decapsulation is performed.
MSD-Value	8 bits	Value of the Maximum SID Depth.
Padding	Variable	Used for 4-octet alignment.

4.3.3.3 SRv6-ERO Subobject

In PCEP, an ERO consists of a series of subobjects that are used to describe paths.[10] With the aim of carrying an SRv6 SID and NAI, or either one of them, PCEP defines a new subobject called SRv6-ERO Subobject.[13] An NAI is associated with an SRv6 SID, which means the NAI can be resolved into an SRv6 SID.

An SRv6-ERO subobject can be carried in a PCRep, PCInitiate, PCUpd, or PCRpt message. Figure 4.17 shows its format.

Table 4.17 describes the fields in the SRv6-ERO subobject.

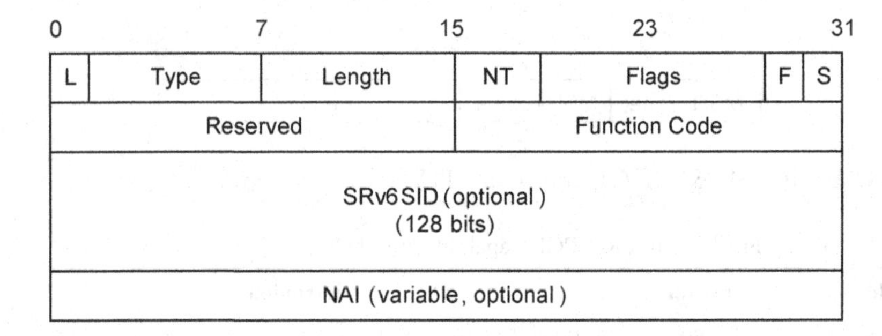

FIGURE 4.17 SRv6-ERO subobject.

TABLE 4.17 Fields in the SRv6-ERO Subobject

Field	Length	Description
L	1 bit	Flag indicating a loose hop.
Type	7 bits	Type of the subobject.
Length	8 bits	Length of the subobject.
NT	4 bits	NAI type: • NT=0: indicates that the NAI is not included in the subobject.[13] • NT=2: indicates that the NAI is the IPv6 address of the node associated with an SRv6 SID. • NT=4: indicates that the NAI is a pair of the global IPv6 addresses of the link associated with an SRv6 End.X SID. • NT=6: indicates that the NAI is a pair of the link-local global IPv6 addresses of the link associated with an SRv6 End.X SID.
Flags	12 bits	Flags: • F-flag: If this flag is set to 1, the NAI value is not included in the subobject. • S-flag: If this flag is set to 1, the SRv6 SID value is not included in the subobject.
Function Code	16 bits	Function associated with an SRv6 SID. This field does not affect the SRH imposed on the packet and is used only for maintainability.
SRv6 SID	128 bits	SRv6 segment.
NAI	Variable	Value of the NAI associated with an SRv6 SID. The format of the NAI is determined by the value of the NT field, and the NAI carries detailed information indicated by the NT field.

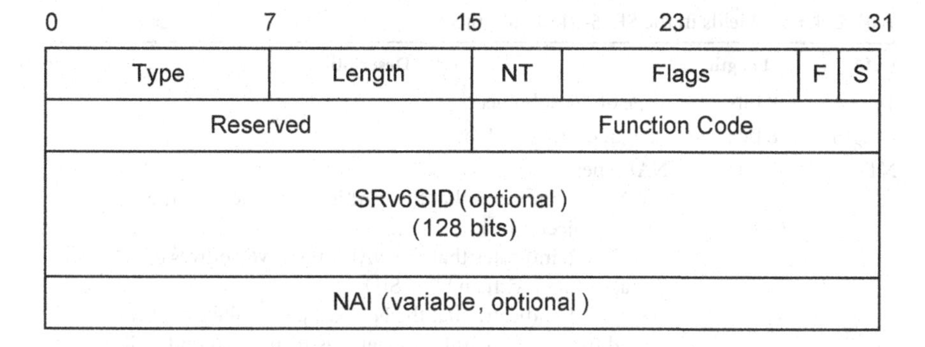

FIGURE 4.18 SRv6-RRO subobject.

4.3.3.4 *SRv6-RRO Subobject*

An RRO is used by a PCC to report the actual path of an LSP when reporting LSP states. In SRv6, this object indicates the SID list of the actual SRv6 path that a PCC uses. To carry SRv6 SID information, PCEP defines the SRv6-RRO subobject,[13] and its format is shown in Figure 4.18. Do note that the SRv6-RRO and SRv6-ERO subobjects share the same format, but the former does not include the L field.

Table 4.18 describes the fields in the SRv6-RRO subobject.

We have now gone over the main points regarding PCEP extensions for SRv6. Put briefly, based on the PCEP extensions, a PCE can receive an SRv6 TE path computation request from a PCC, compute the target path, and return the computation result to the PCC. The combination of the PCE's centralized path computation capability and SRv6 ingress's explicit path programming capability enables TE with flexible controllability, global optimization, and enhanced scalability.

4.4 SRv6 POLICY

SRv6 Policy leverages the source routing mechanism of SR to instruct packet forwarding based on an ordered list of segments encapsulated by the source node. SRv6 uses programmable 128-bit IPv6 addresses to provide diverse network functions, which are expressed using SRv6 instructions that can identify not only forwarding paths but also Value-Added Services (VASs) such as firewalls, application acceleration, and user gateways. Another stand-out feature of SRv6 is its excellent extensibility. This is clearly reflected in the fact that, to support a new network function, all that is needed is to define a new instruction, without changing the

TABLE 4.18 Fields in the SRv6-RRO Subobject

Field	Length	Description
Type	8 bits	Type of the subobject.
Length	8 bits	Length of the subobject.
NT	4 bits	NAI type: • NT=0: indicates that the NAI is not included in the subobject.[13] • NT=2: indicates that the NAI is the IPv6 address of the node associated with an SRv6 SID. • NT=4: indicates that the NAI is a pair of the global IPv6 addresses of the link associated with an SRv6 End.X SID. • NT=6: indicates that the NAI is a pair of the link-local global IPv6 addresses of the link associated with an SRv6 End.X SID.
Flags	12 bits	Flags: • F-flag: If this flag is set to 1, the NAI value is not included in the subobject. • S-flag: If this flag is set to 1, the SRv6 SID value is not included in the subobject.
Function Code	16 bits	Function associated with an SRv6 SID. This field does not affect the SRH imposed on the packet and is used only for maintainability.
SRv6 SID	128 bits	SRv6 segment.
NAI	Variable	Value of the NAI associated with an SRv6 SID. The format of the NAI is determined by the value of the NT field, and the NAI carries detailed information indicated by the NT field.

protocol mechanism or deployment. This advantage significantly shortens the delivery period of innovative network services. As such, the SRv6 Policy mechanism can meet E2E service requirements and is crucial to SRv6 network programming.

By encapsulating a series of SRv6 SIDs into an SRH, a network node can explicitly guide packet forwarding along a planned path, thereby implementing fine-grained E2E control over the forwarding path and meeting SLA requirements such as high reliability, high bandwidth, and low latency. On the network shown in Figure 4.19, source node 1 inserts two SRv6 SIDs (the End.X SID of the link between node 4 and node 5 and the End SID of node 7) into a packet to direct packet forwarding along the specified link and the shortest path that supports load balancing. This is, in short, how SRv6 implements traffic engineering.

To implement SR-TE, the industry put forward SR Policy,[15] a framework that supports both SR-MPLS and SRv6. Focusing on how to use

Segment: instruction telling packets where to go and how to get there

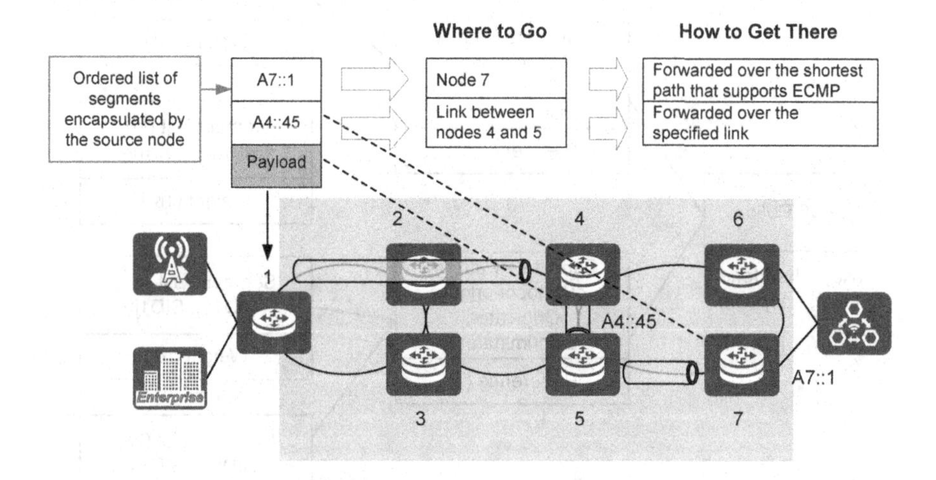

FIGURE 4.19 SRv6 SID encapsulation.

SR Policies to implement SRv6 TE, this section includes the following: SR Policy model definition, control-plane path computation, data-plane forwarding, and protection measures in failure scenarios. The SR Policies used to implement SRv6 TE are called SRv6 Policies in this book.

4.4.1 SRv6 Policy Model

The SRv6 Policy model consists of the following three elements, which are also shown in Figure 4.20.

4.4.1.1 Keys

The SRv6 Policy model uses the following three tuple as the keys to uniquely identify an SRv6 Policy:

- Headend: ingress node of an SRv6 Policy. This node can steer traffic into the corresponding SRv6 Policy.
- Color: attribute used to identify different SRv6 Policies with specified headends and endpoints. It can be associated with a series of service attributes, such as low latency and high bandwidth, and as such can be considered a service requirement template ID. Because no unified coding rule is defined, color values need to be allocated by

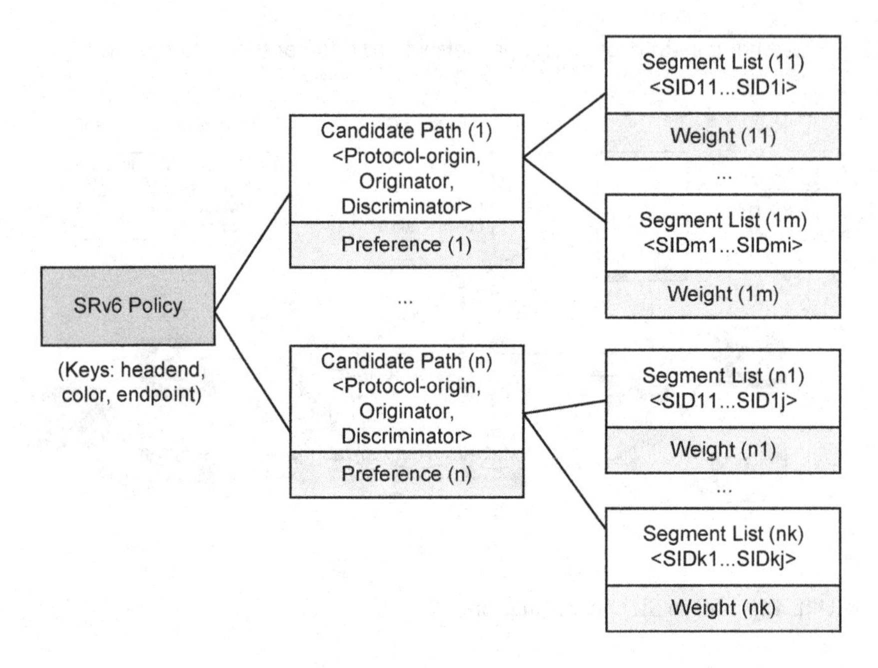

FIGURE 4.20 SRv6 Policy model.

administrators. For example, an administrator can allocate a color value of 100 to an SRv6 Policy whose E2E latency requirement is less than 10 ms. The color attribute enables the association between services and SRv6 Policies.

- Endpoint: destination of an SRv6 Policy.

All the SRv6 Policies delivered to a specific headend have the same headend value. For this reason, the specific headend can use the <Color, Endpoint> tuple to uniquely identify an SRv6 Policy, thereby steering traffic into the SRv6 Policy.

4.4.1.2 Candidate Paths

A candidate path is the basic unit for delivering an SRv6 Policy path to the headend through protocols such as BGP SRv6 Policy and PCEP. Different candidate paths can be delivered using different protocols. An SRv6 Policy may be associated with multiple candidate paths, each of which has a preference. If this is the case, the SRv6 Policy makes the candidate path with the highest preference the primary one.

As shown in the SRv6 Policy model in Figure 4.20, each candidate path of an SRv6 Policy is uniquely identified using the tuple <Protocol-origin, Originator, Discriminator>, where Protocol-origin indicates the protocol or component through which the candidate path is generated, Originator indicates the node that generates the candidate path, and Discriminator helps distinguish candidate paths with the same <Protocol-origin, Originator> tuple. For example, if the originator advertises three candidate paths that belong to SRv6 Policy 1 through BGP, the candidate paths can be distinguished using discriminators.

4.4.1.3 Segment Lists

A segment list identifies a source-routed path through which traffic is sent to the endpoint of the involved SRv6 Policy. One candidate path can be associated with multiple segment lists, each of which is allocated a weight to implement weighted load balancing using Equal-Cost Multiple Path (ECMP) or Unequal-Cost Multiple Path (UCMP).

Furthermore, binding SIDs can be allocated to SRv6 Policies. As basic SR instructions, binding SIDs are used to identify SRv6 Policies and provide diverse functions including traffic steering and tunnel stitching. Packets carrying a binding SID will be steered into the corresponding SRv6 Policy. From this perspective, binding SIDs can be regarded as interfaces that are invoked for SRv6 Policy selection.

Simply put, if we consider SRv6 Policy as a network service, then the binding SID is the interface used to access the service. The SRv6 Policy model is designed to work in publish-subscribe mode, which means that network services are subscribed to as needed and then the network provides service interfaces accordingly. Since the binding SID is considered as the interface, it must meet interface requirements, which include:

- Independence: Any service can invoke such an interface without considering implementation details. Because all implementation details are encapsulated inside SRv6 Policies, services only need to be aware of the interfaces when using SRv6 Policies.
- Reliability: Once released, the interface must be responsible for the contract between the application and network service. This means that SRv6 Policy must guarantee the promised service in the

contract no matter how many applications invoke it. This requires SRv6 Policies to support auto scaling of network resources.

- Stability: Interfaces must be stable. To achieve this, the binding SID of an SRv6 Policy must be kept unchanged throughout the lifecycle of the SRv6 Policy, regardless of network topology, service, and path changes.

4.4.2 SRv6 Policy Path Computation

The candidate paths of an SRv6 Policy can be generated using different methods, including manual configuration, headend computation, and centralized computation. The Protocol-origin and Originator fields can be used to distinguish the generation method of a candidate path.

4.4.2.1 Manual Configuration

In manual configuration, a candidate path is manually planned and configured for an SRv6 Policy through the CLI or NETCONF. As shown in Figure 4.21, an explicit path (A4::45, A7::1, A6::1) is manually configured and then associated with an SRv6 Policy, meaning that a candidate path is manually created for the SRv6 Policy. In manual configuration mode, the

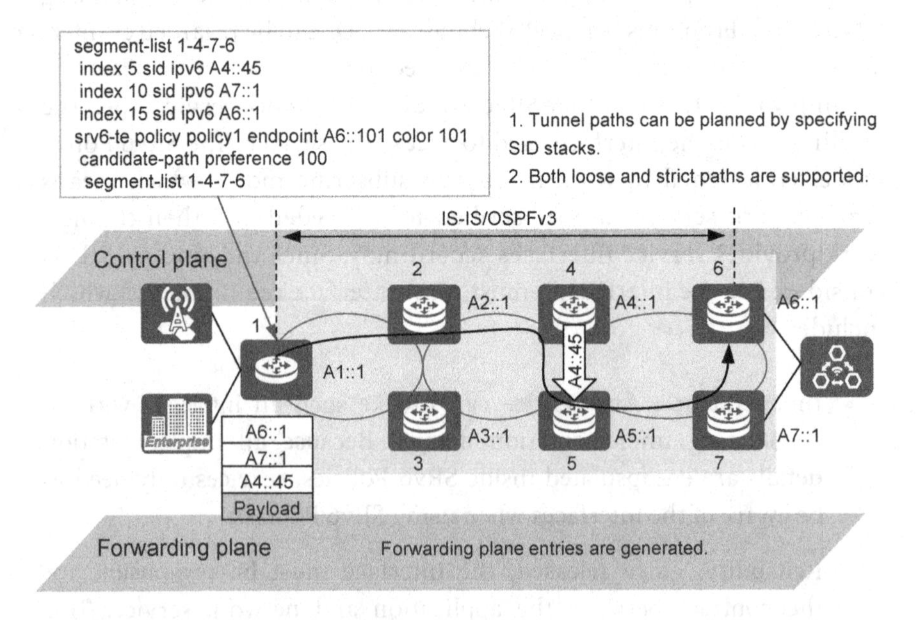

FIGURE 4.21 Manual configuration.

endpoint and color of the SRv6 Policy as well as the preference value and segment list of the candidate path must all be configured. Additionally, the preference value must be unique, and binding SIDs must be allocated within the static range of locators.

Manually configured paths cannot automatically adapt to network topology changes. That is, if a specified link or node fails, SRv6 Policy rerouting cannot be triggered, which leads to continuous traffic interruption. As such, generally, two disjoint paths need to be planned for a manually configured SRv6 Policy, with connectivity detection mechanisms employed to check path connectivity. Thus, when one of the paths fails, traffic can be quickly switched to the other path to ensure network reliability.

4.4.2.2 Headend Computation

Headend computation is similar to RSVP-TE path computation, and its process is illustrated in Figure 4.22. Specifically, the headend uses IGP link-state and TE information to form a TEDB; runs the CSPF algorithm to compute a qualified path according to constraints such as bandwidth, latency, Shared Risk Link Group (SRLG), and disjoint paths; and then installs the corresponding SRv6 Policy to guide traffic forwarding.

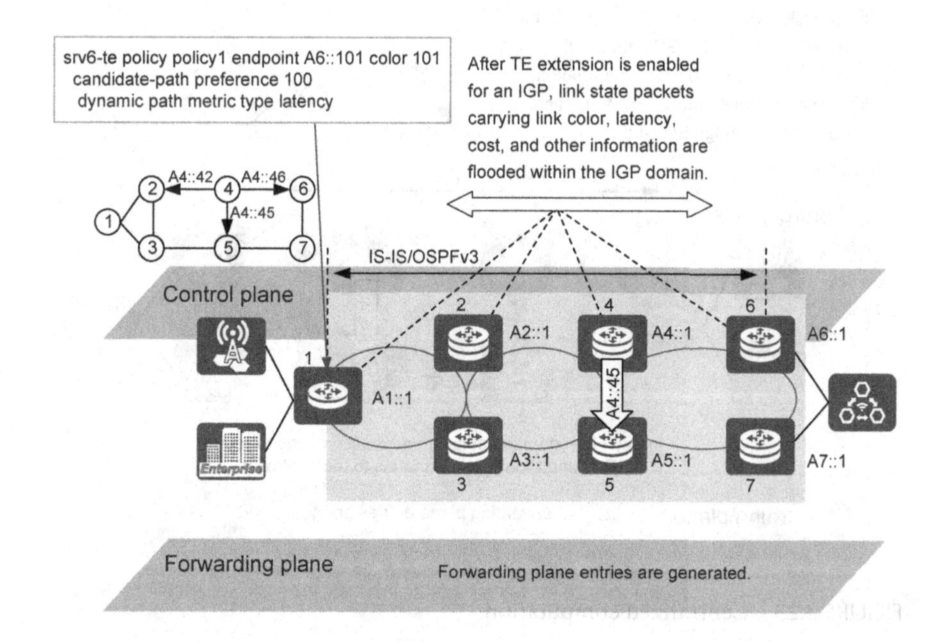

FIGURE 4.22 Headend computation.

Headend computation has the following restrictions:

- Due to the lack of cross-domain topology information, the headend can compute only paths within a single IGP domain, not those that span multiple IGP domains.
- Because SR transit nodes do not maintain the per-flow state, resource preemption and bandwidth reservation are not supported.

4.4.2.3 Centralized Computation

In centralized computation, as shown in Figure 4.23, a controller collects network topology, TE, and SRv6 information through BGP-LS or other protocols; computes a qualified path according to service requirements; and then delivers the corresponding SRv6 Policy to the headend through protocols such as BGP and PCEP. Centralized computation supports global traffic optimization, resource reservation, and E2E cross-domain scenarios.

An SRv6 Policy delivered by a controller to a forwarder does not carry a binding SID. After receiving an SRv6 Policy, the forwarder randomly

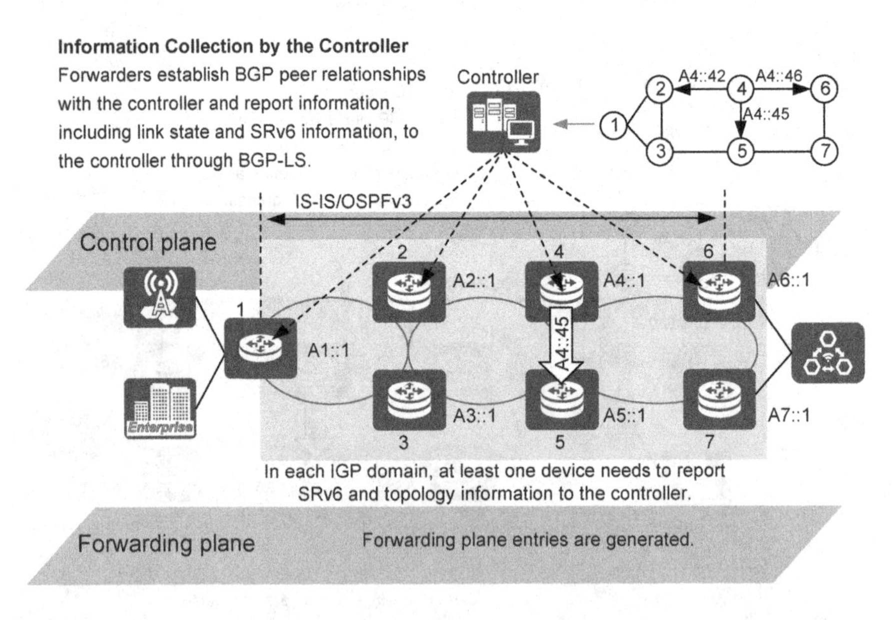

FIGURE 4.23 Centralized computation.

TABLE 4.19 Comparison between the Three Path Computation Methods

TE Application Scenario	Manual Configuration	Headend Computation	Centralized Computation
Manual path planning	Supported	Supported	Supported
Path computation based on the minimum latency	Supported	Supported	Supported
Bandwidth reservation	Not supported	Not supported	Supported
Affinity attribute-based path computation	Supported	Supported	Supported
Hop limit	Supported	Supported	Supported
Global traffic optimization	Not supported	Not supported	Supported
Priority preemption	Not supported	Not supported	Supported
Cross-domain scenario	Supported	Not supported	Supported
SRLG	Not supported	Supported	Supported

allocates a binding SID in the dynamic range of locators and then runs BGP-LS to report the state of the SRv6 Policy carrying the binding SID. In this way, the controller learns the binding SID of the SRv6 Policy and uses this information for SRv6 path programming.

Table 4.19 compares the three path computation methods.

In general, because a controller can obtain global topology, TE, and other relevant information using BGP-LS, it is possible to achieve global traffic optimization through centralized computation. In contrast, headend computation can only compute the optimal path within an IGP domain. Centralized computation also supports bandwidth reservation and priority preemption, which facilitates TE deployment.

4.4.3 Traffic Steering into an SRv6 Policy

After an SRv6 Policy is deployed on the headend, traffic needs to be steered into the SRv6 Policy. This can be done based on either binding SIDs or color values.

4.4.3.1 Binding SID-Based Traffic Steering

As shown in Figure 4.24, an SRv6 Policy with the binding SID B1::100 and segment list <B3::4, B4::6, B5::55> is deployed on node A to which a packet with the destination address of B1::100 is sent from an upstream node.

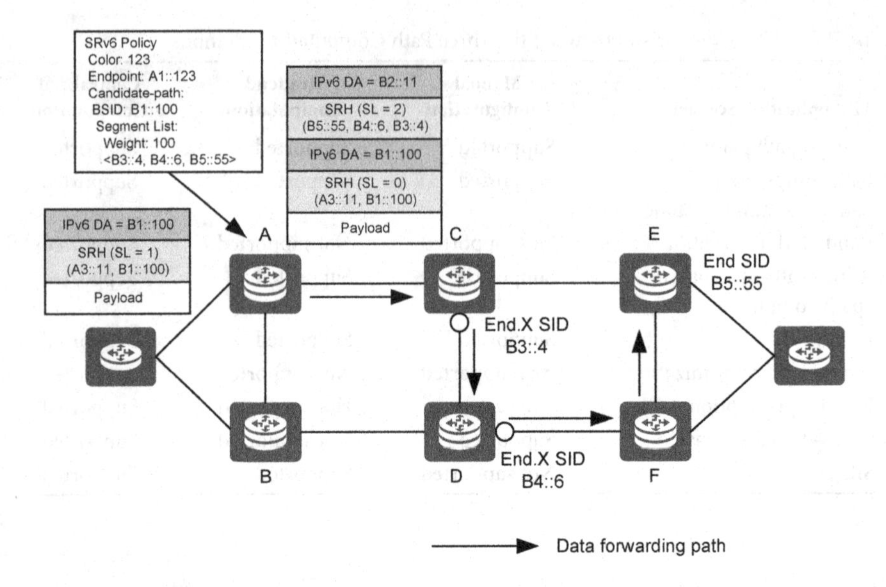

FIGURE 4.24 Binding SID-based traffic steering.

As the destination address is the binding SID of node A's local SRv6 Policy, node A processes the packet as follows:

1. Decrements the SL value in the SRH of the original packet by 1 to point to the next SID.
2. Encapsulates a new IPv6 header for the packet, with the destination address as the first SID (B3::4) of the corresponding SRv6 Policy and the source address as a local interface address.
3. Encapsulates a new SRH carrying the segment list of the corresponding SRv6 Policy.
4. Updates other fields in the packet, searches the forwarding table, and forwards the packet accordingly.

Using this method, traffic can be steered into the SRv6 Policy based on its binding SID. This method is usually used in scenarios such as tunnel stitching and inter-domain path stitching to effectively reduce the SRv6 SID stack depth. It also significantly reduces the inter-domain coupling degree, so that the changes to forwarding paths in one domain do not need to be propagated to others.

4.4.3.2 Color-Based Traffic Steering

The color attribute introduced to SRv6 Policies is used as an extended community of BGP routes. Because of this, the traffic transmitted over a specific route can be steered into the SRv6 Policy configured with the same color value.

Color is an important SRv6 Policy attribute; it is an anchor point for services and tunnels. As shown in Figure 4.25, a color template can be used to define one or more service requirements, including low latency, bandwidth, and affinity attributes. SRv6 Policy paths can then be computed based on the color template. As well as this, services can use the color attribute to define the requirements of network connections. This enables the color attribute of services to be matched with the color of a specific SRv6 Policy, ultimately achieving automatic traffic steering. Hence, during service deployment, only service requirements, not tunnels, need to be defined. The result is the decoupling between service deployment and tunnel deployment.

Figure 4.26 shows the detailed process of color-based traffic steering, the steps of which are as follows:

1. A controller delivers an SRv6 Policy with the color value 123 to node A using BGP or another protocol.

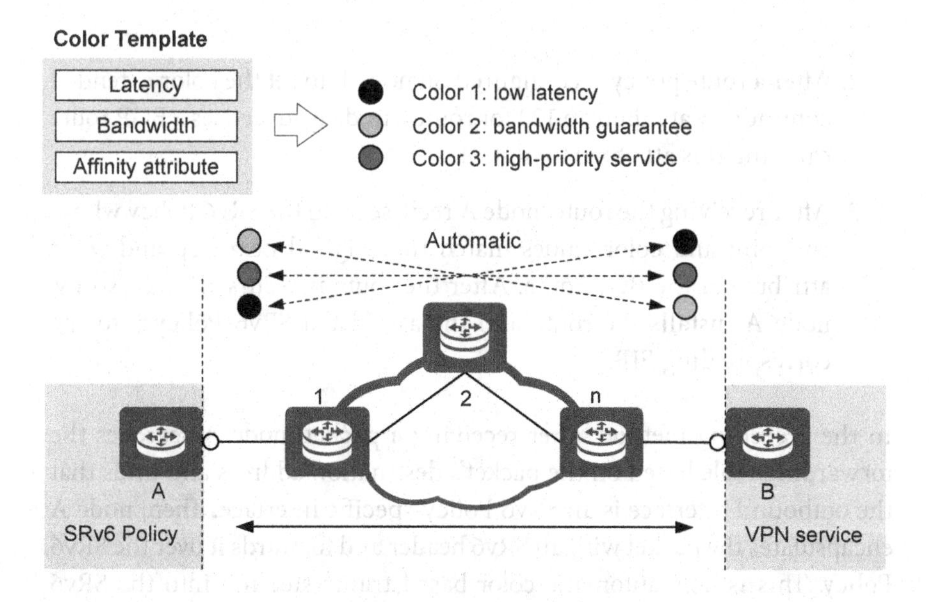

FIGURE 4.25 Schematic diagram for color-based traffic steering.

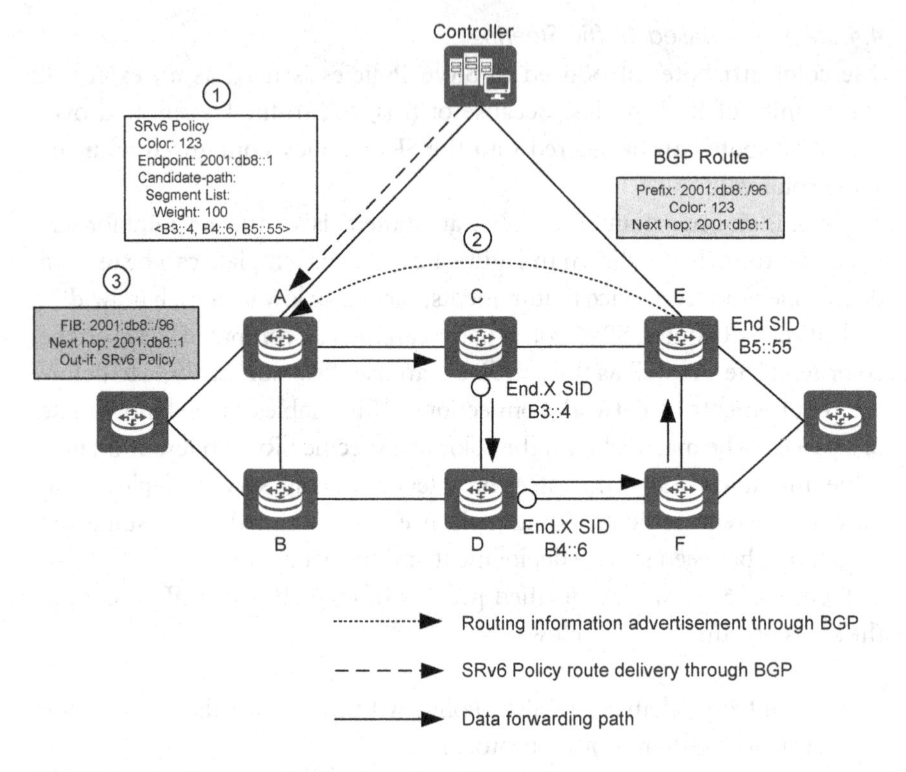

FIGURE 4.26 Process of color-based traffic steering.

2. After a route policy is configured on node E to set the color extended community attribute to 123 for routes, node E advertises a BGP route carrying this attribute to node A.

3. After receiving the route, node A recurses it to the SRv6 Policy whose endpoint and color values match the original next hop and color attribute of the BGP route. After the route is recursed successfully, node A installs the route and the associated SRv6 Policy into the corresponding FIB.

In the preceding method, after receiving a packet, node A searches the forwarding table based on the packet's destination address and finds that the outbound interface is an SRv6 Policy-specific interface. Then, node A encapsulates the packet with an SRv6 header and forwards it over the SRv6 Policy. This is how automatic color-based traffic steering into the SRv6 Policy is implemented.

In this traffic steering mode, route policies are usually used to determine the color values carried by BGP routes. By leveraging flexible coloring policies, color values can be modified as needed on individual nodes, such as the tailend, headend, and even RRs.

4.4.3.3 DSCP-Based Traffic Steering

In addition to binding SIDs and color values, traffic steering can also be performed based on the Differentiated Services Code Point (DSCP) values encapsulated in IP headers. This helps further classify services that match the same route but different origins. For example, an operator can configure multiple SRv6 Policies as a group on the headend, specify the mapping between each SRv6 Policy and DSCP value in the group, and bind services to the group. In this way, after receiving the service traffic, the headend searches for the corresponding SRv6 Policy in the group according to the DSCP value carried in the IP header, and then completes traffic steering accordingly. It is important to note that, in this traffic steering mode, services must be differentiated at the origin and have different DSCP values specified.

Certain scenarios may require both color-based and DSCP-based traffic steering. If this is the case, the color attribute can be configured for the involved group, so that the group is also identified using the <Color, Endpoint> tuple. With a traffic steering policy, based on the next hop and color attribute, a service route is first matched with a group, not an SRv6 Policy. Then, the forwarding plane matches the route with the corresponding SRv6 Policy in the group based on the DSCP value carried in the service packet.

4.4.4 Data Forwarding over an SRv6 Policy

In previous sections, we outlined the methods of computing SRv6 Policies and steering traffic into an SRv6 Policy. This section uses L3VPNv4 services as an example to describe how data is forwarded over an SRv6 Policy. Figure 4.27 illustrates the process, and the steps are detailed here:

1. A controller delivers an SRv6 Policy (color: 123; endpoint: PE2 address 2001:DB8::1) to the ingress PE1. The SRv6 Policy is associated with only one candidate path that contains only one segment list <2::1, 3::1, 4::1>.

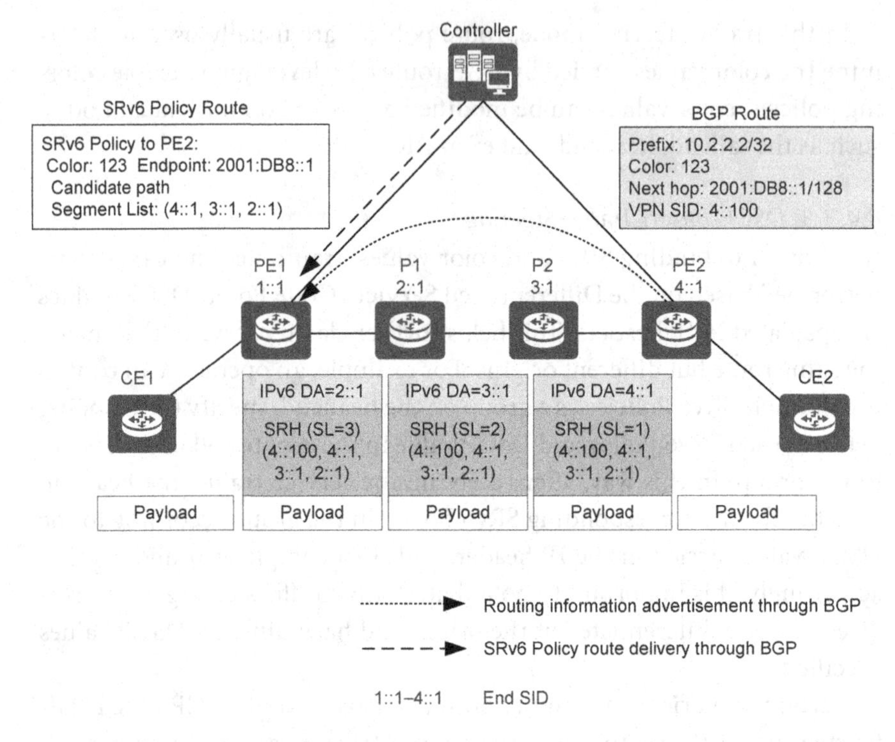

FIGURE 4.27 Data forwarding over an SRv6 Policy.

2. The egress PE2 advertises the BGP VPNv4 route 10.2.2.2/32 (color: 123; next hop: PE2 address 2001:DB8::1/128; VPN SID: 4::100) to PE1.

3. After receiving the BGP route, PE1 recurses it to the corresponding SRv6 Policy based on its color and next hop.

4. After receiving a common unicast packet from CE1, PE1 searches the routing table of the corresponding VPN instance and finds that the corresponding VPN route has been recursed to an SRv6 Policy. Hence, PE1 inserts an SRH carrying the segment list of the SRv6 Policy, encapsulates an IPv6 header into the packet, searches the corresponding routing table, and then forwards the packet accordingly.

5. P1 and P2 forward the packet according to the SIDs encapsulated in the SRH, implementing packet forwarding along the path defined in the SRv6 Policy.

6. After the packet arrives at PE2, PE2 searches the Local SID Table based on the IPv6 destination address 4::1 in the packet and finds a

matching End SID. According to the instruction bound to the SID, PE2 decrements the SL value of the packet by 1 and updates the IPv6 DA field to the VPN SID 4::100.

7. Based on the VPN SID 4::100, PE2 searches the Local SID Table and finds a matching End.DT4 SID. According to the instruction bound to the SID, PE2 decapsulates the packet by removing the SRH and IPv6 header, searches the routing table of the VPN instance corresponding to the VPN SID 4::100 based on the destination address in the inner packet, and then forwards the packet to CE2.

4.4.5 SRv6 Policy Fault Detection

This section details the SRv6 Policy fault detection mechanism.

4.4.5.1 SBFD for SRv6 Policy

Unlike RSVP-TE, SR does not require connections to be established between routers through signaling. So, as long as an SRv6 Policy is delivered to the headend, it can be successfully established. Note that the SRv6 Policy will not go down unless it is withdrawn. As such, some technologies need to be used to quickly detect SRv6 Policy faults and trigger a traffic switchover to the backup path. For example, Seamless Bidirectional Forwarding Detection (SBFD),[16] which is a simplified detection mechanism, can be deployed to achieve flexible path detection. SBFD for SRv6 Policy, specifically, is an E2E fast detection mechanism that can quickly detect faults of the path through which an SRv6 Policy passes.

Figure 4.28 shows the SBFD for SRv6 Policy detection process.

The SBFD for SRv6 Policy detection process is as follows:

1. BFD for SRv6 Policy is enabled on the ingress PE1, and the SBFD destination address and discriminator are configured.

2. PE1 sends an SBFD packet encapsulated with a SID stack corresponding to the SRv6 Policy.

3. After the egress PE2 receives the SBFD packet, it returns an SBFD reply over the IPv6 shortest path.

4. If PE1 receives the SBFD reply, it considers that the segment list in the SRv6 Policy is normal; otherwise, it considers that the segment list fails. If all the segment lists of a candidate path fail, SBFD triggers a switchover to the backup candidate path.

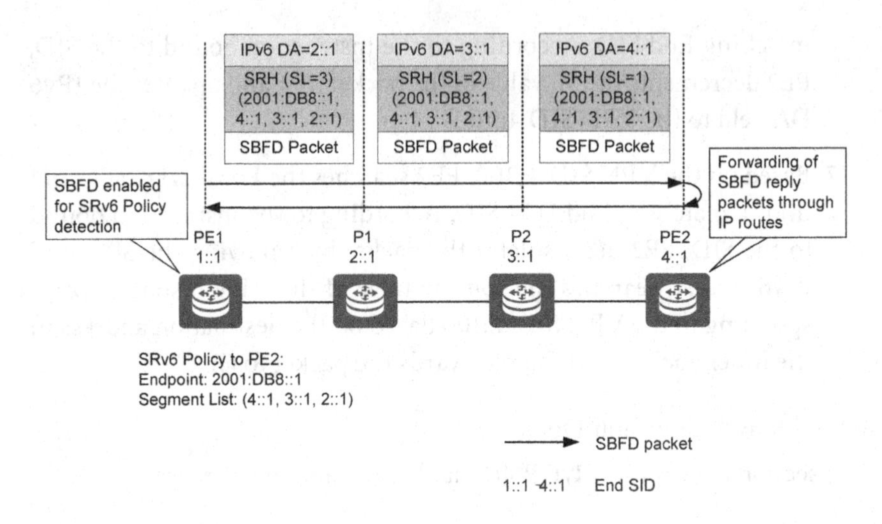

FIGURE 4.28 SBFD for SRv6 Policy.

4.4.5.2 Headend-Based Fault Detection

If SBFD is not supported by the headend or is not allowed in some scenarios, the headend is unable to quickly detect the failure of a candidate path in an SRv6 Policy. So, the only way to update the SRv6 Policy is by using a controller to detect topology changes and then perform rerouting.

If the controller is faulty or fails to manage the network, the SRv6 Policy cannot converge, which may lead to traffic loss. To speed up the traffic switchover when a fault occurs, headend-based fault detection is introduced. With this function enabled, once a segment list fails, the headend sets the segment list to down, triggering a path switchover in the SRv6 Policy or a service switchover.

To detect faults, the headend must be able to collect network topology information and determine the validity of a segment list based on whether the SRv6 SIDs in the segment list exist in the topology and whether routes are reachable.

If both of these are true, the headend sets the state of the segment list to up. But if not, the headend sets the state of the segment list to down and deletes the segment list entry.

When a segment list fails, the headend performs operations according to the following rules:

1. If the preferred candidate path of an SRv6 Policy has multiple segment lists for load balancing and one of them fails, the segment list is

deleted and then removed from the load balancing list. In this case, traffic is load-balanced among the remaining segment lists.

2. If all segment lists in the preferred candidate path of an SRv6 Policy fail and there is a backup candidate path in the SRv6 Policy, traffic is switched to the backup.

3. If both the primary and backup candidate paths of an SRv6 Policy fail, the headend is notified that the SRv6 Policy is in the down state, triggering a service switchover.

To detect the states of SRv6 SIDs and routes, the headend must first obtain SRv6 SID and routing information in the IGP domain. Therefore, headend-based fault detection can be used only for SRv6 Policies in IGP single-domain scenarios.

4.4.6 SRv6 Policy Switchover

As previously described, there are various technologies (including SBFD and headend-based fault detection) for detecting failures of segment lists and candidate paths. If all the candidate paths and segment lists of an SRv6 Policy fail, an SRv6 Policy switchover is required. The following is an example of how the SRv6 Policy switchover mechanism works.

In a VPN dual-homing scenario, if a VPN node fails, so too will all SRv6 Policies destined for that node. To prevent service interruptions, traffic needs to be switched to an SRv6 Policy destined for another VPN node.

As shown in Figures 4.29 and 4.30, the headend and endpoint of SRv6 Policy 1 are PE1 and PE2, respectively, while those of SRv6 Policy 2 are PE1 and PE3, respectively. VPN FRR can be implemented for SRv6 Policy 1 and SRv6 Policy 2, allowing PE2 and PE3 to each function as the other's VPN backup node.

Hot Standby (HSB) protection is enabled for SRv6 Policy 1. In addition, segment list 1 contains the End SIDs of P1, P2, and PE2, and protection mechanisms such as TI-LFA are deployed on network nodes.

For the network shown in Figure 4.29, the SRv6 Policy switchover process is as follows:

- When the link between P1 and P2 fails, TI-LFA local protection takes effect on both P1 and P2. If SBFD for segment list 1 on PE1 detects the failure before TI-LFA local protection takes effect, SBFD

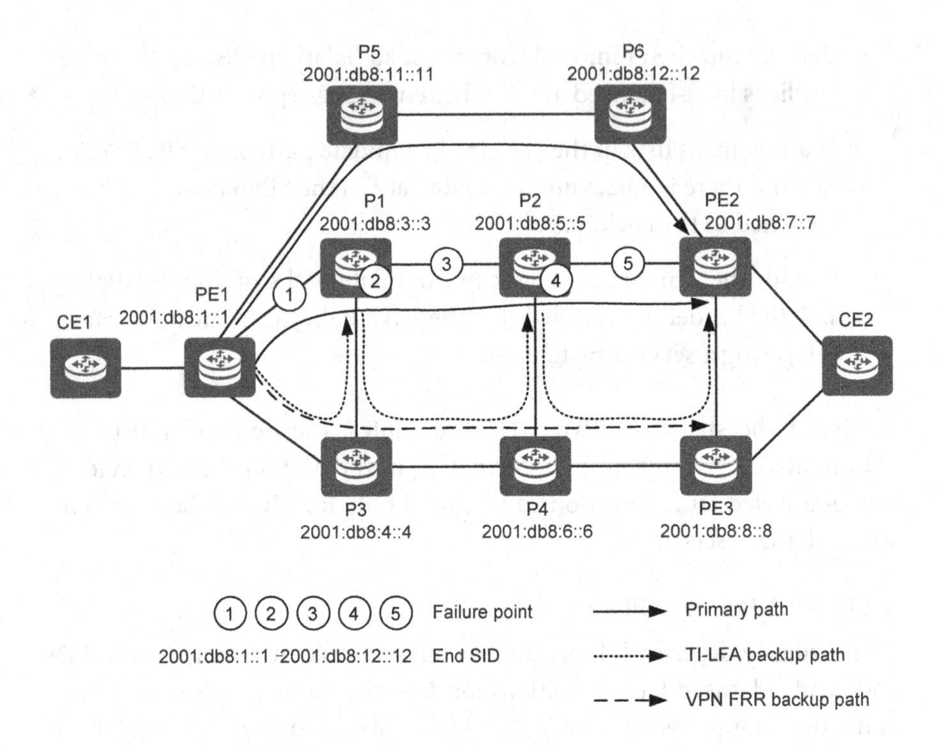

FIGURE 4.29 SRv6 Policy switchover.

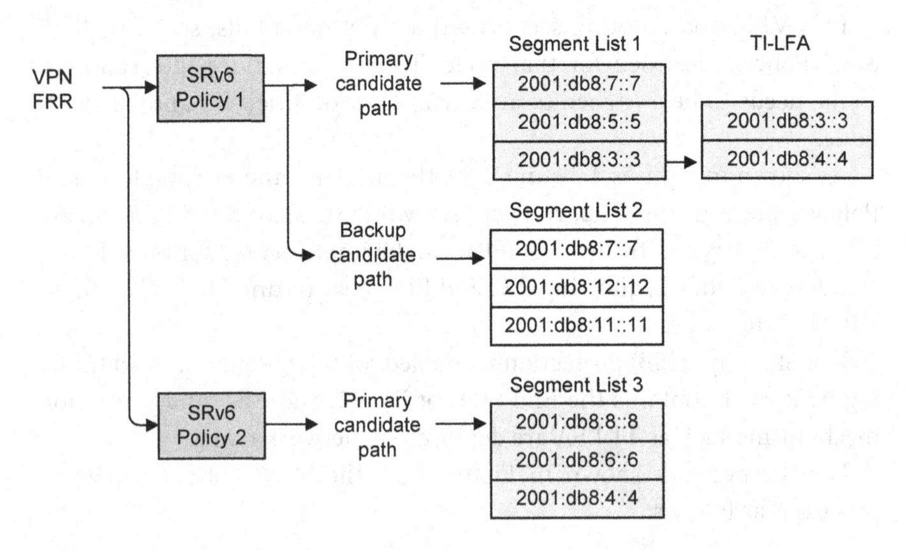

FIGURE 4.30 SRv6 Policy SID stack.

sets segment list 1 to down, which triggers a traffic switchover to segment list 2 of SRv6 Policy 1.

- If P2 fails and does not have TI-LFA configured, SBFD on PE1 detects the failure of P2 and sets segment list 1 to down, which triggers the traffic of SRv6 Policy 1 to be switched to segment list 2.
- If PE2 fails, SBFD can detect that all candidate paths of SRv6 Policy 1 are unavailable. SRv6 Policy 1 is then set to down, which triggers a service switching to SRv6 Policy 2 through VPN FRR.

4.5 BGP SRv6 POLICY

One way to advertise SRv6 Policies is by using BGP. To achieve this, BGP extensions such as Subsequent Address Family Identifier (SAFI), NLRI, and Tunnel Encaps Attribute are defined, as detailed in Table 4.20.

As a newly defined BGP address family, the SAFI is used to identify SR Policy and supports both IPv4 and IPv6 Address Family Identifiers (AFIs). If the AFI is set to IPv6 and the SAFI is set to SR Policy, it indicates that an SRv6 Policy is delivered.

In the Update messages of the IPv6 address family, the NLRI identifies an SR Policy, and the Tunnel Encaps Attribute is used to carry detailed information about the candidate path, including the associated binding SID, preference, and segment list.

After computing an SRv6 Policy, a controller uses a BGP peer relationship to advertise the SRv6 Policy candidate path as a route to the headend on the SRv6 network. If multiple identical candidate path routes are

TABLE 4.20 BGP Extensions for SRv6 Policy

Item	Description
SAFI	The NLRI of the SAFI identifies an SR Policy. If the AFI is set to IPv6, then the NLRI identifies an SRv6 Policy.
Tunnel Encaps Attribute	A new tunnel type identifier and corresponding Tunnel Encaps Attribute are defined for SR Policy. A group of sub-TLVs are added to the Tunnel Encaps Attribute to carry SR Policy information, including binding SID, preference, and segment list information.
Color Extended Community	The Color Extended Community is used to steer traffic into the corresponding SR Policy.

received, the headend performs BGP route selection and delivers the selection result to the SRv6 Policy management module, which then installs the selected route in the data plane.

The following describes BGP extensions for SRv6 Policies in detail.

4.5.1 SRv6 Policy SAFI and NLRI

For SR Policies, BGP defines a new SAFI that has the code of 73 and for which only an IPv4 or IPv6 AFI can be used. For SRv6 Policies, the AFI and SAFI must be set to IPv6 and SR Policy, respectively.

The SRv6 Policy SAFI uses a new NLRI format to describe a candidate path of an SRv6 Policy. The new NLRI format is shown in Figure 4.31.

Table 4.21 describes the fields in the SRv6 Policy NLRI.

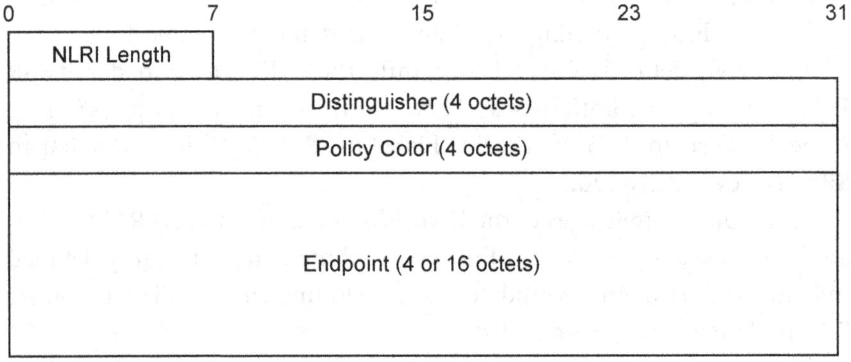

FIGURE 4.31 SRv6 Policy NLRI.

TABLE 4.21 Fields in the SRv6 Policy NLRI

Field	Length	Description
NLRI Length	8 bits	NLRI length. • If the AFI is set to 1 (IPv4), the length value is 12. • If the AFI is set to 2 (IPv6), the length value is 24.
Distinguisher	4 octets	Unique distinguisher of each candidate path in an SRv6 Policy.
Policy Color	4 octets	Color of an SRv6 Policy used together with an endpoint to identify an SRv6 Policy. The color can be matched with the Color Extended Community of the destination route prefix to steer traffic into the corresponding SRv6 Policy.
Endpoint	4 octets or 16 octets	Endpoint of an SRv6 Policy used to identify the destination node of the SRv6 Policy. • If the AFI is set to IPv4, the endpoint is an IPv4 address. • If the AFI is set to IPv6, the endpoint is an IPv6 address.

4.5.2 SR Policy and Tunnel Encaps Attribute

For SR Policies, BGP defines a new tunnel type that has the code of 15. The corresponding Tunnel Encaps Attribute is defined with the value of 23. An SR Policy is encoded as follows:

```
SR Policy SAFI NLRI: <Distinguisher, Policy-Color, EndPoint>
Attributes:
Tunnel Encaps Attribute(23)
 Tunnel Type: SR Policy(15)
       Binding SID
       Preference
       Priority
       Policy Name
       Explicit NULL Label Policy (ENLP)
       Segment List
          Weight
          Segment
          Segment
          ....
       ....
```

We have already looked in detail at the SR Policy SAFI NLRI. The following sections will focus on various sub-TLVs defined for the Tunnel Encaps Attribute of SR Policy, including Binding SID, Preference, Priority, and Segment List sub-TLVs.

4.5.3 Binding SID Sub-TLV

The Binding SID sub-TLV is used to specify the binding SID associated with a candidate path. Figure 4.32 shows the format of this sub-TLV.

Table 4.22 describes the fields in the Binding SID sub-TLV.

4.5.4 Preference Sub-TLV

The Preference sub-TLV is an optional sub-TLV that is used to specify the preference of an SRv6 Policy candidate path. The default preference of a

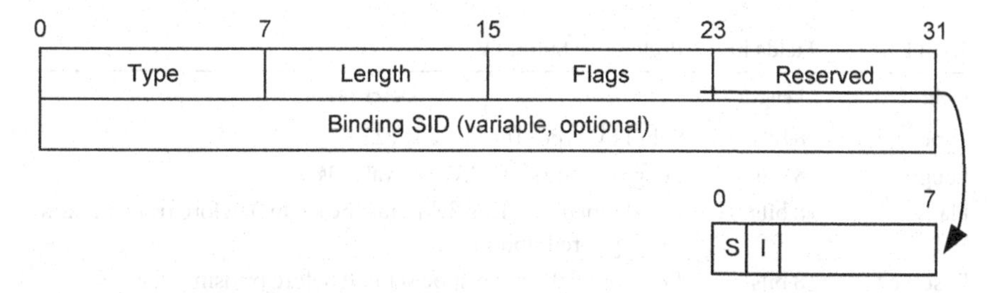

FIGURE 4.32 Binding SID sub-TLV.

TABLE 4.22 Fields in the Binding SID Sub-TLV

Field	Length	Description
Type	8 bits	Sub-TLV type. The value is 13.
Length	8 bits	Length of the sub-TLV. The value can be 2, 6, or 18. For SRv6, the value is 18.
Flags	8 bits	Flags. • S: indicates that the candidate path of an SR Policy must carry a valid binding SID. If no binding SID is carried or the carried binding SID is invalid, the candidate path cannot be used. • I: indicates that the corresponding SR Policy is invalid so that all the received traffic with the binding SID as the destination address is discarded. • Other flags have not yet been defined. They must be set to 0 before transmission and ignored upon receipt.
Reserved	8 bits	Reserved field. It must be set to 0 before transmission.
Binding SID	Variable	Optional field. • If the Length value is 2, no binding SID is carried. • If the Length value is 6, the binding SID is a 4-byte label. • If the Length value is 18, the binding SID is a 16-byte SRv6 SID.

candidate path without this sub-TLV is 100. Figure 4.33 shows the format of this sub-TLV.

Table 4.23 describes the fields in the Preference sub-TLV.

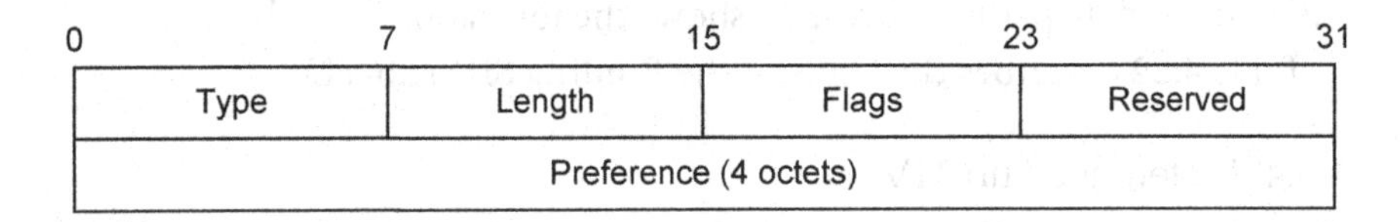

FIGURE 4.33 Preference sub-TLV.

TABLE 4.23 Fields in the Preference Sub-TLV

Field	Length	Description
Type	8 bits	Sub-TLV type. The value is 12.
Length	8 bits	Length of the sub-TLV. The value is 6.
Flags	8 bits	Not defined yet. This field must be set to 0 before transmission and ignored upon receipt.
Reserved	8 bits	Reserved field. It must be set to 0 before transmission.
Preference	32 bits	Preference of an SRv6 Policy candidate path.

4.5.5 Segment List Sub-TLV

The Segment List sub-TLV is used to specify an explicit path to the endpoint of an SRv6 Policy. It contains path information and an optional Weight sub-TLV. The Segment List sub-TLV is optional and may appear multiple times in the SRv6 Policy. Figure 4.34 shows the format of the Segment List sub-TLV.

Table 4.24 describes the fields in the Segment List sub-TLV.

4.5.6 Weight Sub-TLV

The Weight sub-TLV is used to specify the weight of a segment list in a candidate path. If there are multiple segment lists with different weights, weighted load balancing is implemented using UCMP. The default weight of a segment list is 1. Figure 4.35 shows the format of this sub-TLV.

Table 4.25 describes the fields in the Weight sub-TLV.

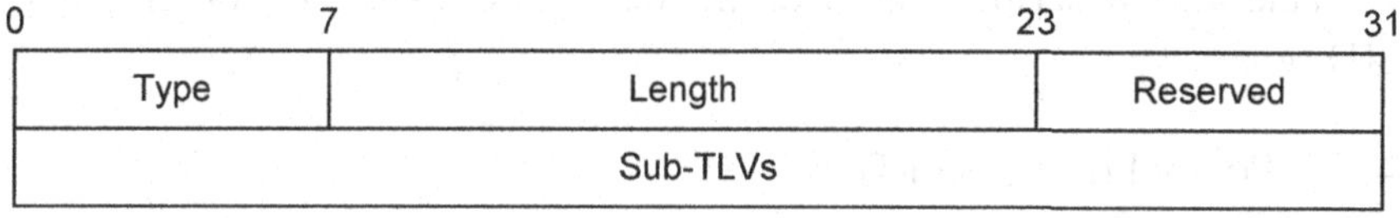

0 7 23 31
Type | Length | Reserved
Sub-TLVs

FIGURE 4.34 Segment List sub-TLV.

TABLE 4.24 Fields in the Segment List Sub-TLV

Field	Length	Description
Type	8 bits	Sub-TLV type. The value is 128.
Length	16 bits	Total length of all sub-TLVs in the Segment List sub-TLV.
Reserved	8 bits	Reserved field. It must be set to 0 before transmission.
Sub-TLVs	Variable	The currently defined sub-TLVs include the optional Weight sub-TLV and one or more Segment sub-TLVs.

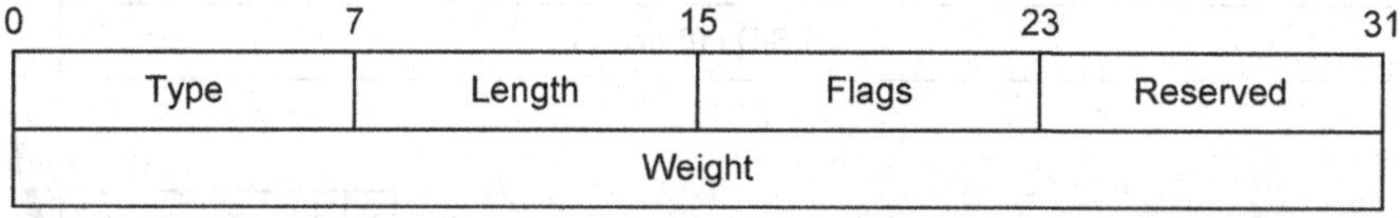

0 7 15 23 31
Type | Length | Flags | Reserved
Weight

FIGURE 4.35 Weight sub-TLV.

TABLE 4.25 Fields in the Weight Sub-TLV

Field	Length	Description
Type	8 bits	Sub-TLV type. The value is 9.
Length	8 bits	Length of the sub-TLV. The value is 6.
Flags	8 bits	Not defined yet. This field must be set to 0 before transmission and ignored upon receipt.
Reserved	8 bits	Reserved field. It must be set to 0 before transmission.
Weight	32 bits	Weight of a segment list in a candidate path. The default value is 1. Traffic can be distributed to different segment lists based on weights.

4.5.7 Segment Sub-TLV

The Segment sub-TLV is used to define a segment in a segment list. The Segment List sub-TLV can contain multiple Segment sub-TLVs. There is a diverse range of Segment sub-TLVs, with the SRv6 SID type being the most commonly used for SRv6 Policies. Figure 4.36 shows the format of the Segment sub-TLV of the SRv6 SID type.

Table 4.26 describes the fields in the Segment sub-TLV of the SRv6 SID type.

4.5.8 Policy Priority Sub-TLV

The Policy Priority sub-TLV is used to specify the path recomputation priorities of SRv6 Policies when a topology change occurs. Figure 4.37 shows the format of the Policy Priority sub-TLV.

Table 4.27 describes the fields in the Policy Priority sub-TLV.

4.5.9 Policy Name Sub-TLV

The Policy Name sub-TLV is an optional sub-TLV that is used to specify a name for an SRv6 Policy candidate path. Figure 4.38 shows the format of this sub-TLV.

Table 4.28 describes the fields in the Policy Name sub-TLV.

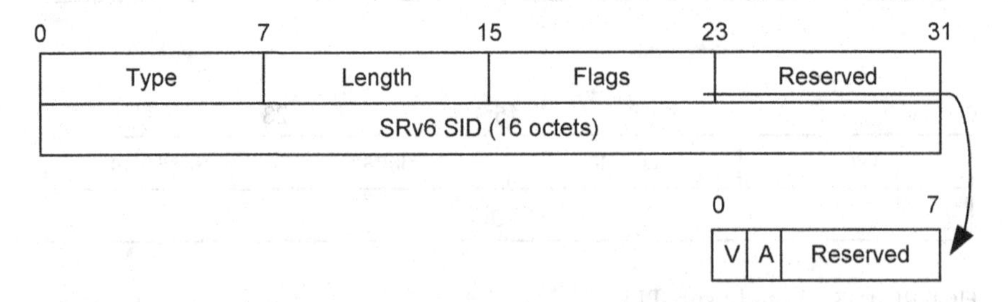

FIGURE 4.36 Segment sub-TLV of the SRv6 SID type.

TABLE 4.26 Fields in the Segment Sub-TLV of the SRv6 SID Type

Field	Length	Description
Type	8 bits	Sub-TLV type. The value is 2.
Length	8 bits	Length of the sub-TLV. The value is 18.
Flags	8 bits	Flags: • V: indicates that the corresponding SID needs to be checked to determine whether it is valid. • A: indicates whether an algorithm ID that exists in the SR Algorithm field (in Segment sub-TLV 3/4/9) is applicable to the current segment. Currently, this flag is valid only for Segment sub-TLV 3/4/9 and must be ignored for Segment sub-TLVs of the SRv6 SID type. • Other flags have not yet been defined. They must be set to 0 before transmission and ignored upon receipt.
Reserved	8 bits	Reserved field. It must be set to 0 before transmission.
SRv6 SID	16 octets	SID in IPv6 address format.

0 7	15	23	31
Type	Length	Priority	Reserved

FIGURE 4.37 Policy Priority sub-TLV.

TABLE 4.27 Fields in the Policy Priority Sub-TLV

Field	Length	Description
Type	8 bits	Sub-TLV type. The value is 15.
Length	8 bits	Length of the sub-TLV. The value is 2.
Priority	8 bits	Priority value.
Reserved	8 bits	Reserved field. It must be set to 0 before transmission and ignored upon receipt.

0 7	23	31
Type	Length	Reserved
Policy Name		

FIGURE 4.38 Policy Name sub-TLV.

TABLE 4.28 Fields in the Policy Name Sub-TLV

Field	Length	Description
Type	8 bits	Sub-TLV type. The value is 129.
Length	16 bits	Length of the sub-TLV.
Reserved	8 bits	Reserved field. It must be set to 0 before transmission and ignored upon receipt.
Policy Name	Variable	Policy name, which is a string of printable ASCII characters excluding the Null terminator.

4.6 STORIES BEHIND SRv6 DESIGN

I. BGP vs. PCEP

During SDN transition, BGP and PCEP become two primary southbound protocols for SDN controllers. Which of these two protocols should be used has long been the subject of debate. In fact, various historical factors and user habits have much effect on the final choice.

IP professionals are familiar with BGP, which therefore becomes their natural choice. Having been around for 30 years, BGP has had time to develop distinct advantages, which can be seen in its diverse functions and excellent interoperability.

PCEP was first developed by experts from the optical transport field, and there were few deployments for the first 10 years after its conception. But in recent years, the development of SDN is giving PCEP a new lease of life. In particular, due to the fact that most IP-based mobile transport networks are originated from Synchronous Digital Hierarchy (SDH), many network O&M personnel with a lot of optical transport experience consider BGP complex, and so tend to opt for PCEP instead. However, the downside of PCEP is that it is a relatively new protocol that has not been thoroughly verified on live networks. It therefore falls short of BGP in many aspects, such as interoperability and performance.

In addition to the impact of historical factors and user habits, BGP and PCEP have essential technical differences which affect the final choice. Specifically, BGP always adopts the Push mode, which enables a BGP RR to push a BGP route received from a peer to other peers; on the contrary, PCEP uses the Pull mode, in which a PCE client requests a PCE server to compute a path, with the PCE server then returning the path computation result to the client. For BGP, this limits its usage in scenarios where it is used for features that need SR Policies to be established on demand. Even though some innovations have been made to request SR Policy establishment through BGP extensions, relevant research efforts are still in the early stages, and as such the effects are yet to be fully observed.[17]

II. SR Policy Design

I always think that Huawei's tunnel policy design in MPLS is very successful. Tunnel policies can be used for VPN instances to select tunnels. Specifically, we can use a tunnel policy to specify an MPLS TE tunnel to a specific destination address and configure multiple tunnels to work in load balancing mode. Later, more new features were added. The tunnel policy is simple, straightforward, and flexible, which leads to high praise from our customers. Since other vendors who first implemented MPLS TE use a different tunnel selection mode, the tunnel policy design was doubted internally, but kept at last.

Personally speaking, I am not in favor of the SR Policy design. Huawei's tunnel policy design is very clear, in that tunnel establishment and tunnel selection are decoupled and do not affect each other. That is, tunnel attributes are configured under tunnels, and tunnel selection policies for VPN instances are configured under tunnel policies. In contrast, the SR Policy design combines tunnel (SR path) establishment, binding SID allocation to tunnels, and tunnel selection, which may lead to the following problems:

- The name "SR Policy" is somewhat misleading. SR Policies involve the setup of SR paths and are quite different from the conventional definition of a policy.
- Due to the coupling design, the data of the tunnel policy that is not changed also needs to be delivered during the update of some tunnel (SR path) attributes (and vice versa), resulting in network resources being wasted.
- Based on prior experience in MPLS TE design, network deployment requires many tunnel attributes and flexible tunnel policies. The current SR Policy mechanism seems simple, but it may become complex as more SR features emerge. This problem is

already emerging, with a large number of drafts having been developed to define protocol extensions for various enhanced SR Policy features. At the IETF meeting, I once explained my concern on the problem and suggested controlling the scope of the SR Policy.

As user-aware policy design is prone to changes, decoupling design takes on particular importance. In addition to the policy function decoupling mentioned above, the internal implementation of IP operating systems must also be decoupled from user interfaces. In essence, the sets of functionalities are similar. If the internal implementation is closely coupled with user interfaces, any changes may noticeably increase the Research and Development (R&D) workload. However, if the internal implementation is decoupled from user interfaces, it can only require a slight adjustment to adapt to user interface changes, thereby minimizing the impacts of the changes.

REFERENCES

[1] Awduche D, Berger L, Gan D, Li T, Srinivasan V, Swallow G. RSVP-TE: Extensions to RSVP for LSP Tunnels[EB/OL]. (2020-01-21)[2020-03-25]. RFC 3209.

[2] Ginsberg L, Previdi S, Giacalone S, Ward D, Drake J, Wu Q. IS-IS Traffic Engineering (TE) Metric Extensions[EB/OL]. (2019-03-15)[2020-03-25]. RFC 8570.

[3] Giacalone S, Ward D, Drake J, Atlas A, Previdi S. OSPF Traffic Engineering (TE) Metric Extensions[EB/OL]. (2018-12-20)[2020-03-25]. RFC 7471.

[4] Gredler H, Medved J, Previdi S, Farrel A, Ray S. North-Bound Distribution of Link-State and Traffic Engineering (TE) Information Using BGP[EB/OL]. (2018-12-20)[2020-03-25]. RFC 7752.

[5] Dawra G, Filsfils C, Talaulikar K, Chen M, Bernier D, Decraene B. BGP Link State Extensions for SRv6[EB/OL]. (2019-07-07)[2020-03-25]. draft-ietf-idr-bgpls-srv6-ext-01.

[6] Filsfils C, Camarillo P, Leddy J, Voyer D, Matsushima S, Li Z. SRv6 Network Programming[EB/OL]. (2019-12-05)[2020-03-25]. draft-ietf-spring-srv6-network-programming-05.

[7] Psenak P, Filsfils C, Bashandy A, Decraene B, Hu Z. IS-IS Extension to Support Segment Routing over IPv6 Dataplane[EB/OL]. (2019-10-04)[2020-03-25]. draft-ietf-lsr-isis-srv6-extensions-03.

[8] Li Z, Hu Z, Cheng D, Talaulikar K, Psenak P. OSPFv3 Extensions for SRv6[EB/OL]. (2020-02-12)[2020-03-25]. draft-ietf-lsr-ospfv3-srv6-extensions-00.

[9] Tantsura J, Chunduri U, Talaulikar K, Mirsky G, Triantafillis N. Signaling MSD (Maximum SID Depth) Using Border Gateway Protocol Link-State [EB/OL]. (2019-08-15)[2020-03-25]. draft-ietf-idr-bgp-ls-segment-routing-msd-05.

[10] Vasseur JP, Le Roux JL. Path Computation Element (PCE) Communication Protocol (PCEP)[EB/OL]. (2020-01-21)[2020-03-25]. RFC 5440.

[11] Crabbe E, Minei I, Medved J, Varga R. Path Computation Element Communication Protocol (PCEP) Extensions for Stateful PCE[EB/OL]. (2020-01-21)[2020-03-25]. RFC 8231.

[12] Crabbe E, Minei I, Sivabalan S, Varga R. PCEP Extensions for PCE-initiated LSP Setup in a Stateful PCE Model[EB/OL]. (2018-12-20)[2020-03-25]. RFC 8281.

[13] Negi M, Li C, Sivabalan S, Kaladharan P, Zhu Y. PCEP Extensions for Segment Routing leveraging the IPv6 data plane[EB/OL]. (2019-10-09)[2020-03-25]. draft-ietf-pce-segment-routing-ipv6-03.

[14] Sivabalan S, Tantsura J, Minei I, Varga R, Hardwick J. Conveying Path Setup Type in PCE Communication Protocol (PCEP) Messages[EB/OL]. (2018-07-24)[2020-03-25]. RFC 8408.

[15] Filsfils C, Sivabalan S, Voyer D, Bogdanov A, Mattes P. Segment Routing Policy Architecture[EB/OL]. (2019-12-15)[2020-03-25]. draft-ietf-spring-segment-routing-policy-06.

[16] Katz D, Ward D. Bidirectional Forwarding Detection (BFD)[EB/OL]. (2020-01-21)[2020-03-25]. RFC 5880.

[17] Li Z, Li L, Chen H, et al. BGP Request for Candidate Paths of SR TE Policies[EB/OL]. (2020-03-08)[2020-03-25]. draft-li-ldr-bgp-request-cp-sr-te-policy-01.

CHAPTER 5

SRv6 VPN

THIS CHAPTER DESCRIBES VPN SERVICES, including SRv6 L3VPN and SRv6 EVPN, which are also fundamental to SRv6 services. VPNs are one of the most important applications on live networks, and they contribute.much to the revenue of carriers. SRv6 carries VPN information in SIDs. With the help of IPv6 route reachability and aggregation, SRv6 VPN services can be provisioned as soon as edge nodes are upgraded to support SRv6, shortening the VPN service provisioning time. This is the SRv6's advantage over SR-MPLS.

5.1 VPN OVERVIEW

A VPN is a virtual private communications network built over the public network of an Internet Service Provider (ISP). Considered to be both private and virtual, logically isolated VPNs can be created over an underlying transport network. Sites of the same VPN are securely connected across and within domains, while VPNs are isolated from each other. In view of this, VPNs can be used to implement interconnection within an enterprise, for example, to interconnect the headquarters with branches.

Not only this, the VPNs can also be used to isolate different services. For example, all employees in a company are granted access to email services, while only R&D department staff are granted access to the code development environment.

Before SRv6 came about, a VPN was generally carried over an MPLS network and such a VPN was called an MPLS VPN. In MPLS VPN, a VPN instance is identified by an MPLS label, also known as a VPN label.

Service data is isolated using VPN labels to prevent VPN resources from being accessed by users who do not belong to the VPN. In SRv6 VPN, a VPN instance can be identified by an SRv6 SID, which we will describe later.

5.1.1 Basic VPN Model

The following devices are deployed on the VPN (Figure 5.1):

- Customer Edge (CE): an edge device connected to an ISP backbone network. The CE can be a router, switch, or host. CEs are unaware of VPNs and do not need special support for VPN transport protocols, such as MPLS or SRv6.
- Provider Edge (PE): an edge device on the ISP backbone network. VPN processing is performed on PEs within a VPN.
- Provider (P): is a backbone device on the ISP backbone network and does not connect to CEs directly. The P is required to provide basic MPLS or IPv6 forwarding capabilities but does not need to maintain VPN information.

5.1.2 VPN Service Types

VPNs can be classified into L3VPN[1,2] and L2VPN,[1,3] differing in the types of services and network characteristics:

- L3VPN: carries Layer 3 services and isolates the services using VPN instances.

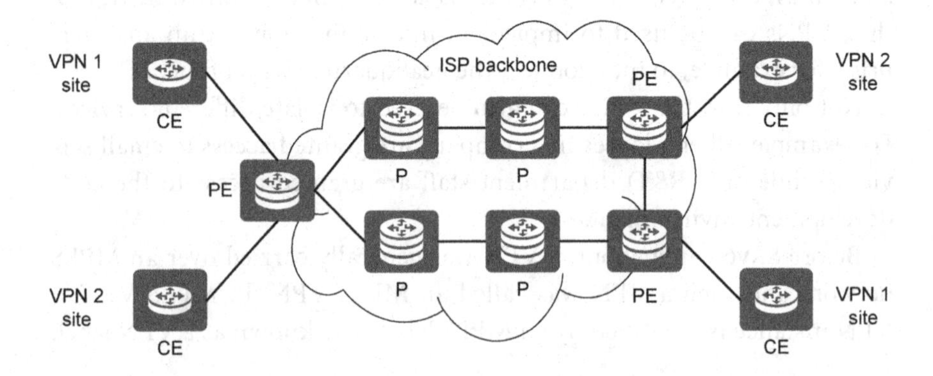

FIGURE 5.1 Basic VPN model.

- L2VPN: carries Layer 2 services, including VPWS and VPLS,[1] and isolates the services using L2VPN instances.

By virtue of evolution to IP, Layer 3 IP services, such as 3G, 4G, 5G, Internet, and Voice over IP (VoIP) services, dominate on live networks. L3VPN accounts for the majority of VPN deployments. For traditional 2G TDM interfaces, the services can only be carried using L2VPN. In addition, L2VPN still takes a big share on enterprise networks or data center networks where many low-speed interfaces or switches are used.

5.1.2.1 L3VPN

L3VPN is the VPN that carries Layer 3 services. Since VPNs manage their own address spaces separately, their address spaces may overlap. For example, both VPN1 and VPN2 can be allocated the address space 10.110.10.0/24. In order to guarantee communication in the two VPN instances, the traffic of the two VPN instances must be isolated.

In MPLS L3VPN, VPNs are isolated using VPN instances, and each VPN instance is identified by an MPLS VPN label. A PE establishes and maintains a dedicated VPN instance for each directly connected site.

In order to guarantee VPN data, the PE separately maintains a routing table and a Label Forwarding Information Base (LFIB) for each VPN instance. A VPN instance typically contains an LFIB, a routing table, interfaces bound to the VPN instance, and VPN instance management information. The VPN instance management information includes a Route Distinguisher (RD), Route Targets (RTs), and a member interface list.

L3VPN uses BGP to redistribute private network routing information of VPN sites and uses MPLS or SRv6 to carry VPN service traffic over a carrier IP network.

5.1.2.2 L2VPN and EVPN

L2VPN is the VPN that carries Layer 2 services. Traditional L2VPN services are classified into VPLS and VPWS.

- VPLS, a type of Multipoint-to-Multipoint (MP2MP) L2VPN service, allows geographically isolated user sites to be connected as a LAN through an intermediate network. VPLS emulates traditional LAN functions to allow VPLS devices to work as if within a LAN.

- VPWS, a type of point-to-point L2VPN service, emulates traditional private line services over an IP network to provide low-cost data services.

However, as the network scale keeps expanding and service requirements keep increasing, traditional L2VPN technologies face the following challenges, as shown in Figure 5.2:

1. Complex service deployment: Specifically, for LDP-based VPLS services, each time a service access point is added, the PE has to establish a remote LDP session with each of the other PEs to create a Pseudo Wire (PW).

Note: Although BGP auto-discovery automatically triggers PW establishment, this BGP extension does not support PW redundancy.[4,5] Back in 2009, researchers proposed solutions for BGP-based VPLS protection. However, the proposed solutions have not been widely implemented or deployed.[6]

2. Limited network scale: PWs are used, no matter which L2VPN technology is used. The PWs, which are point-to-point virtual connections in nature, have to be established between any two points to implement service interworking. As a result, N^2 PWs have to be created to connect any two of N nodes. A larger network may suffer more serious issues induced by full-mesh PW connections. In addition, traditional VPLS learns MAC addresses in the data plane. In case of failures on a VPLS network, all related MAC address entries have to be withdrawn, traffic is flooded, and MAC address entries are learned again, leading to slow convergence.

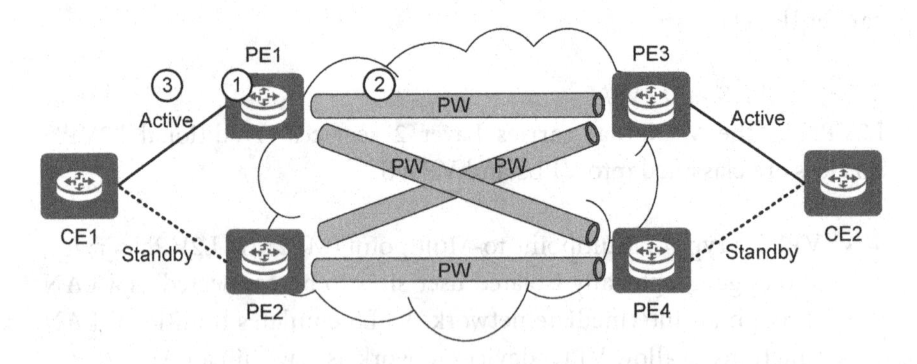

FIGURE 5.2 Challenges faced by traditional L2VPN.

3. Low bandwidth utilization: In a CE dual-homing scenario, traditional L2VPN technologies support active/standby service model, not active-active service model. This makes it impossible to implement flow-based load balancing and efficient use of high bandwidth resources.

   To tackle these L2VPN issues, EVPN technology[7] was proposed in the industry. EVPN was initially designed as BGP- and MPLS-based L2VPN technology. Before EVPN, since 2000, multiple L2VPN technologies were proposed in the industry in an effort to transparently transmit Layer 2 services over IP/MPLS networks. These technologies include LDP VPWS,[8] BGP VPLS,[4] LDP VPLS[9] (that was later optimized to support automatic discovery and simplify deployment[5]), and BGP VPWS.[10]

   EVPN combines the advantages of BGP VPLS and BGP L3VPN. EVPN uses a BGP protocol extension to advertise MAC reachability information, separating the control plane from the data plane for L2VPN services.

   EVPN offers various advantages, such as flexible deployment, efficient bandwidth utilization, provisioning of both Layer 2 and Layer 3 services, and fast convergence. As such, it can be used in a wide variety of scenarios, such as WAN and data center scenarios and trends toward widespread use in the industry as a unified service layer protocol, due to the following advantages. Figure 5.3 illustrates its advantages.

4. Unified service protocol: EVPN serves as a unified service signaling protocol to carry multiple types of L2VPN services, such as Ethernet Line (E-Line), Ethernet LAN (E-LAN), and E-Tree, advertise IP routes to transmit L3VPN services, and use Integrated Routing and Bridging (IRB) routes to support the provisioning of both Layer 2 and Layer 3 services. EVPN L2VPN/L3VPN has been deployed throughout data center and DCI scenarios on a large scale and can carry all cloud-centric services.

5. Simple and flexible deployment: With the BGP RR feature, all PEs in an AS establish BGP peer relationships only with RRs within the AS, implementing interconnection between network devices and eliminating the need for full-mesh PWs. L2VPN PEs can use RTs to automatically discover routes to each other, without the need to establish additional remote LDP sessions. BGP is used to learn remote MAC

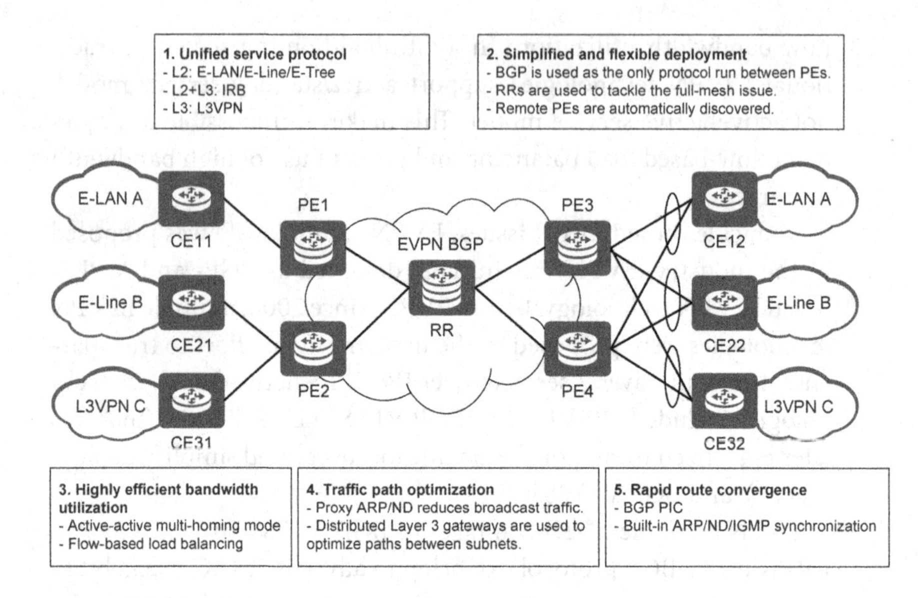

FIGURE 5.3 EVPN advantages.

addresses, and BGP route attributes and policies are flexibly used to control the routing.

6. Efficient bandwidth utilization: All Layer 2 and Layer 3 services can access a transport network in an active-active mode for flow-based load balancing.

7. Traffic path optimization: Proxy Address Resolution Protocol (ARP) and proxy Neighbor Discovery (ND) are used to reduce the volume of broadcast traffic. Distributed gateways are supported, which forward inter-subnet traffic along the optimal paths, eliminating the need for such traffic to detour through a centralized gateway.

8. Rapid route convergence: BGP is used to learn MAC and IP routes to implement Prefix Independent Convergence (PIC) and rapid traffic switchovers if faults occur. The service recovery time becomes independent from the number of MAC/IP addresses. EVPN has the built-in ARP, ND, and Internet Group Management Protocol (IGMP) synchronization mechanisms to allow dual-homing PEs to share information. If a fault occurs on a PE, the other PE does not need to relearn information.

MPLS was used as the forwarding plane when EVPN first came about. By virtue of the separation of the control and forwarding planes in EVPN, the key technologies and advantages in EVPN are inherited by SRv6, as SRv6 replaces MPLS as the forwarding plane. More than that, EVPN is enhanced with SRv6 advantages. For example, EVPN can provide more powerful cloud-network synergy capabilities and higher reliability.

5.2 SRv6 VPN PROTOCOL EXTENSIONS

As mentioned before, a VPN instance on an MPLS network can be identified by an MPLS VPN label. A VPN instance on an SRv6 network can be identified by an SRv6 SID. As described in Chapter 2, the function type of a VPN SID can be End.DX4, End.DT4, End.DX6, or End.DT6, which is determined by the VPN service type. For example, an L3VPN carrying IPv6 packets uses End.DT6 or End.DX6 SIDs, and an EVPN carrying Layer 2 packets uses End.DT2U or End.DT2M SIDs.

To support VPN services, PEs have to advertise SRv6 SIDs to identify VPN instances. The advertisement process involves extensions to control plane protocols. Details about such extensions will be described later in this section.

After a PE's interface bound to a VPN instance receives a private network packet sent by a CE, the PE searches the forwarding table of the VPN instance for an entry matching the destination IP address. The entry contains the SRv6 SID allocated to a VPN route, outbound interface name, and next-hop IP address. If SRv6 Best Effort (BE) is used, the found SRv6 SID is used as the IPv6 destination address when the packet is encapsulated with the IPv6 header. If SRv6 TE is used, the found SRv6 SID and the segment list of the path to which the route recurses are added into the packet for forwarding.

5.2.1 SRv6 Services TLV

Like MPLS VPN, SRv6 VPN supports BGP protocol extensions and uses BGP to advertise VPN reachability information identified by an SRv6 SID, such as IPv4, IPv6, and MAC addresses.

An IETF document defines specific extensions of SRv6 VPN, including the BGP processing procedure and protocol extensions for SRv6-based L3VPN, EVPN, and Internet services.[11] In terms of protocol extensions, the IETF defines two BGP Prefix-SID attribute TLVs to carry service-specific SRv6 SIDs[11]:

- SRv6 L3 Service TLV: carries Layer 3 service SRv6 SIDs, such as End.DX4, End.DT4, End.DX6, and End.DT6 SIDs.
- SRv6 L2 Service TLV: carries Layer 2 service SRv6 SIDs, such as End.DX2, End.DX2V, End.DT2U, and End.DT2M SIDs.

The BGP Prefix-SID attribute is a BGP path attribute[12] specially defined for SR. It is an optional transitive attribute and its type code is 40. The Value field in the BGP Prefix-SID attribute contains one or more TLVs used to implement various functions, such as the multiple types of SRv6 Services TLVs mentioned in this book. Figure 5.4 shows the format of the SRv6 Services TLV.

Table 5.1 describes the fields in the SRv6 Services TLV.

Figure 5.5 shows the format of the SRv6 Service Sub-TLV.

Table 5.2 describes the fields in the SRv6 Service Sub-TLV.

0–7	8–23	24–31
TLV Type	TLV Length	Reserved
SRv6 Service Sub-TLVs		

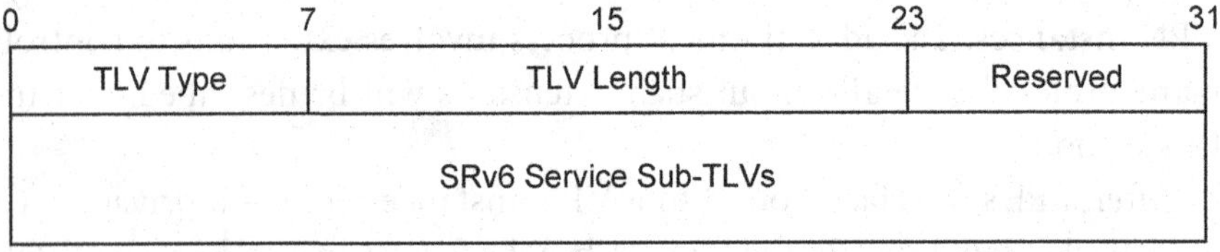

FIGURE 5.4 SRv6 Services TLV.

TABLE 5.1 Fields in the SRv6 Services TLV

Field	Length	Description
TLV Type	8 bits	TLV type. This field is set to 5 for the SRv6 L3 Service TLV and 6 for the SRv6 L2 Service TLV.
TLV Length	16 bits	Total length of the TLV Value field.
Reserved	8 bits	Reserved field.
SRv6 Service Sub-TLVs	Variable	Extended sub-TLVs, which carry more specific SRv6 service information.

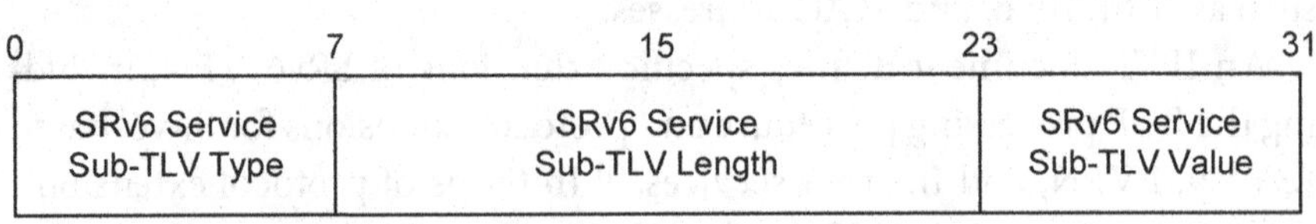

0–7	8–23	24–31
SRv6 Service Sub-TLV Type	SRv6 Service Sub-TLV Length	SRv6 Service Sub-TLV Value

FIGURE 5.5 SRv6 Service Sub-TLV.

TABLE 5.2 Fields in the SRv6 Service Sub-TLV

Field	Length	Description
SRv6 Service Sub-TLV Type	8 bits	Sub-TLV type. Value 1 indicates the SRv6 SID Information Sub-TLV.
SRv6 Service Sub-TLV Length	16 bits	Length of the sub-TLV Value field.
SRv6 Service Sub-TLV Value	Variable	Sub-TLV value. It may contain other optional SRv6 service attributes.

5.2.2 SRv6 SID Information Sub-TLV

A specific type of SRv6 Service Sub-TLV describes a particular service and is in a specific format. Currently, a single SRv6 Service Sub-TLV, that is, SRv6 SID Information Sub-TLV, is defined to carry SID information. Figure 5.6 shows the format of the SRv6 SID Information Sub-TLV.

Table 5.3 describes the fields in the SRv6 SID Information Sub-TLV.

5.2.3 SRv6 SID Structure Sub-Sub-TLV

SRv6 Service Data Sub-Sub-TLVs carry optional service information. Each type of sub-sub-TLV carries a specific type of information.

Currently, only a single type of SRv6 Service Data Sub-Sub-TLV is defined, that is, SRv6 SID Structure Sub-Sub-TLV of type 1. The SRv6 Service Data Sub-Sub-TLV contains the SID format description. Figure 5.7 shows the format of the SRv6 Service Data Sub-Sub-TLV.

0 7	15 23	31
SRv6 Service Sub-TLV Type=1	SRv6 Service Sub-TLV Length	Reserved1
SRv6 SID Value (16 bytes)		
SRv6 SID Flags	SRv6 Endpoint Behavior	Reserved2
SRv6 Service Data Sub-Sub-TLVs		

FIGURE 5.6 SRv6 SID Information Sub-TLV.

TABLE 5.3 Fields in the SRv6 SID Information Sub-TLV

Field	Length	Description
SRv6 Service Sub-TLV Type	8 bits	Sub-TLV type. Value 1 indicates the SRv6 SID Information Sub-TLV.
SRv6 Service Sub-TLV Length	16 bits	Length of the sub-TLV Value field.
Reserved1	8 bits	Reserved and set to 0 by the sender.
SRv6 SID Value	16 bytes	Encodes an SRv6 SID value.[13]
SRv6 SID Flags	8 bits	Reserved flags and not defined.
SRv6 Endpoint Behavior	16 bits	SRv6 endpoint behavior.[13] Value 0xFFFF indicates opaque behavior.
Reserved2	8 bits	Reserved and set to 0 by the sender.
SRv6 Service Data Sub-TLV Value	Variable	Carries a series of SRv6 Service Data Sub-Sub-TLVs that are optional.

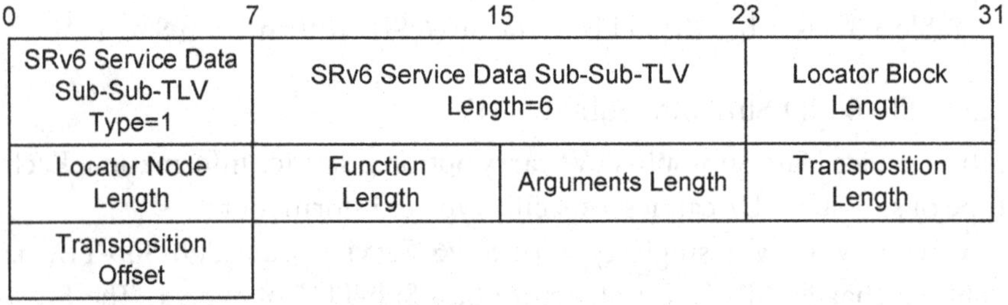

0 7 15 23 31

SRv6 Service Data Sub-Sub-TLV Type=1
SRv6 Service Data Sub-Sub-TLV Length=6
Locator Block Length
Locator Node Length
Function Length
Arguments Length
Transposition Length
Transposition Offset

FIGURE 5.7 SRv6 Service Data Sub-Sub-TLV.

Table 5.4 describes the fields in the SRv6 Service Data Sub-Sub-TLV.

Figure 5.8 shows the Transposition Offset and Transposition Length fields. The value of the Transposition Offset field is 108, and the value of the Transposition Length field is 20.

After the SRv6 SID Structure Sub-Sub-TLV is introduced, the variable parts (function and arguments) in the SRv6 SID can be transposed and stored in the Label field of an existing BGP Update message, and the identical and fixed parts in multiple SRv6 SIDs are still encoded in the SID Information-Sub TLV. However, before such an SRv6 SID is advertised, each length field in the SRv6 SID Structure Sub-Sub-TLV must be correct so that a receiver can correctly restore the original SRv6 SID.

In this way, information carried in the SRv6 SID Structure Sub-Sub-TLV can be repeatedly used by multiple SRv6 SIDs. Then, BGP Update messages used to carry only a single route can now carry multiple routes, greatly reducing the number of BGP Update messages to be sent.

TABLE 5.4 Fields in the SRv6 Service Data Sub-Sub-TLV

Field	Length	Description
SRv6 Service Data Sub-Sub-TLV Type	8 bits	Type of the sub-sub-TLV. Value 1 indicates the SRv6 SID Structure Sub-Sub-TLV.
SRv6 Service Data Sub-Sub-TLV Length	16 bits	Length of the sub-sub-TLV. Currently, the value is fixed at 6 bytes.
Locator Block Length	8 bits	Length of the SRv6 SID Locator Block field.
Locator Node Length	8 bits	Length of the SRv6 SID Locator Node field.
Function Length	8 bits	Length of the SRv6 SID Function field.
Arguments Length	8 bits	Length of the SRv6 SID Arguments field. Arguments are used only for some behaviors, such as End.DT2M. If no arguments are available, the Arguments Length field must be set to 0.
Transposition Length	8 bits	Length of the SID bits that are transposed to the Label field. • The Transposition Length field indicates the number of bits that are transposed to the Label field. • If the Transposition Length value is 0, no bits are transposed, and subsequently, the whole SRv6 SID value is encoded and stored in the SID Information Sub-TLV. In this situation, the Transposition Offset field must be set to 0.
Transposition Offset	8 bits	Positions in the Label field, to which SID bits are transposed. Note that the Label field in a BGP Update message is 24 bits long, and therefore, only 24 bits can be transposed from an SRv6 SID value. The Transposition Offset field specifies the starting point from which bits are transposed. The values of transposed bits must be set to 0 in the SRv6 SID value, and the transposed bits are put into the corresponding most significant bits in the Label field of a BGP Update message. A maximum of 24 bits can be transposed.

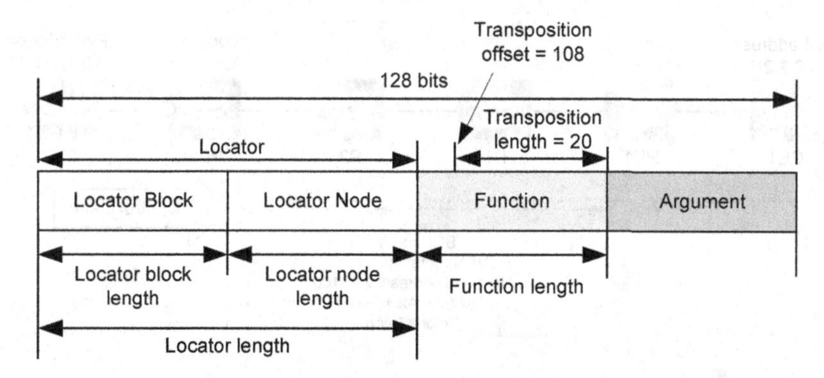

FIGURE 5.8 Transposition Offset and Transposition Length fields.

5.3 SRv6 L3VPN

This section describes SRv6 L3VPN principles. SRv6 is categorized into SRv6 BE and SRv6 TE by service path type. Therefore, the principles of L3VPN over SRv6 BE and L3VPN over SRv6 TE are described separately.

5.3.1 Principles of L3VPN over SRv6 BE

L3VPN over SRv6 BE uses the shortest SRv6 path to carry L3VPN services in a best-effort mode. The implementation of L3VPN over SRv6 BE involves the control plane protocol workflow and data plane forwarding workflow.

5.3.1.1 Workflow of L3VPN over SRv6 BE in the Control Plane

Figure 5.9 shows the workflow of L3VPN over SRv6 BE in the control plane.

1. Basic configuration of network connectivity: IPv6 addresses, an IGP, and BGP are configured on PEs and Ps to implement basic network connectivity.

2. VPN configuration and locator route advertisement: A VPN instance is configured and SRv6 SID parameters are set on PE2. Set the SRv6 locator to A2:1::/64, and the IPv6 address of a loopback interface to

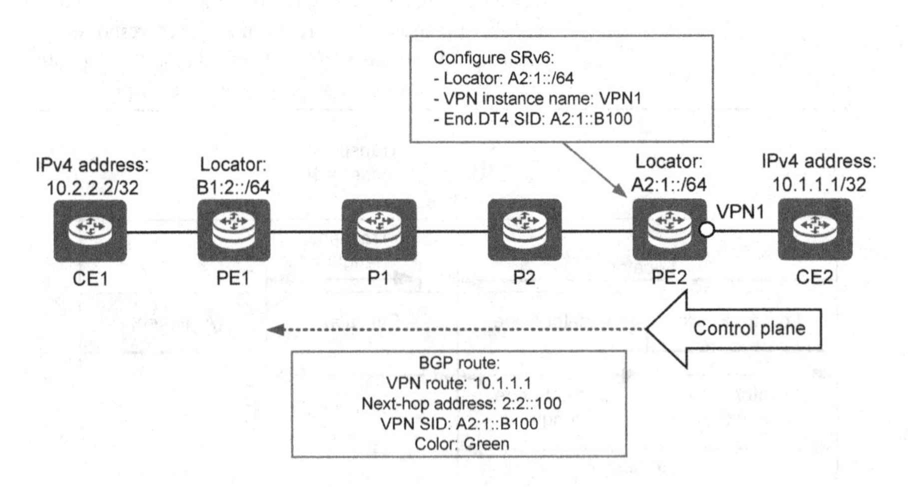

FIGURE 5.9 Workflow of L3VPN over SRv6 BE in the control plane.

A2:2::100 on PE2. VPN1 is configured on PE2 to access IPv4 services sent by CE2. Using per-VPN SID allocation, PE2 allocates End.DT4 SID A2:1::B100 to VPN1. PE2 runs an IGP to advertise the locator prefix A2:1::/64 and the loopback prefix A2:2::100 to each involved node in the domain. Upon receipt of the routes, each node generates forwarding entries.

3. Advertisement of VPN routes and VPN SIDs: After learning a VPN route 10.1.1.1/32 from CE2, PE2 runs BGP to advertise the VPN route and a related VPN SID (carried in the SRv6 L3 Service TLV in the BGP Prefix-SID attribute) to PE1. The route that PE2 advertises to PE1 carries End.DT4 SID A2:1::B100 and a next-hop IPv6 address A2:2::100 that belongs to PE2.

4. Generation of a VPN forwarding table: PE1 learns the VPN route 10.1.1.1/32 from BGP, recurses the route to the outbound interface and next hop through the End.DT4 SID A2:1::B100, and then delivers the route entry with the VPN SID, outbound interface, and next hop to the forwarding table.

5.3.1.2 Workflow of L3VPN over SRv6 BE in the Forwarding Plane

Once the control plane is constructed, L3VPN traffic forwarding is supported. Figure 5.10 shows the workflow of L3VPN over SRv6 BE in the forwarding plane. The process is as follows:

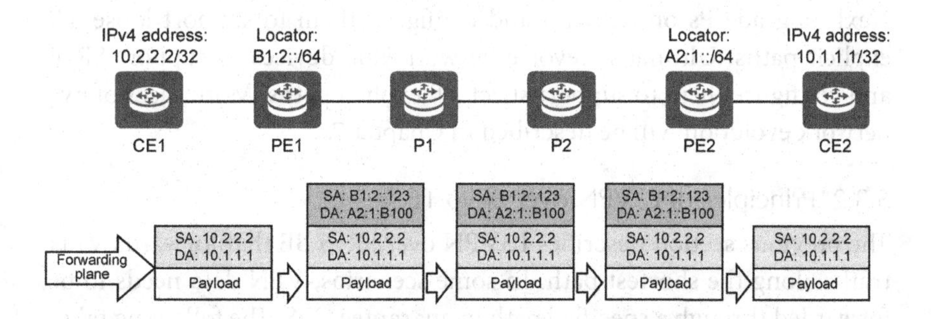

FIGURE 5.10 Workflow of L3VPN over SRv6 BE in the forwarding plane.

1. After PE1 receives a CE1-to-CE2 packet through an inbound interface, PE1 searches the routing table of the VPN instance bound to the inbound interface and obtains the VPN End.DT4 SID. This SID is advertised by PE2 and corresponds to the route 10.1.1.1/32 of CE2. Then PE1 encapsulates an IPv6 header in the original packet and forwards this packet. The destination address of the packet is the End. DT4 VPN SID configured on PE2.

2. P1 and P2 can be SRv6-capable nodes or common IPv6 nodes that do not support SRv6. Because the destination address carried in the packet is not a local SID or interface address, each of P1 and P2 searches the IPv6 forwarding table based on the longest mask matching rule for an entry matching the IPv6 destination address A2:1::B100. Subsequently, the SRv6 locator route A2:1::/64 advertised by an IGP is found. P1 and P2 then forward the packet to PE2 along the shortest path.

3. Upon receipt of the IPv6 packet destined for the End.DT4 SID, PE2 searches the local SID table and finds an entry matching the End.DT4 SID. In accordance with the End.DT4 SID behavior, PE2 removes the outer IPv6 header, searches the routing table of the VPN instance to which the End.DT4 SID is allocated, finds a route to the inner destination IPv4 address 10.1.1.1, and forwards the packet to CE2 over the matching route.

According to the preceding L3VPN over SRv6 BE principles, deploying SRv6 to support VPNs requires merely a few changes on live networks. That is, only PEs need to be upgraded to support SRv6. In evolution to SRv6 on live networks, first PEs can be upgraded to support SRv6 VPN. Next, upgrade Ps on demand and configure them to support loose TE explicit paths. Ultimately, evolve networkwide devices to support SRv6 and configure them to support strict TE explicit paths. More cases of live network evolution will be described in Chapter 7.

5.3.2 Principles of L3VPN over SRv6 TE

The previous section describes L3VPN over SRv6 BE that forwards VPN traffic along the shortest path. In some scenarios, VPN data needs to be forwarded through a specified path to guarantee QoS. The following takes L3VPN over SRv6 Policy as an example to introduce how to forward VPN traffic along a specified SRv6 TE path. For ease of understanding, a loose SRv6 Policy containing only a single End.X SID is used as an example.

5.3.2.1 Workflow of L3VPN over SRv6 Policy in the Control Plane

The workflow of L3VPN over SRv6 Policy in the control plane is as follows, as shown in Figure 5.11:

1. Basic configuration of network connectivity: Configure IPv6 addresses, an IGP, and BGP on PEs and Ps to implement basic network connectivity.
2. SRv6 locator configuration and advertisement: Set SRv6 SID parameters on PE3. Set the SRv6 locator to A2:1::/64, and the IPv6 address of the loopback interface to A2:2::124. PE3 runs an IGP to advertise the locator prefix A2:1::/64 and the loopback prefix A2:2::124 to each involved node in the domain.

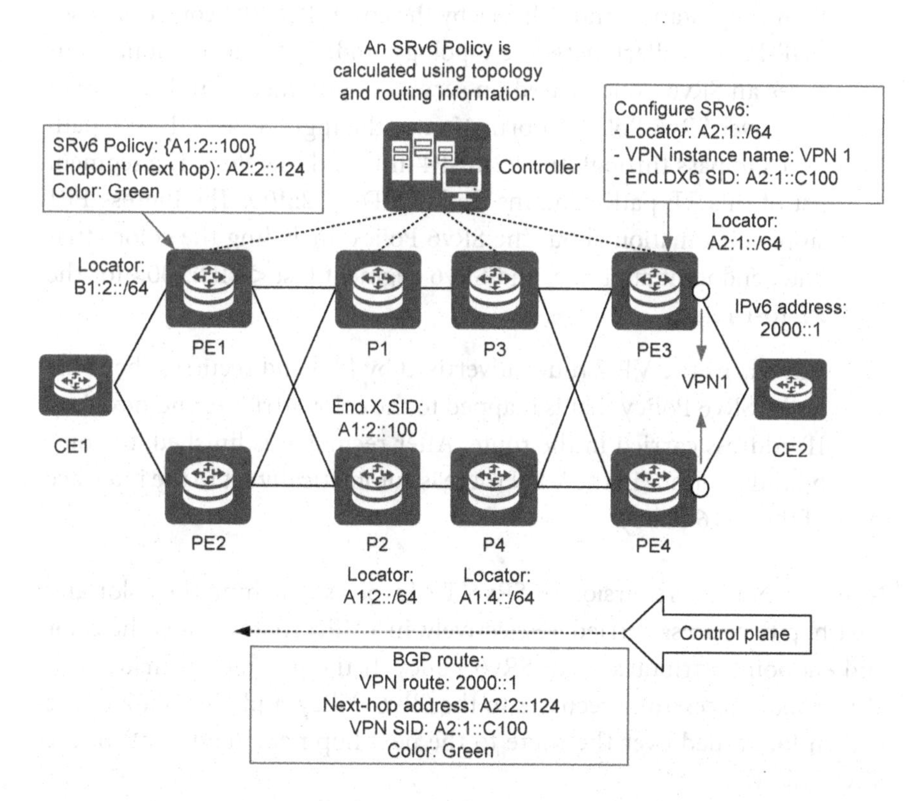

FIGURE 5.11 Workflow of L3VPN over SRv6 Policy in the control plane.

3. VPN configuration and SID advertisement: Configure VPN1 on PE3 to access IPv6 services sent by CE2, and set an End.DX6 SID to A2:1::C100 for a Layer 3 adjacency connected to CE2.

4. VPN route advertisement: PE3 learns the VPN route 2000::1 from CE2 and runs BGP to advertise the VPN route carrying the End.DX6 SID A2:1::C100, next-hop IPv6 address A2:2::124, and optional color attribute (matching different service requirements) to remote PEs.

5. SRv6 SID configuration and advertisement: Set P2's SRv6 locator route to A1:2::/64. P2 runs an IGP to advertise the locator route. After other nodes learn the route, they generate route forwarding entries with A1:2::/64 as the destination. Set an End.X SID to A1:2::100 on P2 for the P2-to-P4 link and use BGP-LS to advertise the End.X SID to the controller.

6. Path computation and delivery by the controller: The controller uses BGP-LS to collect network topology and SID information, computes an SRv6 Policy based on service requirements, and runs BGP to deliver SRv6 Policy information to the ingress PE1. The TE path needs to pass through the P2-to-P4 link, and therefore, the segment list of this TE path contains End.X SID A1:2::100. The ingress PE1 stores information about the SRv6 Policy, including the color attribute, endpoint attribute, and SRv6 Segment List <A1:2::100> for the P2-to-P4 link.

7. PE1 learns the VPN route advertised by PE3 and recurses the route to the SRv6 Policy that is mapped to the color attribute and next-hop IP address carried in the route. After recursion is finished, the outbound interface in the VPN route is automatically set to the interface of the SRv6 Policy.

Note: VPN route recursion in SRv6 TE involves matching the color and next-hop IP address carried, respectively, in a VPN route against the color and endpoint attributes of an SRv6 Policy. If the parameters match, the VPN route successfully recurses to the SRv6 Policy, and VPN data traffic is then forwarded over the route to the next-hop node (endpoint) of the SRv6 Policy.

5.3.2.2 *Workflow of L3VPN over SRv6 Policy in the Forwarding Plane*

After the control plane is constructed, L3VPN over SRv6 Policy forwarding is supported. The workflow of L3VPN over SRv6 Policy in the forwarding plane is as follows, as shown in Figure 5.12:

1. After PE1 receives a CE1-to-CE2 packet through an inbound interface, PE1 searches the routing table of a VPN instance to which the inbound interface is bound and finds a matching End.DX6 SID of a remote VPN instance. PE1 finds a forwarding entry and determines that the outbound interface in the VPN route is an SRv6 Policy, encapsulates an SRH and an IPv6 header into the original packet, and forwards the packet. The SRH carries two SRv6 SIDs, including End.X SID A1:2::100 of the P2-to-P4 link and End.DX6 SID A2:1::C100, which is allocated to the VPN route 2000::1, in VPN1 that is configured on the destination PE3. In the IPv6 packet sent from PE1, the index in the SRH points to the next-hop IPv6 address A1:2::100, and A1:2::100 is used as the outer destination address of the IPv6 packet. PE1 searches the forwarding table for an entry

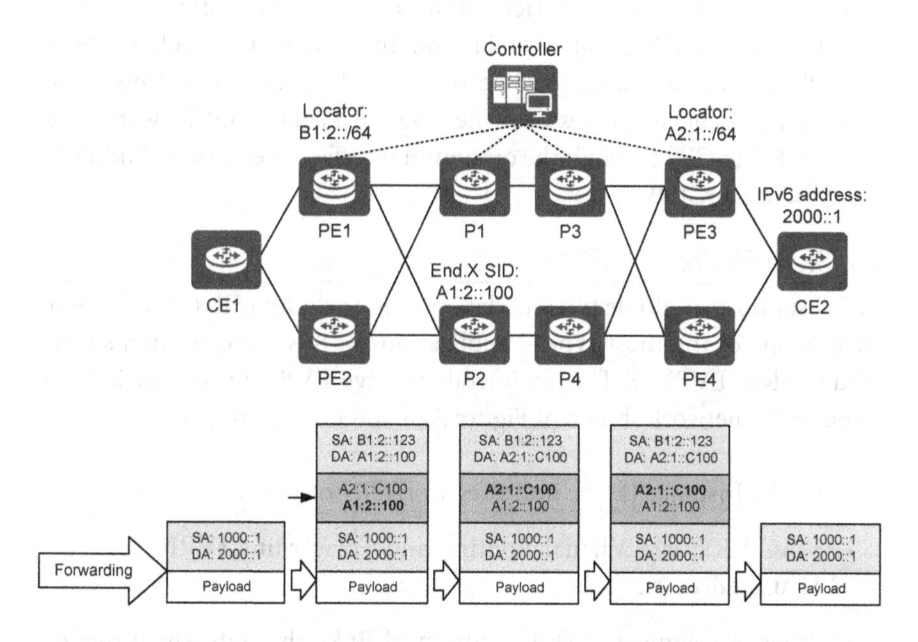

FIGURE 5.12 Workflow of L3VPN over SRv6 Policy in the forwarding plane.

matching A1:2::100 based on the longest mask matching rule and forwards the IPv6 packet to P2.

2. After receiving the IPv6 packet, P2 searches the local SID table for a matching End.X SID. P2 performs the End.X behavior: P2 processes the IPv6 header carrying the SRH, sets the index to the offset that is the next-hop IPv6 address A2:1::C100, replaces the outer destination address in the IPv6 header with A2:1::C100, and forwards the packet to P4 along the P2-to-P4 link through the outbound interface specified by End.X SID A1:2::100.
3. P4 can be a common IPv6 node. When it receives the IPv6 packet destined for End.DX6 SID A2:1::C100, P4 finds that the destination address is not a local SID or interface address. P4 searches the IPv6 forwarding table for an entry corresponding to the IPv6 destination address A2:1::C100 based on the longest mask matching rule. Then P4 finds the matching SRv6 locator route A2:1::/64 advertised by an IGP and forwards the IPv6 packet over this route to PE3 along the shortest path.
4. After receiving the IPv6 packet destined for End.DX6 SID A2:1::C100, PE3 searches the local SID table and finds a matching End.DX6 SID. PE3 then removes the IPv6 header from the packet according to the End.DX6 behavior, restores the original packet, and forwards the packet to CE2 through the outbound interface specified by End.DX6 SID A2:1::C100.

5.4 SRv6 EVPN

EVPN supports multiple types of data planes, including SRv6. EVPN over SRv6 is one of the most typical applications of SRv6 and is often shortened to SRv6 EVPN. Before we introduce SRv6 EVPN principles, let's go through the network shown in Figure 5.13 and learn some EVPN terms.

- EVPN Instance (EVI): indicates an EVPN instance.
- MAC-VRF: is a Virtual Routing and Forwarding (VRF) table for MAC addresses.
- Ethernet Segment (ES): is a group of links through which one or multiple PEs are connected to the same CE.

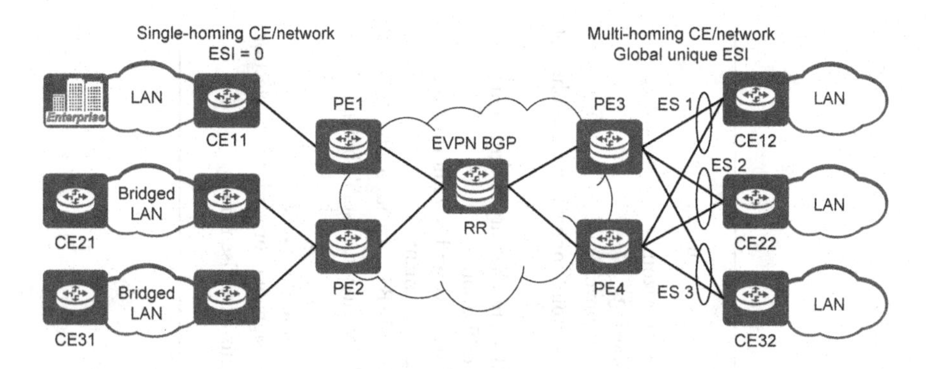

FIGURE 5.13 Typical SRv6 EVPN networking.

- Ethernet Segment Identifier (ESI): specifies the ID of an Ethernet segment. Value 0 indicates that a CE is single-homed to a PE. A non-0 value indicates that a CE is multi-homed to PEs.
- Ethernet Tag: identifies a broadcast domain, for example, a VLAN domain. An EVPN instance may contain multiple broadcast domains, meaning a single EVI corresponds to multiple Ethernet tags.

Unlike traditional L2VPN, EVPN runs BGP to advertise MAC addresses, which means the MAC address learning and advertisement is shifted from the data plane to the control plane. This mechanism is similar to the BGP/MPLS IP VPN mechanism of L3VPN.

EVPN defines new NLRIs, which are used to carry routing information, including routes of Types 1 through 4 listed in Table 5.5. EVPN also defines Type 5 routes to carry EVPN IP prefixes.[14] Besides, more route types exist, such as Type 6 through Type 8 routes,[15] which are used for EVPN multicast and can be directly used in SRv6 EVPN. This section describes how Types 1 through 5 routes (Table 5.5) are used in SRv6 EVPN. For details about the other types of routes, please refer to the related IETF documents.

To support SRv6-based EVPN, one or more SRv6 service SIDs must be advertised when Type 1, Type 2, Type 3, and Type 5 routes are advertised. Such SIDs include End.DX4, End.DT4, End.DX6, End.DT6, End.DX2, End.DX2V, and End.DT2U SIDs. The SRv6 service SID of each route type is encoded in the SRv6 L2 Service TLV or SRv6 L3 Service TLV of the BGP Prefix-SID attribute. The purpose of advertising SRv6 service SIDs is to

TABLE 5.5 Five Types of SRv6 EVPN Routes

Route Type	Full Name	Short Name	Related Standards	Application Scenario	Purpose
Type 1	Ethernet auto-discovery route	A-D route	RFC 7432	Multi-homing	• Split horizon • Aliasing • Fast convergence
			RFC 8214	EVPN VPWS (E-Line)	Point-to-point connection establishment
Type 2	MAC/IP advertisement route	MAC/IP route	RFC 7432 RFC 8365	EVPN E-LAN IRB	• MAC reachability information notification • Host ARP and ND information advertisement • Host IRB information advertisement
Type 3	Inclusive multicast Ethernet tag route	IMET route	RFC 7432 RFC 8365	EVPN E-LAN	BUM flooding tree establishment
Type 4	Ethernet segment route	ES route	RFC 7432	Multi-homing	• ES auto-discovery • Designated Forwarder (DF) election
Type 5	IP prefix route	Prefix route	draft-ietf-bess-evpn-prefix-advertisement	EVPN L3VPN	IP prefix advertisement

bind EVPN instances to SRv6 service SIDs so that EVPN traffic can be forwarded through the SRv6 data plane.

Based on the CE access modes and PE mutual access requirements, EVPN can be deployed in one of the following modes:

- EVPN E-LAN: applies to Layer 2 VPLS access scenarios. In EVPN E-LAN, Type 2 MAC/IP routes must be used to advertise MAC address reachability information, and Type 3 IMET routes may be used to establish a BUM replication list between PEs. Type 1 Ethernet A-D routes and Type 4 ES routes are primarily used to implement PE automatic discovery and rapid route convergence in CE multi-homing scenarios.
- EVPN E-Line: applies to point-to-point Layer 2 VPWS access scenarios. Typically, in EVPN E-Line, Type 1 Ethernet A-D routes are used to establish point-to-point connections.
- EVPN L3VPN: applies to Layer 3 VPN access scenarios. In EVPN L3VPN, Type 5 IP prefix routes are used to advertise IP prefix or host routes.

5.4.1 Principles of EVPN E-LAN over SRv6

In a typical EVPN E-LAN service, a PE must have an EVI and a Bridge Domain (BD) configured, Attach Circuit (AC) interfaces are bound to the BD, and the BD is bound to the EVI. Moreover, RTs must be properly set for all PEs in the EVPN instance to accept and advertise MAC/IP routes, implementing MP2MP interconnection. The local import and export RTs must match the remote export and import RTs, respectively.

The following takes unicast and multicast services as examples to introduce the principles of EVPN over SRv6 BE.

5.4.1.1 MAC Address Learning and Unicast Forwarding

Before forwarding EVPN data, an EVPN instance needs to advertise Type 2 MAC/IP routes through the control plane to create a MAC-VRF table. On each PE of the network shown in Figure 5.14, EVI 1 is created, both the import and export RTs are set to 100:1, and BD 1 is configured. An AC interface on each PE is configured with VLAN 10 and connects to a CE. The process of learning a MAC/IP route on a PE is as follows:

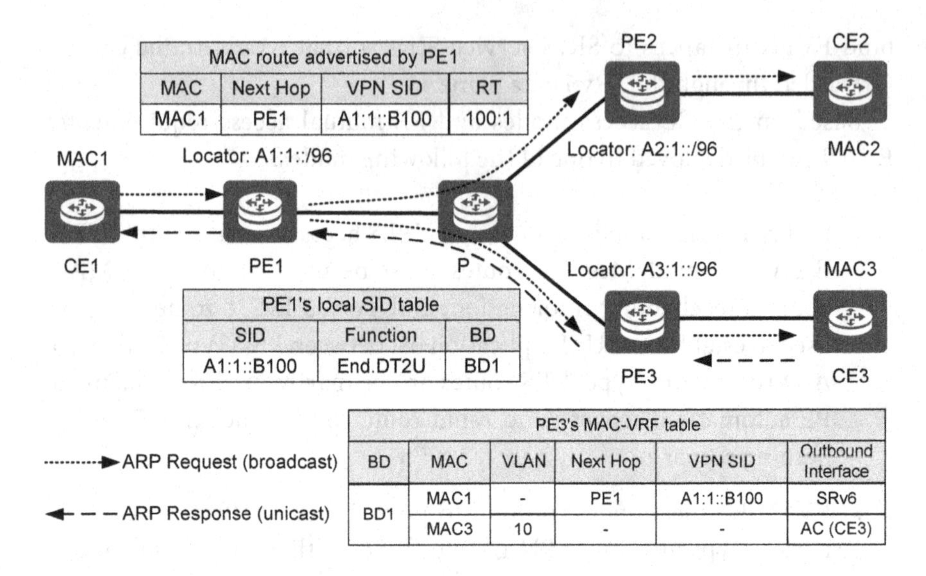

FIGURE 5.14 MAC learning in EVPN E-LAN over SRv6.

1. CE1 sends an ARP Request message to query CE3's MAC address.

2. Upon receipt of the message through an AC interface, PE1 performs the following actions:

   a. Learns CE1's MAC address MAC1.

   b. Broadcasts the ARP Request message to the other PEs.

   c. Advertises CE1's MAC1 through an EVPN MAC/IP route carrying the VPN End.DT2U SID A1:1::B100.

3. PE2 and PE3 receive the ARP message broadcast by PE1 and broadcast the message to their AC interfaces.

4. PE2 and PE3 learn MAC1 in the BGP control plane, use End.DT2U A1:1::B100 as a next-hop IPv6 address to implement route recursion, and generate a forwarding entry with MAC1 as the destination address.

5. Upon receipt of the ARP Request message, CE3 replies with an ARP Response message to CE1. The destination MAC address in the ARP Response message is CE1's MAC1. PE3 receives the ARP Response message through the AC interface, learns CE3's MAC3, and forwards the ARP Response message to PE1 in unicast mode.

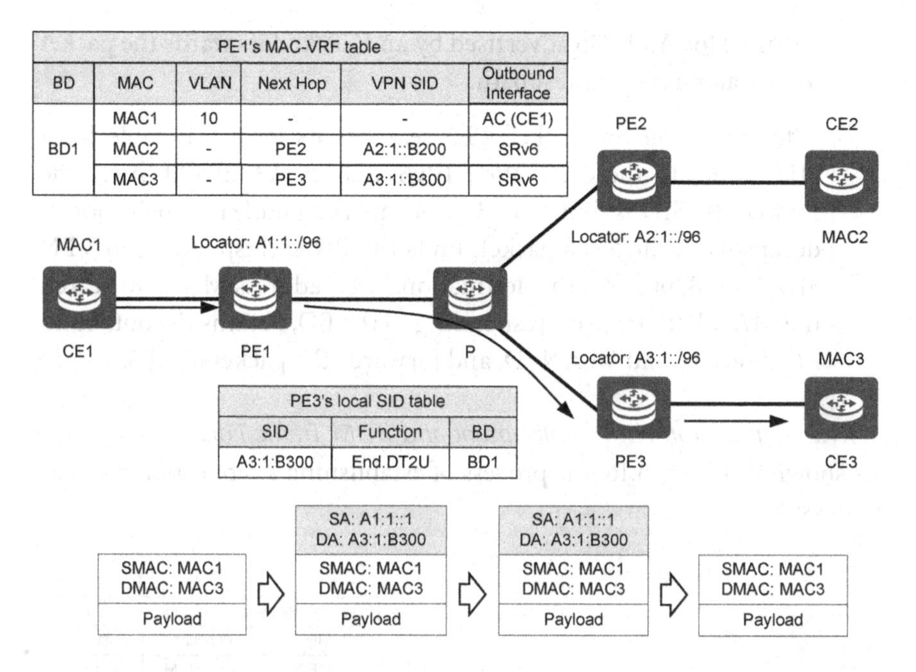

FIGURE 5.15 Unicast traffic forwarding in EVPN E-LAN over SRv6.

On the network shown in Figure 5.15, the process of forwarding unicast traffic from CE1 to CE3 is as follows:

1. After PE1 receives a CE1-to-CE3 packet through a BD interface, PE1 performs the following actions:

    a. Searches a BD MAC entry matching the destination MAC3 and finds VPN End.DT2U SID A3:1::B300 bound to MAC3.

    b. Encapsulates the packet for SRv6 by setting the source IPv6 address to its loopback IPv6 address and the destination IPv6 address to VPN SID A3:1::B300.

    c. Searches the IPv6 forwarding table for a matching outbound interface bound to VPN SID A3:1::B300 that is the destination address, and forwards the packet through the outbound interface.

2. After receiving the packet, the P forwards the packet in native IPv6 mode. Specifically, the P searches the IPv6 forwarding table based on the longest mask matching rule, finds the SRv6 locator route

destined for A3:1::/96 advertised by an IGP, and forwards the packet to PE3 along the shortest path.

3. After receiving the packet, PE3 searches its local SID table for a SID bound to the destination IPv6 address A3:1::B300. After the End.DT2U SID is found, PE3 performs the End.DT2U behavior: It decapsulates the SRv6 packet, finds the BD corresponding to VPN SID A3:1::B300, uses the destination MAC address MAC3 to search the MAC-VRF table corresponding to the BD, obtains the outbound AC interface and VLAN ID, and forwards the packet to CE3.

5.4.1.2 Replication List Establishment and BUM Traffic Forwarding

As shown in Figure 5.16, the process of establishing a replication list is as follows:

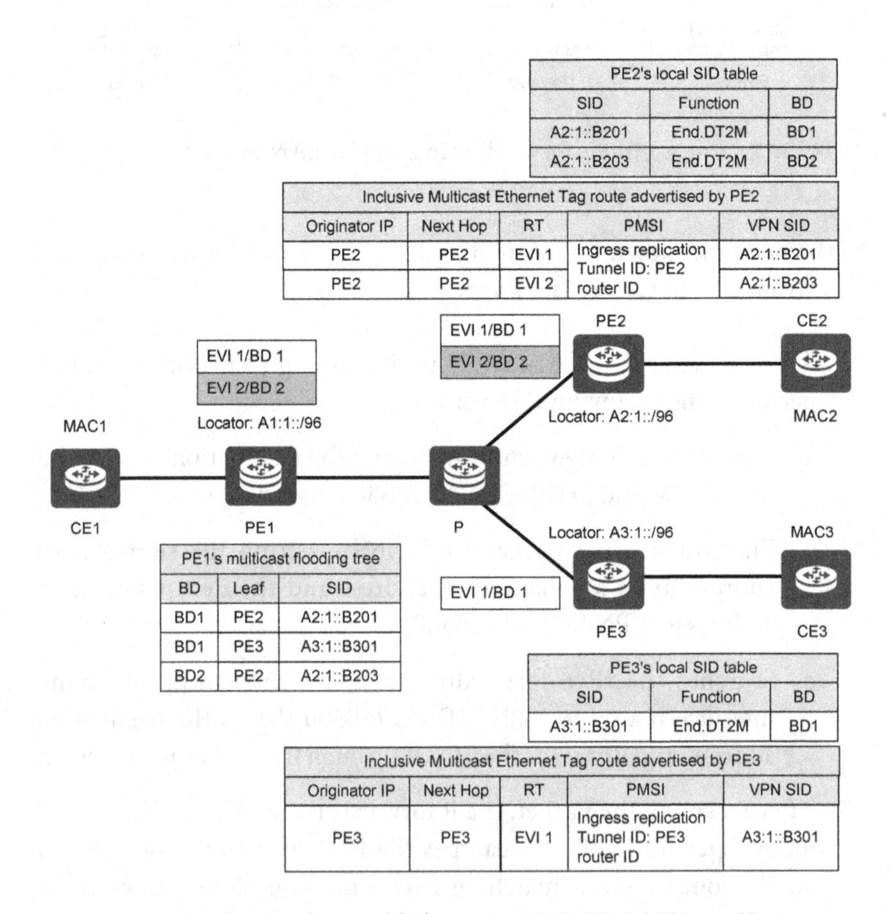

FIGURE 5.16 Replication list establishment in EVPN E-LAN over SRv6.

1. After a BD is bound to an EVPN instance on each PE, a PE advertises a Type 3 IMET route for the BD, an Originator-IP, a Provider Multicast Service Interface (PMSI) Tunnel attribute, and an End. DT2M SID to the other PEs. The PE generates an End.DT2M SID forwarding entry locally. The behavior of the End.DT2M SID is to decapsulate the SRv6 packet and broadcast the original packet in the BD corresponding to the End.DT2M SID.

2. Each PE receives Type 3 IMET routes from the other PEs, finds the EVPN instance and BD matching the RT, and establishes a replication list based on the Originator-IP with the VPN SID. When BUM traffic of an EVPN instance arrives, a PE replicates BUM traffic to all the other PEs.

On the network shown in Figure 5.17, once the replication list is created, BUM traffic can be forwarded. The forwarding process is as follows:

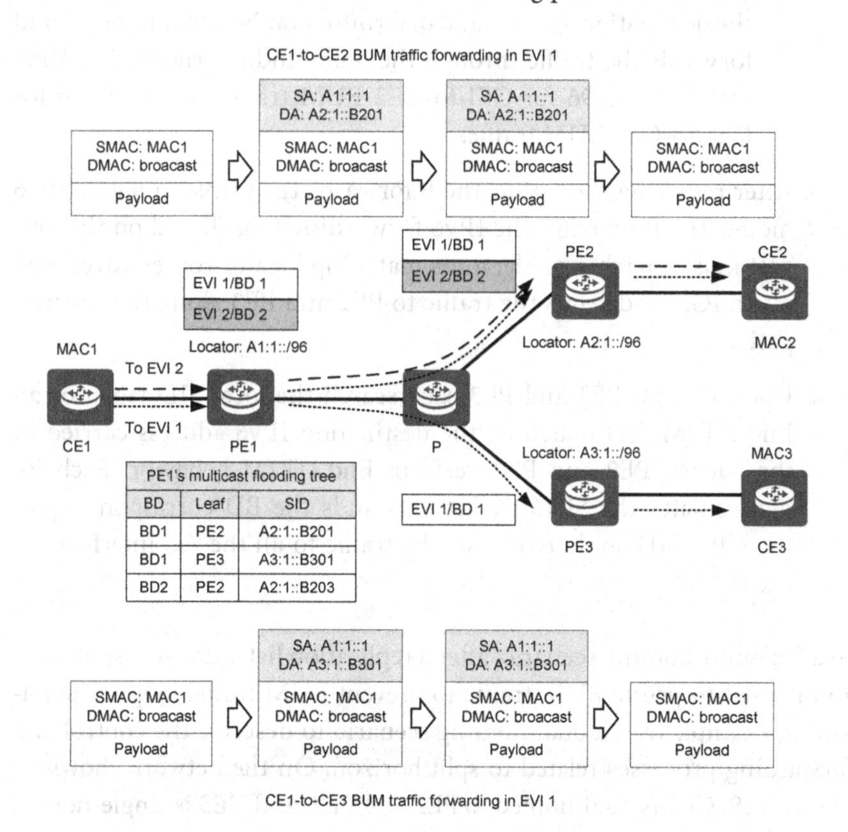

FIGURE 5.17 BUM traffic forwarding in EVPN E-LAN over SRv6.

1. CE1 sends a Layer 2 broadcast packet to PE1. The destination MAC address of the packet can be a broadcast MAC address, a multicast MAC address, or an unknown unicast MAC address. A broadcast destination MAC address is used in Figure 5.17.

2. PE1 receives BUM traffic through a BD interface and replicates and sends the BUM traffic to all leaf PEs according to the established replication list.

    - PE1 encapsulates SRv6 information in BUM traffic by setting the source IPv6 address to a loopback IPv6 address and the destination IPv6 address to a VPN SID contained in the replication list forwarding table. The SRv6 VPN SID is A2:1::B201 for CE1-to-CE2 BUM traffic and A3:1::B301 for CE1-to-CE3 BUM traffic.

    - PE1 searches the IPv6 forwarding table with each VPN SID as the destination address and determines outbound interfaces, and forwards the traffic through the outbound interfaces. The VPN SID is A2:1::/96 for CE1-to-CE2 BUM traffic and A3:1::/96 for CE1-to-CE3 BUM traffic.

3. After receiving the traffic, the P forwards the traffic in native IPv6 mode. The P searches the IPv6 forwarding table based on the longest mask matching rule, finds matching locator routes advertised by an IGP, and sends the traffic to PE2 and PE3 along the shortest paths.

4. Upon receipt, PE2 and PE3 each search the local SID table for an End.DT2M SID matching the destination IPv6 address carried in the packet. PE2 and PE3 perform End.DT2M behavior: Each PE decapsulates the SRv6 packet and finds the BD corresponding to the VPN SID, and broadcasts the traffic to all the AC interfaces in the BD.

In a CE multi-homing scenario, after a replication list is created, split horizon is used to prune BUM traffic to prevent BUM traffic loops. The following example uses a dual-homing scenario to describe the control and forwarding processes related to split horizon. On the network shown in Figure 5.18, CE2 is dual-homed to PE2 and PE3, and CE3 is single-homed to PE3. CE2 sends a copy of BUM traffic.

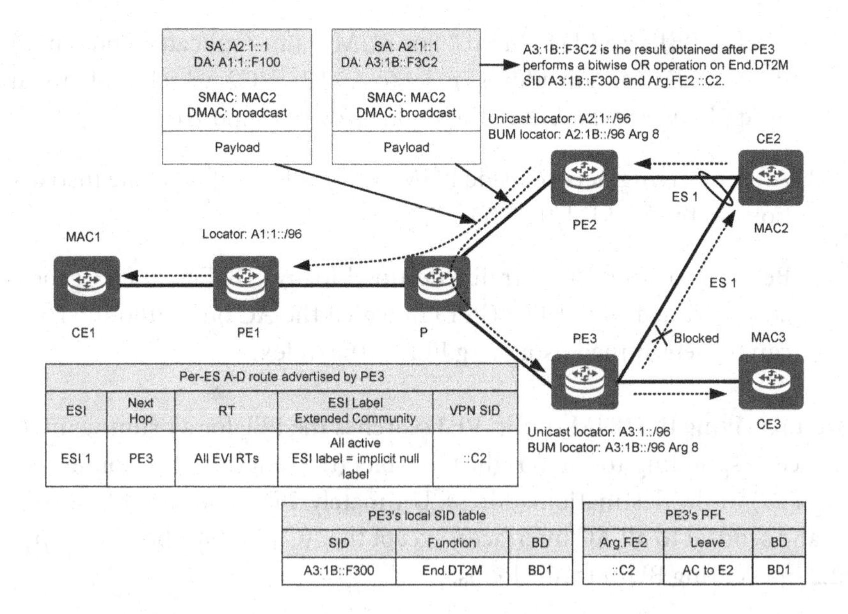

FIGURE 5.18 BUM traffic forwarding in EVPN E-LAN over SRv6 in a CE dual-homing scenario.

The process of creating control entries is as follows:

Both PE2 and PE3 to which CE2 is dual-homed allocate a locally unique Arg.FE2 identifier to the AC interface connected to the CE and advertise the Arg.FE2 identifier to the other PEs. For example, PE3 advertises a per-ES Ethernet A-D route carrying the Arg.FE2 identifier to PE1 and PE2.

5. PE1 does not have the same ESI as PE3 and, therefore, does not process the route after receiving it.

6. When PE2 receives the route and finds that it has the same ESI, PE2 saves the Arg.FE2 (::C2) information advertised by PE3. PE2 performs a bitwise OR operation on the Arg.FE2 (::C2) and PE3's End.DT2M SID (A3:1B::F300), and changes the destination address of the local replication list to A3:1B::F3C2.

The data forwarding process is as follows:

1. When the ingress PE2 receives BUM traffic, it replicates the traffic and sends it to the other PEs according to an EVI-based replication list. For BUM traffic replicated and sent to PE1, the destination address

is set to PE1's End.DT2M SID. For BUM traffic replicated and sent to PE3, the destination address is set to A3:1B::F3C2, which is obtained by applying bitwise OR to the End.DT2M SID and Arg.FE2.

2. After receiving network side BUM traffic, PE1 performs the instructions in the End.DT2M SID.

3. Before receiving BUM traffic destined for A3:1B::F3C2, PE3 generates a Pruned-Flood-List (PFL) in which the AC outbound interface can be determined using Arg.FE2 as the index.

After receiving the BUM traffic, PE3 searches the PFL for a pruning interface corresponding to the 8 right-most bits (the length of the configured Arg.FE2) in the destination address. Ultimately, PE3 replicates BUM traffic and sends it to all AC interfaces, except the AC interface bound to Arg. FE2, preventing BUM traffic loops.

5.4.2 Principles of EVPN E-Line over SRv6

EVPN E-Line is a P2P L2VPN service solution provided based on the EVPN service architecture. In a typical EVPN E-Line service, an AC interface is bound to an Ethernet Virtual Private Line (EVPL) instance. In the EVPL instance, a local AC ID and a remote AC ID are set on PEs on each end. These AC IDs must be the same as the remote and local AC IDs on the remote PE, respectively. In this way, a point-to-point service model is formed. EVPN E-Line over SRv6 can be deployed in single-homing, multi-homing single-active, and multi-homing active-active scenarios. The following example uses the single-homing scenario.

As shown in Figure 5.19, the process of establishing an EVPN E-Line over SRv6 BE path is as follows:

1. An EVPL instance and an EVPN VPWS instance are configured on both PE1 and PE2. The EVPL instance is bound to an AC interface and the EVPN VPWS instance and is allocated a local AC ID and a remote AC ID. After the configuration is complete, each PE generates a forwarding entry associating the AC interface with the EVPL instance.

2. PE1 and PE2 send each other a Type 1 Ethernet A-D route with an End.DX2 VPN SID. The Type 1 Ethernet A-D route carries the local AC ID that is configured on the local end and is corresponding to

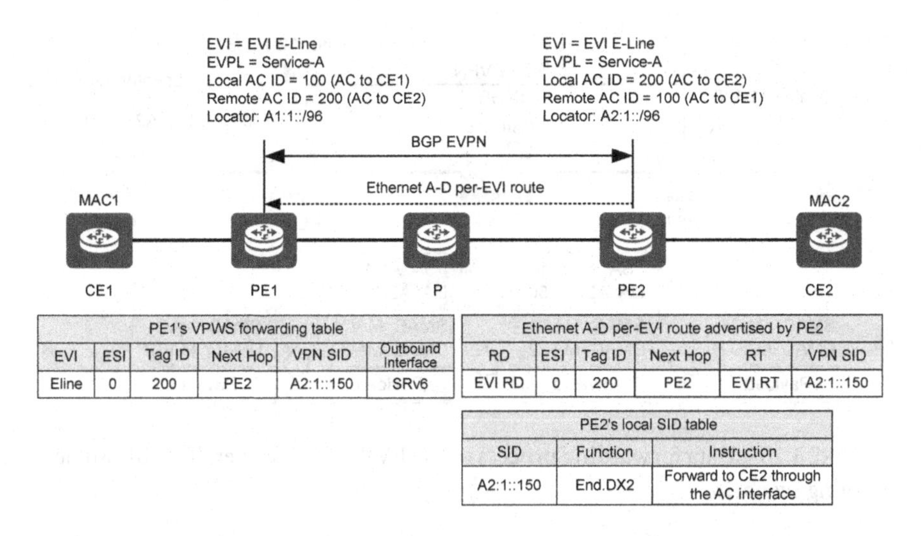

FIGURE 5.19 Route advertisement in an EVPN E-Line over SRv6 BE single-homing scenario.

the Ethernet Tag ID carried in the packet. Each PE generates an End.DX2 SID forwarding entry locally. The behavior of the SID is to decapsulate SRv6 packet and forward the original packet through the AC interface.

3. After PE1 and PE2 receive the Type 1 Ethernet A-D route, both PEs find that the received export and import RTs match the local import and export RTs, respectively. Then, both PEs match the remote AC ID configured for the EVPL instance against the Ethernet Tag ID carried in the Ethernet A-D route. If they match, PE1 and PE2 each install an entry into the EVPN E-Line forwarding table. The forwarding entry contains the VPN SID advertised by the peer. When the E-Line service is transmitted, the packets are encapsulated with the VPN SID before being forwarded.

As shown in Figure 5.20, the data forwarding process in the EVPN E-Line over SRv6 BE single-homing scenario is as follows:

1. Upon receipt of the traffic sent by CE1, PE1 searches the EVPN E-Line forwarding table for an SRv6 VPN SID matching the E-Line service to which the inbound interface is bound, and encapsulates the original packet with SRv6 VPN SID A2:1::150 as the destination

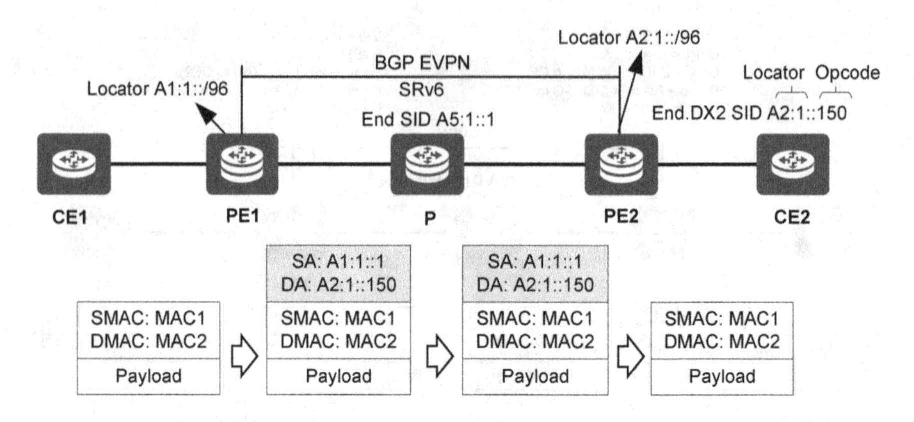

FIGURE 5.20 Data forwarding process in the EVPN E-Line over SRv6 BE single-homing scenario.

address of the outer IPv6 header. Then, PE1 searches the IPv6 forwarding table for an entry with the SRv6 VPN SID as the destination address, determines the outbound interface, and forwards the traffic through the outbound interface.

2. After receiving the traffic, the P forwards the traffic in native IPv6 mode. The P searches the IPv6 forwarding table based on the longest mask matching rule, finds the locator route destined for A2:1::/96 advertised by an IGP, and forwards the traffic to PE2 along the shortest path.

3. Upon receipt, PE2 searches the local SID table based on the destination IPv6 address of the packet and finds the End.DX2 SID forwarding entry. PE2 performs the behavior of End.DX2 SID: It decapsulates the SRv6 packet and finds the EVPL instance to which the End.DX2 SID is allocated. Ultimately, PE2 forwards the traffic to CE2 through the AC interface bound to the EVPL instance.

5.4.3 Principles of EVPN L3VPN over SRv6

EVPN L3VPN over SRv6 forwards data in the same way as L3VPN over SRv6. The only difference is that EVPN L3VPN over SRv6 uses a different route type. L3VPN over SRv6 uses the conventional BGP VPNv4 and VPNv6 address families to advertise End.DT4 and End.DT6 SIDs, respectively. EVPN over SRv6 uses the BGP EVPN address family to send EVPN Type 5 IP prefix routes and advertises VPN SIDs. Due to the similarity with the L3VPN over SRv6 forwarding process, the principles of EVPN L3VPN over SRv6 are not described in detail.

5.4.4 SRv6 EVPN Protocol Extensions

After describing the SRv6 EVPN principles in previous sections, in this section, we cover the protocol extensions and advertisement process of the major types of SRv6 EVPN routes.

5.4.4.1 Ethernet A-D Route

EVPN Type 1 Ethernet A-D routes are used to implement various functions, such as split horizon, rapid route convergence, and aliasing.[7] These routes are also used to advertise P2P service IDs in EVPN VPWS and EVPN VPLS scenarios.

Figure 5.21 shows the NLRI of the EVPN Type 1 route.

The values of the preceding fields vary according to the per-ES and per-EVI routes to be advertised.

- Per-ES Ethernet A-D route: A PE advertises an Ethernet A-D route with an Ethernet Tag ID 0xFFFFFFFF for each ES. In CE multi-homing scenarios, such a route carries the ESI label extension community which is used to distinguish the PE from other PEs connected to the same CE.[7] Such routes are used for split horizon and fast convergence.
- Per-EVI Ethernet A-D route: A PE advertises an Ethernet A-D route with a non-0xFFFFFFFF Ethernet Tag ID for each EVI. Such routes

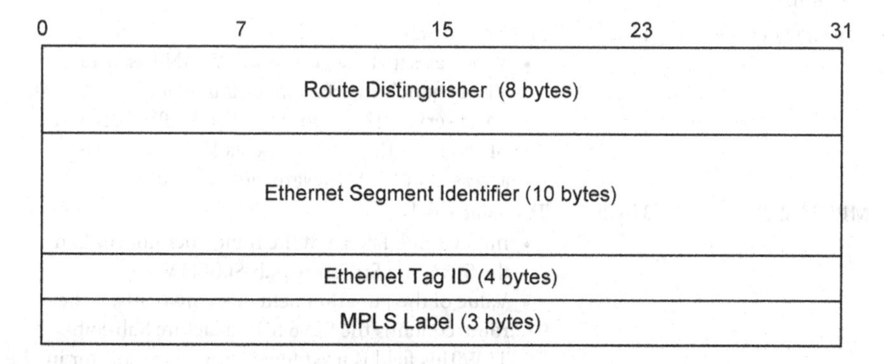

FIGURE 5.21 NLRI of the EVPN Type 1 route.

are used in the aliasing function to implement load balancing or redundancy in E-LAN scenarios or are used to carry P2P service IDs in VPWS scenarios.[7,16]

Table 5.6 describes the fields in the per-ES Ethernet A-D route.

Before a per-ES Ethernet A-D route is advertised, a service SID must be added to the SRv6 L2 Service TLV of the BGP Prefix-SID attribute. The behavior corresponding to the SID is defined by the originator of the advertisement, and the actual function of the SID is to specify the Arg.FE2 of the End.DT2M SID to prune BUM traffic in multi-homing scenarios, preventing BUM traffic loops.

Table 5.7 describes the fields in the per-EVI Ethernet A-D route.

TABLE 5.6 Fields in the Per-ES Ethernet A-D Route

Field	Length	Description
RD	8 bytes	RD value. It contains a PE IP address, for example, X.X.X.X:0, in a per-ES route.
Ethernet Segment Identifier	10 bytes	ES ID, identifying an ES.
Ethernet Tag ID	4 bytes	For per-ES routes, the maximum value of this field is MAX-ET (0xFFFFFFFF).
MPLS Label	3 bytes	The value is fixed at 0 in a per-ES A-D route.

TABLE 5.7 Fields in the Per-EVI Ethernet A-D Route

Field	Length	Description
Route Distinguisher	8 bytes	RD value configured for an EVPN instance.
Ethernet Segment Identifier	10 bytes	ES ID, identifying an ES.
Ethernet Tag ID	4 bytes	The value can be: • 0: indicates that the port-based, VLAN-based, or VLAN bundle interface mode is used for VPLS access. • Local service ID: applies to EVPN VPWS scenarios. • BD tag: identifies a BD when a VPLS service is accessed in VLAN-aware bundle mode.
MPLS Label	3 bytes	The value can be: • Implicit null label 3, if the route does not contain the SRv6 SID Structure Sub-Sub-TLV. • Value of the Function field in an SRv6 SID, if the route contains the SRv6 SID Structure Sub-Sub-TLV. This field is used together with the locator in the SRv6 SID Structure Sub-Sub-TLV to compose a complete 128-bit SID.

Similarly, a service SID must be carried in the SRv6 L2 Service TLV of the BGP Prefix-SID attribute when a per-EVI Ethernet A-D route is to be advertised. The behavior in this service SID can be defined by the originator of the advertisement. The behavior is usually an End.DX2, End.DX2V, or End.DT2U.

5.4.4.2 MAC/IP Advertisement Route

EVPN Type 2 MAC/IP advertisement routes are used to advertise the MAC/IP address reachability for unicast traffic to all other edge devices in the same EVPN instance.

Figure 5.22 shows the NLRI of the EVPN Type 2 route.

Table 5.8 describes the fields in the EVPN Type 2 route.

Service SIDs are carried in MAC/IP advertisement routes. The service SID is mandatory in the SRv6 L2 Service TLV of the BGP Prefix-SID attribute but is optional in the SRv6 L3 Service TLV.

MAC/IP advertisement can be classified into two types based on whether IP addresses are advertised:

1. MAC only: MAC/IP advertisement routes contain only MPLS Label1. In this scenario, a service SID must be carried in the SRv6 L2 Service TLV of the BGP Prefix-SID attribute that is advertised along

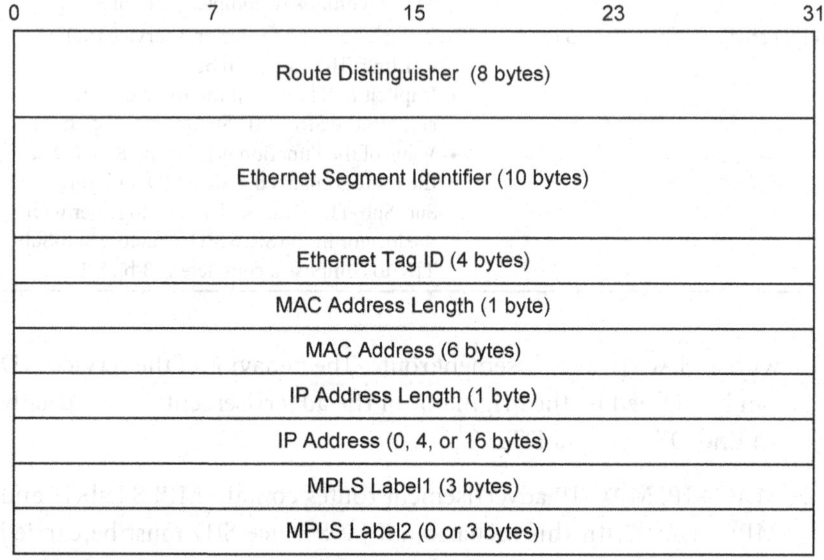

0 7 15 23 31
Route Distinguisher (8 bytes)
Ethernet Segment Identifier (10 bytes)
Ethernet Tag ID (4 bytes)
MAC Address Length (1 byte)
MAC Address (6 bytes)
IP Address Length (1 byte)
IP Address (0, 4, or 16 bytes)
MPLS Label1 (3 bytes)
MPLS Label2 (0 or 3 bytes)

FIGURE 5.22 NLRI of the EVPN Type 2 route.

TABLE 5.8 Fields in the EVPN Type 2 Route

Field	Length	Description
Route Distinguisher	8 bytes	RD of an EVPN instance.
Ethernet Segment Identifier	10 bytes	ES ID, identifying an ES.
Ethernet Tag ID	4 bytes	The value can be: • 0: indicates that the port-based, VLAN-based, or VLAN bundle interface mode is used for VPLS access. • BD tag: identifies a BD when a VPLS service is accessed in VLAN-aware bundle mode.
MAC Address Length	1 byte	MAC address length. The value is fixed at 48, in bits.
MAC Address	6 bytes	Six-octet MAC address.
IP Address Length	1 byte	Length of an IP address mask.
IP Address	0, 4, or 16 bytes	IPv4 or IPv6 address. By default, the IP Address Length field is set to 0, and the IP Address field is omitted from the route.
MPLS Label1	3 bytes	Value of a label used for Layer 2 service traffic forwarding. The value can be: • Implicit null label 3, if the route does not contain the SRv6 SID Structure Sub-Sub-TLV. • Value of the Function field in an SRv6 SID, if the route contains the SRv6 SID Structure Sub-Sub-TLV. This field is used together with the locator in the SRv6 SID Structure Sub-Sub-TLV to compose a complete 128-bit SID.
MPLS Label2	0 or 3 bytes	Value of a label used for Layer 3 service traffic forwarding. The value can be: • Implicit null label 3, if the route does not contain the SRv6 SID Structure Sub-Sub-TLV. • Value of the Function field in an SRv6 SID, if the route contains the SRv6 SID Structure Sub-Sub-TLV. This field is used together with the locator in the SRv6 SID Structure Sub-Sub-TLV to compose a complete 128-bit SID.

with a MAC/IP advertisement route. The behavior of the service SID can be defined by the originator of the advertisement and is usually an End.DX2 or End.DT2U.

2. MAC+IP: MAC/IP advertisement routes contain MPLS Label1 and MPLS Label2. In this scenario, an L2 Service SID must be carried in the SRv6 L2 Service TLV of the BGP Prefix-SID attribute before a

MAC/IP advertisement route is advertised. An L3 Service SID may also be carried in the SRv6 L3 Service TLV of the BGP Prefix-SID attribute advertised along with the route. The behaviors of these SRv6 service SIDs can be defined by the originator of the advertisement. For L2 Service SIDs, this behavior is an End.DX2 or End.DT2U. For L3 Service SIDs, this behavior is usually an End.DX4, End.DX6, End.DT4, or End.DT6.

5.4.4.3 IMET Route

An EVPN Type 3 IMET route consists of a specific EVPN NLRI and PMSI Tunnel attribute and is used to establish a channel for forwarding BUM traffic. After PEs establish BGP peer relationships, they exchange IMET routes that carry the IP prefix (usually the local PE's loopback address) of the local PE and PMSI tunnel information. The PMSI tunnel information contains the tunnel type and label used for BUM packet transmission. BUM traffic includes broadcast, multicast, and unknown unicast traffic. After a PE receives BUM traffic, it forwards the traffic to the other PEs through a channel established using IMET routes in P2MP mode.

Figure 5.23 shows the NLRI of the EVPN Type 3 route.

Table 5.9 describes the fields in the EVPN Type 3 route.

The EVPN Type 3 route carries the PMSI Tunnel attribute to distribute tunnel information. Figure 5.24 shows the format of the PMSI Tunnel attribute.

Table 5.10 describes the fields of the PMSI Tunnel attribute.

When advertising the route, a device must add a service SID to the SRv6 L2 Service TLV in the BGP Prefix-SID attribute. The behavior corresponding to a service SID is defined by the originator of the advertisement and is usually an End.DT2M SID.

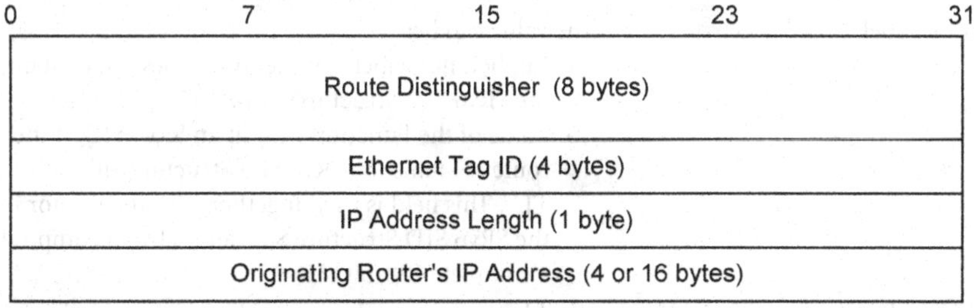

0 7 15 23 31
Route Distinguisher (8 bytes)
Ethernet Tag ID (4 bytes)
IP Address Length (1 byte)
Originating Router's IP Address (4 or 16 bytes)

FIGURE 5.23 NLRI of the EVPN Type 3 route.

TABLE 5.9 Fields in the EVPN Type 3 Route

Field	Length	Description
Route Distinguisher	8 bytes	RD of an EVPN instance.
Ethernet Tag ID	4 bytes	The value can be: • 0: indicates that the port-based, VLAN-based, or VLAN bundle interface mode is used for VPLS access. • BD tag: identifies a BD when a VPLS service is accessed in VLAN-aware bundle mode.
IP Address Length	1 byte	Length of the source IP address configured on a PE.
Originating Router's IP Address	4 or 16 bytes	Source IPv4 or IPv6 address configured on a PE.

0	7	15	23	31
Flags (1 byte)				
Tunnel Type (1 byte)				
MPLS Label (3 bytes)				
Tunnel Identifier (variable)				

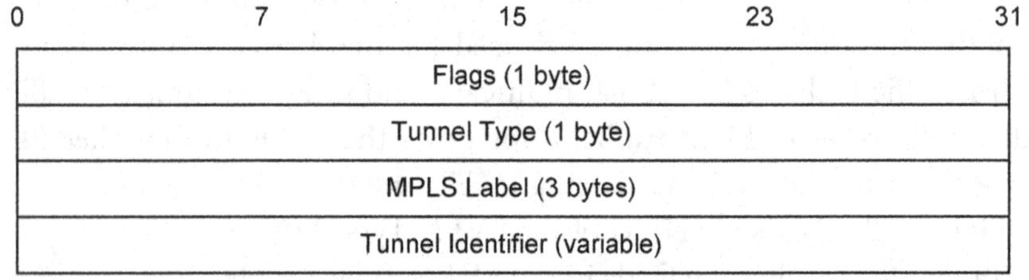

FIGURE 5.24 Format of the PMSI Tunnel attribute.

TABLE 5.10 Fields in the PMSI Tunnel Attribute

Field	Length	Description
Flags	1 byte	Whether leaf node information must be obtained for an existing tunnel. The value is set to 0 in SRv6.
Tunnel Type	1 byte	Tunnel type carried in a route. For SRv6 EVPN, value 6 indicates ingress replication currently.
MPLS Label	3 bytes	The value can be: • Implicit null label 3, if the route does not contain the SRv6 SID Structure Sub-Sub-TLV. • Value of the Function field in an SRv6 SID, if the route contains the SRv6 SID Structure Sub-Sub-TLV. This field is used together with the locator in the SRv6 SID Structure Sub-Sub-TLV to compose a complete 128-bit SID.
Tunnel Identifier	Variable	IP address of a tunnel's egress when ingress replication mode is used.

5.4.4.4 ES Route

EVPN Type 4 ES routes are used by PEs connecting to the same CE to automatically discover each other, and this type of route is also used in DF election. Figure 5.25 shows the NLRI of the EVPN Type 4 route.

Table 5.11 describes the fields in the EVPN Type 4 route.

Note that the BGP Prefix-SID attribute is not used when EVPN Type 4 routes are advertised. This means the route processing on the SRv6 network is the same as that described in RFC 7432.

5.4.4.5 IP Prefix Route

EVPN Type 5 IP prefix routes are used to advertise IP address reachability to all edge devices in an EVPN instance. An IP address may contain a host IP prefix or any specific subnet prefix.

Figure 5.26 shows the NLRI of the EVPN Type 5 route.

Table 5.12 describes the fields in the EVPN Type 5 route.

The SRv6 Service SID corresponding to an EVPN Type 5 route is encoded in the SRv6 L3 Service TLV of the BGP Prefix-SID attribute and

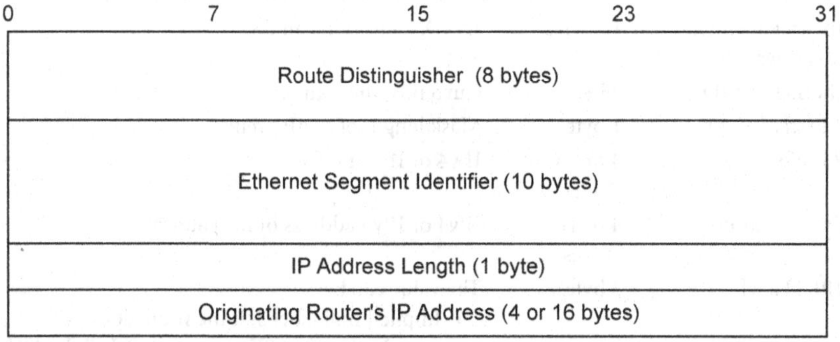

0 7 15 23 31

Route Distinguisher (8 bytes)

Ethernet Segment Identifier (10 bytes)

IP Address Length (1 byte)

Originating Router's IP Address (4 or 16 bytes)

FIGURE 5.25 NLRI of the EVPN Type 4 route.

TABLE 5.11 Fields in the EVPN Type 4 Route

Field	Length	Description
Route Distinguisher	8 bytes	RD of an EVPN instance.
Ethernet Segment Identifier	10 bytes	ES ID, identifying an ES.
IP Address Length	1 byte	Length of the source IP address configured on a PE.
Originating Router's IP Address	4 or 16 bytes	Source IPv4 or IPv6 address configured on a PE.

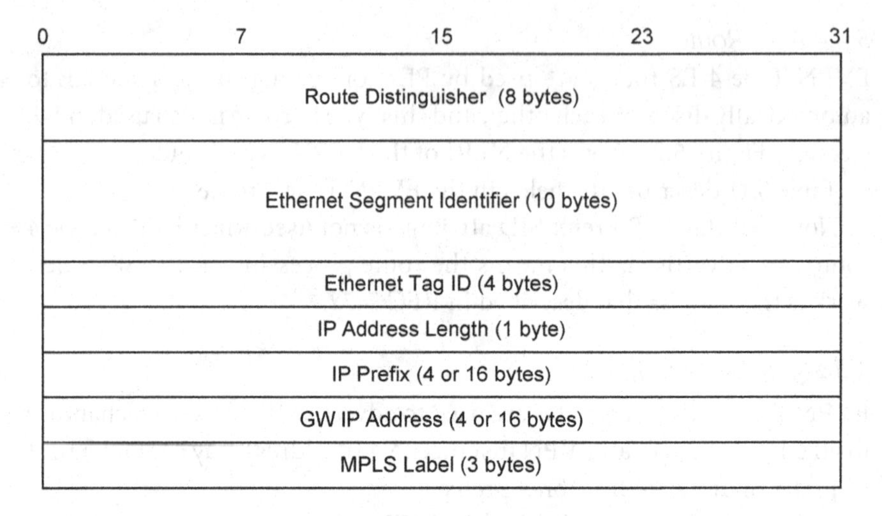

FIGURE 5.26 NLRI of the EVPN Type 5 route.

TABLE 5.12 Fields in the EVPN Type 5 Route

Field	Length	Description
RD	8 bytes	RD of an EVPN instance.
Ethernet Segment Identifier	10 bytes	ES ID, identifying an ES.
Ethernet Tag ID	4 bytes	Currently, the value can only be set to 0.
IP Prefix Length	1 byte	Mask length of an IP prefix.
IP Prefix	4 or 16 bytes	IPv4 or IPv6 prefix.
GW IP Address	4 or 16 bytes	IPv4 or IPv6 address of the gateway.
MPLS Label	3 bytes	The value can be: • Implicit null label 3, if the route does not contain the SRv6 SID Structure Sub-Sub-TLV. • Value of the Function field in an SRv6 SID, if the route contains the SRv6 SID Structure Sub-Sub-TLV. This field is used together with the locator in the SRv6 SID Structure Sub-Sub-TLV to compose a complete 128-bit SID.

advertised together with the NLRI. The behavior of the SID is specified by the originator of the advertisement and is usually an End.DT4, End.DT6, End.DX4, or End.DX6.

We have now covered SRv6 EVPN extensions. For more details, please refer to the SRv6 EVPN standard.[11]

5.5 STORIES BEHIND SRv6 DESIGN

SRv6 VPN Is Promising

In terms of achieving revenue growth, carriers have been facing intense competition with Over the Top (OTT) providers. Technically speaking, OTT cloud computing is adept at virtualizing and pooling infrastructure resources, as well as high O&M automation. For instance, on a data center network, the controller is used to implement network service provisioning with ease; VXLAN, an IP-based VPN technology, only needs to be configured on both ends of a VXLAN tunnel to implement multi-tenant services. As one of the most valuable services for carriers, MPLS VPN is more complex than IP VPN, especially in inter-domain scenarios. By virtue of high automation, it can take minutes to obtain IaaS virtual resources from an OTT provider, whereas as far as we know, it may take carriers months to deploy typical VPNs.

With SRv6 VPN, VPN services are deployed using IPv6 reachability information, which eliminates the complexity of deploying intra-domain MPLS and inter-AS VPN. More precisely, one-hop access can be implemented just after two endpoints are properly configured. In this way, SRv6 makes it easier for carriers to deploy VPNs and the complexity is similar to that for OTT providers to deploy data center VPNs, laying a solid foundation for carriers to engage in rapid VPN deployment.

Although as early as mid-2018, I mentioned how SRv6 is advantageous in terms of network simplification, I was still impressed by this when China Telecom Sichuan deployed SRv6 VPN in early 2019. Video services of China Telecom Sichuan are transmitted between the provincial and municipal data centers across the data center, metro, and national IP backbone networks. If the traditional MPLS VPN is adopted, it is inevitable to coordinate with all kinds of network administrative departments. Thanks to China's efforts at promoting IPv6 deployment, IPv6 reachability has been achieved on IP networks. In light of this, services were quickly provisioned and transmitted after two PEs supporting SRv6 VPN were deployed in the provincial and municipal data centers to establish an SRv6 BE path. I was very glad to learn this simplified deployment. Evidently, technological advancements have been utilized to eliminate the complex coordination issues between departments. I joked with guys from domestic carriers: In November 2017, when Chinese government began to push all carriers' IP networks to support IPv6, Chinese carriers at that time could hardly recognize the value of IPv6 and may be reluctant to execute the policy because most Internet services were still running well on an IPv4 foundation. But the opportune emergence of SRv6 means that IPv6 networks will support new services better and obtain technical advantages over IPv4/MPLS, which saved the carriers and is truly a surprise and bonus. I also believe that, without a doubt, SRv6 will in turn help propel IPv6 application and deployment.

REFERENCES

[1] Andersson L, Madsen T. Provider Provisioned Virtual Private Network (VPN) Terminology[EB/OL]. (2018-12-20) [2020-03-25]. RFC 4026.

[2] Callon R, Suzuki M. A Framework for Layer 3 Provider-Provisioned Virtual Private Networks (PPVPNs)[EB/OL]. (2015-10-14) [2020-03-25]. RFC 4110.

[3] Andersson L, Rosen E. Framework for Layer 2 Virtual Private Networks (L2VPNs)[EB/OL]. (2015-10-14) [2020-03-25]. RFC 4664.

[4] Kompella K, Rekhter Y. Virtual Private LAN Service (VPLS) Using BGP for Auto-Discovery and Signaling[EB/OL]. (2018-12-20) [2020-03-25]. RFC 4761.

[5] Rosen E, Davie B, Radoaca V, Luo W. Provisioning Auto-Discovery and Signaling in Layer 2 Virtual Private Networks (L2VPNs)[EB/OL]. (2015-10-14) [2020-03-25]. RFC 6074.

[6] Kothari B, Kompella K, Henderickx W, Balus F, Uttaro J. BGP Based Multihoming in Virtual Private LAN Service[EB/OL]. (2020-03-12) [2020-03-25]. draft-ietf-bess-vpls-multihoming-05.

[7] Sajassi A, Aggarwal R, Bitar N, Isaac A, Uttaro J, Drake J, Henderickx W. BGP MPLS-Based Ethernet VPN[EB/OL]. (2020-01-21) [2020-03-25]. RFC 7432.

[8] Martini L, Rosen E, El-Aawar N, Smith T, Heron G. Pseudowire Setup and Maintenance Using the Label Distribution Protocol (LDP)[EB/OL]. (2015-10-14) [2020-03-25]. RFC 4447.

[9] Lasserre M, Kompella V. Virtual Private LAN Service (VPLS) Using Label Distribution Protocol (LDP) Signaling[EB/OL]. (2020-01-21) [2020-03-25]. RFC 4762.

[10] Kompella K, Kothari B, Cherukuri R. Layer 2 Virtual Private Networks Using BGP for Auto-Discovery and Signaling[EB/OL]. (2015-10-14) [2020-03-25]. RFC 6624.

[11] Dawra G, Filsfils C, Raszuk R, Decraene B, Zhuang S, Rabadan J. SRv6 BGP Based Overlay services[EB/OL]. (2019-11-04) [2020-03-25]. draft-ietf-bess-srv6-services-01.

[12] Previdi S, Filsfils C, Lindem A, Sreekantiah A, Gredler H. Segment Routing Prefix Segment Identifier Extensions for BGP[EB/OL]. (2019-12-06) [2020-03-25]. RFC 8669.

[13] Filsfils C, Camarillo P, Leddy J, Voyer D, Matsushima S, Li Z. SRv6 Network Programming[EB/OL]. (2019-12-05) [2020-03-25]. draft-ietf-spring-srv6-network-programming-05.

[14] Rabadan J, Henderickx W, Drake J, Lin W, Sajassi A. IP Prefix Advertisement in EVPN[EB/OL]. (2018-05-29) [2020-03-25]. draft-ietf-bess-evpn-prefix-advertisement-11.
[15] Sajassi A, Thoria S, Patel K, Yeung D, Drake J, Lin W. IGMP and MLD Proxy for EVPN[EB/OL]. (2019-11-18) [2020-03-25]. draft-ietf-bess-evpn-igmp-mld-proxy-04.
[16] Boutros S, Sajassi A, Salam S, Drake J, Rabadan J. Virtual Private Wire Service Support in Ethernet VPN[EB/OL]. (2018-12-11)[2020-03-25]. RFC 8214.

CHAPTER 6

SRv6 Reliability

THIS CHAPTER DESCRIBES SRv6 RELIABILITY TECHNOLOGIES, including SRv6 Topology Independent Loop-Free Alternate (TI-LFA),[1] midpoint protection,[2] egress protection,[3] and microloop avoidance. Leveraging the programmability of SRv6 paths and the routability of SRv6 instructions, E2E local protection solutions can be designed with ease for networks that employ a distributed path computation architecture, helping to achieve E2E convergence within 50 ms. In addition, SRv6 can provide various scenario-specific microloop avoidance mechanisms to eliminate transient microloops, further improving network reliability.

6.1 IP FRR AND E2E PROTECTION

Interactive multimedia applications — such as VoIP — are extremely sensitive to packet loss, typically requiring it to be lower than dozens of milliseconds. However, network convergence usually takes hundreds or even thousands of milliseconds to complete when a network link or node fails, which cannot meet service requirements. To minimize packet loss and improve convergence performance, FRR[4] technologies can be used. With FRR, a backup path is pre-installed on relevant nodes, so that the Point of Local Repair (PLR)[1] adjacent to the failure point can quickly switch traffic to the backup path after detecting a failure.

Note: Rerouting is a technology related to FRR. With this mechanism, the ingress converges to a new path after the network topology changes in the case of a network failure. Unlike rerouting, FRR allows a backup path to be pre-computed or to be pre-configured and then pre-installed before a failure occurs. This allows transit nodes to switch traffic to the backup path directly after detecting the failure, without waiting for IGP convergence. FRR gets its name from the fact that it responds to failures faster than rerouting does.

Generally, due to the limited protection scope and application scenarios of traditional FRR technologies, multi-hop Bidirectional Forwarding Detection (BFD)[5] and E2E protection mechanisms have to be introduced to work with FRR. For example, HSB+BFD and VPN FRR+BFD can be used to achieve E2E TE protection and protect PEs against failures, respectively. Yet despite this, there are some drawbacks, for example:

- Hierarchical BFD, which implements hierarchical switching based on different BFD packet sending intervals at different layers, cannot achieve switching within 50 ms.
- For each service connection, multi-hop BFD has to be configured, which increases the complexity of service deployments.
- Network service deployment is limited by the maximum number of BFD sessions supported by devices.

SRv6 can support multiple local protection technologies that achieve E2E local protection switching within 50 ms without using multi-hop BFD and E2E protection mechanisms.

With such technologies, the node adjacent to the failure point switches traffic to a suboptimal path when a failure occurs. IGP convergence, BGP convergence, and TE path re-optimization are then performed in sequence so that the traffic is switched to the new optimal path. These local protection technologies can significantly enhance IP network reliability and expand the FRR protection scope.

6.1.1 TI-LFA Protection

Before introducing TI-LFA, we will describe some traditional FRR technologies, the earliest of which is Loop-Free Alternate (LFA),[6] followed by Remote Loop-Free Alternate (RLFA).[7]

6.1.1.1 LFA

The basic principle of LFA is to find a neighboring node that is not the primary next hop, namely, a next hop that is not on the shortest forwarding path. If the shortest path from the neighboring node to the destination node does not pass through the source node, the neighboring node is considered an LFA next hop.

In link protection scenarios, an LFA next hop can be computed based on the following inequality, where N, D, and S indicate the neighboring node, the destination node, and the source node that performs LFA computation, respectively:

```
Distance_opt(N, D) < Distance_opt(N, S) + Distance_opt(S, D)
```

Distance_opt(N, D) indicates the distance (usually measured by cost) of the shortest path from node N to node D. This inequality means that the shortest-path distance between these two nodes is shorter than the distance from node N to node S and then to node D. We can therefore conclude from this inequality that the shortest path from node N to node D does not pass through node S. Consequently, a neighboring node that complies with the inequality can function as a link-protecting node; specifically, traffic will not be sent back to node S from the neighboring node.

In node protection scenarios, if node S's primary next hop on the shortest path fails, traffic needs to bypass the primary next hop. In this case, a neighboring node that complies with the following inequality can function as a node-protecting node. Here, nodes N, D, and E indicate the neighboring node, destination node, and primary next hop, respectively.

```
Distance_opt(N, D) < Distance_opt(N, E) + Distance_opt(E, D)
```

This inequality means that the distance of the shortest path from node N to node D is shorter than the distance from node N to node E and then to node D. We can therefore conclude from this inequality that the shortest path from node N to node D does not pass through the failed primary next-hop node E. In this case, node N can function as the LFA node when a failure occurs on node E, so that traffic can be forwarded from node S (which performs LFA computation) to node N and then to node D, thereby bypassing node E.

According to the preceding two inequalities, we can see that a node-protecting LFA node can also function as a link-protecting LFA node, whereas a link-protecting LFA node may not function as a node-protecting LFA node.

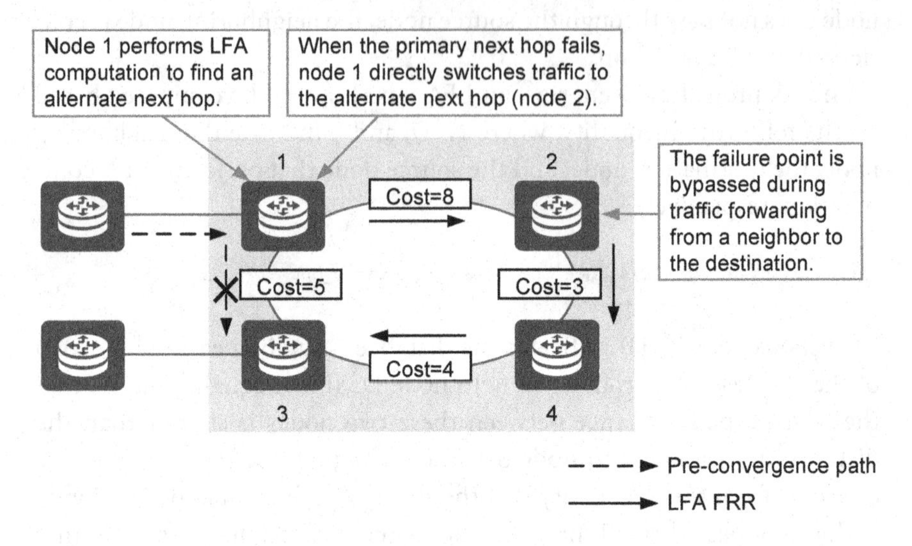

FIGURE 6.1 LFA protection.

Taking the network shown in Figure 6.1 as an example, assume that traffic is forwarded from node 1 to node 3 over the node 1→node 3 path, where node 3 is the primary next hop of node 1. LFA computation is performed to find an LFA next hop. As node 2 is the only LFA next hop available, this node is applied to the preceding two LFA inequalities for computation. The computation results show that node 2 complies with the link protection inequality and therefore meets LFA link protection requirements. So, node 1 preinstalls node 2 as the LFA next hop in its FIB. When a link failure occurs as shown in the figure, node 1 directly switches traffic to node 2 in the forwarding plane, without waiting for control plane convergence.

Despite its benefits, LFA cannot cover all scenarios. Specifically, it is unable to compute an appropriate alternate next hop in many topologies — RFC 6571 states that it can cover only 80%–90% of topologies.[8] For example, on the network shown in Figure 6.2, if the cost of the link between nodes 2 and 4 is 10, LFA protection fails.

6.1.1.2 RLFA

To extend the FRR protection scope, RLFA was proposed and defined in RFC 7490.[7]

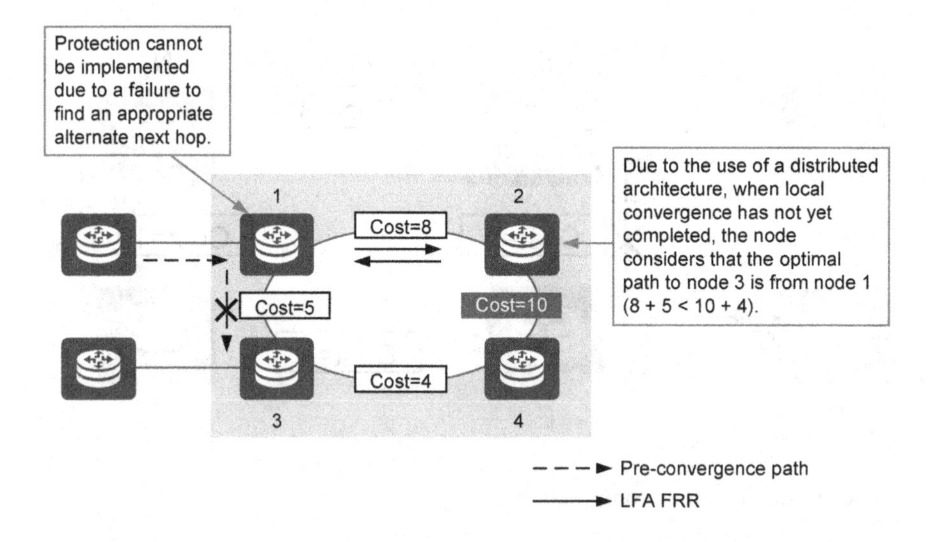

FIGURE 6.2 LFA protection failure.

As an enhancement of LFA, RLFA extends the LFA computation scope to remote nodes besides neighboring nodes, improving the success rate of LFA computation. The basic principle of RLFA is to find a remote LFA node (usually called a PQ node) that meets both of the following conditions:

- The packets sent from the source node to the PQ node are not looped back to the source node.
- The packets sent from the PQ node to the destination node are not looped back to the source node.

To illustrate how a PQ node is computed, we use the network shown in Figure 6.3 as an example. Assume that the S–E link on the network fails. In this case, the computed PQ node must meet the following requirements:

- Node S can reach the PQ node over the shortest path, and the path does not pass through the S–E link.
- The PQ node can reach node E over the shortest path, and the path does not pass through the S–E link.

In order to compute PQ nodes, the P-space and Q-space are introduced, and the intersection of the two spaces is a set of PQ nodes.[7]

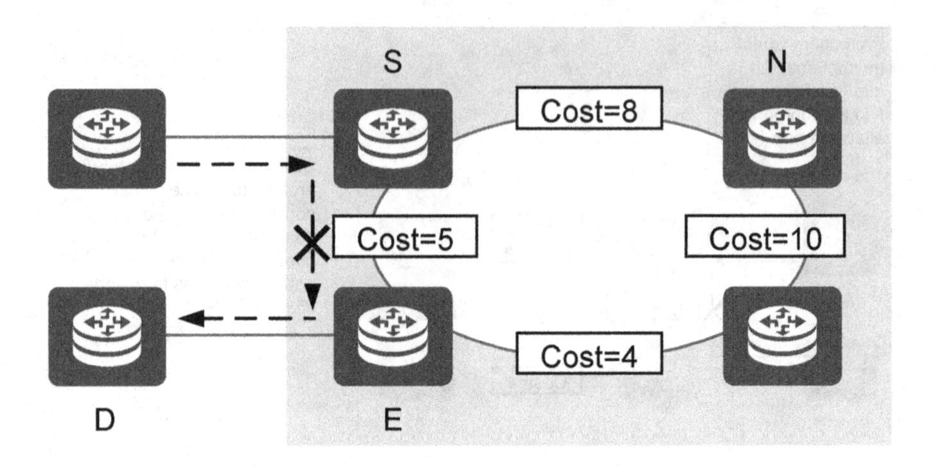

FIGURE 6.3 PQ node computation.

- P-space: As defined in RFC 7490, the P-space of a router with respect to a protected link is the set of routers reachable from that specific router using the pre-convergence shortest paths without any of those paths (including equal-cost path splits) transiting that protected link. For the network shown in Figure 6.3, the P-space of node S with respect to the protected S–E link is the set of nodes that node S can reach over the shortest path without transiting the S–E link, and it can be expressed as P (S, S–E).

  Each node in the P-space is called a P node, which can be computed using the following inequality, where N indicates the neighboring node and S indicates the source node:

```
Distance_opt(N, P) < Distance_opt(N, S) + Distance_opt(S, P)
```

  This means traffic that is forwarded over the shortest path from node N to node P is not sent to node S. That is, this path is loop-free.

- Extended P-space: The extended P-space of the protecting router with respect to the protected link is the union of the P-spaces of the neighbors in that set of neighbors with respect to the protected link.

- Q-space: The Q-space of a router with respect to a protected link is the set of routers from which that specific router can be reached without any path (including equal-cost path splits) transiting that protected link. For the network shown in Figure 6.3, the Q-space of node E with respect to the protected S–E link is the set of nodes that can reach node E over the shortest paths without transiting the S–E link, and it can be expressed as Q (E, S–E).

Each node in the Q-space is called a Q node, which can be computed using the following inequality, where D indicates the destination node involved in Shortest Path Tree (SPT) computation and S indicates the source node that performs LFA computation:

```
Distance_opt(Q, D) < Distance_opt(Q, S) + Distance_opt(S, D)
```

This means traffic that is forwarded over the shortest path from node Q to node D is not sent to node S. That is, this path is loop-free.

For the topology shown in Figure 6.4, LFA cannot compute an appropriate alternate next hop, whereas RLFA computes node 4 as a PQ node through the following process:

1. Computes the extended P-space of source node 1. Specifically, SPTs rooted at source node 1's neighbors (excluding those that are reachable only through the protected link, such as node 3) are computed. The nodes that are reachable from node 1's neighbors over the shortest path without transiting the link between nodes 1 and 3 form a P-space, and the P-spaces of all the neighbors form an extended P-space {Node 2, Node 4}.
2. Computes a reserve SPT rooted at node 3, thereby obtaining the Q-space {Node 4} based on the Q node computation inequality.
3. Computes a PQ node. Because only node 4 exists in both the extended P-space and Q-space, it is a PQ node.

As we can see in Figure 6.4, neither the path from node 2 (a neighbor of the source node) to node 4 nor the path from node 4 to destination node 3 traverses the failure point. This means that node 4 meets PQ node requirements. After computing PQ node 4, RLFA establishes a tunnel (e.g., an LDP tunnel) between nodes 1 and 4, with node 2 as the next hop. Node 1 then preinstalls the tunnel as a virtual LFA next hop in its FIB. This allows node 1 to switch traffic quickly to the next hop if the primary next hop fails, thereby achieving FRR protection.

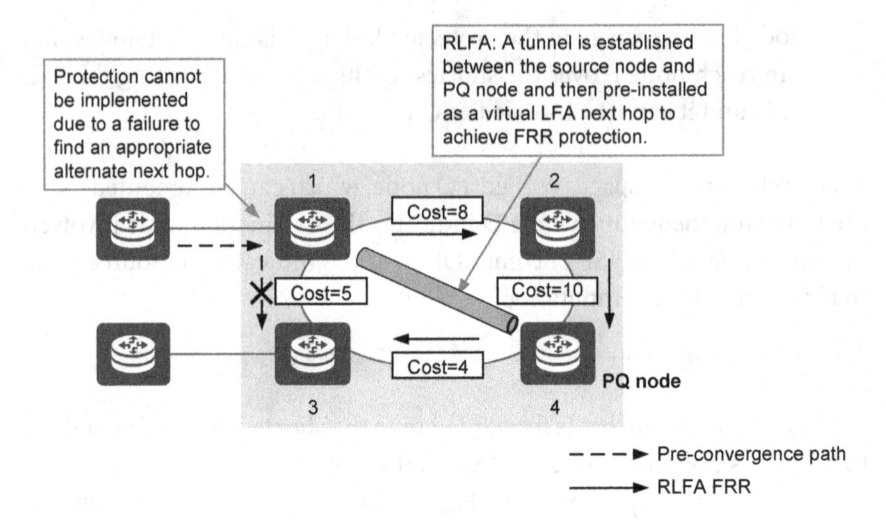

FIGURE 6.4 RLFA protection.

As an enhancement of LFA, RLFA supports more scenarios than LFA. However, it is still limited by topologies because it requires the presence of a PQ node on the involved network. For example, let's change the cost of the node 4→node 3 link to 30, as shown in Figure 6.5. As this cost is now higher than that of the node 4→node 2→node 1→node 3 link (10+8+5=23), traffic from node 4 to node 3 needs to pass through the

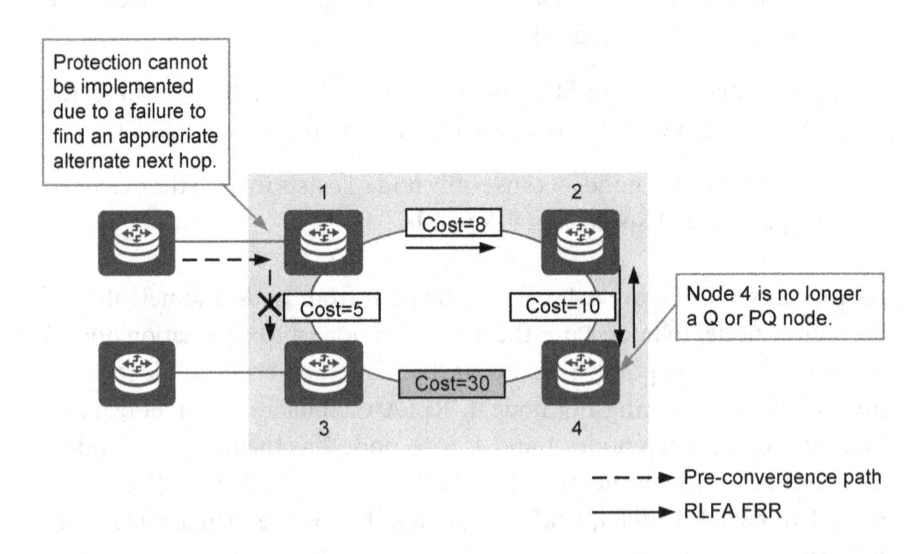

FIGURE 6.5 RLFA protection failure.

TABLE 6.1 Success Rates of LFA and RLFA in Link and Node Protection

Topology No.	LFA Link Protection (%)	RLFA Link Protection (%)	LFA Node Protection (%)	RLFA Node Protection (%)
1	78.5	99.7	36.9	53.3
2	97.3	97.5	52.4	52.4
3	99.3	99.999	58	58.4
4	83.1	99	63.1	74.8
5	99	99.5	59.1	59.5
6	86.4	100	21.4	34.9
7	93.9	99.999	35.4	40.6
8	95.3	99.5	48.1	50.2
9	82.2	99.5	49.5	55
10	98.5	99.6	14.9	14.1
11	99.6	99.9	24.8	24.9
12	99.5	99.999	62.4	62.8
13	92.4	97.5	51.6	54.6
14	99.3	100	48.6	48.6

protected link. In this case, node 4 is no longer a Q node, and the P-space and Q-space do not intersect. As a result, RLFA protection fails.

RFC 7490 compares LFA and RLFA in terms of their success rates in protecting against link and node failures according to the topologies in typical scenarios, as listed in Table 6.1.[7] Although it comes close, RLFA is still unable to achieve 100% link protection.

If no PQ node is found using RLFA, 100% link and node protection cannot be achieved. In addition, RLFA requires targeted LDP sessions[7] to be created, which consequently introduces a large number of tunnel states to be maintained. Because the backup path computation algorithm used by RLFA (and LFA) is not SPF, the computed backup path may not be the shortest one. When this is the case, traffic needs to be switched from the backup path to the shortest path after convergence is complete.

6.1.1.3 TI-LFA

Neither LFA nor RLFA is able to achieve 100% failure protection, and both are limited by topologies. So, is there any protection technology that can achieve 100% failure protection without topology limitations? The answer is yes: TI-LFA.[1]

TI-LFA — an FRR protection mechanism — is an SR-based enhancement of LFA. It has the following characteristics:

- TI-LFA can provide protection in any topology, regardless of whether a PQ node exists. During TI-LFA backup path computation, if a PQ node exists, a repair segment list is inserted on the PLR to direct packet forwarding to the PQ node. If no PQ node exists, a repair segment list between the PLR's furthest P node and nearest Q node on the backup path is computed to direct packet forwarding from the P-space to the Q-space. In this way, a loop-free backup path can be computed in any topology.
- TI-LFA computes an FRR backup path based on the post-convergence topology. This ensures that the FRR backup path is the same as the post-convergence rerouting path, eliminating the need to switch the traffic forwarding path again. Specifically, the FRR backup path is computed by excluding the failed node or link.
- Because SRv6 supports repair segment list insertion on the PLR to explicitly specify a forwarding path, SRv6 TI-LFA does not require the creation of targeted LDP sessions, reducing the number of tunnel states that nodes need to maintain.

To describe how TI-LFA provides protection when no PQ node exists on the network, let's use an example topology that cannot be protected using RLFA, as shown in Figure 6.6.

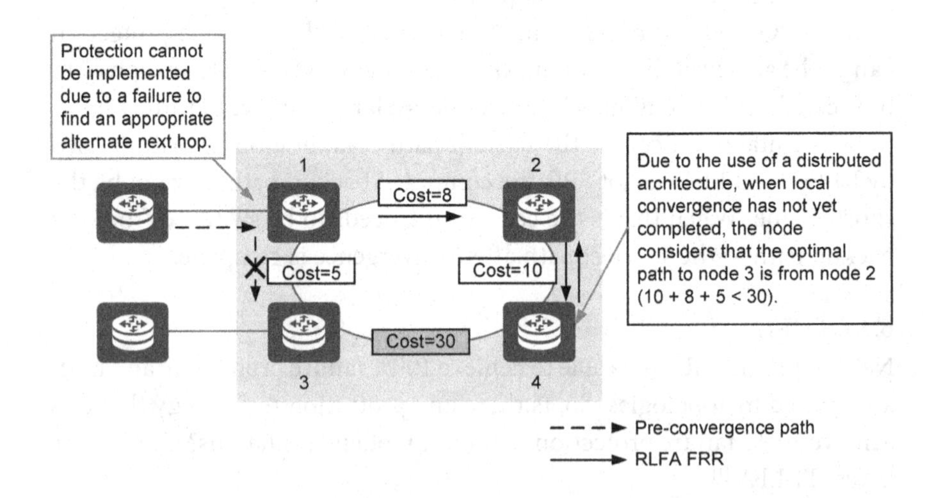

FIGURE 6.6 Topology where RLFA cannot provide protection.

TI-LFA first assumes that the link between nodes 1 and 3 fails, and then computes the post-convergence shortest path: node 1→node 2→node 4→node 3. According to the P-space and Q-space computation methods, node 4 is a P node rather than a Q node, and only node 3 exists in the Q-space. This means that there is no PQ node. FRR is achieved if traffic can be forwarded from the P-space to the Q-space, so the key is how to forward traffic from node 4 in the P-space to node 3 in the Q-space. SRv6 TI-LFA can compute a repair segment list from node 4 to node 3 and use it to direct packet forwarding, thereby achieving FRR protection in this topology. In this example, an SRv6 End.X SID can be used to instruct the network to forward traffic from node 4 to node 3.

Figure 6.7 illustrates how SRv6 TI-LFA works. In this figure, the shortest path from node A to node F is A→B→E→F, and node B needs to compute a backup path to node F as follows:

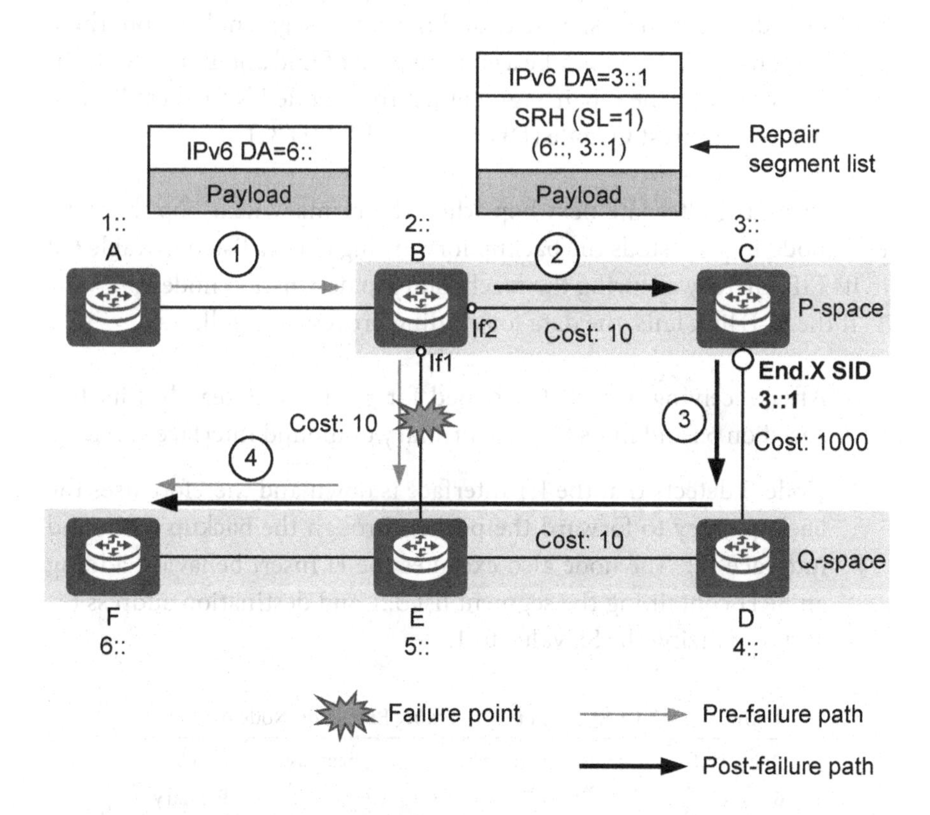

FIGURE 6.7 SRv6 TI-LFA protection.

1. Excludes the primary next hop (B–E link) and computes the post-convergence shortest path B→C→D→E→F.

2. Computes the P-space. In this case, nodes B and C are in the P-space.

3. Computes the Q-space. In this case, nodes D, E, and F are in the Q-space.

4. Computes a backup path expressed using a repair segment list. Any path can be represented using the source node→P node→Q node→destination node format. The path from the source node to the P node and the path from the Q node to the destination node are both loop-free. If a PQ node exists, traffic can be directly forwarded to the PQ node, and the entire path is loop-free. In this case, the repair segment list can be composed of the PQ node's End SID. If no PQ node exists, a loop-free forwarding path from the P node to the Q node needs to be specified, and the repair segment list from the P node to the Q node may be a combination of End and End.X SIDs. In this example, the repair segment list from node B's furthest P node (C) to its nearest Q node (D) can be End.X SID 3::1.

To activate the alternate next hop when the primary next hop fails, the PLR (node B) preinstalls the backup forwarding entries listed in Table 6.2 in its FIB, thereby ensuring the reachability to destination node F.

If the B–E link fails, the data forwarding process is as follows:

1. After receiving a packet destined for 6::, node B searches its FIB based on 6:: and finds that the primary outbound interface is If1.

2. Node B detects that the If1 interface is down and therefore uses the backup entry to forward the packet through the backup outbound interface If2. The node also executes the H.Insert behavior: adding an SRH containing the segment list 3::1 and destination address 6::, and initializing the SL value to 1.

TABLE 6.2 TI-LFA Backup Forwarding Entries on Node B

Route Prefix	Outbound Interface	Segment List	Role
6::	If1	-	Primary
	If2	3::1	Backup

3. After receiving the packet sent from node B, node C finds that 3::1 is an End.X SID and therefore executes the instruction corresponding to the End.X SID. Specifically, node C decrements the SL value by 1, updates the outer IPv6 address to 6::, and forwards the packet to node D along the C–D link based on the outbound interface and next hop that are bound to 3::1. Because the SL value is now 0, node C can remove the SRH as instructed by the PSP flavor.

4. After receiving the packet, node D searches its IPv6 routing table based on the destination address 6:: and forwards the packet to destination node F over the shortest path.

In conclusion, in addition to 100% topology protection, TI-LFA offers the following benefits:

- In most cases, the TI-LFA backup path is consistent with the post-convergence shortest path due to TI-LFA performing computation based on the post-convergence shortest path. This reduces the number of forwarding path switchovers.
- TI-LFA backup path computation depends on IGP SRv6, reducing the number of protocols introduced for reliability technology.
- TI-LFA uses existing End, End.X, or a combination of both SIDs to establish a backup path, without needing to maintain additional forwarding states.

6.1.2 SRv6 Midpoint Protection

In SRv6 TE scenarios, we usually need to constrain the forwarding path of data packets by specifying the nodes or links they need to traverse. For example, on the network shown in Figure 6.8, node A is required to send a packet to node F over a forwarding path that passes through node E. Because the backup path also needs to pass through node E even if this node fails, TI-LFA cannot provide protection in this case.

To address this issue, SRv6 midpoint protection is introduced. As we described in Chapter 2, endpoint nodes need to decrement the SL value by 1 and copy the next SID to the DA field in the IPv6 header during SRv6 packet processing. If an endpoint node fails, a forwarding failure occurs because this node cannot execute the instruction bound to the corresponding SID.

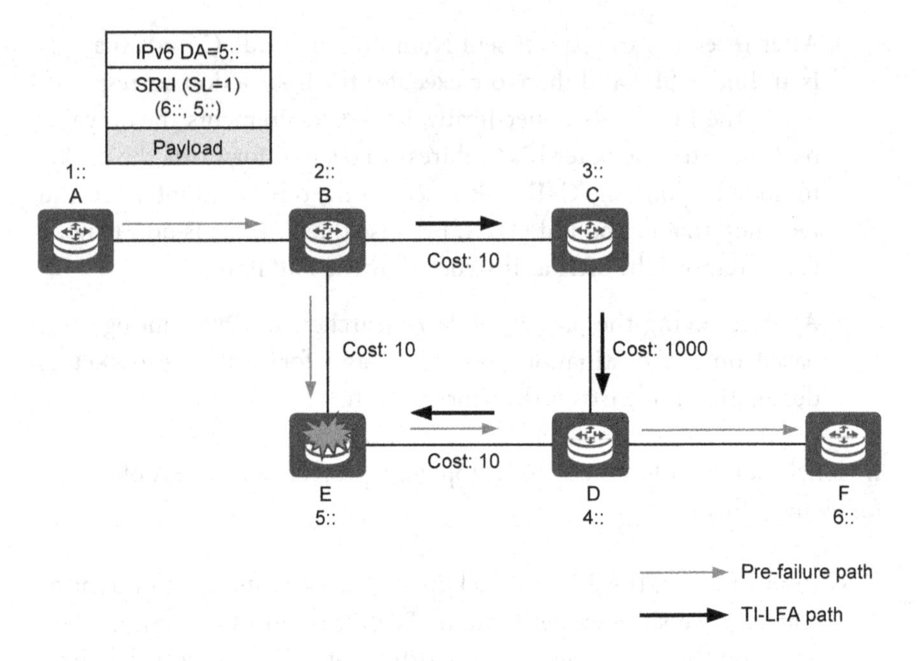

FIGURE 6.8 TI-LFA protection failure.

With SRv6 midpoint protection, a proxy forwarding node (a node upstream to the failed endpoint) takes over from the failed endpoint to complete the forwarding. Specifically, when the proxy forwarding node detects that the next-hop interface of the packet fails, the next-hop address is the destination address of the packet, and the SL value is greater than 0, the proxy forwarding node performs the End behavior on behalf of the endpoint. The behavior involves decrementing the SL value by 1, copying the next SID to the DA field in the outer IPv6 header, and then forwarding the packet according to the instruction bound to the SID. In this way, the failed endpoint is bypassed, achieving SRv6 midpoint protection.

Taking the network shown in Figure 6.9 as an example, the following describes the SRv6 midpoint protection process:

1. Node A forwards a packet to destination node F, with an SRv6 SRH instructing the packet to pass through node E.

2. If node E fails, node B detects that the next-hop interface of the received packet fails. It also learns that the next-hop address is the current destination address 5:: of the packet, and the SL value is

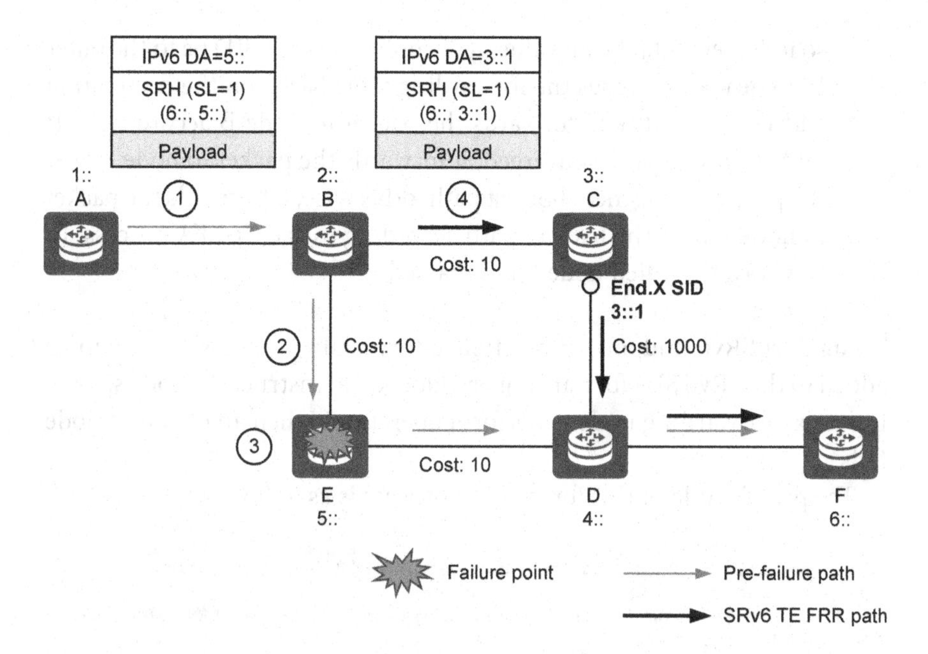

FIGURE 6.9 SRv6 midpoint protection.

greater than 0. In this case, node B functions as a proxy forwarding node, decrementing the SL value by 1 and copying the next SID 6:: to the DA field in the outer IPv6 header. Because the SL value is now 0, node B can remove the SRH and then search the forwarding table to forward the packet based on the destination address 6::.

3. The primary next hop to the destination address 6:: is still node E. Because the SL value is 0 or no SRH exists, node B can no longer perform proxy forwarding. Instead, it switches the packet to a backup path according to the normal TI-LFA process, with the repair segment list on the backup path as <3::1>. As such, node B executes the H.Insert behavior: encapsulating the segment list <3::1> into the packet, adding an SRH, and then forwarding the packet to node F over the backup path.

4. After node A detects that node E fails and that IGP convergence is completed, it deletes the forwarding entry towards node E. Consequently, when node A searches the forwarding table based on 5::, it cannot find any matching route. In this case, node A needs to function as a proxy forwarding node to perform proxy forwarding,

so it decrements the SL value by 1, copies the next SID 6:: to the outer IPv6 header, searches the forwarding table based on the destination address 6::, and then forwards the packet to node B accordingly. If node B has already converged, it forwards the packet to node F over the post-convergence shortest path; otherwise, it forwards the packet to node F over the backup path according to the TI-LFA process. In this way, the failed node E is bypassed.

To support SRv6 midpoint protection, a forwarding process[2] needs to be added to the SRv6 SID forwarding pseudocode to instruct the corresponding proxy forwarding node to perform operations when an endpoint node fails.

The pseudocode of this forwarding process is as follows:

```
IF the primary outbound interface used to forward the packet failed
 IF NH = SRH && SL != 0,
   IF the failed endpoint is directly connected to the PLR THEN //Note 1:
SRv6 midpoint protection
     SL--; update the IPv6 DA with SRH[SL];
     FIB lookup on the updated DA;
     forward the packet according to the matched entry;
  ELSE
     forward the packet according to the backup nexthop;    //Note 2: SRv6
TI-LFA protection
ELSE // There is no FIB entry for forwarding the packet
  IF NH = SRH && SL != 0 THEN
 SL--; update the IPv6 DA with SRH[SL];
 FIB lookup on the updated DA;
 forward the packet according to the matched entry;
  ELSE
 drop the packet;
ELSE
 forward accordingly to the matched entry;
```

It may be difficult at first to understand the difference between TI-LFA and midpoint protection. In essence, the difference lies in whether the next-hop node is a transit node or an endpoint node.

On the network shown in Figure 6.10, node A sends a packet carrying the segment list <E, F>. Because TI-LFA computes a backup path based on the destination address of the packet, the backup path passes through node E. If node E fails, TI-LFA cannot provide protection. In contrast, in an SRv6 midpoint protection scenario, a backup path is computed based on the next SID to be processed. This ensures that the failed endpoint E is bypassed, thereby implementing SRv6 TE midpoint protection.

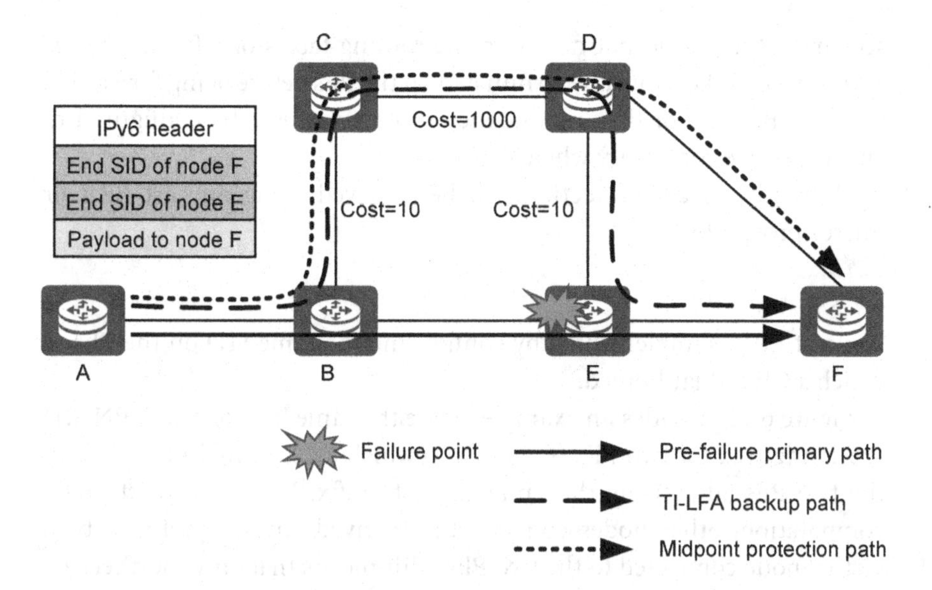

FIGURE 6.10 Difference between TI-LFA and midpoint protection.

6.1.3 Egress Protection

On an SRv6 network, nodes on each forwarding path have their own roles, such as source nodes, transit nodes, endpoint nodes, and egress nodes. In previous sections, we already introduced the protection technologies for transit and endpoint nodes. In this section, let's take a look at the FRR technology used for egress protection.

Figure 6.11 shows an SRv6 L3VPN scenario where node A is an ingress PE, and nodes F and G are egress PEs to which a CE is dual-homed. In this

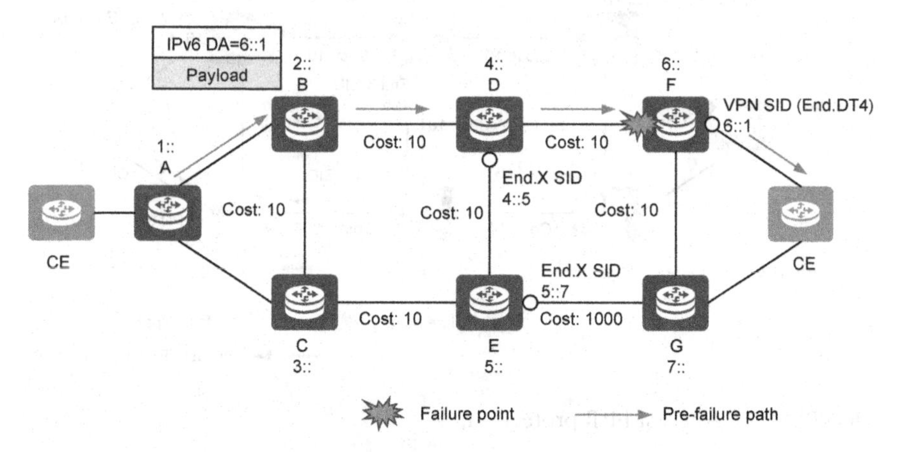

FIGURE 6.11 SRv6 L3VPN protection.

scenario, P nodes do not maintain the routing tables of VPN instances, and service packets must be terminated on the PEs before being forwarded to a CE. Because of this, redundancy protection must be configured to ensure service continuity when a PE fails.

Currently, egress protection can be achieved using anycast FRR or mirror protection.

6.1.3.1 Anycast FRR

Anycast FRR is implemented by configuring the same SID on the PEs to which a CE is dual-homed.

Figure 6.12 provides an example where the same locator and VPN SID are configured for two PEs (F and G) to which a CE is dual-homed, with the two PEs advertising the same anycast prefix. During anycast prefix computation, other nodes consider the received anycast prefix to be a virtual node connected to the two PEs. This means that only the alternate next hop to the virtual node needs to be computed, and it is the same as the alternate next hop to the anycast prefix. In this example, node D pre-installs a backup path destined for 6::, forming primary and backup FRR entries. The primary and backup paths are D→F and D→E→G, respectively. The computed repair segment list is <5::7>.

When node F is working properly, node D forwards the received packet to node F, which then executes the instruction bound to the configured

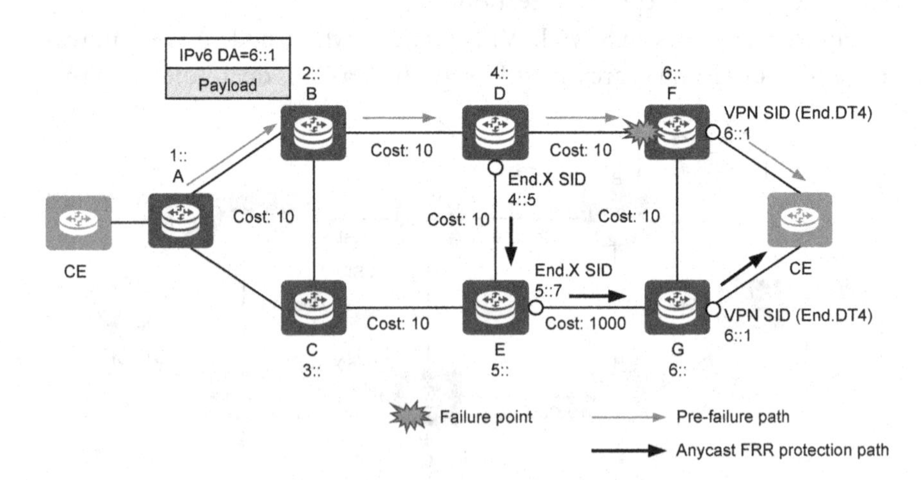

FIGURE 6.12 Anycast FRR protection.

VPN SID and forwards the packet to the connected CE through a VPN interface. If node F fails, however, node D performs a fast switchover to the backup path after detecting the failure, encapsulates the repair segment list <5::7> into the packet, and then forwards the packet to node G. Because nodes G and F are configured with the same VPN SID, node G is also capable of processing the VPN SID 6::1. As such, node G searches its Local SID Table for VPN SID 6::1, executes the instruction bound to the VPN SID, and then forwards the packet to the connected CE through a VPN interface accordingly.

Although anycast FRR can provide protection against PE failures, it has the following drawbacks:

- For two PEs configured with the same VPN instance, anycast FRR requires the same VPN SID to be configured. To achieve this, VPN SIDs must be manually configured, which could result in a lot of configuration work if a large number of VPN instances exist.
- To implement anycast FRR, two PEs need to advertise routes with the same prefix. However, only IGP route selection (not VPN route selection) can be performed. For example, if VPN services need to be load-balanced between nodes F and G or the route advertised by node F needs to be preferentially selected, VPN route selection cannot be performed if the route advertised by node G is preferentially selected through an IGP on the path to 6::.
- If there is a VPN interface failure, such as a failure on the link between node F and its connected CE, traffic is still forwarded to node F and then to node G, resulting in a traffic loop that cannot be eliminated.
- In a scenario where CEs are multi-homed to different PE groups, different locators need to be planned, complicating deployment.

6.1.3.2 Mirror Protection

As an alternative to anycast FRR, mirror protection[3] is recommended to achieve egress protection. If a CE is dual-homed to PEs, we can consider that the two PEs provide the same VPN forwarding service. Therefore, the PEs can be configured as a mirror group, and the corresponding mirror

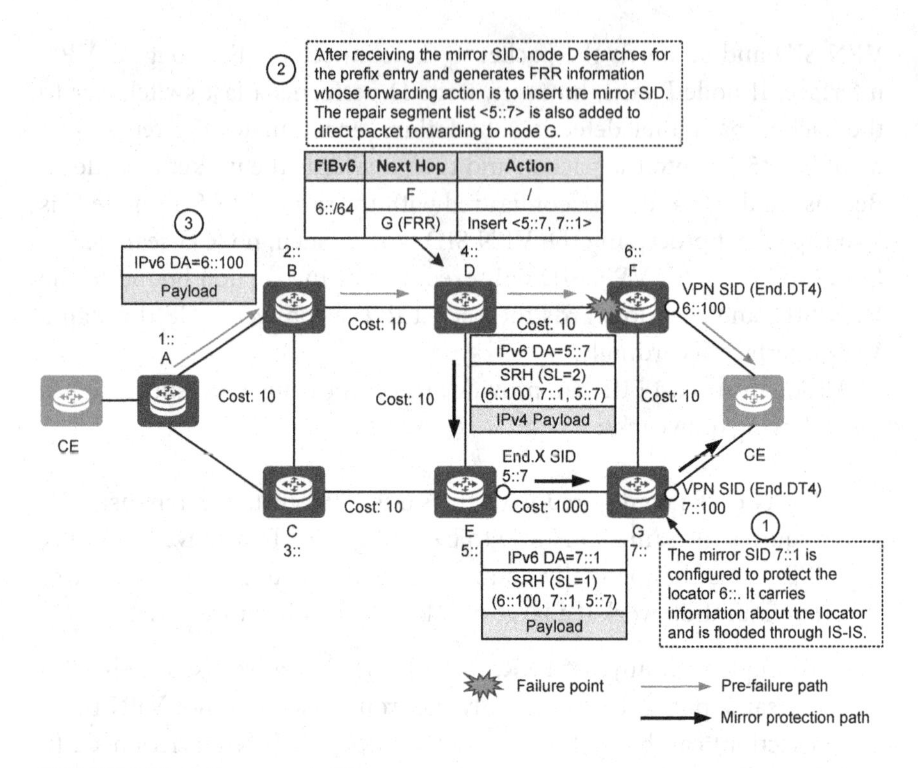

FIGURE 6.13 Mirror protection.

relationship can then be propagated throughout the network using an IGP. Then, if one of the nodes in the mirror group fails, traffic can be forwarded to the other node in the mirror group through FRR, achieving fast convergence.

Figure 6.13 shows an SRv6 L3VPN scenario where a CE is dual-homed to nodes F and G. The locator 6::/64 and VPN SID 6::100 are configured for node F, and the locator 7::/64 and VPN SID 7::100 for node G. On node G, the mirror SID 7::1 is configured for the locator 6::/64 of node F. In this case, after receiving a VPN route advertised by node F, node G generates the mirror entry <7::1, 6::100>:<7::100> for the VPN SIDs of the two PEs based on the mirror SID. This mirror entry indicates that <7::1, 6::100> is equivalent to 7::100.

After being configured with a mirror SID, node G advertises a Mirror SID TLV carrying protected locators to propagate the mirror SID using an IGP. Figure 6.14 shows the format of the IS-IS-defined Mirror SID TLV.[3]

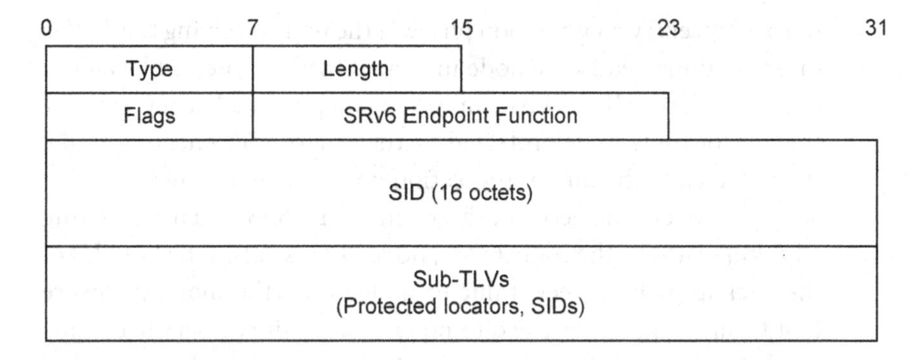

FIGURE 6.14 Format of the Mirror SID TLV.

TABLE 6.3 Fields in the Mirror SID TLV

Field	Length	Description
Type	8 bits	TLV type.
Length	8 bits	TLV length.
Flags	8 bits	Flags. No flags are currently defined.
SRv6 Endpoint Function	16 bits	SRv6 endpoint function type of the mirror SID.
SID	16 octets	SRv6 mirror SID to be advertised.
Sub-TLVs (variable)	Variable	Sub-TLVs in the Mirror SID TLV used to encapsulate the locator information of the node to be protected by the mirror SID.

Table 6.3 describes the fields in the Mirror SID TLV.

The Mirror SID TLV is defined as a sub-TLV of the SRv6 Locator TLV. On top of that, the forwarding behavior of the End.M SID is also defined for egress protection. The pseudocode of this forwarding behavior is as follows:

```
End.M: Mirror protection
When N receives a packet destined to S and S is a local End.M SID,
N does:
IF NH=SRH and SL = 1 ;;
 SL--
 Map to a local VPN SID based on Mirror SID and SRH[SL] ;;
 forward according to the local VPN SID ;;
ELSE
 drop the packet
```

After an IGP node receives the TLV, the following operations are performed:

In the backup entry computation phase, if the node receiving the TLV is directly connected to the node indicated by the protected locator, it uses the mirror SID as a backup when computing a backup path for the locator route to be protected by the mirror SID, enabling traffic to be steered to the mirror protection node (where the mirror SID is configured). For the network shown in Figure 6.13, when computing a backup path for the route 6::/64, node D uses 7::1 as the last SID of the backup path to steer traffic to node G. Furthermore, to ensure that traffic can be forwarded to node G, a repair segment list <5::7> needs to be added. Thus, the complete repair segment list computed by node D is <5::7, 7::1>.

In the data forwarding phase, generally, node A functions as an ingress PE and preferentially selects the route advertised by node F according to the cost of the IGP route to the next hop. Therefore, when forwarding a packet, node A sets the destination address in the packet to the VPN SID 6::100 and then forwards the packet to node F. After receiving the packet, node F searches the routing table of the VPN instance associated with 6::100 based on the inner destination IP address, and forwards the packet to the corresponding CE accordingly.

If node F fails, node D activates the backup path after detecting the failure and then encapsulates the repair segment list <5::7, 7::1> into the packet to direct packet forwarding to node G over a loop-free path. Upon receipt, node G searches its Local SID Table based on 7::1 and finds a matching mirror SID. According to the instruction bound to the mirror SID, node G searches the mapping table based on the next SID 6::100 and finds that the SID is mapped to the local SID 7::100. Then, based on the IP address in the inner packet, node G searches the routing table of the local VPN instance associated with 7::100 and forwards the packet to the corresponding CE.

Mirror protection has the following advantages over anycast FRR:

- In mirror protection, node protection pairs are configured to achieve egress protection, and mapping entries are automatically created based on BGP synchronization information. As such, VPN SIDs can be dynamically generated, without needing to configure the same

SID for identical VPN instances on the two PEs to which a CE is dual-homed.

- Mirror protection does not require the two PEs to be configured with the same locator, making VPN route selection possible.
- If a PE's VPN interface fails, the ingress can detect the failure and then leverage the mirror protection mechanism to switch traffic to the other PE through VPN FRR. With anycast FRR, a loop will occur in such a scenario, and convergence cannot be implemented.

Although mirror protection can solve some problems of anycast FRR, some other problems are shared. For example, protection groups also need to be manually planned, and scenarios where CEs are multi-homed to different PE groups are not supported.

In this section, we covered FRR and E2E protection technologies applicable to an SRv6 network, including TI-LFA, midpoint protection, and egress protection. In the following section, we will introduce solutions to the microloop problem that may occur during the process from failure occurrence to recovery.

6.2 MICROLOOP AVOIDANCE

This section describes microloops that form when failures occur and then are rectified on an SRv6 network, as well as how such microloops can be removed.

6.2.1 Microloop Cause

Due to the distributed nature of IGP LSDBs on IP networks, devices may converge at different times after a failure occurs, presenting the risk of microloops. Take the link state of IS-IS/OSPF as an example. Each time the network topology changes, relevant devices need to update their FIBs based on the new topology and then converge accordingly. The issue here is that, because the devices converge at different times, their LSDBs are asynchronous for a short period of time. The period may be a matter of milliseconds or seconds, depending on specific device conditions, such as the device capability, parameter configurations, and service scale. During this period, the states of devices along the involved packet forwarding path may be different. More specifically, some of the devices may be in the pre-convergence state, whereas others are in the post-convergence state. In this case, forwarding routes may become inconsistent, leading to microloops which disappear after all the devices along the forwarding path have converged. This type of transient loop is called a microloop. Microloops may result in a series of issues, including packet loss, jitter, and out-of-order packets, and as such must be avoided.

As shown in Figure 6.15, node A sends a packet to node D over the A→B→E→D path. If the B–E link fails, node B converges first, causing node G to become the next hop. In this case, if node G has not yet converged, its next hop to node D remains node B. This results in a microloop forming between nodes B and G, and this microloop lasts for a period of time.

Microloops may occur with any topology change, for example, a link or node goes up or down, or the metric value of a link changes. As such, microloops are very common on IP networks. To better eliminate them, the industry had already carried out a lot of relevant research before SR emerged and developed relevant technologies, such as ordered FIB[9] and ordered metric. The basic principle of all these technologies is to control the sequence in which network nodes converge (e.g., allowing the node

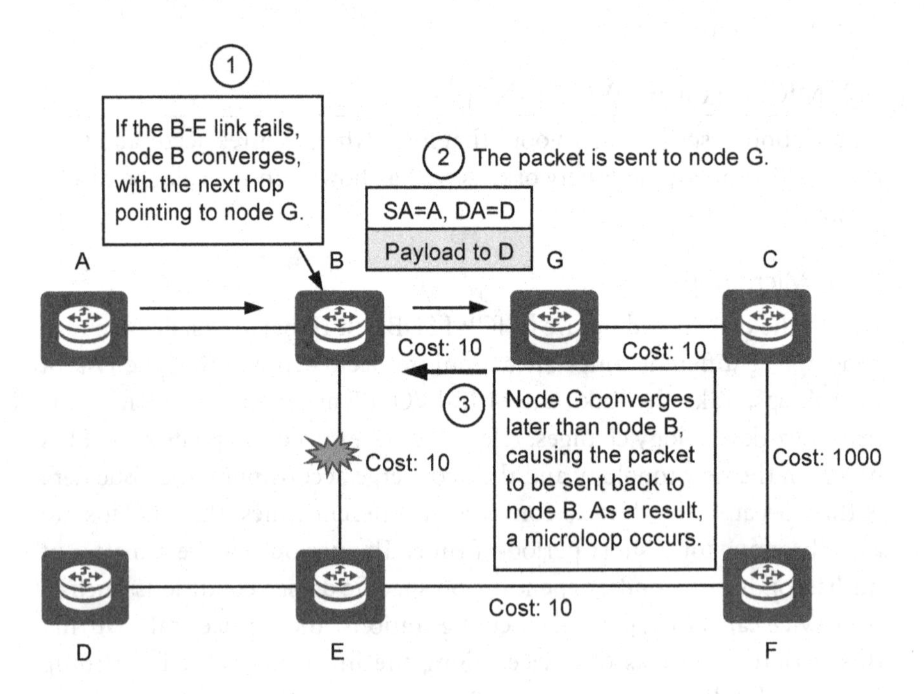

FIGURE 6.15 Microloop.

farthest from the failure point to converge first and the nearest to converge last in a switchover scenario; or allowing the node nearest to the failure point to converge first and the farthest to converge last in a switchback scenario). With this method, network nodes can converge in the correct sequence to eliminate loops. The downside of this method is that it complicates the convergence process of the entire network and prolongs the overall convergence time. For these reasons, it is not used on live networks.

Note: Typically, switchover refers to the process in which traffic is switched from the primary path to the backup path after a network failure occurs; switchback refers to the process in which traffic is switched back from the backup path to the post-convergence primary path after the network failure is rectified.

SR provides a method to help avoid potential loops while minimizing impacts on the network. Specifically, if a network topology change may cause a loop, SR first allows network nodes to insert loop-free SRv6 segment lists to steer traffic to the destination address. Then normal traffic forwarding is restored only after all the involved network nodes converge. This can effectively avoid loops on the network.

6.2.2 SRv6 Local Microloop Avoidance in a Traffic Switchover Scenario

In a traffic switchover scenario, a local microloop may be formed when a node adjacent to the failed node converges earlier than the other nodes on the network. On the network shown in Figure 6.16, SRv6 TI-LFA is deployed on all nodes. If node B fails, node A undergoes the following process to perform convergence for the route to node C:

1. Node A detects the failure and enters the TI-LFA FRR process, during which node A inserts the SRv6 repair segment list <5::1> into the packet to direct the packet to the TI-LFA-computed PQ node (node E). So the packet is forwarded to the next-hop node D first and the SIDs encapsulated into the packet are <5::1, 3::1>.

2. After performing route convergence, node A searches for the route to node C and forwards the packet to the next-hop node D through the route. In this case, the packet does not carry any SRv6 repair segment list and is directly forwarded based on the destination address 3::1.

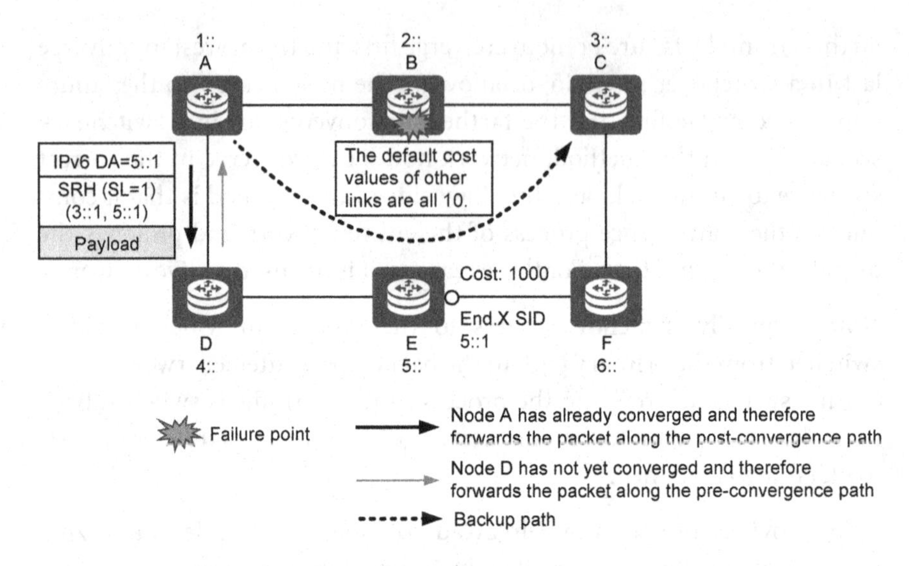

FIGURE 6.16 SRv6 local microloop in a traffic switchover scenario.

3. If node D has not yet converged when receiving the packet, node A is considered as the next hop of the route from node D to node C. As a result, node D forwards the packet back to node A, which in turn causes a microloop between the two nodes.

According to the preceding convergence process, we can see that the microloop occurs when node A converges, quits the TI-LFA FRR process, and then implements normal forwarding before other nodes on the network converge. The issue is that node A converges earlier than the other nodes, so by postponing its convergence, the microloop can be avoided. As TI-LFA backup paths are loop-free, the packet can be forwarded along a TI-LFA backup path for a period of time. Node A can then wait for the other nodes to complete convergence before quitting the TI-LFA FRR process and performing convergence, thereby avoiding the microloop.

After microloop avoidance is deployed on the network shown in Figure 6.17, the convergence process is as follows:

1. After node A detects the failure, it enters the TI-LFA FRR process, encapsulating the SRv6 repair segment list <5::1> into the packet and forwarding the packet along the TI-LFA backup path, with node D as the next hop.

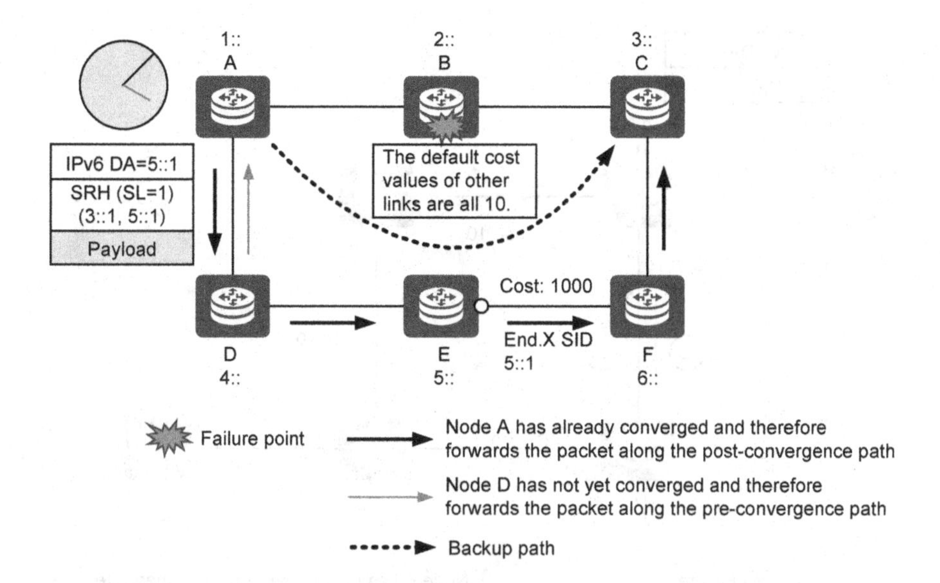

FIGURE 6.17 SRv6 local microloop avoidance in a traffic switchover scenario.

2. Node A starts the timer T1. During T1, node A does not respond to topology changes, the forwarding table remains unchanged, and the TI-LFA backup path continues to be used for packet forwarding. Other nodes on the network converge properly.

3. When T1 expires, other nodes on the network have already converged. Node A can now converge and quit the TI-LFA FRR process to forward the packet along a post-convergence path.

The preceding solution can protect only a PLR against microloops in a traffic switchover scenario. This is because only a PLR can enter the TI-LFA FRR process and forward packets along a TI-LFA backup path. In addition, this solution applies only to single point of failure scenarios. In a multiple points of failure scenario, the TI-LFA backup path may also be adversely affected and therefore cannot be used for packet forwarding.

6.2.3 SRv6 Microloop Avoidance in a Traffic Switchback Scenario

Besides traffic switchover scenarios, microloops may also occur in traffic switchback scenarios. The following uses the network shown in Figure 6.18 as an example to describe how a microloop occurs in a traffic switchback scenario. The process is as follows:

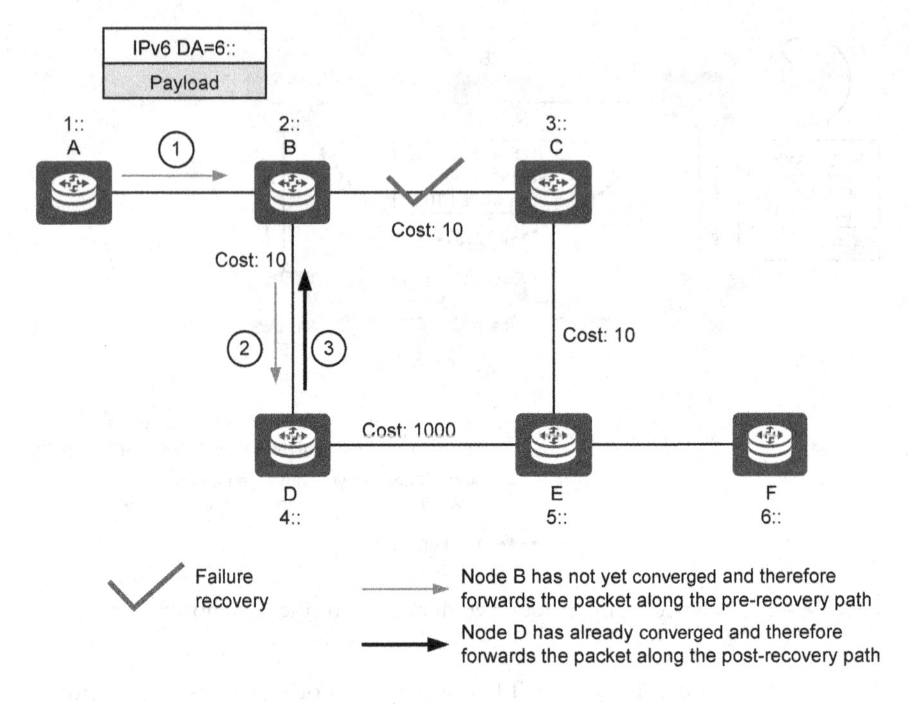

FIGURE 6.18 SRv6 microloop in a traffic switchback scenario.

1. Node A sends a packet to destination node F along the path A→B→C→E→F. If the B–C link fails, node A sends the packet to destination node F along the post-convergence path A→B→D→E→F.
2. After the B–C link recovers, a node (for example, node D) first converges.
3. When receiving the packet sent from node A, node B has not yet converged and therefore still forwards the packet to node D along the pre-recovery path.
4. Because node D has already converged, it forwards the packet to node B along the post-recovery path, resulting in a microloop between the two nodes.

Traffic switchback does not involve the TI-LFA FRR process. Therefore, delayed convergence cannot be used for microloop avoidance in such scenarios.

According to the preceding process, we can see that a transient loop occurs when node D converges earlier than node B during recovery. Node D is unable to predict link up events on the network and so is unable to precompute any loop-free path for such events. To avoid loops that may occur during traffic switchback, node D needs to be able to converge to a loop-free path.

On the network shown in Figure 6.19, after node D detects that the B–C link goes up, it computes the D→B→C→E→F path to destination node F.

Since the B–C link up event does not affect the path from node D to node B, it can be proved that the path is loop-free. So, to construct the loop-free path from node D to node F, it is not necessary to specify the path from node D to node B.

NOTICE

Topology changes triggered by a link up event affect only the post-convergence forwarding path that passes through the link. As such, if the post-convergence forwarding path from node D to node B does not pass through the B–C link, it is not affected by the B–C link up event. Similarly, topology changes triggered by a link down event affect only the pre-convergence forwarding path that passes through the link.

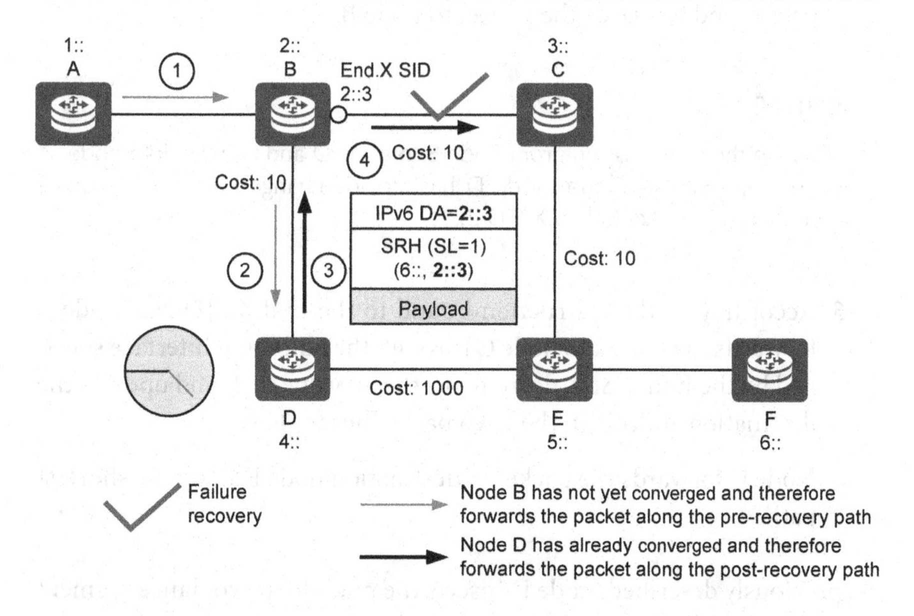

FIGURE 6.19 SRv6 microloop avoidance in a traffic switchback scenario.

Similarly, because the path from node C to node F is not affected by the B–C link up event, it is definitely loop-free. In this scenario, only the path from node B to node C is affected. Given this, to compute the loop-free path from node D to node F, only a path from node B to node C needs to be specified. According to the preceding analysis, a loop-free path from node D to node F can be formed by inserting only an End.X SID that instructs packet forwarding from node B to node C into the post-convergence path of node D.

After microloop avoidance is deployed, the convergence process is as follows:

1. After the B–C link recovers, a node (e.g., node D) first converges.
2. Node D starts the timer T1. It computes a microloop avoidance segment list <2::3>, which is used for the packet destined for node F before T1 expires.
3. When receiving the packet sent from node A, node B has not yet converged and therefore still forwards the packet to node D along the pre-recovery path.
4. Node D inserts the microloop avoidance segment list <2::3> into the packet and forwards the packet to node B.

NOTICE

Although the packet is sent from node B to node D and then back to node B, no loop occurs because node D has already changed the destination address of the packet to End.X SID 2::3.

5. According to the instruction bound to the End.X SID 2::3, node B forwards the packet to node C through the outbound interface specified by the End.X SID, decrements the SL value by 1, and updates the destination address in the IPv6 packet header to 6::.
6. Node C forwards the packet to destination node F along the shortest path.

As previously described, node D inserts the microloop avoidance segment list <2::3> into the packet, avoiding loops.

When T1 of node D expires, other nodes on the network have already converged, allowing node A to forward the packet along the post-convergence path A→B→C→E→F.

6.2.4 SRv6 Remote Microloop Avoidance in a Traffic Switchover Scenario

In a traffic switchover scenario, a remote microloop may also occur between two nodes on a packet forwarding path if the node close to the failure point converges earlier than one farther from the point. The following uses the network shown in Figure 6.20 as an example to describe how a remote microloop occurs in a traffic switchover scenario. The process is as follows:

1. After detecting a C–E link failure, node G first converges, whereas node B has not yet converged.
2. Nodes A and B forward the packet to node G along the path used before the failure occurs.
3. Because node G has already converged, it forwards the packet to node B according to the next hop of the corresponding route, resulting in a microloop between the two nodes.

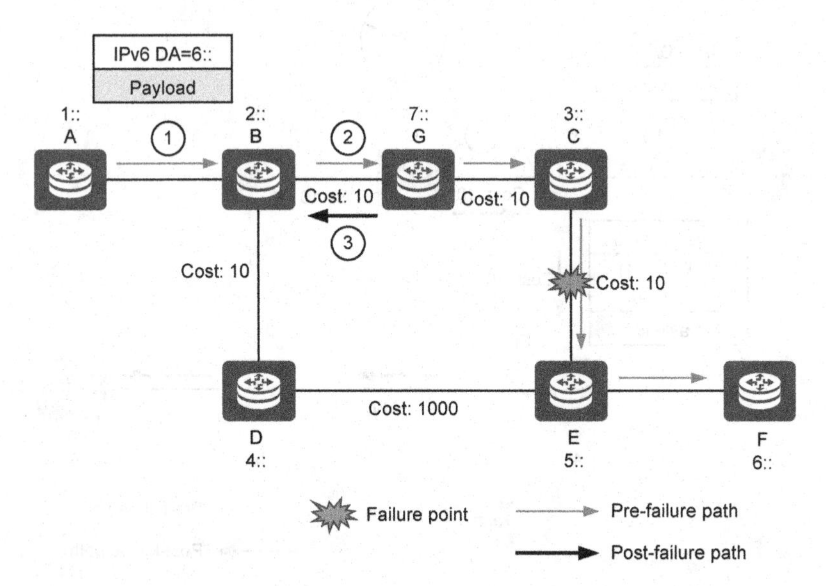

FIGURE 6.20 SRv6 remote microloop in a traffic switchover scenario.

To minimize computation workload, a network node typically precomputes a loop-free path only when a directly connected link or node fails. That is, no loop-free path can be precomputed against any other potential failure on the network. Given this, the microloop can be avoided only by installing a loop-free path after node G converges.

As mentioned in the previous section, topology changes triggered by a link down event affect only the pre-convergence forwarding path that passes through the link. Therefore, if the path from a node to the destination node does not pass through the failed link before convergence, it is absolutely not affected by the link failure. According to the topology shown in Figure 6.20, we can see that the path from node G to node D is not affected by the C–E link failure. Therefore, this path does not need to be specified for computing a loop-free path from node G to node F. Similarly, the path from node E to node F is not affected by the C–E link failure and therefore does not need to be specified, either. Because only the path from node D to node E is affected by the C–E link failure, only the End.X SID 4::5 needs to be specified to identify the path from node D to node E to determine the loop-free path, as shown in Figure 6.21.

After microloop avoidance is deployed, the convergence process is as follows:

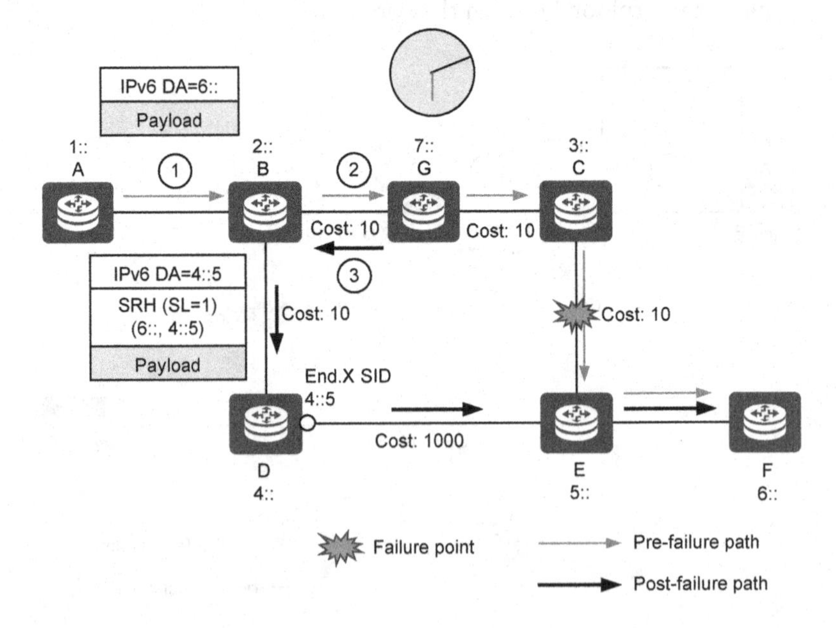

FIGURE 6.21 SRv6 remote microloop avoidance in a traffic switchover scenario.

1. After detecting a C–E link failure, node G first converges.
2. Node G starts the timer T1. It computes a microloop avoidance segment list <4::5>, which is used for the packet destined for node F before T1 expires.
3. When receiving the packet sent from node A, node B has not yet converged and therefore still forwards the packet to node G along the pre-recovery path.
4. Node G inserts the microloop avoidance segment list <4::5> into the packet and forwards the packet to node B.

NOTICE

Although the packet is sent from node B to node G and then back to node B, no loop occurs because node G has already changed the destination address of the packet to End.X SID 4::5.

5. Node B searches the routing table for the route to destination address 4::5 and forwards the packet to node D through the route.
6. According to the instruction bound to the End.X SID 4::5, node D forwards the packet to node E through the outbound interface specified by the End.X SID, decrements the SL value by 1, and updates the destination address in the IPv6 packet header to 6::.
7. Node E forwards the packet to destination node F along the shortest path.

As previously described, node G inserts the microloop avoidance segment list <4::5> into the packet, avoiding loops.

When T1 of node G expires, other nodes on the network have already converged, allowing node A to forward the packet along the post-convergence path A→B→D→E→F.

6.3 STORIES BEHIND SRv6 DESIGN

We established FRR targeting 100% IP network coverage as a key research area in 2010 but could not make any breakthrough in determining loop-free backup paths. Later, I encountered an expert in the industry who gave me a hint: "Don't forget that when the topology of the network is determined, it is impossible to find a backup path in all scenarios using mathematics." Pondering these words for a long time, I tried to figure out: what would be the alternative if there was no way with mathematics? Finally, I understood that the expert was hinting to me that the issue can only be solved by using the engineering method to create new possible paths. As such, we started relevant research into achieving this using multi-topology LDP.

There have been multiple engineering solutions for FRR targeting 100% network coverage, including:

- Multi-topology LDP combined with Maximally Redundant Trees (MRT-FRR)[10]
- LFA series: LFA, RLFA, and SR-based TI-LFA
- LDP path protection using RSVP-TE LSPs[11]

In essence, searching for backup paths in IP FRR is a traffic engineering issue. Also, solutions 2 and 3 both use TE paths to protect the shortest paths. In fact, there should have been a chance for traditional MPLS to solve the issue well. In the past, LDP has shown it can support LSP setup over the shortest path and MPLS TE LSP setup through Constraint-based Routing Label Distribution Protocol (CR-LDP) extensions. So, with a single protocol, LDP LSPs can be protected using the MPLS TE LSPs. Moreover, CR-LDP is more scalable than RSVP-TE and can be used to establish numerous MPLS TE LSPs. Leveraging these advantages, traditional MPLS could have provided a perfect solution. However, the IETF MPLS Working Group decided in the early years to focus on RSVP-TE extensions for MPLS TE and to make no new efforts in CR-LDP.[12] This caused MPLS to be divided into LDP and RSVP-TE. Against the backdrop of protocol simplification, IGP-based SR replaces LDP and RSVP-TE, and SR TI-LFA ends up as the winner of the FRR solutions.

REFERENCES

[1] Litkowski S, Bashandy A, Filsfils C, Decraene B, Francois P, Voyer D, Clad F, Camarillo P. Topology Independent Fast Reroute using Segment Routing[EB/OL]. (2019-09-06)[2020-03-25]. draft-ietf-rtgwg-segment-routing-ti-lfa-01.

[2] Chen H, Hu Z, Chen H. SRv6 Proxy Forwarding[EB/OL]. (2020-03-16)[2020-03-25] draft-chen-rtgwg-srv6-midpoint-protection-01.
[3] Hu Z, Chen H, Chen H, Wu P. SRv6 Path Egress Protection[EB/OL]. (2020-03-18)[2020-03-25]. draft-ietf-rtgwg-srv6-egress-protection-00.
[4] Shand M, Bryant S. IP Fast Reroute Framework[EB/OL]. (2015-10-14)[2020-03-25]. RFC 5714.
[5] Katz D, Ward D. Bidirectional Forwarding Detection (BFD)[EB/OL]. (2020-01-21)[2020-03-25]. RFC 5880.
[6] Atlas A, Zinin A. Basic Specification for IP Fast Reroute: Loop-Free Alternates[EB/OL]. (2020-01-21)[2020-03-25]. RFC 5286.
[7] Bryant S, Filsfils C, Previdi S, Shand M, So N. Remote Loop-Free Alternate (LFA) Fast Reroute (FRR)[EB/OL]. (2015-10-14)[2020-03-25]. RFC 7490.
[8] Filsfils C, Francois P, Shand M, Decraene B, Uttaro J, Leymann N, Horneffer M. Loop-Free Alternate (LFA) Applicability in Service Provider (SP) Networks[EB/OL]. (2015-10-14)[2020-03-25]. RFC 6571.
[9] Shand M, Bryant S, Previdi S, Filsfils C, Francois P, Bonaventure O. Framework for Loop-Free Convergence Using the Ordered Forwarding Information Base (oFIB) Approach[EB/OL]. (2018-12-20)[2020-03-25]. RFC 6976.
[10] Atlas A, Bowers C, Enyedi G. An Architecture for IP/LDP Fast Reroute Using Maximally Redundant Trees (MRT-FRR)[EB/OL]. (2016-06-30)[2020-03-25]. RFC 7812.
[11] Esale S, Torvi R, Fang L, et al. Fast Reroute for Node Protection in LDP-based LSPs[EB/OL]. (2017-03-13)[2020-03-25]. draft-esale-mpls-ldp-node-frr-05.
[12] Andersson L, Swallow G. The Multiprotocol Label Switching (MPLS) Working Group decision on MPLS signaling protocols[EB/OL]. (2003-02-28)[2020-03-25]. RFC 3468.

CHAPTER 7

SRv6 Network Evolution

COMPATIBILITY AND SMOOTH EVOLUTION CAPABILITY are key to the success of new IP technologies. In light of this, compatibility with live networks was fully taken into account when SRv6 was designed. This guarantees incremental deployment and smooth evolution by being capable of upgrading some nodes on demand on an IPv6 or SR-MPLS network. In this chapter, we describe the challenges encountered during SRv6 network evolution, how incremental SRv6 deployment and SRv6 smooth evolution are performed, how SRv6 is compatible with legacy devices, and how SRv6 network security is ensured.

7.1 CHALLENGES FACED BY SRv6 NETWORK EVOLUTION

Typically, the following challenges are encountered during evolution to SRv6:

1. Network devices must support IPv6.
2. SRv6 must be compatible with legacy devices.
3. Security challenges arise.

7.1.1 Network Upgrade to Support IPv6

Networks with IPv6 capabilities are a prerequisite of supporting SRv6. Although IPv4/MPLS networks are more prevalent, IPv6 is becoming increasingly popular with the development and maturity of IP technologies. This popularity is also increasing due to the joint efforts of carriers, content providers, and equipment vendors. According to Google's statistics,[1] the percentage of users who access the Internet over IPv6 was approximately 30% in 2019, and IPv6 adoption continued to accelerate. It is therefore foreseeable that the all-IPv6 network is getting closer to us.

Actually, most devices on live networks can support the IPv4/IPv6 dual-stack, laying a solid foundation for SRv6 evolution.

7.1.2 SRv6 Compatibility with Legacy Devices

Compatibility is the key to the success of new IP protocols, so does SRv6. This is because the investment in upgrading a network is high, especially when network devices have to be replaced. Examples of this include upgrading from IPv4 to IPv6, and from SR-MPLS to SRv6.

To support SRv6, the software and hardware of network devices must be upgraded. The software upgrading involves supporting SRv6 control plane protocols, such as IGP and BGP protocol extensions. Its impact on live networks can be under control.

In contrast, it is considerably challenging to upgrade hardware to support SRv6. To put it into perspective, SRv6 introduces more semantics to IPv6 addresses, SRH processing, and encoding of TLVs in SRHs in the forwarding plane. These technologies bring various network programming capabilities while imposing higher requirements on forwarding processing capabilities of network devices. To support high-performance packet processing and forwarding, SRHs and TLVs are generally processed using hardware. In addition, to support traffic engineering, SFC, and IOAM, hardware devices must be capable of processing more SRv6 SIDs. All these factors pose challenges to live network devices. Figure 7.1 illustrates preliminarily assessed requirements on the SRv6 SID stack processing capability of network devices.

From the above figure, we can see that the SRH is not required in the scenario of L3VPN over SRv6 BE. To support the scenario of L3VPN over SRv6 BE+TI-LFA, a device should be capable of processing the SRv6 SID stack with up to four SIDs. To support L3VPN over SRv6 TE, a device is required to support the processing of more SIDs, even up to ten SIDs. To support SRv6 SFC and IOAM, a higher SRv6 SID stack processing capability is required.

7.1.3 Security Challenges Faced by SRv6

Network security is linked to everyone's interests and is one of the topics that carriers are most concerned about. Security must be considered when

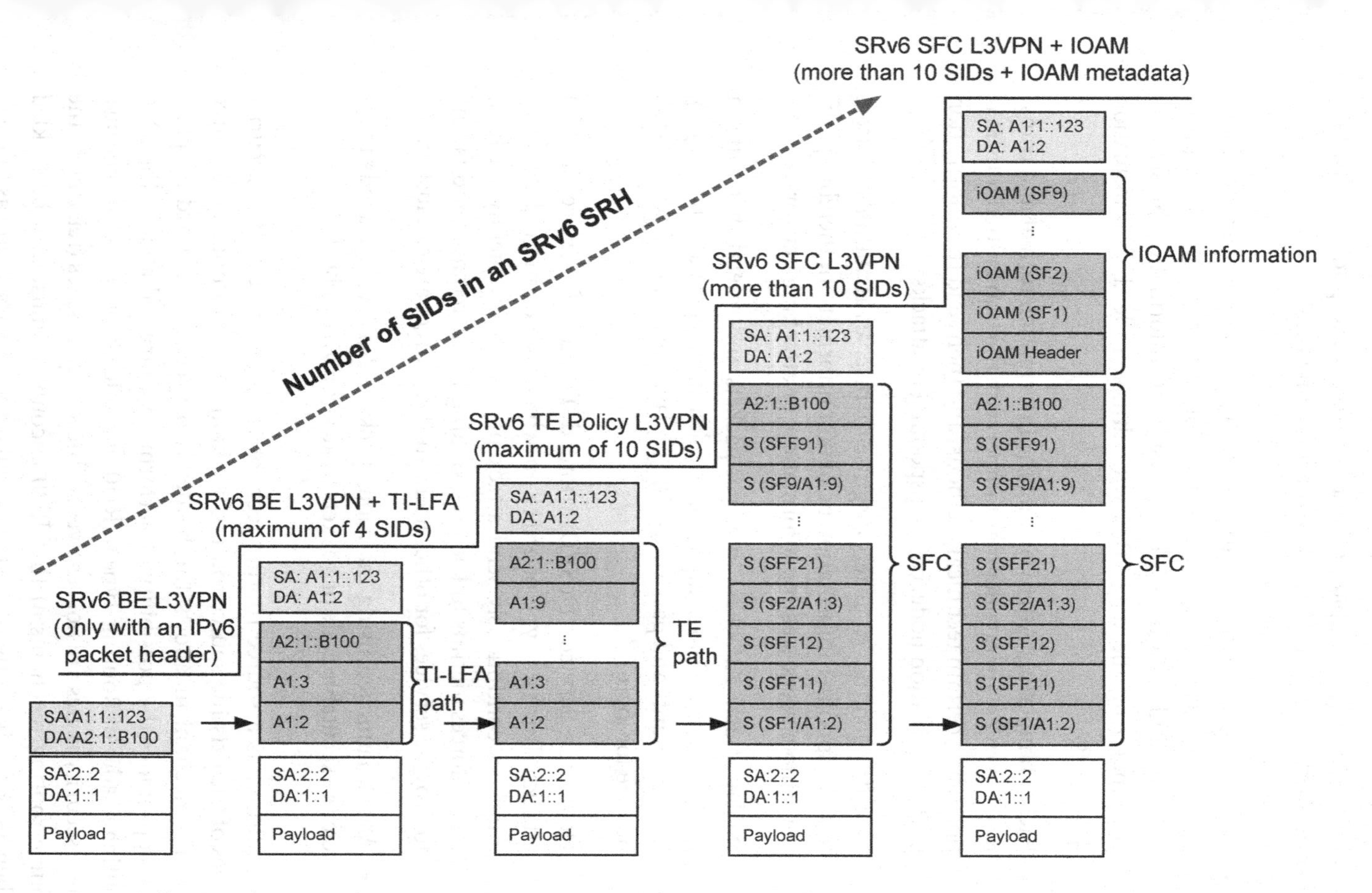

FIGURE 7.1 Preliminarily assessed requirements on the SRv6 SID stack processing capability of network devices.

a new IP technology is deployed, and SRv6 is no exception to this. Like other networks, SRv6 networks face the following threats[2]:

- Eavesdropping: If packets are not encrypted, an intruder may intercept transmitted data on the devices or links through which packets pass, and then obtain the transmitted information by stealing and analyzing the data. This will ultimately lead to information disclosure.
- Packet tampering: After learning the information format and transmission rules, an intruder may tamper with the transmitted information using certain technical methods, and then send the information to the destination node, causing spoofing or attacks.
- Identity spoofing: Attackers disguise themselves as authorized users to send spoofing information or obtain information, thereby spoofing remote users, stealing information, or attacking networks.
- Replay attack: Attackers intercept packets and resend them after a period of time. For example, an attacker intercepts packets carrying login accounts and passwords, and replays the packets after a period of time for login.
- Denial of Service (DoS)/Distributed Denial of Service (DDoS) attacks: This is a type of attack in which one or more distributed attackers continuously send attack packets. Subsequently, such an attack forces victims to keep responding to the requests and failing to provide services for other users after resources are exhausted.
- Attack launched using malicious packets: An attacker sends malicious executable code or packets to attack a victim host.

In addition, as a source routing technology, SRv6 inherits the security issues of the existing source routing mechanism. For example, the source routing mechanism allows the headend to specify a forwarding path, which provides the possibility of performing a targeted attack. The IPv6 routing header of Routing Type 0 (RH0) is made obsolete mainly because these security issues are not resolved.[3] The security issues (relating to the source routing mechanism) that RH0 encounters must also be tackled when SRv6 SRHs are used. These particular security issues are as follows:

- Remote network discovery: Information about nodes along the shortest path is obtained using existing tracing technologies. Then, a specified node address can be added to RH0 in a packet, allowing the packet to be forwarded along a specified path and then other network nodes are discovered. The process repeats, ultimately helping an attacker discover the networkwide topology.
- Bypassing a firewall: An attacker can bypass a firewall by exploiting the vulnerability in a 5-tuple-based-only matching policy. For example, to bypass a firewall that discards packets originating from node A and destined for node D, an attacker only needs to insert node B's address into RH0 for the packets to pass through the firewall and then reach node B. After the packets arrive at node B, the attacker can then change the destination address to node D's address, ultimately bypassing the firewall.
- DoS attack: Because RH0 allows duplicate addresses to be inserted, if multiple groups of duplicate SIDs are inserted into the RH0, forwarding data back and forth on the network amplifies attacks and ultimately exhausts device resources, leading to DoS attacks. For example, if 50 pairs of node A's and node B's addresses are inserted into RH0, 50 times the amount of attack traffic can be created between node A and node B.

These are the major security issues induced by RH0, and the common security issues of source routing. As such, these issues must be resolved when SRv6 SRHs are used, and this requires us to design the corresponding security solution.

7.2 INCREMENTAL DEPLOYMENT FOR SRv6 NETWORKS

Although some challenges are faced during evolution to SRv6, compatibility was fully taken into account at the beginning of the design process to ensure that live networks smoothly evolve to SRv6.

SRv6 is constructed based on IPv6. From the encapsulation perspective, without SRHs, an SRv6 packet that uses an SRv6 SID as the destination address is the same as a regular IPv6 packet. SRHs do not need to be introduced if L3VPN services are transmitted over SRv6 BE paths. For VPN services, if Ps already support IPv6 forwarding, only PEs need to be upgraded to support SRv6 VPN service.

Hardware devices with higher processing capabilities are needed only when more advanced SRv6 functions (such as L3VPN over SRv6 TE, SFC, and IOAM) are to be introduced. As network technologies improve and SRv6 gains traction, devices with higher capabilities will emerge, and the issues of hardware processing capabilities will be resolved.

Since SRv6 is compatible with IPv6, some nodes can be upgraded on demand to support SRv6, without the need to upgrade all the nodes on the network. The IPv6 nodes that are not upgraded to support SRv6 are transparent and agnostic to the traffic that originates from or is destined for the nodes that have already been upgraded to support SRv6.

7.2.1 SRv6 Evolution Paths

Figure 7.2 shows two typical SRv6 evolution paths. Path 1 is for evolution from IPv4/MPLS to SRv6 networks; path 2 is for evolution from IPv4/MPLS to SR-MPLS and then to SRv6 networks.

Path 1: Evolving IPv4/MPLS networks to SRv6 networks directly. The key to achieving this lies in upgrading an IPv4 network to an IPv6 network. If a network already supports the IPv4/IPv6 dual-stack, this evolution path will be easy and simple to deploy. Once a network has been upgraded to IPv6, in the initial phase, SRv6 can be easily deployed by introducing SRv6 capabilities to related network nodes. As for highlights, on-demand incremental upgrade is one of the main advantages of SRv6. For example, for the scenario of L3VPN over SRv6 BE, SRv6 is deployed only on edge nodes, and transit devices just need to support IPv6 forwarding, rather than being upgraded to SRv6.

Moreover, to support more advanced network functions (e.g., SRv6 TE), only some key nodes such as Autonomous System Boundary Routers

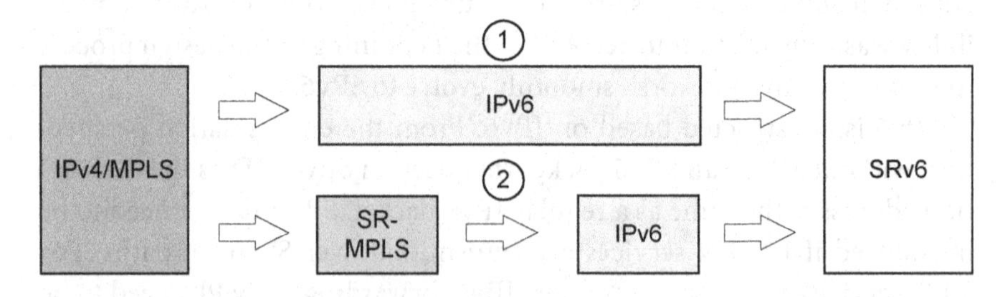

FIGURE 7.2 Typical SRv6 evolution paths.

(ASBRs) and PEs need to be upgraded to SRv6. In other words, network-wide or E2E upgrade does not need to be performed. All these benefits are brought about by the SRv6 smooth evolution capability.

From the network evolution perspective, there is a consensus across the industry that IPv6 is the future of IP networks. Taking this into account, SRv6 evolution path 1 is recommended.

Path 2: Evolving IPv4/MPLS networks to SR-MPLS networks and then to SRv6 networks. Currently, SR-MPLS outshines SRv6 in terms of standards maturity and industry readiness. In addition, most live networks are IP/MPLS networks, and carriers have abundant experience in IP/MPLS network O&M. In view of this, some carriers may choose to evolve IPv4/MPLS networks to SR-MPLS networks, and then to SRv6 networks, provided that SRv6 grows in popularity.

Evolution from SR-MPLS to SRv6 also requires a network upgrade to support IPv6. When IPv6 is ready, networks can be gradually and smoothly upgraded to support SRv6. That being said, the networks have to support both SR-MPLS and SRv6 in the long term during evolution. In addition to support for both SR-MPLS and SRv6, some scenarios involve SRv6 and SR-MPLS interworking, meaning dedicated protocols and standards must be defined. The outcome of this is that the complexity of network protocols, devices, and network O&M increases accordingly.

From the network evolution perspective, there is a consensus across the industry that IPv6 is the future of IP networks. Within the context of the industry currently advocating for simplified networks, it can be affirmed that evolution path 2 is more complex than evolution path 1 (direct evolution from IPv4/MPLS networks to SRv6 networks). Taking these into account, evolution path 1 is recommended.

7.2.2 SRv6 Deployment Process

Incremental deployment is the key to smooth SRv6 evolution. In this section, we provide a typical SRv6 deployment roadmap according to the past experience which has been adopted in SRv6 deployment practices.

First, perform an upgrade to support IPv6: An IP transport network is generally upgraded to support the IPv4/IPv6 dual-stack, as shown in Figure 7.3.

Second, upgrade PEs on demand to support SRv6 features, such as SRv6 VPN, as shown in Figure 7.4.

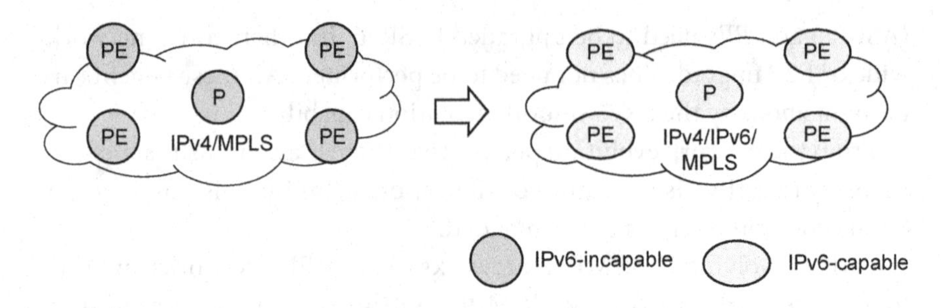

FIGURE 7.3 Upgrade to IPv6.

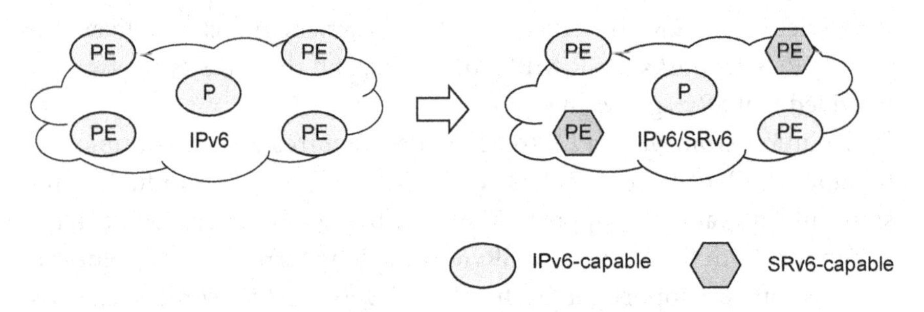

FIGURE 7.4 On-demand upgrade of edge devices.

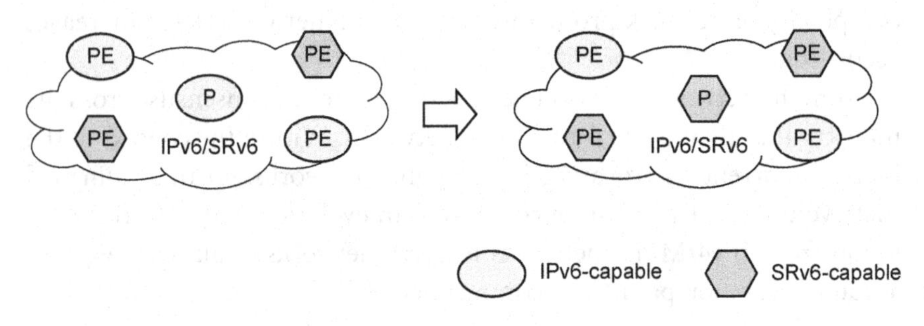

FIGURE 7.5 On-demand upgrade of transit devices.

Third, upgrade some Ps on demand to support SRv6. In this way, SRv6 TE paths can be established to implement path optimization or advanced network functions, such as SFC, on the network shown in Figure 7.5.

Ultimately, upgrade the other nodes to support SRv6, implementing an E2E SRv6 network. After this process, the network programming capability of SRv6 is fully utilized. Figure 7.6 shows E2E support for SRv6 on the network.

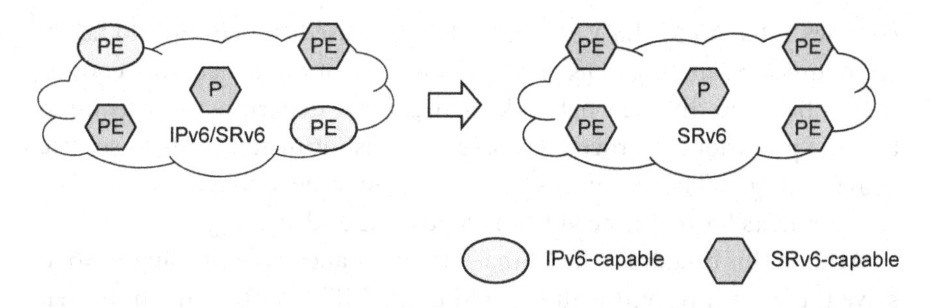

FIGURE 7.6 E2E support for SRv6.

7.2.3 SRv6 Evolution Practices

SRv6 has been deployed on the networks of multiple carriers around the world, with China Telecom and China Unicom being among the first to introduce it. In this section, we describe the SRv6 deployment practices of these two carriers to ease the understanding of SRv6's smooth evolution capability.

7.2.3.1 SRv6 Deployment Practice of China Telecom

Figure 7.7 shows SRv6 deployment in a video backhaul project of a China Telecom branch. In this project, SRv6 has been introduced to transmit videos collected by municipal DCs in a province to the provincial DC in real time and dynamically steer video data in all these DCs. For example, the provincial DC pushes video data from one or more municipal Internet Data Centers (IDCs) to other municipal IDCs as needed.

As such, a solution that can rapidly set up, modify, or delete VPN connections is needed. Traditionally, the MPLS VPN solution that involves

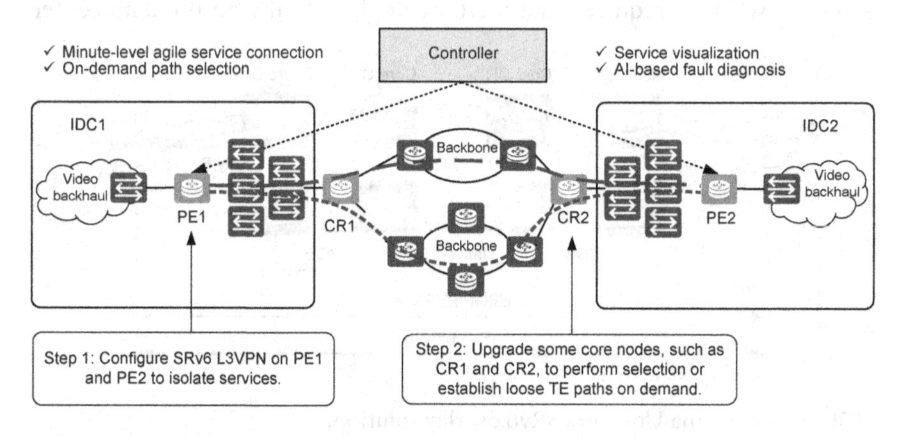

FIGURE 7.7 Video backhaul project of a China Telecom branch.

complex inter-AS MPLS VPN technologies is used, which leads to heavy configurations and high O&M costs. Not to mention the fact that coordination between different network management departments is required and the provisioning process is complex. This ultimately leads to service provisioning that takes months to complete, making it difficult to meet the requirements for real-time video data pushing and steering.

Against this backdrop, the China Telecom branch aims at using SRv6 to solve the issues involved with the traditional MPLS VPN solution. In that regard, since China Telecom's ChinaNet and CN2 networks are already IPv6 capable, to support SRv6-based VPN, it only needs to introduce the SRv6 L3VPN function on egress devices (such as PE1 and PE2 in Figure 7.7) of each data center. With the help of a controller, services can be fast provisioned in a matter of minutes. The advantage of this is the fact that transit devices on the network do not need to be aware of such changes, and coordination between different network management departments is avoided. If SRv6 is introduced on some transit nodes, such as ChinaNet/CN2 edge nodes (CR1 and CR2), loose TE paths can be supported, and the paths can be selected on demand to steer traffic through the CN2 or ChinaNet network based on service requirements, facilitating network path optimization.

7.2.3.2 SRv6 Deployment Practice of China Unicom

Drawing on the SRv6 technology, China Unicom has successfully built a cross-domain cloud private line network across the China169 backbone network from city A to city B. This network can provide flexible and fast cross-province private line access services for enterprise users.

As shown in Figure 7.8, China Unicom has adopted the SRv6 overlay solution, where it requires that SRv6 be deployed only on the data center

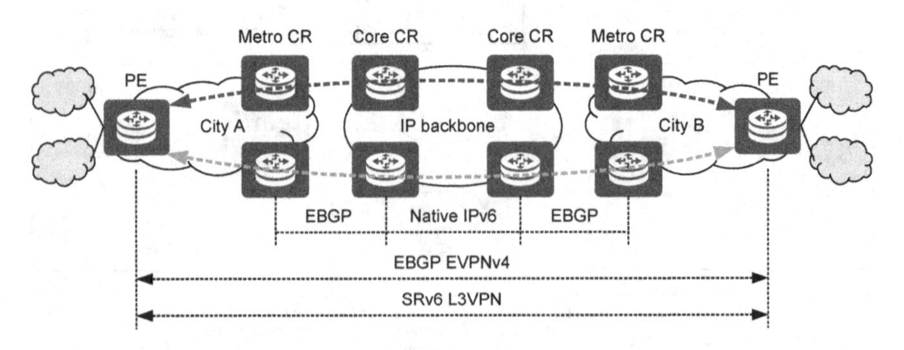

FIGURE 7.8 China Unicom's SRv6 overlay solution.

egress devices (e.g., PE) in both cities A and B. The SRv6 VPN is deployed across the China169 backbone network, quickly constructing cross-province cloud private lines between cities A and B.

7.3 SRv6 COMPATIBILITY WITH LEGACY DEVICES

Compatibility with IPv6 enables SRv6 traffic to easily pass through existing network devices using regular IPv6 routing. That is, it will facilitate smooth network evolution and protect investments as SRv6 is compatible with devices on live networks. Currently, the industry has proposed some solutions for supporting SRv6 development by upgrading the software of legacy devices or through improvements based on of limited hardware capability resources of legacy devices. This involves two approaches: reducing the depth of the SID stack in an SRv6 SRH so that existing hardware can support SRv6; supporting the E2E traffic steering by combining some existing traffic steering methods (e.g., FlowSpec) without requiring an E2E upgrade. Next, we will introduce the two approaches. The related document[4] delves into more details.

7.3.1 Using Binding SIDs to Reduce the Depth of the SRv6 SID Stack

Binding SIDs can be used to reduce the depth of the SRv6 SID stack. With binding SIDs, an SRv6 TE path can be divided into multiple shorter paths, which are called SRv6 sub-paths.[5,6] The sub-paths are concatenated using binding SIDs to shield the details of each sub-path's SID stack, thereby lowering the requirement for the SID stack depth on devices. Figure 7.9 is an example of using binding SIDs to reduce the depth of an SRv6 SID stack.

As shown in Figure 7.9, on an explicit path <R2, R3, R4, R5, R6> from R1 to R6, a five-SID list <A2::1, A3::1, A4::1, A5::1, A6::1> must be supported if no binding SID is used. In the SID list, A2::1, A3::1, A4::1, A5::1, and A6::1 represent R2, R3, R4, R5, and R6, respectively. However, if R1 only supports a maximum of three SIDs for processing, R1 cannot process a SID stack with more than three SIDs.

After binding SIDs are introduced, the controller uses IGP/BGP-LS to collect the SRv6 MSD processing capability of each node on a network. Then, the controller computes SRv6 paths that do not exceed the MSD processing capability of each node on the path. For R1 (which can process a maximum of three SIDs), a path <A6::1, A2::B1> that contains only two

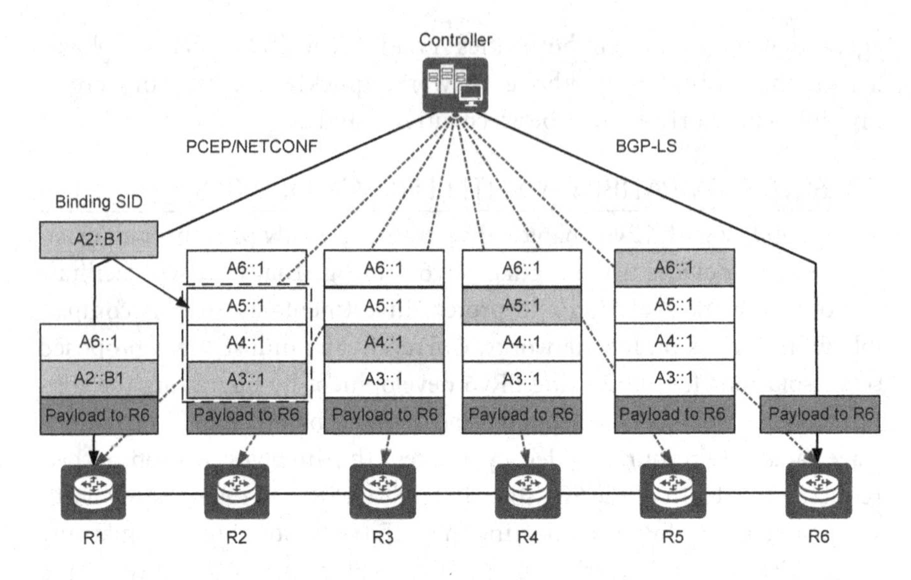

FIGURE 7.9 Using binding SIDs to reduce the depth of the SRv6 SID stack.

SIDs is computed. A2::B1 is a binding SID, representing a sub-path with the SID list <R3, R4, R5>. In this way, an SRv6 path that contains two SIDs is delivered to R1, the sub-path <R3, R4, R5> is delivered to R2, and mapping between the sub-path and A2::B1 is formed.

For a packet destined for R6 along the explicit path <R2, R3, R4, R5, R6>, R1 only needs to add a two-SID list <A6::1, A2::B1> to the SRH in the packet. After the packet is forwarded to R2, R2 adds another SRH based on SID A2::B1. The SID list in the SRH is <A3::1, A4::1, A5::1>. The new SRH is used to forward the R2-to-R5 packet through R3 and R4. If PSP is enabled, the SRH is removed on R4. The packet arriving at R5 has only the inner SRH, and the SL pointing to A6::1. SID A6::1 is used to direct packets to the destination R6.

7.3.2 Applying FlowSpec to SRv6

Currently, the IETF defines two FlowSpec-related standards and solutions: BGP FlowSpec[7] and PCEP FlowSpec.[8] The former introduces a new BGP NLRI to carry flow processing rules[9] for anti-DDoS and traffic filtering, whereas the latter defines how to automatically steer traffic to a specified tunnel. PCEP FlowSpec inherits the BGP FlowSpec idea. The difference is that PCEP FlowSpec uses PCEP to deliver FlowSpec information. After path computation, the PCE delivers both path and FlowSpec information to the ingress PCC of a path. In this way, path/tunnel information and FlowSpec rules are combined to dynamically steer traffic, as well as simplify network configuration and management.

A FlowSpec rule includes a series of flow filter criteria and associated processing rules. These rules include traffic filtering, rate limiting, traffic redirect, re-marking, traffic sampling, and traffic monitoring. Therefore, FlowSpec can be applied to a network to control related nodes, enabling traffic to be processed on demand.

In this section, we elaborate on how to use SRv6 together with the FlowSpec traffic redirect function to implement E2E traffic steering. The key idea involves using SRHs to steer traffic on SRv6-capable nodes and using FlowSpec redirect on SRv6-incapable nodes.

As shown in Figure 7.10, an E2E SRv6 TE path needs to be established from R1 to R4 through R2, R6, R7, and R3. R1 and R4 support SRv6, whereas the other devices do not. In this scenario, the controller can

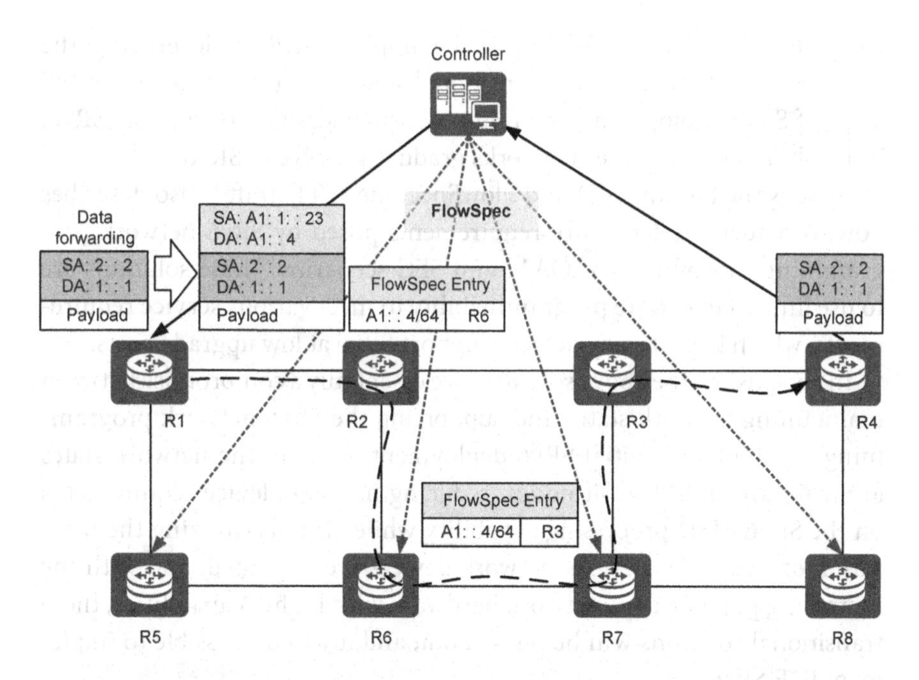

FIGURE 7.10 Applying FlowSpec to SRv6.

leverage FlowSpec to steer traffic on those SRv6-incapable devices. For example:

1. On R2, a FlowSpec rule is installed to redirect traffic destined for A1::4/64 (this destination address is contained in the outer IPv6 header, representing R4) to R6, indicating that the packet destined for A1::4/64 is sent through R6.

2. If R6's next hop to A1::4/64 is R7, the controller does not need to deliver a FlowSpec route to R6. Instead, R6 sends packets to R7 along the shortest path.

3. There are two paths from R7 to R4. One is R7→R8→R4, and the other is R7→R3→R4. To direct traffic along the latter path, the controller can deliver a FlowSpec rule to R7 for redirecting traffic destined for A1::4/64 to R3. Then packets are sent to R3.

4. After R3 receives the packets, it forwards them to R4 along the shortest path.

As such, E2E traffic steering can be implemented by leveraging the FlowSpec function. This is a transitional solution intended for the initial stage of SRv6 deployment when numerous devices do not support SRv6. This solution helps carrier networks gradually evolve to SRv6.

Besides the binding SID and FlowSpec, the IETF draft[4] also describes how to reduce the hardware requirements posed by SRv6 network programming capabilities in IOAM and SFC scenarios. These solutions are to use limited hardware programmability to meet various service requirements, which helps legacy devices support SRv6 at low upgrade costs.

In conclusion, the above solutions are essentially a compromise between maintaining network states and supporting the SRv6 network programming capability. In initial SRv6 deployment, some of the network states are maintained on transit nodes, reducing network device requirements on the SRv6 MSD processing capability while also maximizing the utilization of legacy devices. As network devices are upgraded, and with the increasing popularity of network hardware with higher capabilities, these transitional solutions will be phased out, and it will be possible to implement E2E SRv6.

7.4 SRv6 NETWORK SECURITY

In Section 7.1, we described the SRv6 network security risks, such as eavesdropping, packet tampering, identity spoofing, replay attacks, DoS/DDoS attacks, and attacks launched using malicious packets. In general, these attacks are initiated by sending attack packets with forged sources or intercepting, tampering with, or forging SRv6 packets during packet transmission. In order to ensure the security of SRv6 network, the following requirements must be met:

1. The communication sources must be trustworthy.
2. Packets must not be tampered with during transmission.

The overall SRv6 security solution will be designed based on these two requirements.

SRv6 is an IPv6-based source routing mechanism. Therefore, SRv6 inherits security risks from IPv6 and source routing accordingly. In this context, the SRv6 security solution needs to be designed from the aspects of IPv6 and source routing to tackle the preceding issues, ensuring that the communication sources are trustworthy and preventing packets from being tampered with during transmission. In the following sections, we will describe the SRv6 security solution in terms of IPv6 and source routing.

7.4.1 IPv6 Security Measures

IPv6 uses Internet Protocol Security (IPsec), which is a protocol suite defined by the IETF for IP data encryption and authentication, as a basic protocol for network security.[10]

IPsec consists of three parts:

1. Authentication Header (AH)[11]: an IP extension header that provides functions, such as integrity check, data origin verification, and anti-replay protection for IPv4 and IPv6 packets.
2. Encapsulating Security Payload (ESP)[12]: an IP extension header that provides encryption, data origin verification, integrity check, and anti-replay protection for IPv4 and IPv6 packets.

3. Secure Association (SA): a unidirectional security association established between two communicating parties. The SA contains the specified algorithms and parameters needed in IPsec. The Internet Key Exchange (IKE) protocol is always used to establish SAs.

IPsec AH and ESP headers can be used separately or together. AH verifies packet integrity but does not support encryption. AH verifies an entire packet, including the header and payload. ESP verifies packet integrity and encrypts packets, and adds a check value in the ESP Integrity Check Value (ICV) field. It verifies packet content from the ESP header to the ESP trailer. ESP encrypts packet content from the field next to the ESP header to ESP trailer.

IPsec works in transport or tunnel mode based on the position where an IPsec header is added:

- In transport mode, the AH or ESP header is inserted after the original IP header. Then the whole IP packet participates in AH authentication digest calculation, and ESP encrypts only transport layer packets.
- In tunnel mode, a whole user IP packet is transmitted as payload, following the outer tunnel headers. The AH, ESP, or both headers are inserted after the outer IP header. Subsequently, the whole user IP packet participates in AH authentication digest calculation and ESP encryption.

Figures 7.11 and 7.12 show IPsec encapsulation in transport and tunnel modes, respectively. Note that we do not cover their detailed mechanisms in this section.

In Figure 7.11, the ESP field is the ESP header, containing security-related parameters. The ESP Trailer field contains subfields, such as Padding and Next Header. The ESP ICV field carries the ICV value.[11,12]

IPsec can be used to verify the communication source of IP packets as well as encrypt and verify IP packets to block those that have been modified during transmission. Therefore, IPsec is one of the most important security measures at the network protocol layer.

IPv6-based SRv6 can also use IPsec to encrypt and authenticate packets, improving network security. However, to provide more flexible

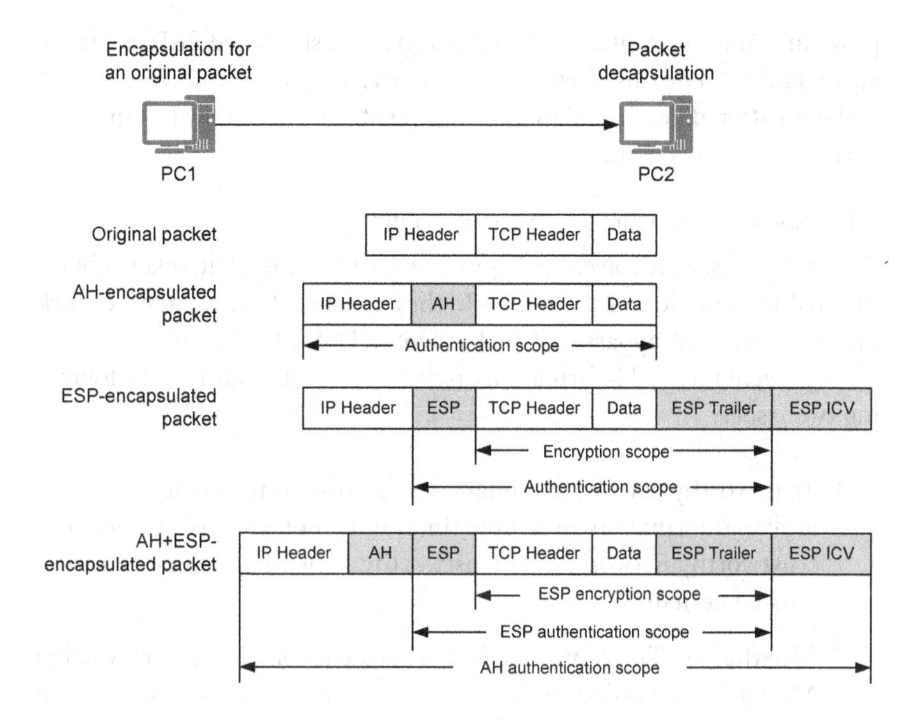

FIGURE 7.11 IPsec encapsulation in transport mode.

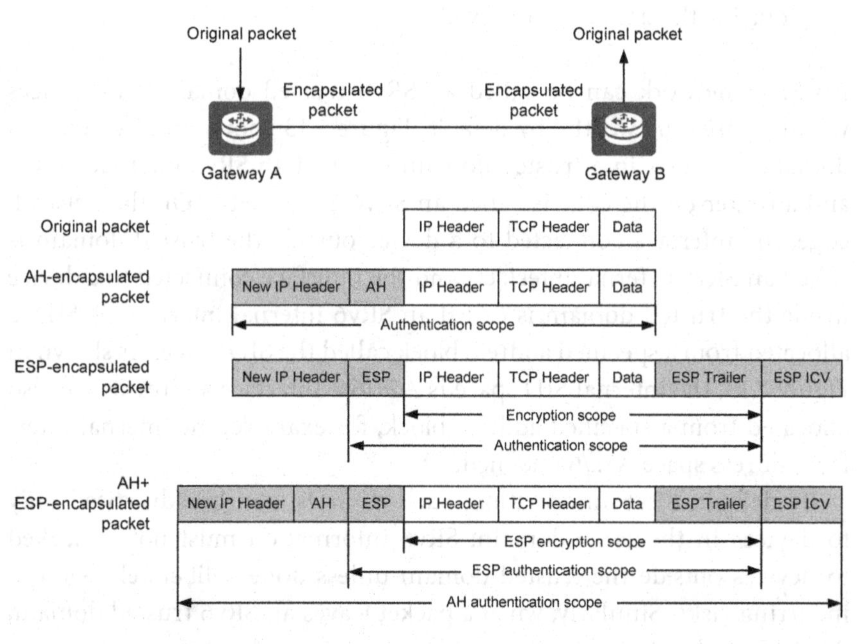

FIGURE 7.12 IPsec encapsulation in tunnel mode.

programming capabilities, it is defined that the stipulated SRH fields are all mutable.[13] For this reason, such an SRH needs to be ignored in AH authentication digest calculation. Subsequently, an SRH may be tampered with even if AH is used.

7.4.2 Security Measures for Source Routing

The security issue involved with source routing is that if attackers obtain internal information of the network, they can use it for remote network discovery, firewall-bypassing attacks, DoS/DDoS attacks, and so on.

To prevent internal information disclosure, we must address the following two issues:

1. Trustworthy network boundary, also known as information boundary[14]: Information in a domain should not be spread outside a trustworthy network. This trustworthy network domain is called a trusted domain.
2. Whether traffic from outside the trusted domain is trustworthy: Methods are needed to determine that the communication source is trustworthy, and the traffic is not tampered with when the traffic outside the domain is received.

An SRv6 network can be called an SRv6 trusted domain, and devices within it are trustworthy by default. Figure 7.13 shows an SRv6 trusted domain. A router in a trusted domain is called an SRv6 internal router, and a router on the edge is called an SRv6 edge router. On the network edge, the interface connected to a device outside the trusted domain is called an SRv6 external interface, and the interface connected to a device inside the trusted domain is called an SRv6 internal interface. A SID is allocated from a specified address block called the SID space. As shown in Figure 7.13, the internal SID space is A2::/64. Interface addresses are also allocated from a specified address block, for example, the internal interface address space A1::/64 defined.

By default, SRv6 information, such as SIDs, can be advertised only to devices in the same domain. SRv6 information must not be leaked to devices outside the trusted domain unless done deliberately without incurring risks. Similarly, when a packet leaves an SRv6 trusted domain, SRv6 SID-related information must not be carried outside the domain.

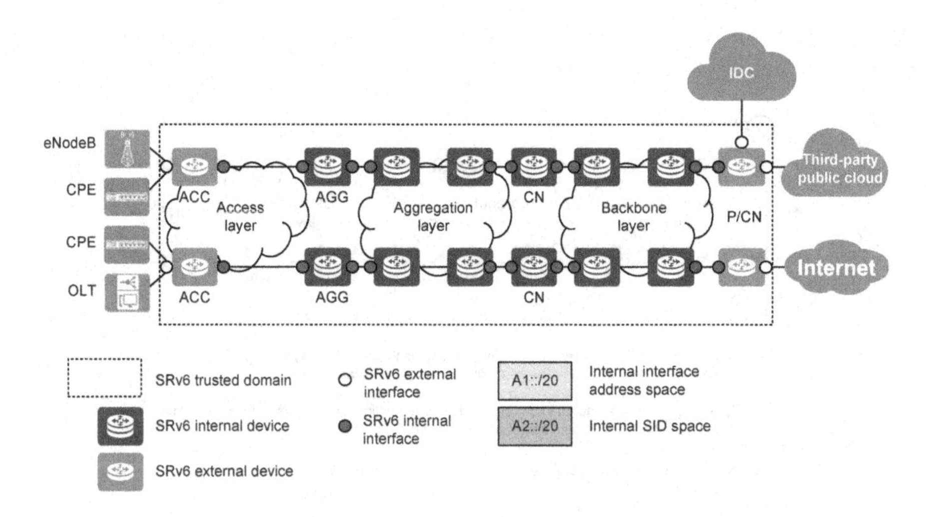

FIGURE 7.13 SRv6 trusted domain.

Theoretically, devices outside the SRv6 trusted domain cannot obtain source routing information from inside. The benefit of this is that it prevents attackers from obtaining intra-domain source routing information while also ensuring that the communication sources are trustworthy.

In addition, in some special cases, SIDs need to be advertised outside the domain. To ensure security and trustworthiness for traffic received from outside the domain, we need to deploy filtering criteria (e.g., ACLs) or verification mechanisms (e.g., HMAC) on the edge and inside the network to discard the traffic that carries the information to the domain.

7.4.3 SRv6 Security Solution

Currently, SRv6 source routing security solutions are classified into basic and enhanced solutions.[6]

- Basic solution: Traffic is filtered based on ACLs, and traffic for unauthorized access to internal information is discarded.
- Enhanced solution: HMAC[15] is used to verify the identity of the communication source and authenticate SRv6 packets, preventing attacks caused by packet tampering.

As stated in RFC 7855 and RFC 8402, the network boundary and trusted domain of the SR network must be defined, and devices in the

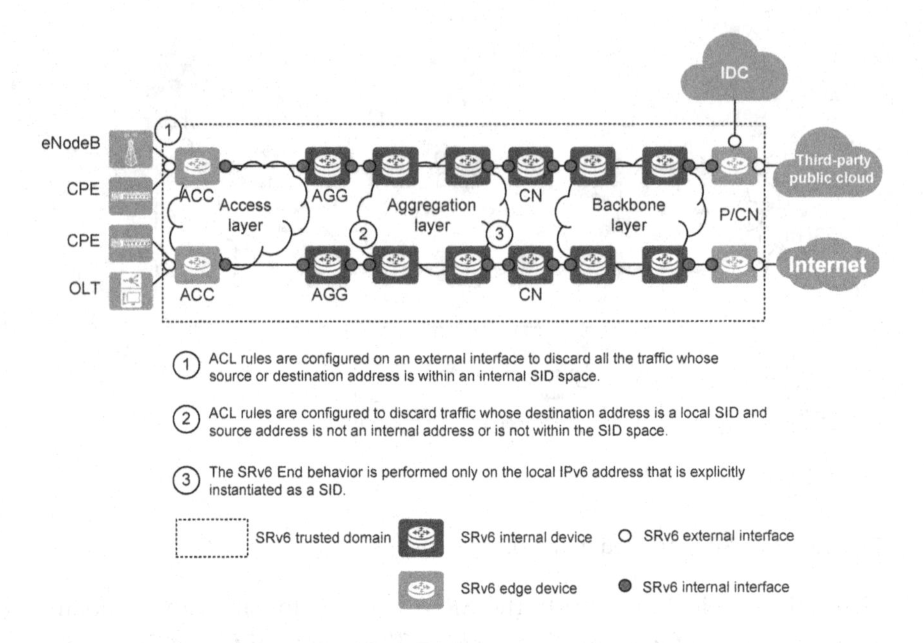
IDC
eNodeB
CPE
CPE
OLT
ACC
Access layer
AGG
Aggregation layer
CN
Backbone layer
P/CN
Third-party public cloud
Internet
ACC
AGG
CN
1
2
3
ACL rules are configured on an external interface to discard all the traffic whose source or destination address is within an internal SID space.
ACL rules are configured to discard traffic whose destination address is a local SID and source address is not an internal address or is not within the SID space.
The SRv6 End behavior is performed only on the local IPv6 address that is explicitly instantiated as a SID.
SRv6 trusted domain
SRv6 internal device
SRv6 external interface
SRv6 edge device
SRv6 internal interface

FIGURE 7.14 Basic SRv6 security solution.

trusted domain are deemed to be secure. Based on the prerequisite of the trusted domain, the basic solution involves deploying an ACL policy to prevent packets of unauthorized users from entering the trusted domain.[5,14] As shown in Figure 7.14, the basic solution consists of three parts.

1. ACL rules must be configured on an SRv6 external interface. Then the rules can be used to discard the SRv6 packets whose source or destination addresses are within the address blocks used to allocate SIDs. This is because the internal SID should not be leaked out of the domain in most cases. If the internal SID is obtained by a device outside the domain, the packet carrying the internal SID as the source or destination address is considered to be an attack packet.

   # Configure an ACL rule on an external interface.

```
1. IF DA or SA in internal address or SID space:
2.     drop the packet
```

2. ACL rules must be configured on both external and internal interfaces. An SRv6 packet is discarded based on the ACL rules if its destination address is a local SID and the source address is not an internal

SID or internal interface address. This is because SRv6 packets can only originate at devices within the trusted domain. Therefore, when the destination address is a SID, the source address must be an internal SID or internal interface address.

Configure ACL rules on external and internal interfaces.

```
1. IF (DA == LocalSID) && (SA!= internal address or SID space)
2.     drop the packet
```

3. The SRv6 End behavior is performed only on the local IPv6 interface whose address is explicitly instantiated as a SID. If the local IPv6 interface address that is not advertised as a SID is inserted into the segment list, a matching interface address route (not a SID) will be found, and as a result, SID processing is not triggered. If the local IPv6 interface address is the last SID (SL=0), the node skips SRH processing; if it is a middle SID (SL>0), the packet is discarded.

Theoretically, information such as internal SIDs cannot be leaked out of an SRv6 trusted domain, based on which the preceding security policies secure the SRv6 network to some extent.

In some cases, however, the SIDs in a domain may be leaked to an external domain. For example, a binding SID is leaked outside an SRv6 domain for TE path selection. For these types of exceptions, add an ACL on the external interface of an edge device to allow traffic carrying the binding SID to pass. On security grounds, binding SIDs are leaked only within a limited scope. For example, they are leaked to some other SRv6 trusted domains, which does not cause high security risks.

To add on to that, SRv6 uses the HMAC mechanism to verify SRHs, which improves security and prevents risks caused by SID disclosure to external domains. This ensures that SRv6 packets from external domains come from trustworthy data sources. These security measures are called enhanced SRv6 security solutions, as shown in Figure 7.15.

In some scenarios, such as cloud data centers, an administrator may delegate the SRH encapsulation capability from routers to hosts to support host-based autonomous route selection and application-level fine-tuning of traffic. In this case, the HMAC mechanism must be adopted on the access device to enhance security for the received SRv6 packets.[13]

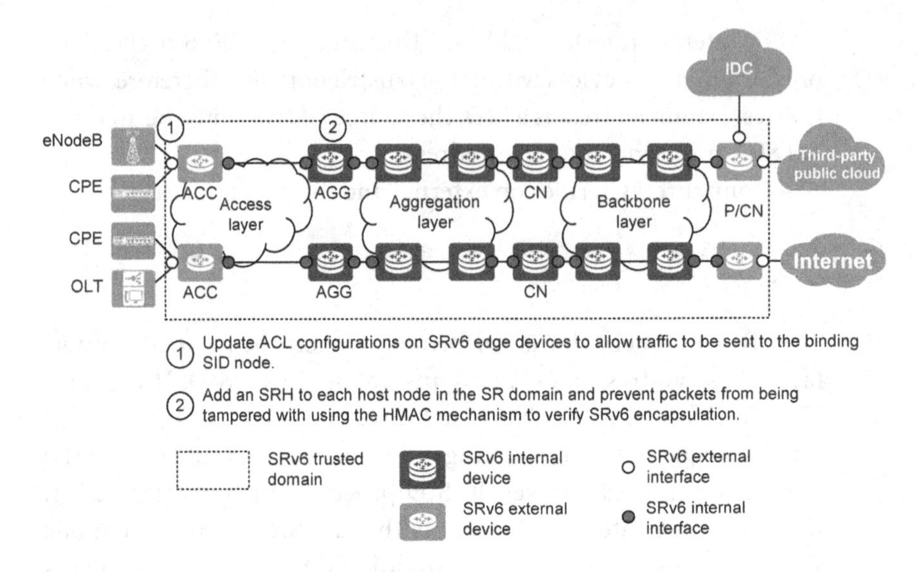

FIGURE 7.15 Enhanced SRv6 security solution.

HMAC is typically used for the following purposes:

1. To verify SRHs on an edge router.
2. To prevent data, such as an SRH, from being tampered with.
3. To verify the identity of a data source.

In view of this, in the enhanced solution, an HMAC policy must be configured on edge routers. Once a packet carrying an SRH enters an SRv6 trusted domain, HMAC processing is triggered. The packet is discarded if the SRH does not carry the HMAC TLV or if the HMAC check fails. Note that the packets can pass only if the HMAC check is successful. Figure 7.16 shows the format of the HMAC TLV in an SRH.

Table 7.1 describes the fields in the HMAC TLV of an SRH.

When processing HMAC, a router first checks whether the destination address field in the IPv6 header matches the SID that an SL value in the segment list points to if the D-bit is not set. If they match, the router verifies that the SL value is greater than the Last Entry field value, and concludes the process by calculating and verifying the checksum based on the preceding fields. We do not delve into further details on the specific checksum calculation method as it is already defined in RFC 2104.[15]

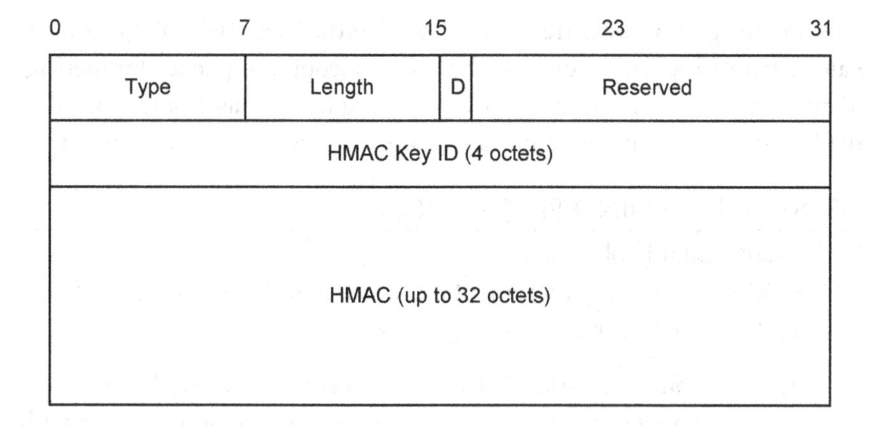

FIGURE 7.16 Format of the HMAC TLV in an SRH.

TABLE 7.1 Fields in the HMAC TLV

Field	Length	Description
Type	8 bits	TLV type.
Length	8 bits	Length of the Value field.
D	1 bit	When this bit is set to 1, destination address verification is disabled. This field is mainly used in reduced SRH mode.
Reserved	15 bits	Reserved. This field is set to 0 before transmission and ignored upon receipt.
HMAC Key ID	32 bits	HMAC key ID that identifies the pre-shared key and algorithm used for HMAC computation.
HMAC	Variable (up to 256 bits)	HMAC computation result. This parameter is optional. It is valid only when the HMAC key ID is not 0. The value of the HMAC field is the checksum digest of the following fields computed by the algorithm identified by the HMAC key ID: 1. IPv6 Header: Source Address field 2. SRH: Last Entry field 3. SRH: Flags field 4. SRH: HMAC Key-id field 5. SRH: all addresses in the segment list

The HMAC mechanism ensures that SRv6 packets are sent by trustworthy hosts and that data, such as SRv6 SRHs, is not tampered with during transmission, effectively protecting the SRv6 network.

IPsec provides a certain level of network security assurance. In addition, the basic SRv6 security solution with ACL policies configured and the HMAC check-based enhanced security solution enhance SRv6 domain security by ensuring that the communication source in the SRv6 domain is trustworthy and packets are not tampered with. These two solutions mitigate network security issues, such as eavesdropping, packet tampering, identity spoofing, and DoS/DDoS, providing us with assurance for SRv6 deployment. For more information, please refer to the related document.[2]

7.5 STORIES BEHIND SRv6 DESIGN

I. Incremental Evolution

The reasons why incremental evolution for IP networks is of vital importance to carriers are as follows:

1. As public networks that carry converged services, IP networks will have a great impact, and therefore, a gradual evolution path is more fitting.

2. Carriers have to protect their investments, and incremental evolution allows them to reuse existing resources.

3. The mixed networking solution is inevitably used, because both old and new IP devices coexist and need to interwork with each other.

Our implementation of incremental network evolution depends on protocol compatibility, which is also an important lesson learned from the IPv6 design. At the beginning, IPv6 designers used 128-bit IPv6 addresses, which are completely incompatible with IPv4 addresses. The IPv6 address format seems simple, but it means all IPv4-based protocols have to be upgraded to support IPv6, and complex IPv4/IPv6 transition technologies have to be adopted. The consequence is the process of migration to IPv6 has been severely affected.

We discussed the extensibility of MPLS and IPv6 in the previous sections. In that context, if history could repeat itself, perhaps there would be a more compatible option for IPv6 to extend some identifiers on the basis of IPv4 addresses. Along this line of reasoning, MPLS could also be considered to be an IPv4 identifier extension. Taking MPLS VPN into account, it can be seen to use the MPLS label extended from the inner IPv4 address to identify a specific VPN instance, and then performs lookup in the IP routing table for IPv4 forwarding. However, MPLS introduces the "shim" layer, which has to require a networkwide upgrade.

These lessons were fully considered, and SRv6 was designed to be compatible with IPv6. As such, the SRv6 compatibility design is an important aspect in which SRv6 has an upper hand, and it plays an essential role in SRv6 deployment, as we already covered in many sections of this document.

II. Trading State for Space

In essence, the protocol design is a type of engineering work, and tradeoff and balance are the important design rules. In fact, it also tortures us to implement reasonable tradeoff and balance in the process of designing and developing protocols. Natural science usually involves definite results, whereas engineering technology always presents us with multiple choices. The different design objectives may cause different choices and the objectives may also change with time. All this makes it difficult for us to come up with a final design.

From the content covered in this chapter, we can see the example of tradeoff in the protocol design. We can take the solution which has to add the network state in order to reduce the requirement on the hardware capability. RSVP-TE needs to maintain many network states and has poor scalability, but it has the lowest hardware requirement as only one layer of label needs to be processed. On the flip side, SR has fewer network states and good scalability but has high hardware requirements for processing a SID stack with over ten SIDs. Essentially, using binding SIDs and FlowSpec routes is a tradeoff between RSVP-TE and SR. The passport and postcard modes of IOAM in Chapter 9 and the stateful and stateless SRv6 SFCs in Chapter 11 are also the reflection of "trading the control plane network state for the forwarding plane programming space." That being said, there is no such thing as a free lunch. Striking a balance between the control and forwarding planes, hardware and software, as well as network state and programming space, the design adopts various methods due to different objectives and constraints. Having said that, we also need to control the number of such options. Otherwise, they will increase system complexity and hinder industry focus.

REFERENCES

[1] Google. Statistics – IPv6 Adoption. (2020-09-30) [2020-10-31].

[2] Li C, Li Z, Xie C, Tian H, Mao J. Security Considerations for SRv6 Networks[EB/OL]. (2019-11-04) [2020-03-25]. draft-li-spring-srv6-security-consideration-03.

[3] Abley J, Savola P, Neville-Neil G. Deprecation of Type 0 Routing Headers in IPv6[EB/OL]. (2015-10-14) [2020-03-25]. RFC 5095.

[4] Peng S, Li Z, Xie C, Cong L. SRv6 Compatibility with Legacy Devices[EB/OL]. (2019-07-08) [2020-03-25]. draft-peng-spring-srv6-compatibility-01.

[5] Filsfils C, Previdi S, Insberg L, Decraene B, Litkowski S, Shakir, R. Segment Routing Architecture[EB/OL]. (2018-12-19) [2020-03-25]. RFC 8402.

[6] Filsfils C, Camarillo P, Leddy J, Voyer D, Matsushima S, Li Z. SRv6 Network Programming[EB/OL]. (2019-12-05) [2020-03-25]. draft-ietf-spring-srv6-network-programming-05.

[7] Loibl C, Hares S, Raszuk R, et al. Dissemination of Flow Specification Rules[EB/OL]. (2020-09-30) [2020-03-25]. draft-ietf-idr-rfc5575bis-17.

[8] Dhody D, Farrel A, Li, Z. PCEP Extension for FlowSpecification[EB/OL]. (2020-09-30) [2020-03-25]. draft-ietf-pce-pcep-flowspec-07.

[9] Marques P, Sheth N, Raszuk R, Greene B, Mauch J, Mcpherson D. Dissemination of Flow Specification Rules[EB/OL]. (2020-01-21) [2020-03-25]. RFC 5575.

[10] Kent S, Atkinson R. Security Architecture for the Internet Protocol[EB/OL]. (2013-03-02) [2020-03-25]. RFC 2401.

[11] Kent S, Atkinson R. IP Authentication Header[EB/OL]. (2013-03-02) [2020-03-25]. RFC 2402.

[12] Kent S. IP Encapsulating Security Payload[EB/OL]. (2020-01-21) [2020-03-25]. RFC 4303.

[13] Filsfils C, Dukes D, Previdi S, Leddy J, Matsushima S, Voyer D. IPv6 Segment Routing Header (SRH)[EB/OL]. (2020-03-14) [2020-03-25]. RFC 8754.

[14] Previdi S, Filsfils C, Decraene B, Litkowski S, Horneffer M, Shakir R. Source Packet Routing in Networking (SPRING) Problem Statement and Requirements[EB/OL]. (2020-01-21) [2020-03-25]. RFC 7855.

[15] Krawczyk H, Bellare M, Canetti R. HMAC: Keyed-Hashing for Message Authentication[EB/OL]. (2020-01-21) [2020-03-25]. RFC 2104.

CHAPTER 8

SRv6 Network Deployment

THIS CHAPTER DESCRIBES SRv6 NETWORK DEPLOYMENT, covering IPv6 address planning, IGP route design, BGP route design, tunnel design, VPN service design, and network evolution design. SRv6 can be deployed on a typical single-AS network (such as an IP backbone network, metro network, mobile transport network, or data center network) or on an E2E network (such as an inter-AS VPN or carrier's carrier network). Before SRv6 is deployed, IPv6 address planning is needed for SID allocation. IGP and BGP designs are then implemented for network nodes, and the corresponding SIDs are advertised for services, such as TE and VPN.

8.1 SRv6 SOLUTION

SRv6 can be deployed on a single-AS network (such as an IP backbone network, metro network, mobile transport network, or data center network) or on an E2E network (such as an inter-AS VPN or carrier's carrier network).

8.1.1 Single-AS Network

This section describes SRv6 deployment on a typical single-AS network, such as an IP backbone network, metro network, mobile transport network, or data center network.

8.1.1.1 IP Backbone Network

Figure 8.1 shows a typical IP backbone network. PEs are service access nodes that connect to CEs or a metro/data center network, Ps are responsible for connectivity between PEs, RRs are used to reflect BGP routes, and Internet Gateways (IGWs) interconnect with the Internet.

On an IP backbone network, SRv6 can be used to carry all types of services, including Internet, private line, and voice services. SRv6 BE paths or SRv6 TE tunnels can be used to carry different services depending on their different SLA requirements.

- SRv6 BE paths can be used to carry Internet and private line services without high SLA requirements.
- SRv6 TE tunnels can be used to carry voice and high-value private line services.

In addition, some Internet services can also use SRv6 TE tunnels for fine-grained scheduling, meeting requirements of scenarios such as traffic scrubbing and high-value service (such as online gaming) guarantee.

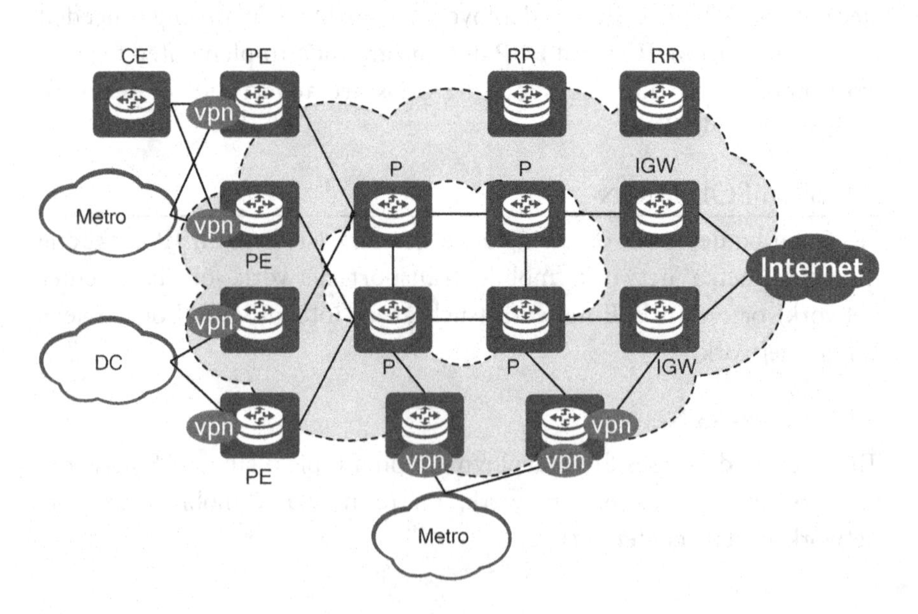

FIGURE 8.1 Typical IP backbone network.

On an IP backbone network, different services are carried over different types of VPNs.

- Internet services can be carried using SRv6 L3VPN, IP over SRv6 (public network services are carried over SRv6 tunnels), or native IP.
- Enterprise Layer 3 private line services can be carried using SRv6 L3VPN.
- Enterprise Layer 2 private line services can be carried using SRv6 EVPN E-Line/SRv6 EVPN E-LAN/SRv6 EVPN E-Tree.
- Voice services can be carried using SRv6 L3VPN.

It can be seen that SRv6 can be used to carry all services on an IP backbone network and is advantageous in scenarios such as high-value service SLA guarantee, fine-grained traffic optimization, and traffic scrubbing. MPLS-incapable IP backbone networks especially benefit from SRv6 in terms of carrying VPN services.

8.1.1.2 Metro Network

Metro networks are also an important scenario for SRv6 deployment. Figure 8.2 shows a typical metro network, which is connected to the access network and IP backbone network. Typically, a metro network includes two layers of devices: Aggregation (AGG) and Broadband Network Gateway (BNG)/Metro Core (MC). AGGs are used for fixed broadband and enterprise private line services. BNGs or MCs are used to aggregate all services and connect them to the IP backbone network. Services, such as fixed broadband and enterprise private line services, on a metro network can be carried using SRv6.

- Fixed broadband: SRv6 BE can be used to carry services without high SLA requirements, and SRv6 TE can be used to provide SLA guarantee for high-value services (such as online gaming).
- Broadcast TV (BTV) in fixed broadband: Bit Index Explicit Replication IPv6 Encapsulation (BIERv6) can be used to carry BTV in fixed broadband services. For details about BIERv6, refer to Chapter 12.

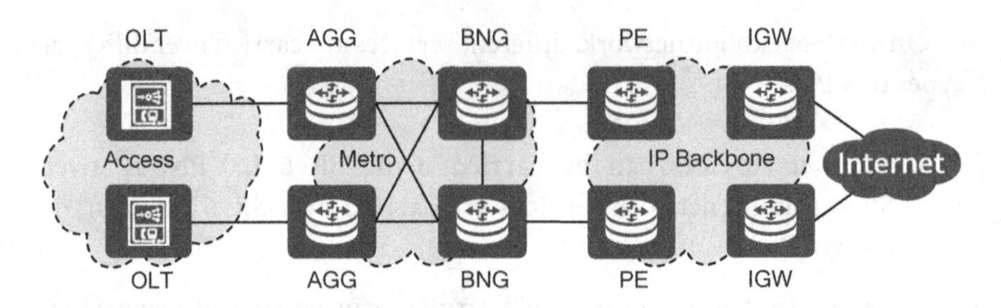

FIGURE 8.2 Typical metro network.

- Enterprise private line service: SRv6 BE can be used to carry private line services without high SLA requirements, and SRv6 TE can be used to carry high-value private line services.

Services on a metro network are usually carried over L2VPN or L3VPN. The SRv6 solutions used for specific VPN services are as follows:

- SRv6 EVPN E-LAN/SRv6 EVPN E-Tree can be used to carry fixed broadband services on the access side (from AGGs to BNGs).
- SRv6 L3VPN can be used to carry fixed broadband services on the network side (from BNGs to PEs).
- SRv6 L3VPN can be used to carry enterprise Layer 3 private line services.
- SRv6 EVPN E-Line/SRv6 EVPN E-LAN/SRv6 EVPN E-Tree can be used to carry enterprise Layer 2 private line services.

To summarize, SRv6 can be used to carry all services on a metro network and features tremendous advantages in terms of high-value service SLA guarantee and simplifying multicast service deployment.

8.1.1.3 Mobile Transport Network

Mobile transport networks are an important scenario for SRv6 deployment. Figure 8.3 shows a typical mobile transport network, which typically involves three roles:

- Cell Site Gateway (CSG): is located at the access layer and is responsible for base station access.

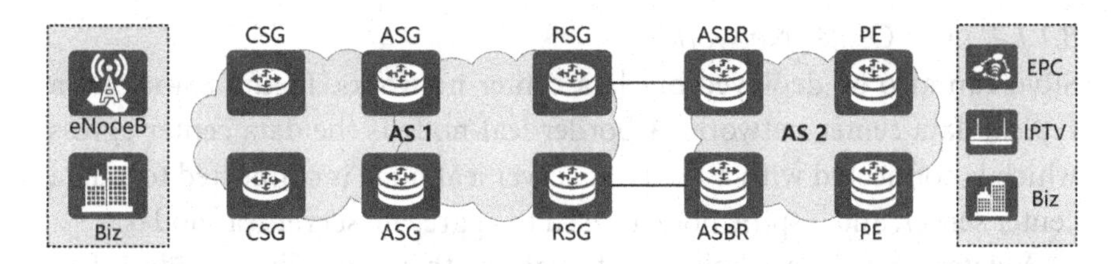

FIGURE 8.3 Typical mobile transport network.

- Aggregation Site Gateway (ASG): is located at the aggregation layer and is responsible for aggregating CSG service flows.

- Radio Network Controller Site Gateway (RSG): functions as the egress of the mobile transport network and is interconnected with an IP backbone network.

A mobile transport network mainly carries wireless voice and Internet services. Some mobile transport networks also carry enterprise private line services. All these services can be carried using SRv6.

- Wireless voice service: This type of service has high SLA requirements and is typically carried using SRv6 TE.

- Internet service: This type of service is usually carried using SRv6 BE. High-value services, such as online gaming, can be carried using SRv6 TE.

- Enterprise private line service: Private line services without high SLA requirements are carried using SRv6 BE, and high-value private line services are carried using SRv6 TE.

Typically, wireless services on a mobile transport network are carried using SRv6 EVPN L3VPN, and enterprise private line services are carried using SRv6 EVPN E-Line/SRv6 EVPN E-LAN/SRv6 EVPN E-Tree.

To summarize, SRv6 can be used to carry all services on a mobile transport network and offers tremendous advantages in high-value scenarios, such as high-value private line SLA guarantee and voice transport.

8.1.1.4 Data Center Network

SRv6 can also be deployed on data center networks. Figure 8.4 shows a typical data center network. A border leaf node is the data center egress which is connected with a WAN. A server leaf node is connected to a data center server, and a spine node is used to aggregate server leaf nodes.

A data center is a whole set of facilities that centrally provide information services for external systems. A data center involves three parts: computing, storage, and networking. Currently, data center networks mainly use VXLAN to connect tenants and isolate traffic between tenants.

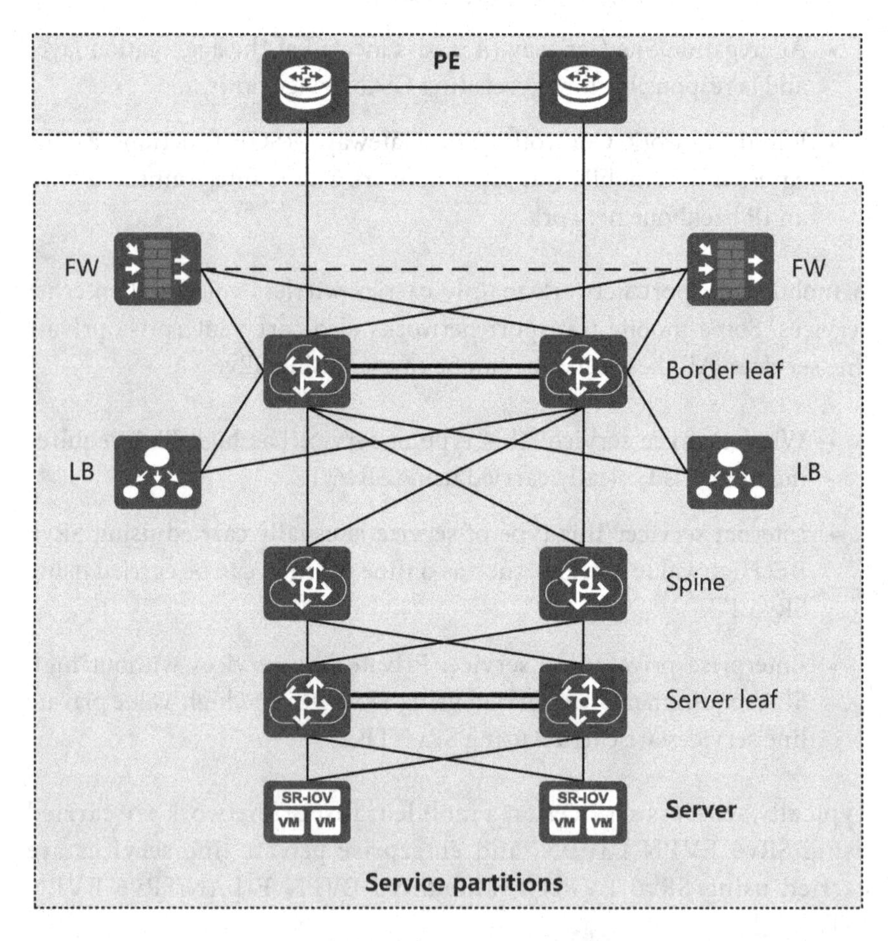

FW: firewall
LB: load balancer

FIGURE 8.4 Data center network.

However, VXLAN only provides best-effort IP forwarding and falls short in terms of traffic engineering functions, such as forwarding along a specified path.

In contrast, SRv6 EVPN does not just implement multi-tenant isolation like VXLAN does, but also forwards tenant traffic along specified paths with SRv6 TE. Therefore, data center networks are also a typical scenario for SRv6 deployment. For example, currently, SRv6 has been deployed in Japan Line's data center network.[1]

8.1.2 E2E Network

E2E network transport is an important application of SRv6. We briefly introduced the technical advantages of SRv6 over MPLS and SR-MPLS in Chapter 2, and we will go into more detail in this section. Typical E2E network applications include inter-AS VPN and carrier's carrier.

8.1.2.1 Inter-AS VPN

Traditionally, if a network is deployed across multiple ASs, three inter-AS VPN solutions (Option A, Option B, and Option C) are available to carry services. These solutions, however, are complex to deploy.

Inter-AS Option A is currently the most widely used option. Figure 8.5 shows the typical deployment of an E2E private line service in Option A mode.

Note: Inter-AS VPN Option A is a VRF-to-VRF mode. Specifically, the two nodes connected across ASs serve as each other's PE or CE and learn each other's specific VPN routes, and packets are forwarded based on IP addresses.

In the preceding figure, PW+L3VPN is used to carry services at the access and aggregation layers, and Option A is used to deploy inter-AS VPN between the aggregation and backbone layers. In this inter-AS scenario, eight configuration points (the eight nodes in Figure 8.5) are needed

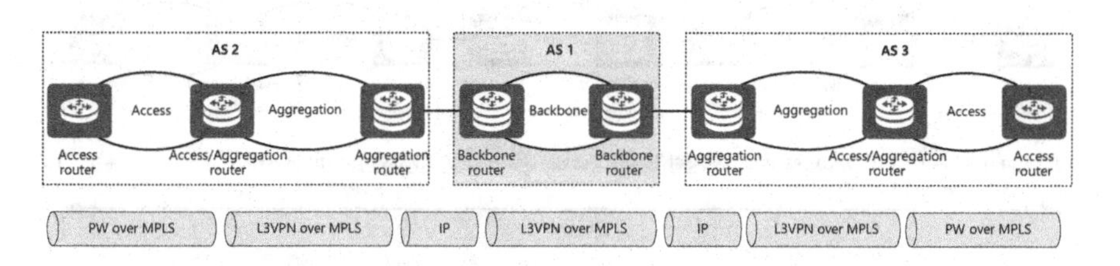

FIGURE 8.5 Inter-AS VPN Option A.

for provisioning a service, and networkwide resources (such as VLAN IDs) need to be uniformly allocated and managed on the IP backbone network, making service deployment extremely complex.

Seamless MPLS is a technology used to achieve inter-AS interworking and is commonly used for inter-AS service transport[2] on the live network. Seamless MPLS leverages the principles of inter-AS VPN Option C. The key idea of seamless MPLS is to provide E2E connectivity between any two network nodes and to allow VPN services to be deployed only on both ends, other than on border nodes. Seamless MPLS reduces the number of service configuration points compared with inter-AS VPN Option A, but it still struggles to support large-scale networks because it depends on the use of BGP to establish inter-AS LSPs.

As shown in Figure 8.6, in order to establish an E2E BGP LSP, the loopback address (/32 route) of the device on one end needs to be advertised to the device on the other end, and the corresponding labels need to be distributed. All loopback routes on the network need to be exchanged, and this creates extremely high scalability requirements on the control plane and forwarding plane of network nodes (especially edge nodes). If there are 100,000 nodes on the network, they need to support at least 100,000 routes and LSPs for the network to be scalable with seamless MPLS.

During seamless MPLS deployment, a common way to reduce the pressure on access nodes is to configure a route policy on border nodes during service creation, so that loopback routes that are leaked from an external domain to the local domain can be distributed to access nodes on demand. This deployment mode, however, complicates configuration.

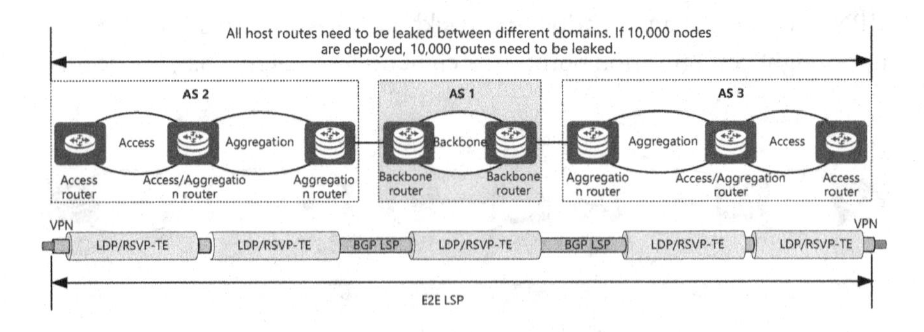

FIGURE 8.6 Inter-AS seamless MPLS.

In contrast to inter-AS VPN and seamless MPLS, SRv6 is based on IPv6 forwarding and inherently supports inter-AS connectivity. Furthermore, SRv6 supports IPv6 route aggregation, thereby reducing the number of routes received on access nodes. The use of the E2E SRv6 VPN technology in inter-AS scenarios not only reduces the number of service configuration points but also enables service deployment through route aggregation, significantly reducing the scalability requirements on network nodes. Figure 8.7 shows an E2E SRv6 network.

During IPv6 address planning for an E2E SRv6 network, each network domain is configured with a network prefix for locator allocation to devices in this domain, allowing advertisement of only an aggregated locator route to devices outside the domain. If no IPv6 loopback interface has been configured on the network, the locator and loopback address with the same network prefix can be allocated so that only the aggregated route shared by the locator and loopback address needs to be advertised, thereby reducing the number of routes. As shown in Figure 8.7, a separate network prefix is allocated to the access and aggregation layers, and another separate network prefix is allocated to the IP core layer. Only an aggregated IPv6 route (locator and loopback address) is advertised between the aggregation and IP core layers. SRv6 service nodes only need to learn the aggregated route and the specific routes in the local domain to carry E2E SRv6 services. In addition, the number of service configuration points is reduced to two: ingress and egress. As such, the specific routes of a domain are not flooded to other domains. In addition, route changes, such as route flapping, in one domain do not cause frequent route changes in another domain. This enhances security and stability within the network and makes SRv6 a clear winner over other technologies in inter-AS VPN transport.

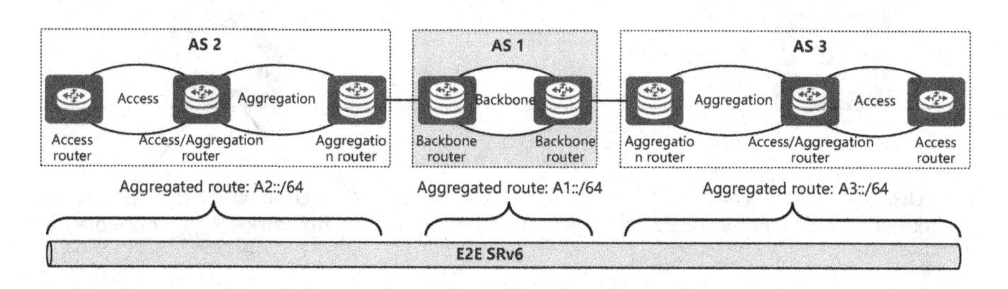

FIGURE 8.7 E2E SRv6 network.

8.1.2.2 Carrier's Carrier

A user of a VPN service provider may themselves be a service provider. In this case, the VPN service provider is called a provider carrier or first carrier, and the user is called a customer carrier or second carrier, as shown in Figure 8.8. This networking model is called carrier's carrier, where the customer carrier is a VPN customer of the provider carrier.

As shown in Figure 8.9, to ensure good scalability, provider carrier CEs (also the customer carrier PEs) advertise only the public network routes, instead of the VPN routes, of the customer carrier network to provider carrier PEs. That is, the VPN routes of the customer carrier network will not be advertised to the provider carrier network. As such, MPLS (LDP or labeled BGP) needs to be deployed between provider carrier PEs (PE1 and PE2) and provider carrier CEs (CE1 and CE2) and between provider carrier CEs (CE1 and CE2) and customer carrier PEs (PE3 and PE4). In addition, Multiprotocol Extensions for BGP (MP-BGP) needs to be deployed between customer carrier PEs (PE3 and PE4), leading to very complex service deployment.

In contrast, SRv6 VPN can forward traffic based on native IP. If SRv6 VPN is used on the customer carrier network, MPLS does not need to be deployed between the provider carrier network and customer carrier network. Instead, only the locator and loopback routes of the customer

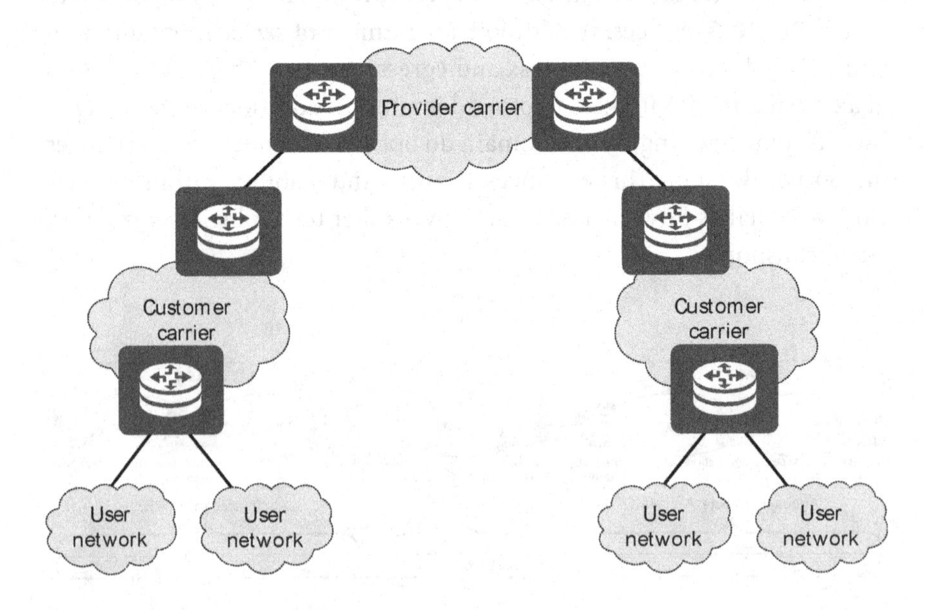

FIGURE 8.8 Carrier's carrier network.

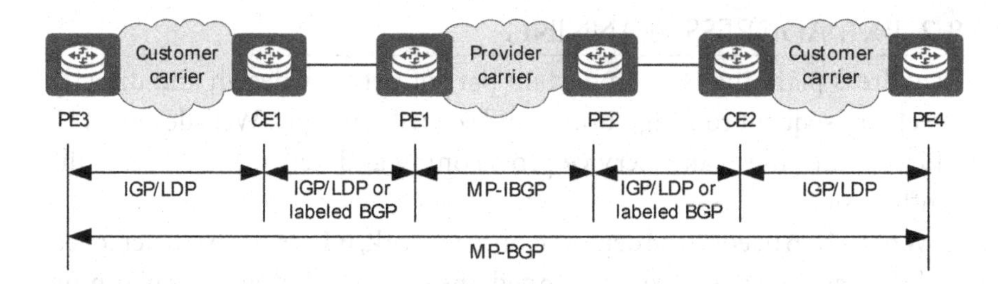

FIGURE 8.9 Conventional service model of carrier's carrier.

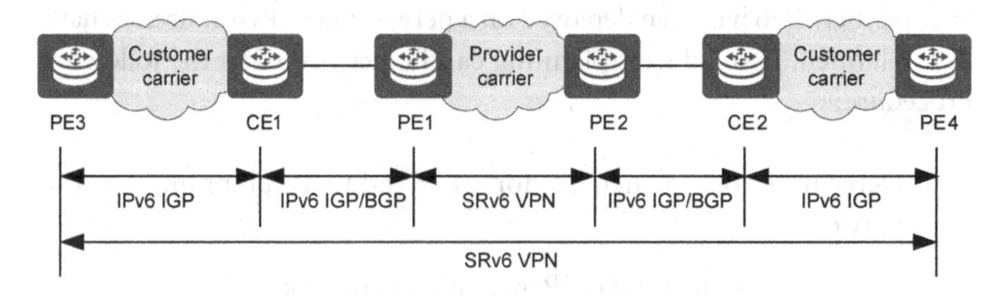

FIGURE 8.10 Carrier's carrier through SRv6.

carrier network need to be advertised over the provider carrier network to establish connectivity. In this way, the customer carrier can provide E2E VPN services based on IPv6 connectivity. As shown in Figure 8.10, CE1 and CE2 on the provider carrier network are the same as the CEs in common VPN scenarios, greatly simplifying VPN service deployment and reducing the workload.

MPLS is not required for SRv6-based carrier's carrier. Only IGP or BGP needs to be deployed between provider carrier PEs and provider carrier CEs in order to advertise the routes of the customer carrier network. The provider carrier network learns IPv6 routes from the customer carrier network, leaks them to a VPN instance, and advertises VPN routes between sites of the provider carrier. After IPv6 routes of the customer carrier network are advertised, Internal Border Gateway Protocol (IBGP) peer relationships are directly established between customer carrier PEs, and SRv6 VPN is established. After that, the customer carrier's service routes can be advertised and user services can be carried. Therefore, the use of SRv6 in carrier's carrier reduces the number of service configuration points to two (ingress and egress) and eliminates the need to maintain multi-segment MPLS networks, significantly simplifying deployment.

8.2 IPv6 ADDRESS PLANNING

IP address planning is an important part of network design and directly affects subsequent routing, tunnel, and security designs. Well-designed IP address planning makes service provisioning and network O&M significantly easier.

When SRv6 needs to be deployed on a network, if IPv6 has been deployed and IPv6 addresses have been planned, the original IPv6 address planning does not need to be modified, and we only need to select a reserved network prefix and use it to allocate SRv6 locators.

If neither IPv6 has been deployed on a network nor IPv6 addresses have been planned, IPv6 address planning can be performed in the following procedure:

1. Determine the principles for IPv6 address planning on the network.
2. Determine the method of IPv6 address allocation.
3. Hierarchically allocate IPv6 addresses.

8.2.1 Principles for IPv6 Address Planning

On an SRv6 network, IPv6 address planning must comply with the following rules:

1. Uniformity: All IPv6 addresses on the entire network are planned in a unified manner, including service addresses (addresses allocated to end users), platform addresses (such as addresses of the IPTV platforms and DHCP servers), and network addresses (addresses used for network device interconnection).
2. Uniqueness: Each address is unique throughout the entire network.
3. Separation:
   - Service and network addresses are planned separately to facilitate route control and traffic security control on border nodes.
   - SRv6 locators, loopback addresses, and link addresses are planned in different address blocks to facilitate route control and management.

4. Hierarchy and aggregatability: Address planning (such as SRv6 locator and loopback address planning) must support route aggregation in an IGP or BGP domain to facilitate route import. It is recommended that:

    - A separate network prefix should be allocated to each backbone network.
    - A separate network prefix should be allocated to each metro network.
    - Each pair of metro aggregation devices should be allocated a separate subnet prefix from the metro network prefix. This facilitates route aggregation between metro aggregation domains.
    - Each metro access domain should be allocated a separate subnet prefix from the network prefix of the metro aggregation devices. This facilitates route aggregation during route import between metro access domains.
    - Each access device should be allocated a separate subnet prefix from the network prefix of the metro access domain.

    Figure 8.11 illustrates a typical address allocation example. A separate network prefix is allocated for each of the two backbone networks, for all devices in the metro domain, for all devices connected to each pair of aggregation devices, and for each IGP access domain, respectively.

5. Security: Source tracing can be quickly performed on addresses, and traffic can be easily filtered by address.

6. Evolvability: A certain amount of address space needs to be reserved in each address block for future service development.

8.2.2 IPv6 Address Allocation Methods

IPv6 addresses are most commonly allocated via sequential, sparse, best-fit, or random allocation.[3]

1. Sequential allocation

    Addresses are allocated from right to left (values from low to high) in an address block based on the same mask. This method of allocation

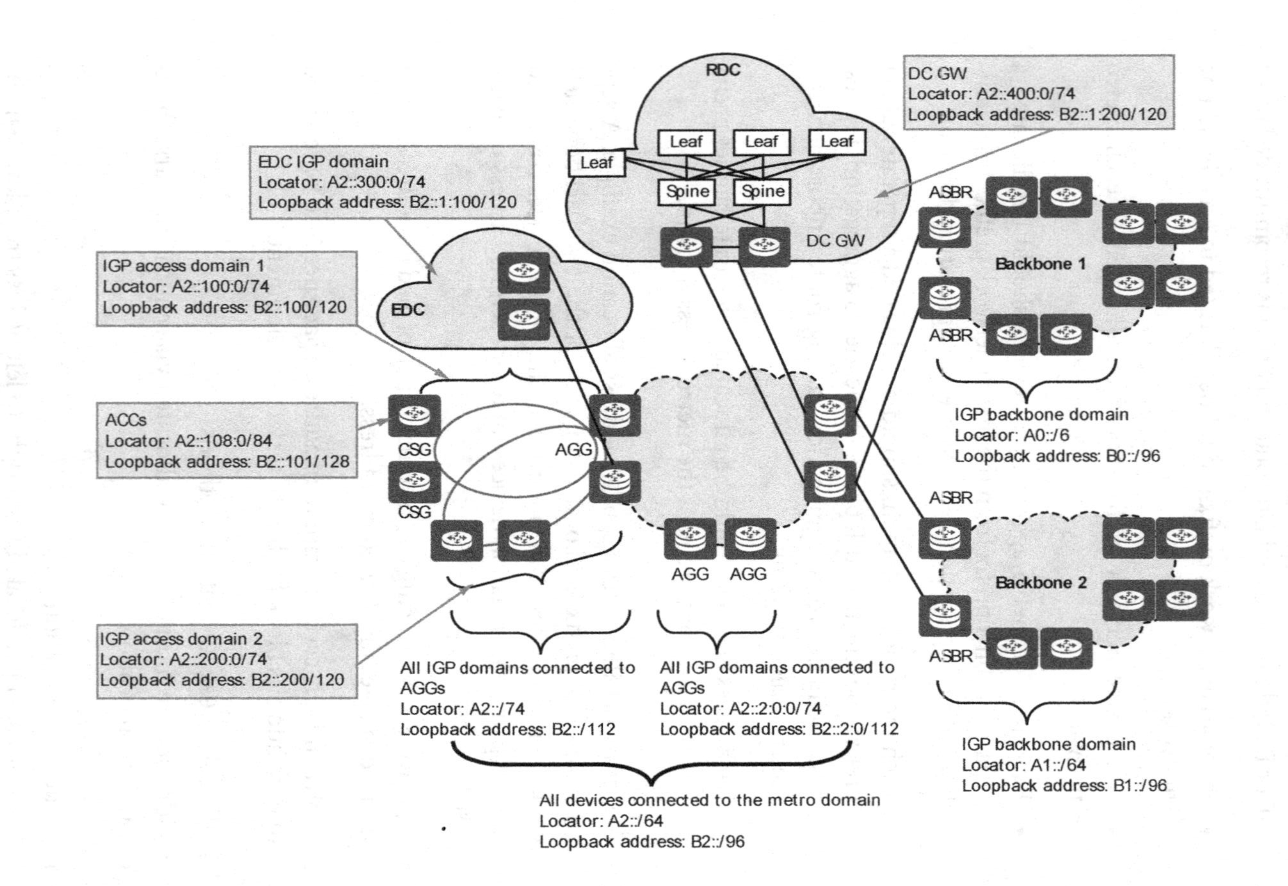

FIGURE 8.11 IPv6 network address planning.

is simple, but sufficient addresses must be reserved for scalability to prevent a large number of routes from having to be advertised due to the failure of route aggregation caused by insufficient addresses. Figure 8.12 uses the source address block 2001:db8:1a::/48 as an example. If the source address block needs to be divided into multiple address blocks with a 52-bit mask and the sequential address allocation method is used, the four bits from 49 to 52 are allocated in ascending order, starting from 0000 (e.g., 0000, 0001, 0010, and so on). In this way, the first address block allocated is 2001:db8:1a::/52, and the second is 2001:db8:1a:1000::/52. Note that the value is in hexadecimal notation and the first digit in 1000 is 0001 in binary (the 49th bit to the 52nd bit).

2. Sparse allocation

   The addresses in an address block are allocated from left to right based on the same mask. The initially allocated data blocks are discrete, but the addresses can still be aggregated after address extension. Figure 8.13 uses the source address block 2001:db8:1a::/48 as an example. If the source address block needs to be divided into multiple address blocks with a 52-bit mask and the sparse allocation method is used, the four bits from 49 to 52 are allocated from left to right, starting from 0000 (e.g., 0000, 1000, 0100, 1100, and so on). In this way, the first address block allocated is 2001:db8:1a::/52, and the second is 2001:db8:1a:8000::/52. Note that the value is in hexadecimal notation and the first digit in 8000 is 1000 in binary (the 49th bit to the 52nd bit). This allocation method needs to reserve sufficient address space for future extensions.

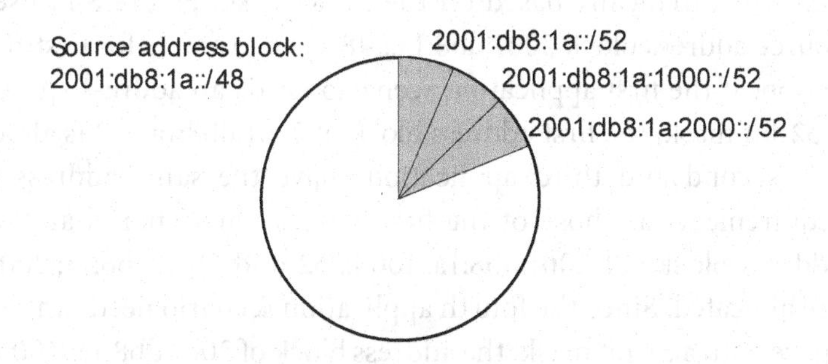

FIGURE 8.12 Sequential IPv6 address allocation.

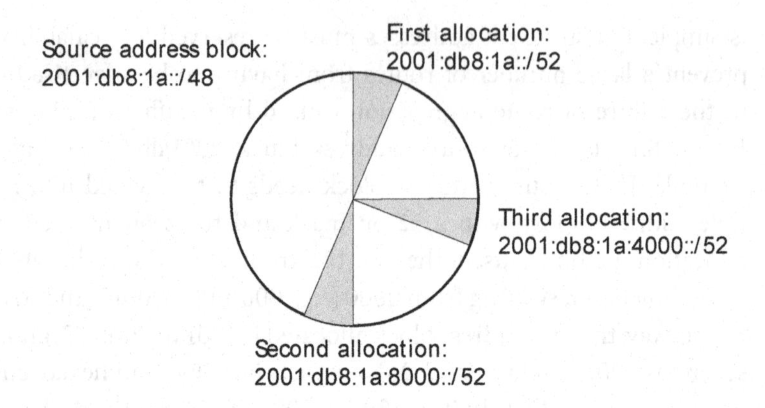

FIGURE 8.13 Sparse IPv6 address allocation.

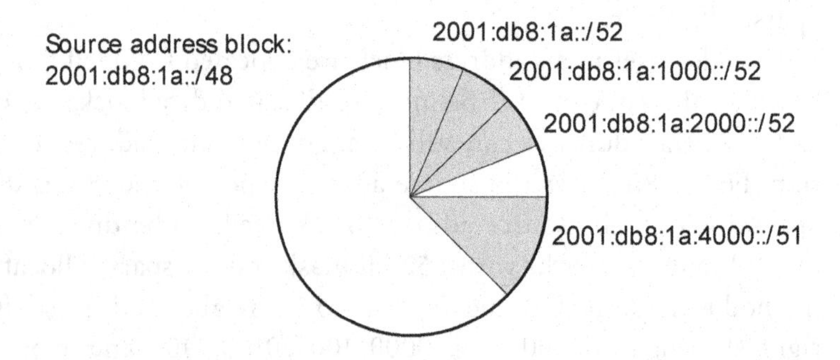

FIGURE 8.14 Best-fit IPv6 address allocation.

3. Best-fit allocation

   Similar to IPv4 Classless Inter-Domain Routing (CIDR), IPv6 addresses are allocated sequentially into appropriate available address blocks according to requirements, rather than being allocated hierarchically based on the same mask. Figure 8.14 uses the source address block 2001:db8:1a::/48 as an example for best-fit allocation. If the first application scenario needs an address space with a 52-bit mask, the first address block of 2001:db8:1a::/52 is allocated. The second and third applications have the same address space requirements as those of the first application scenario, and so the address blocks of 2001:db8:1a::1000::/52 and 2001:db8:1a::2000::/52 are allocated. Since the fourth application scenario needs an address space with a 51-bit mask, the address block of 2001:db8:1a::4000/51 is allocated accordingly.

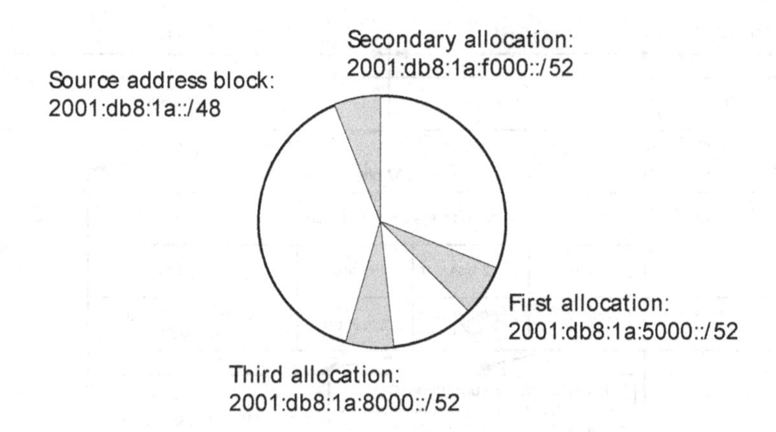

FIGURE 8.15 Random IPv6 address allocation.

4. Random allocation

   As the name suggests, address blocks are randomly allocated. Figure 8.15 uses the source address block 2001:db8:1a::/48 as an example where a random number ranging from 0 to 15 is generated during each allocation. If an address block is still available, it will be allocated. For example, the first random number is 5, the second 15, and the third 8. Random allocation is suitable for scenarios with high-security requirements.

These methods for address allocation have their own advantages and disadvantages, and are therefore used in different scenarios. During address planning, the recommended address allocation method varies according to the network type. For instance, sequential allocation is adopted for an IP backbone network, whereas best-fit allocation is adopted for a metro network.

8.2.3 Hierarchical IPv6 Address Allocation

According to the defined rules and methods, IPv6 addresses can be allocated hierarchically. Figure 8.16 shows an example of IPv6 address planning.

Table 8.1 describes the related fields involved in the IPv6 address planning.

Address planning can also be performed hierarchically, as shown in Figure 8.17.

First up, 27-bit network prefixes are allocated from the 24-bit fixed prefix to addresses with different attributes. If addresses in one 27-bit network

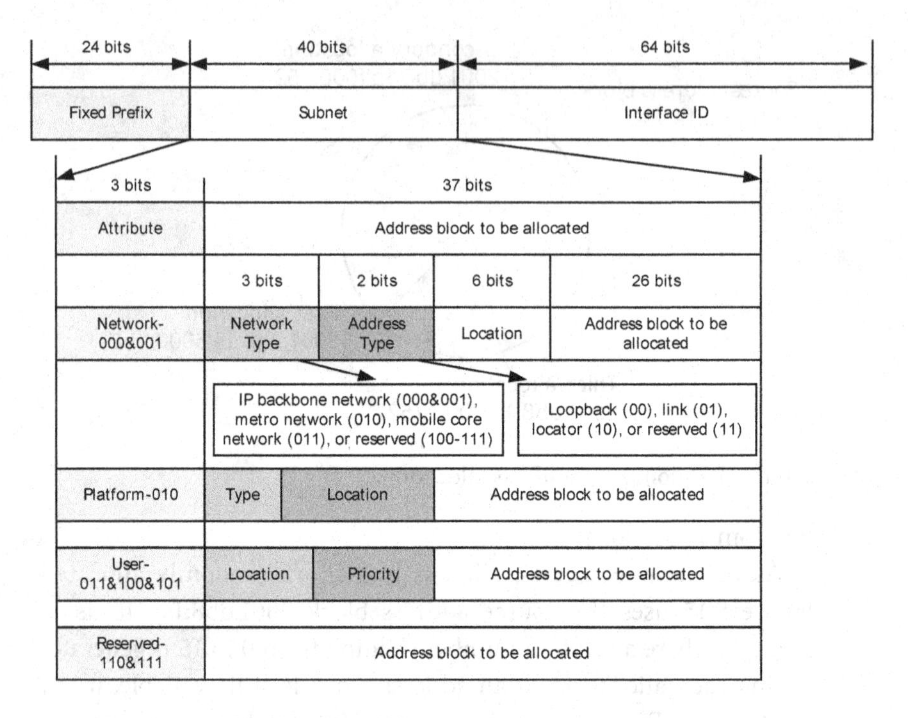

FIGURE 8.16 Example of IPv6 address planning.

prefix are insufficient, multiple 27-bit network prefixes can be allocated. As shown in Figure 8.17, three 27-bit network prefixes are allocated to user addresses. Two 27-bit network prefixes are reserved for future use.

- For network addresses, multiple 30-bit network prefixes can be allocated from a 27-bit network prefix to different network types or only one network type. Multiple 32-bit network prefixes are further allocated to different address types according to specific network types. From there, multiple 38-bit network prefixes are further allocated to different locations according to specific address types.
- For platform addresses, multiple 29-bit network prefixes can be allocated from a 27-bit network prefix to different platforms, and multiple 35-bit addresses can be further allocated to different locations.
- For user addresses, multiple 33-bit network prefixes can be allocated from a 27-bit network prefix to different locations, and multiple 35-bit addresses can be further allocated to users with different priorities in each location.

TABLE 8.1 Fields Involved in IPv6 Address Planning

Field	Description
Fixed Prefix	Indicates the prefix with a fixed length, which is allocated by a regional Internet registry. In this example, the length of the fixed prefix is 24 bits.
Subnet	Indicates an IPv6 network prefix used for a carrier's address planning. A subnet includes: • Attribute: indicates the first-level classification of a subnet. It is used to distinguish the address type, such as a network address (used for network device interconnection), a platform address (used for a service platform), or a user address (allocated to users, such as home or enterprise users). The length of this field typically ranges from 1 to 4 bits. In Figure 8.16, 3 bits are used to identify an attribute. 000/001 indicates a network address, 010 a platform address, and 011/100/101 a user address. 110 and 111 are reserved for future use. • Network address (000/001): used for interconnection between network devices. It can be further classified into: o Network Type: indicates the IP backbone network, metro network, or mobile core network. The length of this field typically ranges from 1 to 3 bits. In Figure 8.16, 3 bits are used to identify a network type. 000 indicates an IP backbone network, 010 a metro network, and 011 a mobile core network. 100, 101, 110, and 111 are reserved for future use. o Address Type: indicates the type of an address, such as a loopback address, link address, or locator address. In Figure 8.16, 2 bits are used to identify an address type. 00 indicates a loopback address, 01 a link address, and 10 an SRv6 locator address. 11 is reserved for future use. o Location: identifies a location on the network. The length of this field typically ranges from 3 to 6 bits. In Figure 8.16, 6 bits are used to identify a location. • Platform address (010): is similar to a network address and can be further decomposed based on the platform type and location. • User address (011/100/101): can be further decomposed based on the user type and priority. • Reserved address block (110/111): indicates an address block that can be further allocated at a location.
Interface ID	Identifies a link. An interface ID is the last 64 bits of an IPv6 address and can be generated based on the link-layer address or manually configured. An interface ID is mainly used for non-P2P interconnection interfaces.

Note: In real-world network deployment, it is recommended that IPv6 addresses be planned in 4-bit increments to facilitate subsequent O&M.

Hierarchically allocating IPv6 addresses based on specified principles and methods delivers an appropriate IPv6 address planning result and paves the way for subsequent IPv6 and SRv6 network design.

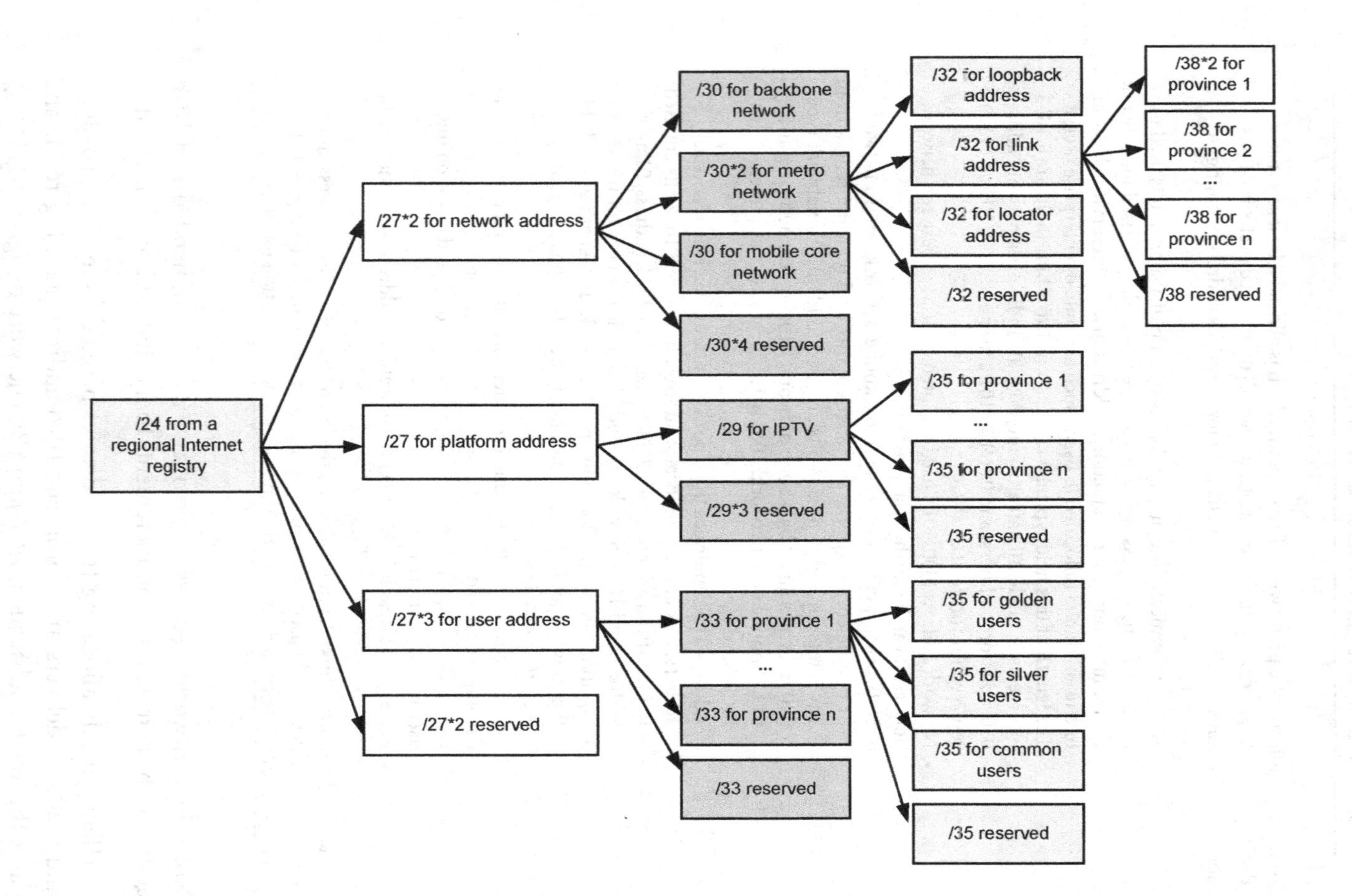

FIGURE 8.17 Hierarchical address planning.

8.3 SRv6 NETWORK DESIGN

The previous sections have discussed SRv6 application on the IP backbone network, metro network, and mobile transport network as well as the basics of IPv6 address planning. In the following sections, we will move on to SRv6 design in real-world deployment, covering basic SRv6 configuration, IGP design, BGP design, tunnel design, and VPN service design.

8.3.1 Basic SRv6 Configuration

Interface IPv6 addresses need to be configured prior to SRv6 configuration. The following is an example of interface IPv6 address configuration:

```
<HUAWEI> system-view
[~HUAWEI] interface GigabitEthernet 1/0/0
[~HUAWEI-GigabitEthernet1/0/0] ipv6 enable
[*HUAWEI-GigabitEthernet1/0/0] ipv6 address 2001:db8::1 127
[*HUAWEI-GigabitEthernet1/0/0] commit
[~HUAWEI-GigabitEthernet1/0/0] quit
```

In this example, IPv6 is enabled on an interface, and IPv6 address 2001:db8::1/127 is configured for this interface.

Basic SRv6 configuration includes the following operations:

- Enable SRv6.
- Configure the source address to be encapsulated into the IPv6 header of an SRv6 packet.
- Configure locators. During locator configuration, the prefix, prefix length, and the bits reserved for static Function IDs are specified. The following is an example of basic SRv6 configuration:

```
[~HUAWEI] segment-routing ipv6
[*HUAWEI-segment-routing-ipv6] encapsulation source-address 1::1
[*HUAWEI-segment-routing-ipv6] locator SRv6_locator ipv6-prefix A1:: 64
  static 32
[*HUAWEI-segment-routing-ipv6] commit
[~HUAWEI-segment-routing-ipv6] quit
```

In this example, the source address in the IPv6 header is **1::1**, the locator name is **SRv6_locator**, the corresponding IPv6 prefix is **A1::/64**, and **static 32** indicates that 32 bits are reserved for Function IDs in static SIDs.

Note: Bits need to be reserved for static Function IDs when SIDs are manually configured. End SIDs and End.X SIDs often need to be specified manually for convenient O&M, so that they are not changed after device restart or link flapping. When this is the case, static SIDs can be configured. Dynamic Function IDs are used when the system automatically allocates SIDs, for example, when the system automatically allocates a SID to a VPN instance on the local device.

After the configuration is completed, the locator information can be viewed using the following command:

```
[~HUAWEI] display segment-routing ipv6 locator verbose

                    Locator Configuration Table
                    ---------------------------

LocatorName  : SRv6_locator                    LocatorID   : 1
IPv6Prefix   : A1::                           PrefixLength: 64
StaticLength : 32                             Reference   : 4
ArgsLength   : 0
AutoSIDPoolID: 8193
AutoSIDBegin : A1::1:0:0
AutoSIDEnd   : A1::FFFF:FFFF:FFFF:FFFF

Total Locator(s): 1
```

The command output shows the configured locator name, IPv6 prefix, prefix length, bits reserved for static Function IDs, as well as start and end dynamic SIDs.

8.3.2 IGP Design

After the local configuration is completed on network nodes, information such as the interface address and SRv6 SID needs to be advertised to the network through an IGP to achieve network connectivity. IS-IS IPv6 or OSPFv3 can be used as an IGP on an SRv6 network.

Figure 8.18 shows typical IGP planning on a single-AS network. Different IGP domains are deployed at different network layers.

As shown in Figure 8.18, the entire network is in a single AS, and the access, aggregation, and IP core layers belong to different IGP domains. The following methods are available for IGP deployment on this network:

Method 1: IS-IS Level-2 or OSPFv3 area 0 is deployed at the IP core layer, IS-IS Level-1 or OSPF area X at the aggregation layer, and a separate IS-IS or OSPFv3 process at the access layer. This method is commonly used.

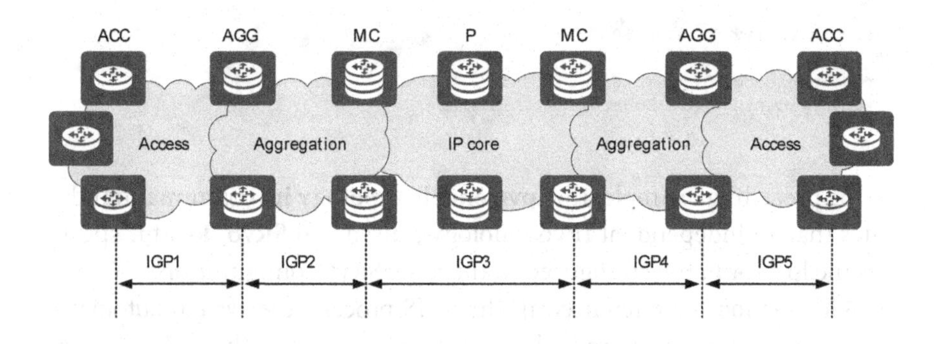

FIGURE 8.18 IGP planning.

Method 2: Different IGP processes are deployed at the access, aggregation, and IP core layers.

Regardless of the deployment method, only aggregated routes are advertised on demand between IGP domains of different IS-IS levels/OSPF areas or processes. This effectively reduces the number of IGP routes that need to be maintained by each node, relieves device pressure, and speeds up route convergence. In addition, route aggregation can effectively reduce the size of forwarding tables of border nodes.

SRv6 replaces MPLS labels with IPv6 addresses, making full use of the routable and aggregatable features of IPv6 addresses. Locators can be advertised across the entire network using only a few routes to form an E2E reachable SRv6 BE path.

Taking IS-IS as an example, the mandatory basic configurations include network-entity, level, and cost-style. To enable IS-IS to support SRv6, IPv6 needs to be configured in the IS-IS process. To prevent IPv4 and IPv6 IS-IS neighbors from impacting each other, it is recommended that IPv6 IS-IS use an independent topology and the SRv6 locator configured in basic SRv6 configuration be referenced. The following is an example of these configurations:

```
[~HUAWEI] isis 1
[~HUAWEI-isis-1] display this
#
isis 1
 is-level level-2
 cost-style wide
```

```
 network-entity 01.0000.0000.0007.00
 #
 ipv6 enable topology ipv6
 segment-routing ipv6 locator SRv6_locator
#
```

In the preceding example, the **ipv6 enable topology ipv6** command indicates that an independent IPv6 topology is used and **SRv6_locator** specifies the locator name configured in the basic SRv6 configuration.

With the locator referenced in the IS-IS process, the system automatically allocates End SIDs and advertises locator routes in the IS-IS process. The following output shows two End SIDs allocated by the system. The SID whose Flavor field is **PSP** is used for PSP, and the other SID is used for USP.

```
[~HUAWEI] display segment-routing ipv6 local-sid end forwarding

                   My Local-SID End Forwarding Table
                   ---------------------------------

SID        : A1::1:0:72/128                         FuncType : End
Flavor     : PSP
LocatorName: SRv6_locator                            LocatorID: 1

SID        : A1::1:0:73/128                         FuncType : End
Flavor     : --
LocatorName: SRv6_locator                            LocatorID: 1

Total SID(s): 2
```

The following lists information about the locator routes advertised in the IS-IS process.

```
[~HUAWEI] display ipv6 routing-table A1::
Route Flags: R - relay, D - download to fib, T - to vpn-instance, B - black
hole route
------------------------------------------------------------------------------
Routing Table : _public_
Summary Count : 1

Destination : A1::                          PrefixLength : 64
NextHop     : ::                           Preference   : 15
Cost        : 0                           Protocol     : ISIS-L2
RelayNextHop : ::                           TunnelID     : 0x0
Interface    : NULL0                        Flags        : DB
```

In addition to the configuration of the IS-IS process, the interface configuration is the same as that of a common IS-IS IPv6 interface. Typical configuration includes enabling IS-IS IPv6, configuring the IS-IS IPv6 cost, and configuring the IS-IS circuit-type. The following is an example of IS-IS IPv6 interface configurations:

```
[~HUAWEI] interface gigabitethernet 1/0/1
[~HUAWEI-GigabitEthernet1/0/1] display this
#
interface GigabitEthernet1/0/1
 undo shutdown
 ipv6 enable
 ipv6 address 2001:db8::1/127
 isis ipv6 enable 1
 isis circuit-type p2p
 isis ipv6 cost 10
#
```

In the preceding output, IPv6 IS-IS process 1 is enabled on the interface, circuit-type is set to P2P, and the IS-IS IPv6 cost is set to 10.

After IS-IS IPv6 is enabled on the interface, the system automatically generates End.X SIDs for each interface. The following is an example of the generated End.X SIDs. This example includes two interfaces, each of which is allocated two End.X SIDs. The SID whose Flavor field is **PSP** is used for PSP, and the other SID is used for USP.

```
[~HUAWEI] display segment-routing ipv6 local-sid end-x forwarding

           My Local-SID End.X Forwarding Table
           ------------------------------------

SID        : A1::1:0:74/128                     FuncType : End.X
Flavor     : PSP
LocatorName: SRv6_locator                        LocatorID: 1
NextHop    :                 Interface :        ExitIndex:
FE80::82B5:75FF:FE4C:2B1A      GE1/0/1            0x0000001d

SID        : A1::1:0:75/128                     FuncType :End.X
Flavor     : --
LocatorName: SRv6_locator                        LocatorID: 1
NextHop    :                 Interface :        ExitIndex:
FE80::82B5:75FF:FE4C:2B1A      GE1/0/1            0x0000001d

SID        : A1::1:0:76/128                     FuncType :End.X
Flavor     : PSP
LocatorName: SRv6_locator                        LocatorID: 1
NextHop    :                 Interface :        ExitIndex:
FE80::82B5:75FF:FE4C:326A      GE1/0/2            0x0000001e

SID        : A1::1:0:77/128                     FuncType :End.X
Flavor     : --
LocatorName: SRv6_locator                        LocatorID: 1
NextHop    :                 Interface :        ExitIndex:
FE80::82B5:75FF:FE4C:326A      GE1/0/2            0x0000001e

Total SID(s): 4
```

End SIDs and End.X SIDs can also be manually configured. After a locator is configured, the Function field can be configured using the

opcode command. Opcode indicates the Function ID specified on the basis of a configured locator. The following is an example of these configurations:

```
[~HUAWEI] segment-routing ipv6
[~HUAWEI-segment-routing-ipv6] display this
#
segment-routing ipv6
 encapsulation source-address 1::1
 locator SRv6_locator ipv6-prefix A1:: 64 static 32
  opcode ::1 end
  opcode ::2 end-x interface GigabitEthernet1/0/1 nexthop 2001:db8:12::1
#
```

In the preceding output, the Function ID of the End SID is set to ::1 based on the locator A1::/64 and the End SID is eventually A1::1; the Function ID of the End.X SID is set to ::2 for GigabitEthernet 1/0/1 and the End.X SID is eventually A1::2. After being manually configured, the SIDs are directly used by an IGP.

8.3.3 BGP Design

On an SRv6 network, in addition to the conventional route advertisement function, BGP also supports information exchange between forwarders and a controller. Forwarders use BGP-LS to report information, such as the network topology and latency, to the controller for path computation. To support SR, forwarders need to report SR information to the controller through BGP-LS. In addition, the controller uses BGP SR Policy to deliver SR path information. For this reason, on an SRv6 network, BGP design needs to consider not only the IPv6 unicast address family peer design and VPN/EVPN address family peer design, but also the BGP-LS address family peer design and BGP IPv6 SR-Policy address family peer design.

Do note that BGP IPv6 unicast address family peers are not necessary on a single-AS network. As such, locator and loopback routes can be advertised between different IGP domains through mutual route import, instead of through BGP. This is not the case on a multi-AS network as inter-AS locator and loopback routes need to be advertised through BGP. As for the process, the route advertisement using BGP is the same as the conventional mode, which means that IGP routes can be imported to BGP for aggregation and advertisement, or a static aggregated route can be configured and imported to BGP for advertisement.

BGP-LS and BGP IPv6 SR-Policy are mainly used for interaction between forwarders and the controller, and Section 8.3.5 dives into the design details. As for BGP L3VPN/EVPN, it is mainly used to advertise L3VPN/EVPN private network routes, and refer to Section 8.3.6 for design details.

8.3.4 SRv6 BE Design

After the basic network configuration is completed, SRv6 paths can be configured to provide differentiated services. These paths are classified into SRv6 BE and SRv6 TE paths. The former are automatically computed based on the IGP shortest path and BGP optimal route, with no need for a controller, whereas the latter are always planned using a controller. In terms of usage, SRv6 BE applies to services that have no special requirements on path planning, such as Internet and private line services without high SLA requirements, whereas SRv6 TE applies to services that have high SLA requirements.

8.3.4.1 Locator Route Advertisement

On an SRv6 BE path, packets are forwarded along the shortest path based on locator routes. It is worth noting that SRv6 BE paths inherently support ECMP.

To reduce the size of forwarding tables of border nodes through route aggregation, locators on the entire network must be planned and allocated hierarchically. Specifically, a separate large network prefix is allocated to each pair of MCs, and certain addresses are reserved for scalability. On top of that, a subnet prefix is allocated to each pair of AGGs from the network prefix of the MCs and to each access device (ACC) from the network prefix of the AGGs, respectively.

Figure 8.19 shows the process of advertising locator routes in a single AS.

The details of this process are as follows:

1. The locator route A1::8:0/84 is configured on ACC2, imported to an IGP, and then advertised to AGG2.
2. AGG2 imports the locator routes of all ACCs, summarizes them into the route A1::/74, and advertises the aggregated route in the IGP aggregation domain (IGP4). In the opposite direction, AGG2 also

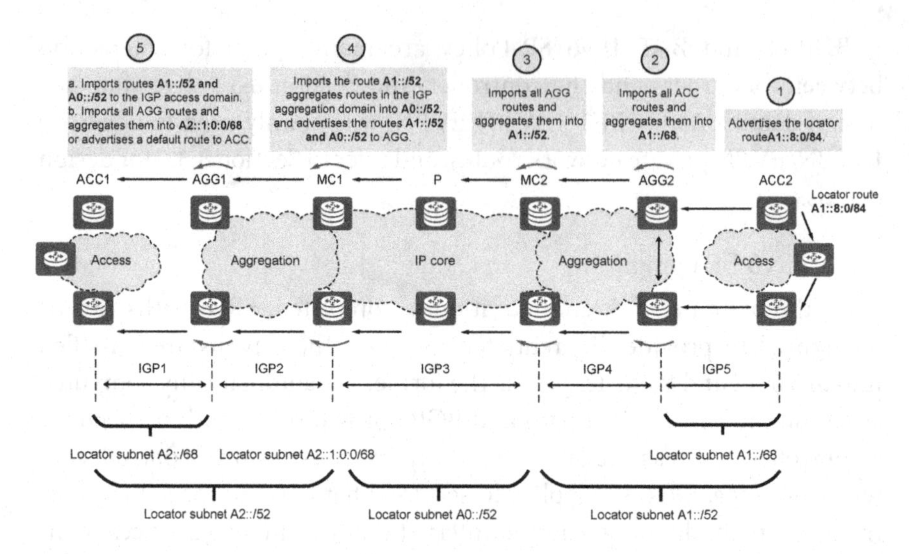

FIGURE 8.19 Locator route advertisement in a single AS.

needs to import locator routes to the IGP access domain (IGP5) and advertise them to ACC2. Therefore, a tag needs to be configured for this route import to prevent routing loops.

3. MC2 imports the locator routes of all AGGs to the IGP core domain (IGP3), summarizes them into A1::/64, and advertises the aggregated route in the IGP core domain (IGP3).

4. MC1 advertises the following two types of routes to the IGP aggregation domain (IGP2):

   a. Locator route A1::/64 leaked by MC2

   b. Aggregated locator route A0::/64 in the IGP core domain (IGP3)

5. AGG1 advertises routes to ACC1 in the IGP access domain (IGP1) using either of the following methods:

   a. Leaks networkwide locator routes. Because routes have been aggregated at each layer, only a limited number of routes need to be leaked.

   b. Advertises a default route.

Both methods can be used to set up SRv6 BE paths.

The locator route advertisement process of ACC1 is similar to that of ACC2.

After all locator routes are advertised, an E2E SRv6 BE path can be established between all nodes on the network. From this process, we can see that locator route advertisement on the entire network requires route import and aggregation in different IGP domains. We describe how to import and summarize IS-IS routes in the following section.

8.3.4.2 IS-IS Route Import and Aggregation

In the following example for configuring IS-IS route import and aggregation, routes are imported from IS-IS 100 to IS-IS 1 and then aggregated. During route import, tag 100 is configured for the routes, and a route policy is configured to reject the routes (with tag 1) imported from IS-IS 1 to IS-IS 100. This prevents routing loops during mutual route import.

```
[~HUAWEI] isis 1
[*HUAWEI-isis-1] ipv6 import-route isis 100 route-policy 100TO1
[*HUAWEI-isis-1] ipv6 summary A1::1:0:0 74
[*HUAWEI-isis-1] quit
[*HUAWEI] route-policy 100TO1 deny node 10
[*HUAWEI-route-policy] if-match tag 1
[*HUAWEI-route-policy] quit
[*HUAWEI] route-policy 100TO1 permit node 20
[*HUAWEI-route-policy] apply tag 100
[*HUAWEI-route-policy] quit
[*HUAWEI] commit
[~HUAWEI] isis 100
[*HUAWEI-isis-1] ipv6 import-route isis 1 route-policy 1TO100
[*HUAWEI-isis-1] ipv6 summary A1::2:0:0 74
[*HUAWEI-isis-1] quit
[*HUAWEI] route-policy 1TO100 deny node 10
[*HUAWEI-route-policy] if-match tag 100
[*HUAWEI-route-policy] quit
[*HUAWEI] route-policy 1TO100 permit node 20
[*HUAWEI-route-policy] apply tag 1
[*HUAWEI-route-policy] quit
[*HUAWEI] commit
```

8.3.4.3 SRv6 BE TI-LFA Protection

We need to design protection and failure recovery solutions on the grounds that we must take network reliability into account during SRv6 solution design.

In that aspect, TI-LFA FRR can be used to implement topology-independent protection for the transit nodes of SRv6 BE paths in an IGP domain. When doing this, microloop avoidance needs to be used together

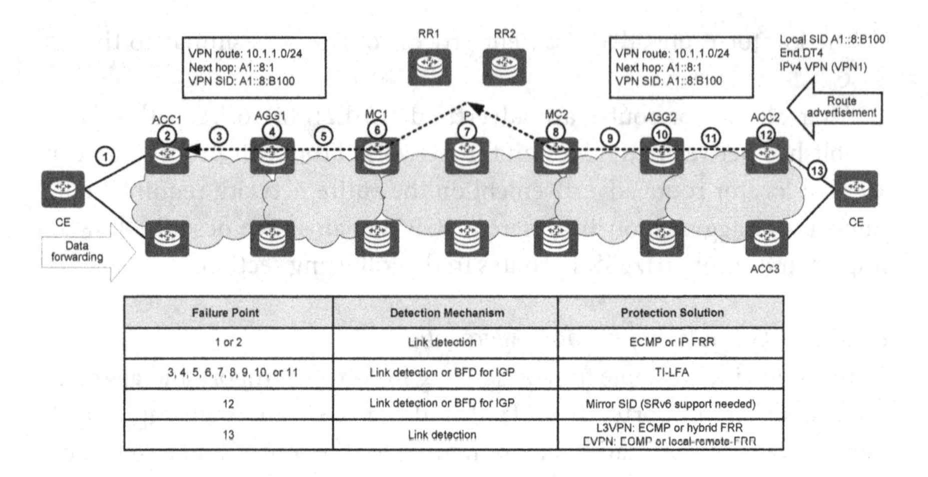

Failure Point	Detection Mechanism	Protection Solution
1 or 2	Link detection	ECMP or IP FRR
3, 4, 5, 6, 7, 8, 9, 10, or 11	Link detection or BFD for IGP	TI-LFA
12	Link detection or BFD for IGP	Mirror SID (SRv6 support needed)
13	Link detection	L3VPN: ECMP or hybrid FRR EVPN: ECMP or local-remote-FRR

FIGURE 8.20 E2E reliability scenarios and technologies.

with TI-LFA to mitigate against microloop scenarios by enabling devices to achieve fast switching through microloop avoidance. During network design, special design considerations are not needed for TI-LFA and microloop avoidance; instead, we simply need to enable them in IGP. The egress can use mirror SIDs to implement fast protection. Figure 8.20 illustrates the E2E reliability scenarios and technologies.

As shown in Figure 8.20, different protection technologies are used for failures at different locations.

- For failure points 1 and 2, link detection can be used to detect failures, and ECMP or IP FRR can be used to protect services.
- For failure points 3–10, TI-LFA FRR can be used to protect services, and link detection or BFD for IGP can be used to detect failures and trigger FRR switching.
- For failure points 11 and 12, mirror SIDs can be used to protect services, and link detection or BFD for IGP can be used to detect failures and trigger FRR switching.
- For failure point 13, link detection can be used to detect failures. For L3VPN, ECMP or hybrid FRR (IP FRR in VPN instance) can be used to protect services. Specifically, when the next hop of the route that originates from ACC2 to its connected CE is unreachable, traffic can

be forwarded to ACC3 through a tunnel and then to the CE through VPN forwarding table lookup. For EVPN, ECMP or local-remote-FRR is used to protect services. More precisely, if the link between ACC2 and the CE fails, the traffic sent to ACC2 is bypassed to ACC3 and then forwarded to the CE in order to reduce packet loss.

The following shows an example of TI-LFA FRR and microloop avoidance configuration:

```
[~HUAWEI] isis 1
[~HUAWEI-isis-1] display this
#
isis 1
 is-level level-2
 cost-style wide
 network-entity 01.0000.0000.0007.00
 #
 ipv6 enable topology ipv6
 segment-routing ipv6 locator SRv6_locator
 ipv6 avoid-microloop segment-routing
 ipv6 avoid-microloop segment-routing rib-update-delay 10000
 ipv6 frr
  loop-free-alternate level-2
  ti-lfa level-2
#
```

TI-LFA is enabled using the **ti-lfa level-2** command; while local microloop avoidance and remote microloop avoidance are enabled using the **avoid-microloop frr-protected** and **ipv6 avoid-microloop segment-routing** commands, respectively. The **ipv6 avoid-microloop segment-routing rib-update-delay** command configures a delay time for updating Routing Information Bases (RIBs) based on the number of routes on the network.

After the configuration is completed, a backup path is generated.

The following shows an example of verifying TI-LFA FRR and microloop avoidance configuration:

```
[~HUAWEI] display isis route ipv6 A1:: verbose

                 Route information for ISIS(1)
                 -----------------------------

                 ISIS(1) Level-1 Forwarding Table
                 --------------------------------

IPV6 Dest  : A1::/128         Cost: 20            Flags: A/-/-/-
Admin Tag  : -            Src Count: 1          Priority: Low
NextHop    :             Interface:         ExitIndex  :
```

```
FE80::82B5:75FF:FE4C:3268           GE1/0/2           0x0000001e
SRv6 TI-LFA:
Interface : GE1/0/1
Nexthop   : FE80::82B5:75FF:FE4C:2B1A IID:0x01000227
Backup sid Stack(Top->Bottom): {A2::5}
Flags: D-Direct, A-Added to URT, L-Advertised in LSPs, S-IGP Shortcut,
      U-Up/Down Bit Set, LP-Local Prefix-Sid
```

In short, this section explains SRv6 BE design, which covers route advertisement and protection solution deployment. In the following section, we expand on SRv6 TE design.

8.3.5 SRv6 TE Design

For services with high SLA requirements, a controller is needed to compute paths based on constraints and SRv6 TE tunnels need to be deployed to meet service requirements.

A controller's SRv6 TE path computation result can be either a strict explicit path (each hop has an outbound link specified) or a loose explicit path (only part of the nodes have an outbound link specified).

In loose explicit path scenarios, an unspecified node only needs to support IPv6 forwarding, as opposed to SRv6. This is one of the aspects that SRv6 transcends SR-MPLS in, making it easier for conventional IP/MPLS networks to evolve to SRv6.

8.3.5.1 SRv6 Policy

SRv6 Policy is a way of establishing SRv6 TE tunnels. It not only unifies tunnel models but also supports load balancing and primary/backup path protection, thereby improving reliability. Beyond that, SRv6 Policy uses the color attribute to define application-level network SLA policies. This attribute can be seen as an SLA template ID that corresponds to the requirements of a group of SLA indicators, such as network latency and bandwidth. Based on this very attribute, the controller can uniformly plan path constraints such as network latency and bandwidth constraints. As we covered in Chapter 4, the color attribute can also be carried in the BGP routes to be advertised. Network nodes can compare the color attribute of a BGP route with that of an SRv6 Policy to associate services with tunnels.

As described in Chapter 4, the process involved in SRv6 Policy path computation is as follows:

1. The controller uses BGP-LS to collect network topology and TE link information.
2. The controller computes a path that meets the SLA requirements based on the preset color, headend, and endpoint information.
3. The controller uses BGP extension to deliver an SRv6 Policy that carries the color attribute and endpoint information to the headend.
4. The BGP VPN routes to be advertised carry the color attribute and endpoint information (next hop).
5. After receiving a BGP VPN route, the headend recurses the route to an SRv6 Policy based on the color attribute and next hop.
6. The controller monitors the SLAs of services or paths in real time. If SLA performance is reduced, it performs path computation again and delivers a new SRv6 Policy to the headend.

8.3.5.2 BGP-LS and BGP SRv6 Policy

As we mentioned in Chapter 4, the controller needs to collect link state information through BGP-LS. It also computes paths based on the link state information and service requirements, and delivers the path computation result through BGP IPv6 SR-Policy. Figure 8.21 serves as a design example.

To delve into more details, BGP-LS is used to report topology, TE, and SRv6 information to the controller. One recommendation for reducing the number of BGP peers for the controller involves establishing BGP-LS peer relationships between the controller and RRs, and then between RRs and each node.

Regarding BGP IPv6 SR-Policy, it is used by the controller to deliver path information to forwarders. As for reducing the number of BGP peers for the controller, the recommendation is to establish BGP IPv6 SR-Policy peer relationships between the controller and RRs and then between RRs and each node. A BGP IPv6 SR-Policy message takes effect only on a specified network node. As such, to reduce the advertisement scope of BGP IPv6 SR-Policy messages, RRs can be configured to forward the BGP IPv6 SR-Policy messages delivered by the controller to only the specified network nodes.

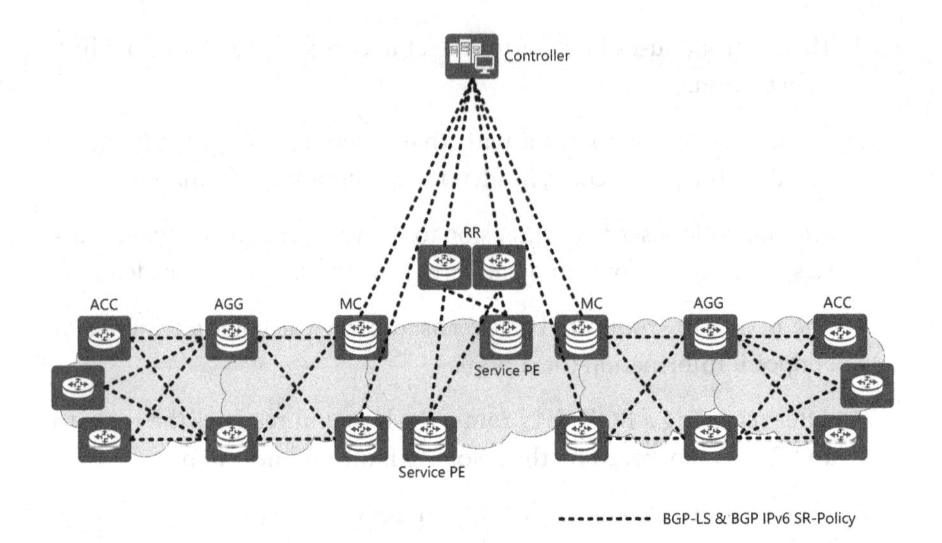

FIGURE 8.21 BGP SRv6 Policy peer relationship design.

The following is an example of BGP-LS configuration:

```
[~HUAWEI] bgp 100
[*HUAWEI-bgp] peer 100::100 as-number 100
[*HUAWEI-bgp] link-state-family unicast
[*HUAWEI-bgp-af-ls] peer 100::100 enable
[*HUAWEI-bgp-af-ls] quit
[*HUAWEI-bgp] quit
[*HUAWEI] commit
```

The following is an example of BGP-SRv6 Policy configuration:

```
[~HUAWEI] bgp 100
[~HUAWEI-bgp] ipv6-family sr-policy
[*HUAWEI-bgp-af-ipv6-srpolicy] peer 100::100 enable
[*HUAWEI-bgp-af-ipv6-srpolicy] quit
[*HUAWEI-bgp] quit
[*HUAWEI] commit
```

8.3.5.3 SRv6 Policy Path Computation

To meet SLA requirements, SRv6 Policy path computation needs to be performed based on constraints, including the priority, bandwidth, affinity attribute, explicit path, latency threshold, primary and backup paths, as well as path disjointness constraints.

When these constraints are met, the computed optimal path may be the one with the minimum cost, minimum latency, or balanced bandwidth.

SRv6 Policy path computation can be centrally performed on the controller, which computes a path that meets constraints based on the preset color, headend, and endpoint information, as well as delivers the path computation result to the headend through BGP IPv6 SR-Policy. In addition to being delivered through the controller, SRv6 Policies can also be statically configured using commands.

The following is an example of SRv6 Policy configuration:

```
[~HUAWEI] segment-routing ipv6
[~HUAWEI-segment-routing-ipv6] segment-list list1
[*HUAWEI-segment-routing-ipv6-segment-list-list1] index 5 sid ipv6 A2::1:0:0
[*HUAWEI-segment-routing-ipv6-segment-list-list1] index 10 sid ipv6 A3::1:0:0
[*HUAWEI-segment-routing-ipv6-segment-list-list1] commit
[~HUAWEI-segment-routing-ipv6-segment-list-list1] quit
[~HUAWEI-segment-routing-ipv6] srv6-te-policy locator SRv6_locator
[*HUAWEI-segment-routing-ipv6] srv6-te policy policy1 endpoint 3::3 color 101
[*HUAWEI-segment-routing-ipv6-policy-policy1] binding-sid A1::100
[*HUAWEI-segment-routing-ipv6-policy-policy1] candidate-path preference 100
[*HUAWEI-segment-routing-ipv6-policy-policy1-path] segment-list list1
[*HUAWEI-segment-routing-ipv6-policy-policy1-path] quit
[*HUAWEI-segment-routing-ipv6-policy-policy1] quit
[*HUAWEI-segment-routing-ipv6] quit
[*HUAWEI] commit
```

In this example, the segment list named **list1** is configured, with two nodes (A2::1:0:0 and A3::1:0:0) that the path needs to pass through specified. Following this, an SRv6 Policy is configured, including the endpoint, color, and binding SID. To wind up, a candidate path with the priority of 100 is configured for the SRv6 Policy, and the segment list **list1** is referenced by the candidate path.

8.3.5.4 SRv6 Policy Reliability

SRv6 BE and SRv6 TE share common ground in that they both require reliability solution design. As shown in Figure 8.22, TI-LFA can be used to protect the segment lists of the SRv6 Policy computed by the controller. To go a step further and ensure reliability in extreme scenarios, it is recommended that an SRv6 BE path be used as the best-effort path of an SRv6 Policy. In other words, if an SRv6 Policy fails, services are switched to the SRv6 BE path for best-effort forwarding. Another similarity between SRv6 BE and SRv6 TE is their use of mirror SIDs to protect services on the egress.

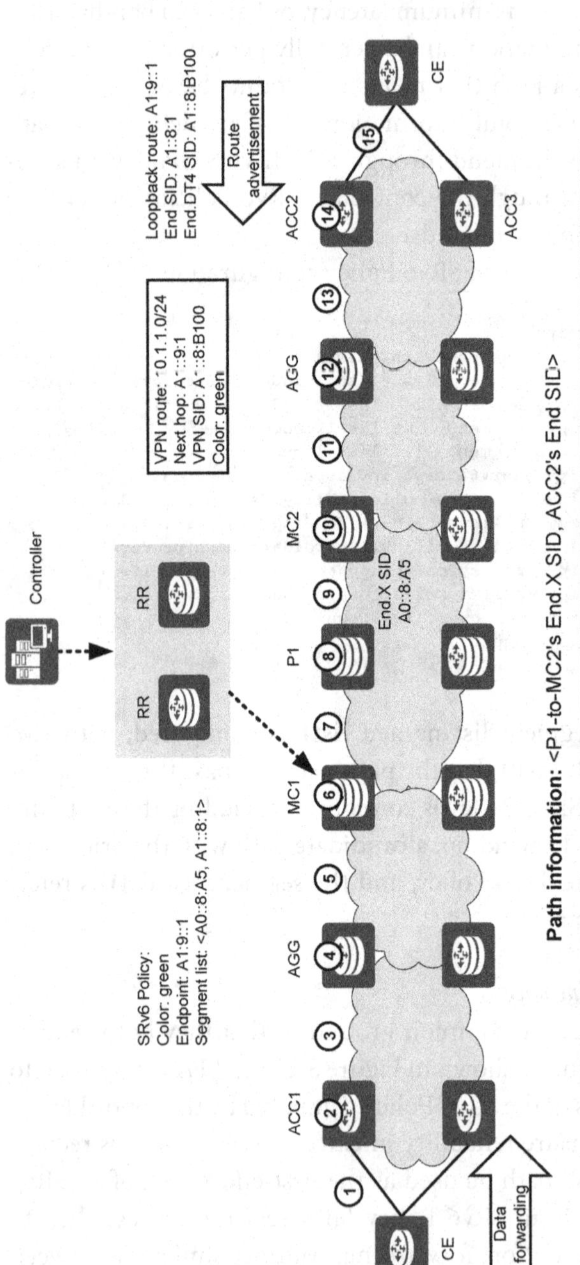

Failure Point	Detection Mecehanism	Protection Solution
1 or 2	Link detection	ECMP or IP FRR (L3VPN)
3, 4, 5, 6, 7, 10, 11, or 12	Link detection or BFD for IGP	TI-LFA
8 or 9	Link detection or BFD for IGP	TI-LFA protection for transit nodes
13 or 14	Link detection or BFD for IGP	Mirror SID (SRv6 support needed)
15	Link detection	L3VPN: ECMP or hybrid FRR EVPN: ECMP or local-remote-FRR

FIGURE 8.22 SRv6 Policy reliability design.

As shown in Figure 8.22, different protection technologies are used for failures at different locations.

- For failure points 1 and 2, link detection can be used to detect failures, and ECMP or IP FRR can be used to protect services.
- For failure points 3–12 excluding 8 and 9, TI-LFA FRR can be used to protect services, and link detection or BFD for IGP can be used to detect failures and trigger FRR switching.
- For failure points 8 and 9, TI-LFA FRR can be used to protect transit nodes, and link detection or BFD for IGP can be used to detect failures and trigger FRR switching.
- For failure points 13 and 14, mirror SIDs can be used to protect services, and link detection or BFD for IGP can be used to detect failures and trigger FRR switching.
- For failure point 15, link detection technology can be used to detect failures. For L3VPN, ECMP or hybrid FRR (IP FRR in VPN instance) can be used to protect services. In more details, when the next hop of the route that originates from ACC2 to its connected CE is unreachable, traffic can be forwarded to ACC3 through a tunnel and then to the CE through VPN forwarding table lookup. For EVPN, ECMP or local-remote-FRR is used to protect services. Specifically, if the link between ACC2 and the CE fails, the traffic sent to ACC2 is bypassed to ACC3 and then forwarded to the CE in order to reduce packet loss.

Looking at the above SRv6 BE design and SRv6 TE design side by side, it is noted that the latter adds TE path calculation and deployment, as well as midpoint protection for high availability.

8.3.6 VPN Service Design

SRv6 VPN services can use BGP as the unified signaling control plane to provide both Layer 2 and 3 service connections, with no need for deploying MPLS LDP.

Suggestions on VPN service deployment are as follows:

1. Layer 2 services: EVPN is used as the transport protocol.

2. L2/L3 hybrid services (both Layer 2 and Layer 3 services exist in a VPN): EVPN is used as the transport protocol.

3. Layer 3 services: Conventional L3VPN or EVPN L3VPN can be used.

As a recommendation, EVPN can be used to carry both L3VPN and L2VPN services in SRv6, thereby simplifying protocols.

The VPN deployment modes on the live network are classified into E2E VPN and hierarchical VPN. Among them, hierarchical VPN is widely deployed on MPLS networks to reduce the number of routes on access devices at network edges. It involves a large number of service access points, and transit nodes need to be aware of services. E2E VPN is recommended for SRv6 networks because only service access points, instead of transit nodes, need to be configured. Also, transit nodes do not need to be aware of services, and this in turn facilitates both deployment and maintenance.

Table 8.2 lists E2E VPN deployment scenarios.

Figure 8.23 shows a typical VPN service deployment model.

In the preceding model, a network is divided into three layers: access, aggregation, and IP core. The model also includes three levels of data

TABLE 8.2 E2E VPN Deployment Scenarios

Service Type	Technology	Application Scenario
L3VPN	• L3VPNv4/L3VPNv6 • EVPN L3VPNv4/L3VPNv6	• 4G/5G access • Internet service • VoIP • Video on Demand (VOD) • Enterprise Layer 3 private line
L2 E-Line	EVPN VPWS[4]	• Enterprise Layer 2 P2P private line
L2 E-LAN	EVPN[5]	• Enterprise Layer 2 MP2MP private line
L2 E-Tree	EVPN E-Tree[6]	• Enterprise Layer 2 P2MP private line • Home broadband access
L2/L3 hybrid	EVPN IRB[7]	• Telco cloud • DCI

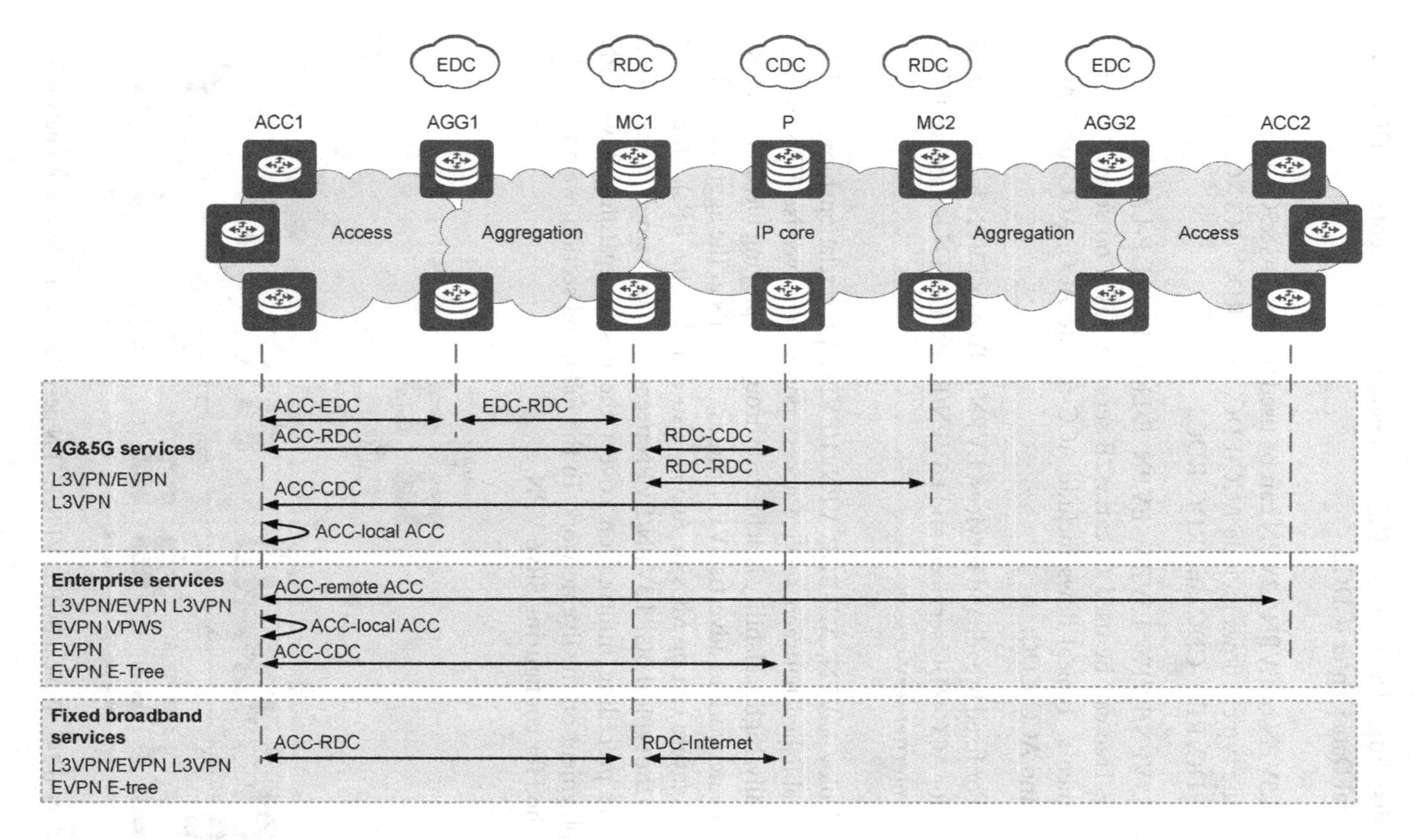

FIGURE 8.23 Typical VPN service deployment model.

centers: Edge Data Center (EDC), Regional Data Center (RDC), and Central Data Center (CDC).

- L3VPN or EVPN L3VPN can be used to carry 4G&5G services, and the typical flows include ACC-EDC, ACC-RDC, ACC-CDC, EDC-RDC, RDC-CDC, and RDC-RDC.
- L3VPN/EVPN L3VPN, EVPN E-Line, EVPN E-LAN, or EVPN E-Tree can be used to carry 2B services based on service requirements. Typical flows include ACC-remote ACC, ACC-local ACC, and ACC-CDC.
- For fixed broadband services, EVPN E-Tree or EVPN E-Line is used for ACC-RDC services, and L3VPN/EVPN L3VPN is used for RDC-Internet services.

To deploy a VPN, MP-BGP VPNv4/EVPN peer relationships need to be established between service access points to advertise VPN routes. Typically, high scalability is achieved through deploying hierarchical RRs on the network to advertise VPN routes. The RRs at the aggregation and access layers can be MCs or AGGs (also called inline RRs). Figure 8.24 shows the typical BGP L3VPN/EVPN peer design on a single-AS network.

The preceding information covers the basics regarding SRv6 VPN deployment on the live network. In the following section, we present the method for configuring SRv6 VPN.

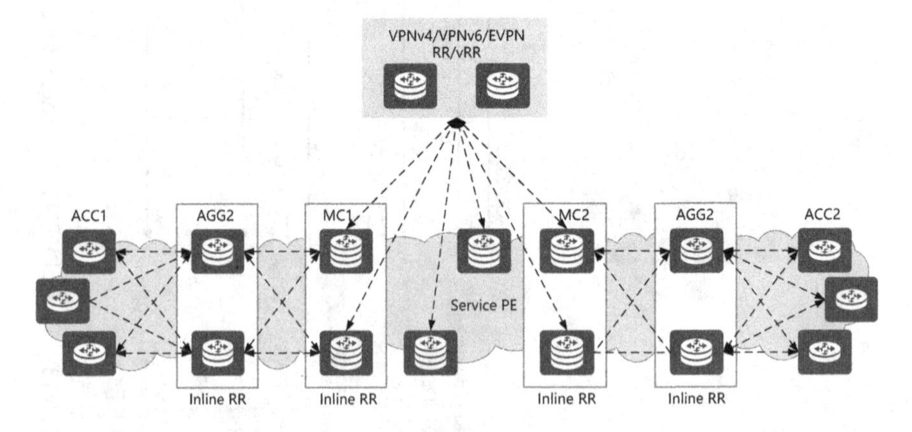

FIGURE 8.24 Typical BGP L3VPN/EVPN peer design on a single-AS network.

8.3.6.1 SRv6 EVPN L3VPN

In real-world deployment, SRv6 EVPN L3VPN is a common and recommended SRv6 VPN technology, and its configuration procedure is as follows:

1. Configure an EVPN L3VPN instance and bind an interface to the EVPN L3VPN instance. The following shows an example configuration:

```
[~HUAWEI] ip vpn-instance srv6_vpn2
[~HUAWEI-vpn-instance-srv6_vpn2] display this
#
ip vpn-instance srv6_vpn2
 ipv4-family
  route-distinguisher 100:2
  vpn-target 100:2 export-extcommunity evpn
  vpn-target 100:2 import-extcommunity evpn
#
[~HUAWEI] interface GigabitEthernet1/0/0.2
[~HUAWEI-GigabitEthernet1/0/0.2] display this
#
interface GigabitEthernet1/0/0.2
 vlan-type dot1q 2
 ip binding vpn-instance srv6_vpn2
 ip address 10.78.2.2 255.255.255.0
#
```

2. Configure BGP IPv6 peer relationships and enable the BGP IPv6 peers in the EVPN address family view. The following shows an example configuration:

```
[~HUAWEI] bgp 100
[~HUAWEI-bgp] display this
#
bgp 100
 peer 2::2 as-number 100
 peer 2::2 connect-interface LoopBack0
 #
 l2vpn-family evpn
  policy vpn-target
  peer 2::2 enable
  peer 2::2 advertise encap-type srv6
#
```

   For SRv6 application, BGP peer relationships need to be established using IPv6 addresses, and the **peer 2::2 advertise encap-type srv6** command needs to be run on BGP peers to enable SRv6 encapsulation.

3. In the BGP VPN instance view, configure the device to recurse VPN routes to SRv6 BE paths. The following shows an example configuration:

```
[~HUAWEI] bgp 100
[~HUAWEI-bgp] ipv4-family vpn-instance srv6_vpn1
[~HUAWEI-bgp-srv6_vpn1] display this
#
ipv4-family vpn-instance srv6_vpn1
 import-route direct
 advertise l2vpn evpn
 segment-routing ipv6 locator SRv6_locator
 segment-routing ipv6 best-effort
 peer 10.78.1.1 as-number 65002
#
```

The **segment-routing ipv6 locator** command specifies a locator. After the locator is configured, an End.DT4 SID is dynamically allocated from the locator to the EVPN L3VPN instance. The **segment-routing ipv6 best-effort** command configures an SRv6 BE path for VPN service transport.

The following command can be run to check End.DT4 SID entries in the Local SID Table:

```
[~HUAWEI] display segment-routing ipv6 local-sid end-dt4 forwarding

             My Local-SID End.DT4 Forwarding Table
             -------------------------------------

SID         : A1::1:0:9B/128                            FuncType  : End.DT4
VPN Name    : srv6_vpn1                                 VPN ID    : 2
LocatorName: SRv6_locator                                LocatorID: 1

SID         : A1::1:0:9C/128                            FuncType  : End.DT4
VPN Name    : srv6_vpn2                                 VPN ID    : 5
LocatorName: SRv6_locator                                LocatorID: 1

Total SID(s): 2
```

4. End.DT4 SIDs used by the VPN can also be manually configured. The following shows an example configuration:

```
[~HUAWEI] segment-routing ipv6
[~HUAWEI-segment-routing-ipv6] locator SRv6_locator
[~HUAWEI-segment-routing-ipv6-locator] display this
#
locator SRv6_locator ipv6-prefix A1:: 64 static 32
 opcode ::80 end-dt4 vpn-instance srv6_vpn1
#
```

The following command can be run to check the static End.DT4 SIDs:

```
[~HUAWEI] display segment-routing ipv6 local-sid end-dt4 forwarding

            My Local-SID End.DT4 Forwarding Table
            -------------------------------------

SID        : A1::80/128                      FuncType : End.DT4
VPN Name   : srv6_vpn1                        VPN ID   : 2
LocatorName: SRv6_locator                      LocatorID: 1

SID      : A1::1:0:9C/128                     FuncType : End.DT4
VPN Name   : srv6_vpn2                        VPN ID   : 5
LocatorName: SRv6_locator                      LocatorID: 1

Total SID(s): 2
```

The command output shows that after static SIDs are configured for the VPN, the system automatically uses the static SIDs and dynamic allocation is not needed.

After the configuration is completed, routing entries on the peer PE can be checked. The command output shows that the route carries the prefix SID A1::80.

```
[~HUAWEI] display bgp vpnv4 vpn-instance srv6_vpn2 routing-table 10.7.7.0
BGP local router ID : 10.37.112.122
Local AS number :100

VPN-Instance srv6_vpn2 Router ID 10.37.112.122:
Paths:  1 available, 1 best, 1 select, 0 best-external, 0 add-path
BGP routing table entry information of 10.78.1.0/24:
Route Distinguisher: 100:1
Remote-Cross route
Label information (Received/Applied): 3/NULL
From: 2::2 (10.37.112.119)
Route Duration: 0d00h29m01s
Relay IP Nexthop: FE80::E45:BAFF:FE28:7258
Relay IP Out-Interface: GigabitEthernet1/0/2
Relay Tunnel Out-Interface:
Original nexthop: 1::1
Qos information : 0x0
Ext-Community: RT <100 : 1>
Prefix-sid: A1::80
AS-path Nil, origin incomplete, MED 0, local preference 100, pref-val 0,
valid, internal, best, select, pre 255, IGP cost 20
Originator: 10.37.112.117
Cluster List: 10.37.112.119
Advertised to such 1 peers:
 10.79.1.1
```

8.3.6.2 SRv6 EVPN E-Line

The procedure for EVPN E-Line over SRv6 configuration is as follows:

1. Configure an EVPN instance. The following shows an example configuration:

```
[~HUAWEI] evpn vpn-instance srv6_vpws vpws
[~HUAWEI-vpws-evpn-instance-srv6_vpws] display this
#
evpn vpn-instance srv6_vpws vpws
 route-distinguisher 100:2
 segment-routing ipv6 best-effort
 vpn-target 100:2 export-extcommunity
 vpn-target 100:2 import-extcommunity
#
```

2. Configure an EVPL instance (EVPN E-Line instance) and specify the SRv6 mode. The following shows an example configuration:

```
[~HUAWEI] evpl instance 1 srv6-mode
[~HUAWEI-evpl-srv6-1] display this
#
evpl instance 1 srv6-mode
 evpn binding vpn-instance srv6_vpws
 local-service-id 100 remote-service-id 200
 segment-routing ipv6 locator SRv6_locator
#
```

The **segment-routing ipv6 locator** command specifies the locator used by the EVPL instance and dynamically allocates an End.DX2 SID from the locator to the EVPN E-Line instance. The following command can be run to view SID information.

```
[~HUAWEI] display segment-routing ipv6 local-sid end-dx2 forwarding

            My Local-SID End.DX2 Forwarding Table
            -------------------------------------

SID     : A1::82/128                      FuncType : End.DX2
EVPL ID  : 1
LocatorName: SRv6_locator                   LocatorID: 1

Total SID(s): 1
```

SIDs can also be manually configured for the EVPL instance. The following shows an example configuration:

```
[~HUAWEI] segment-routing ipv6
[~HUAWEI-segment-routing-ipv6] locator SRv6_locator
[~HUAWEI-segment-routing-ipv6-locator] display this
```

```
#
 locator SRv6_locator ipv6-prefix A1:: 64 static 32
  opcode ::82 end-dx2 evpl-instance 1
#
```

3. Bind an interface to the EVPL instance. The following shows an example configuration:

```
[~HUAWEI] interface GigabitEthernet1/0/0.100 mode l2
[~HUAWEI-GigabitEthernet1/0/0.100] display this
#
interface GigabitEthernet1/0/0.100 mode l2
 encapsulation dot1q vid 100
 rewrite pop single
 evpl instance 1
#
```

After the configuration is completed, run the following command to view A-D routes on the peer PE. The command output shows that the A-D route carries the prefix SID A1::82.

```
[~HUAWEI] display bgp evpn vpn-instance srv6_vpws routing-table ad-route 0000
.0000.0000.0000.0000:100
BGP local router ID : 10.37.112.122
Local AS number : 100

EVPN-Instance srv6_vpws:
Number of A-D Routes: 1
BGP routing table entry information of 0000.0000.0000.0000.0000:100:
Route Distinguisher: 100:2
Remote-Cross route
Label information (Received/Applied): 3/NULL
From: 2::2 (10.37.112.119)
Route Duration: 0d06h07m06s
Relay IP Nexthop: FE80::82B5:75FF:FE4C:326D
Relay IP Out-Interface: GigabitEthernet1/0/2
Relay Tunnel Out-Interface:
Original nexthop: 1::1
Qos information : 0x0
Ext-Community: RT <100 : 2>, EVPN L2 Attributes <MTU:1500 C:0 P:1 B:0>
AS-path Nil, origin incomplete, localpref 100, pref-val 0, valid, internal,
best, select, pre 255, IGP cost 20
Originator: 10.37.112.117
Cluster list: 10.37.112.119
Prefix-sid: A1::82
Route Type: 1 (Ethernet Auto-Discovery (A-D) route)
ESI: 0000.0000.0000.0000.0000, Ethernet Tag ID: 100
 Not advertised to any peer yet
```

The EVPN E-LAN over SRv6 configuration is similar to the EVPN E-Line over SRv6 configuration. More details are provided in Huawei product documentation.

8.3.6.3 EVPN SRv6 Policy

If EVPN traffic needs to be forwarded over SRv6 TE tunnels, the following steps need to be performed:

1. Configure the device to recurse EVPN routes to an SRv6 Policy.
2. Configure a tunnel policy.
3. Apply the tunnel policy to a VPN instance.
4. Add the color attribute to routes.

Detailed configurations are as follows:

1. Configure the device to recurse EVPN routes to an SRv6 Policy.

   The following is an example of EVPN L3VPN configuration:

```
[~PE1] bgp 100
[*PE1-bgp] ipv4-family vpn-instance srv6_vpn2
[*PE1-bgp-srv6_vpn2] segment-routing ipv6 traffic-engineering evpn
[*PE1-bgp-srv6_vpn2] quit
[*PE1-bgp] quit
[*PE1] commit
```

   The following is an example of EVPN E-Line configuration:

```
[~PE1] evpn vpn-instance srv6_vpws vpws
[*PE1-vpws-evpn-instance-srv6_vpws] segment-routing ipv6 traffic-engineering
[*PE1-vpws-evpn-instance-srv6_vpws] quit
[*PE1] commit
```

2. Configure a tunnel policy. The following is an example of tunnel policy configuration:

```
[~PE1] tunnel-policy p1
[*PE1-tunnel-policy-p1] tunnel select-seq ipv6 srv6-te-policy load-balance-
number 1
[*PE1-tunnel-policy-p1] quit
[*PE1] commit
```

3. Apply the tunnel policy to the VPN instance.

   The following is an example of EVPN L3VPN configuration:

```
[~PE1] ip vpn-instance srv6_vpn2
[*PE1-vpn-instance-srv6_vpn2] ipv4-family
[*PE1-vpn-instance-srv6_vpn2-af-ipv4] tnl-policy p1 evpn
[*PE1-vpn-instance-srv6_vpn2-af-ipv4] quit
[*PE1-vpn-instance-srv6_vpn2] quit
```

The following is an example of EVPN E-Line configuration:

```
[*PE1] evpn vpn-instance srv6_vpws vpws
[*PE1-vpws-evpn-instance-srv6_vpws] tnl-policy p1
[*PE1-vpws-evpn-instance-srv6_vpws] quit
[*PE1] commit
```

4. Add the color attribute to routes and ensure that it is consistent with the color attribute in SRv6 Policy so that traffic can be steered into the SRv6 Policy. The color attribute can be added to the import route policy on the local PE or the export route policy on the peer PE (PE2). The following example uses the configuration of an export route policy on the peer PE (PE2).

```
[~PE2] route-policy color100 permit node 1
[*PE2-route-policy] apply extcommunity color 0:100
[*PE2-route-policy] quit
[*PE2] commit
[~PE2] bgp 100
[*PE2-bgp] l2vpn-family evpn
[*PE2-bgp-af-evpn] peer 2::2 route-policy color100 export
[*PE2-bgp-af-evpn] quit
[*PE2-bgp] quit
[*PE2] commit
```

After the configuration is completed, run the following command to check the EVPN L3VPN forwarding table. The command output shows that the remote VPN route (10.7.7.0) has recursed to an SRv6 Policy.

```
[~PE1] display ip routing-table vpn-instance srv6_vpn2 10.7.7.0 verbose
Route Flags: R - relay, D - download to fib, T - to vpn-instance, B - black
hole route
------------------------------------------------------------------------------
Routing Table : srv6_vpn2
Summary Count : 1

Destination: 10.7.7.0/24
    Protocol: IBGP             Process ID: 0
   Preference: 255                  Cost: 0
     NextHop: 2::2             Neighbour: 2::2
      State: Active Adv Relied        Age: 00h03m15s
        Tag: 0                  Priority: low
      Label: 3                   QoSInfo: 0x0
  IndirectID: 0x10000E0           Instance:
 RelayNextHop: 0.0.0.0           Interface: SRv6-TE Policy
    TunnelID: 0x000000003400000001    Flags: RD
```

Run the following command to check the EVPN E-Line state. The command output shows that routes in the EVPL instance have recursed to an SRv6 Policy.

```
[~PE1] display bgp evpn evpl
Total EVPLs: 1    1 Up    0 Down

EVPL ID : 1
State : up
Evpl Type : srv6-mode
Interface : GigabitEthernet1/0/0.100
Ignore AcState : disable
Local MTU : 1500
Local Control Word : false
Local Redundancy Mode : all-active
Local DF State : primary
Local ESI : 0000.0000.0000.0000.0000
Remote Redundancy Mode : all-active
Remote Primary DF Number : 1
Remote Backup DF Number : 0
Remote None DF Number : 0
Peer IP : 2::2
 Origin Nexthop IP : 2::2
 DF State : primary
 Eline Role : primary
 Remote MTU : 1500
 Remote Control Word : false
 Remote ESI : 0000.0000.0000.0000.0000
 Tunnel info : 1 tunnels
  NO.0  Tunnel Type : srv6te-policy, Tunnel ID : 0x000000003400000001
Last Interface UP Timestamp : 2019-8-14 3:21:34:196
Last Designated Primary Timestamp : 2019-8-14 3:23:45:839
Last Designated Backup Timestamp : --
```

To sum up, this section explains SRv6 configuration for VPN services, including how to configure SRv6 EVPN L3VPN and SRv6 EVPN E-Line, as well as providing the related configuration examples.

8.4 EVOLUTION FROM MPLS TO SRv6

The following example uses MPLS L3VPN to describe how L3VPN services evolve from MPLS to SRv6.

Figure 8.25 shows MPLS network deployment in which the entire network is divided into two IGP domains, with AGGs at the border. LDP/RSVP-TE tunnels and BGP LSPs are deployed in both of the domains, and E2E BGP VPNv4 peer relationships are established between ACCs and MCs to advertise VPNv4 routes. Traffic is encapsulated into MPLS tunnels for forwarding.

After network nodes are upgraded to support SRv6, L3VPN services can be migrated from MPLS to SRv6 based on the following procedure:

1. Configure interface IPv6 addresses and locators.
2. Configure IS-IS IPv6 and enable SRv6, and then configure the forwarders to advertise locator routes.

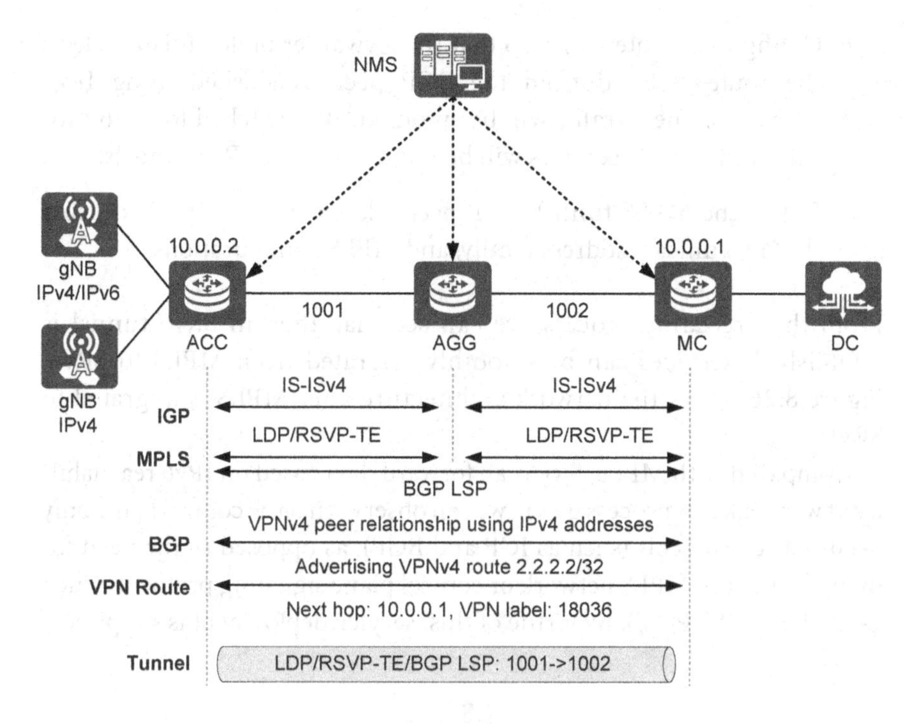

FIGURE 8.25 Evolution from MPLS to SRv6.

3. Establish BGP peer relationships between the controller and forwarders using the IPv6 unicast address family, and enable BGP-LS and BGP IPv6 SR-Policy. The controller delivers SRv6 Policies, and SRv6 TE tunnels are established on forwarders.

4. On forwarders, establish BGP VPNv4 peer relationships using IPv6 addresses so that BGP VPNv4 peers advertise VPN routes to each other. The color attribute of the VPN routes is consistent with that of SRv6 Policies to ensure that VPN routes can recurse to the SRv6 Policy.

5. In this case, each forwarder has two routes with the same prefix, one carrying the MPLS VPN label received from the BGP peer established using IPv4 addresses and the other carrying the VPN SID received from the BGP peer established using IPv6 addresses. If the two routes have the same attributes, a forwarder by default preferentially selects the route received from the BGP peer established using IPv4 addresses, and services can still be carried over MPLS tunnels.

6. Configure a route policy so that the forwarder preferentially selects the route received from the BGP peer established using IPv6 addresses. Then, traffic will be automatically switched to SRv6 tunnels, and L3VPN services will be migrated to the SRv6 tunnels.

7. Delete the MPLS tunnel, BGP peer relationships established using the IPv4 unicast address family, and MPLS configurations.

From the preceding process, we can see that after an SRv6 tunnel is established, services can be smoothly migrated from MPLS to SRv6. Figure 8.26 shows the network architecture after MPLS is migrated to SRv6.

Compared with MPLS, SRv6 can forward data based on IPv6 reachability. If we break this process down, we can observe that the control plane only needs basic protocols (such as IGP and BGP), as opposed to the need for maintaining the MPLS network or control plane signaling protocols (such as LDP and RSVP-TE). By virtue of this, service deployment is simplified.

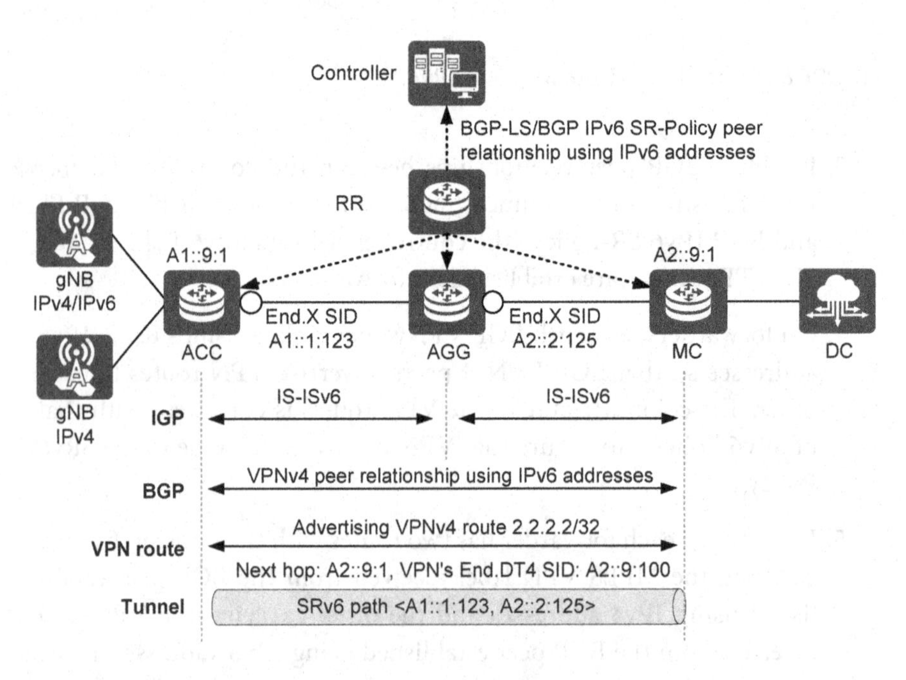

FIGURE 8.26 Network architecture after MPLS is migrated to SRv6.

8.5 STORIES BEHIND SRv6 DESIGN

Protocols are the base and an important part of IP network solutions.

Protocol requirements vary with network scenarios, and network solutions must be able to integrate protocol features to meet customer requirements well. In the previous chapter, we mentioned the difficulty in choosing the appropriate protocol design. In order to solve the issues, when we select protocols during network design, the customer requirements and application scenarios must be viewed through the same pair of lenses. For example, ring network protection is developed for a mobile transport network that often uses ring networking, and the spine-leaf architecture of data center networks triggers load balancing design and optimization.

Network solutions involve the comprehensive application of multiple protocols, which need to form an organic whole. This chapter provides such an example, in which multiple protocols, such as IGP, BGP, and PCEP, are integrated into the SRv6 network design.

Network solutions encompass multiple aspects, some of which may not involve interworking and go beyond the scope of protocols in a strict sense. A typical case in point is that certain local policies, such as route-policies, can be flexibly designed. In IETF standards, such policies are often simply described as "Out of Scope." To unify these policies and to allow these policies to be distributed between the controller and devices as SDN proceeds on a development trajectory, the IETF has standardized many protocols.

Network solution design, protocol design, and protocol implementation involve similar functions and design roadmaps. These similarities represent the architecture's quality attributes, such as scalability, reliability, security, usability, and maintainability. It is worth noting that the quality attributes of a network solution and those involved during protocol design and implementation complement each other. For example, the maximum number of allowable tunnels on a 5G transport network is relevant to the scalability of protocol design and implementation. From the protocol design perspective, SR has better scalability than RSVP-TE. From the protocol implementation perspective, the distributed architecture of the controller outshines the centralized architecture to support more SR Policies.

REFERENCES

[1] LINE Developers. LINE Data Center Networking with SRv6. (2019-09-20)[2020-10-31].

[2] Leymann N, Decraene B, Filsfils C, Konstantynowicz M, Steinberg D. Seamless MPLS Architecture[EB/OL]. (2015-10-14)[2020-03-25]. draft-ietf-mpls-seamless-mpls-07.

[3] Rooney T. IPv6 Address Planning: Guidelines for IPv6 address allocation[EB/OL]. (2013-09-24)[2020-03-25].

[4] Boutros S, Sajassi A, Salam S, Drake J, Rabadan J. Virtual Private Wire Service Support in Ethernet VPN[EB/OL]. (2018-12-11)[2020-03-25]. RFC 8214.

[5] Sajassi A, Aggarwal R, Bitar N, Isaac A, Uttaro J, Drake J, Henderickx W. BGP MPLS-Based Ethernet VPN[EB/OL]. (2020-01-21)[2020-03-25]. RFC 7432.

[6] Sajassi A, Salam S, Drake J, Uttaro J, Boutros S, Rabadan J. Ethernet-Tree (E-Tree) Support in Ethernet VPN (EVPN) and Provider Backbone Bridging EVPN (PBB-EVPN)[EB/OL]. (2018-01-31)[2020-03-25]. RFC 8317.

[7] Sajassi A, Salam S, Thoria S, Drake J, Rabadan J. Integrated Routing and Bridging in EVPN[EB/OL]. (2019-10-04)[2020-03-25]. draft-ietf-bess-evpn-inter-subnet-forwarding-08.

III

SRv6 2.0

CHAPTER 9

SRv6 OAM and On-Path Network Telemetry

THIS CHAPTER DESCRIBES the key technologies of SRv6 OAM and data-plane on-path network telemetry. SRv6 can be easily extended based on the existing IPv6 OAM mechanism to support Fault Management (FM) and Performance Measurement (PM). In terms of PM, this chapter covers its basic concepts including those of on-path network telemetry as well as the framework of In-situ Flow Information Telemetry (IFIT) that is used to implement on-path network telemetry. On-path network telemetry achieves higher accuracy than active PM methods without needing to generate extra packets for measurement. The IFIT framework supports data plane encapsulation for multiple on-path network telemetry technologies, enabling them to be used extensively on IP networks.

9.1 SRv6 OAM

9.1.1 OAM Overview

OAM is a general term that refers to a toolset for fault detection and isolation, and for performance measurement. In terms of functionality, it provides both FM and PM, as shown in Figure 9.1[1,2]

- FM
    - Continuity Check (CC): detects address reachability by using mechanisms such as IP ping,[3] BFD,[4] and LSP ping.

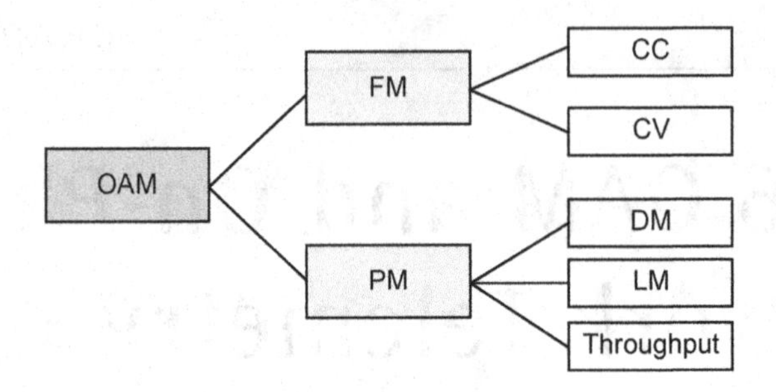

FIGURE 9.1 Classification of OAM functions.

- Connectivity Verification (CV): verifies paths and locates faults by using mechanisms such as IP traceroute,[3] BFD, and LSP traceroute.

- PM

  - Delay Measurement (DM): measures metrics such as delay and jitter.

  - Loss Measurement (LM): measures metrics such as the number of lost packets and packet loss rate.

  - Throughput measurement: measures metrics such as interface bandwidth, link bandwidth, and the packet processing capability per second.

The OAM mechanism varies according to networks. For example, IP ping and traceroute are used to implement FM functions such as CC and CV on an IP network, while LSP ping/traceroute is used to check LSP continuity and connectivity on an MPLS network. Additionally, BFD can be used for fast CC and CV on both IP and MPLS networks.

OAM implements PM in one of three ways: active, passive, or hybrid,[5] differing in whether OAM packets need to be proactively sent.

- Active PM: proactively generates OAM packets, measures the OAM packet performance, and uses the measurement result to infer the network performance. Two-Way Active Measurement Protocol (TWAMP) is a typical active PM method.[6]

- Passive PM: monitors the undisturbed and unmodified data packets to obtain performance parameters. Unlike active PM, passive PM observes the data packets directly, without modifying them or generating extra OAM packets, and can therefore accurately reflect the performance.

- Hybrid PM: combines both active and passive PM. It measures network performance by modifying only certain fields in data packets (e.g., it colors certain fields in packet headers) and does not send additional OAM packets to the network for measurement. IP Flow Performance Measurement (IP FPM) is a typical hybrid PM method, which colors the data packets in measurement.[7] Without generating additional OAM packets on the network, hybrid PM achieves a fair measurement accuracy with passive PM.

As mentioned earlier, OAM consists of FM and PM. The following sections describe the OAM functions of SRv6 based on them.

9.1.2 SRv6 FM

FM includes CC and CV. On an IP network, CC is implemented primarily through IP ping based on Internet Control Message Protocol (ICMP). Because SRv6 forwarding is based on IPv6, IPv6 OAM functions can be applied to SRv6. For example, IPv6 ping can be directly used on an SRv6 network to detect the reachability of an IPv6 address or an SRv6 SID.

9.1.2.1 Classic IP Ping

As alluded to earlier, SRv6 implements forwarding based on the IPv6 data plane. This means that ICMPv6 ping can be directly applied to an SRv6 network to detect the reachability of common IPv6 addresses,[3] without any changes to hardware or software. ICMPv6 ping supports packet forwarding to a destination address over the shortest path to detect the reachability of the destination address.

To ensure that ICMPv6 Echo messages are forwarded to an IPv6 destination address through a specified path (referred to as a path-specific ping operation), we can add an SRH carrying the specified path (segment list) to the IPv6 header. As shown in Figure 9.2, the ICMPv6 Echo message sent by H1 carries the segment list <R3, R4, R5> in the SRH. As a result, the message is forwarded to H2 through the R3→R4→R5 path to ping H2.

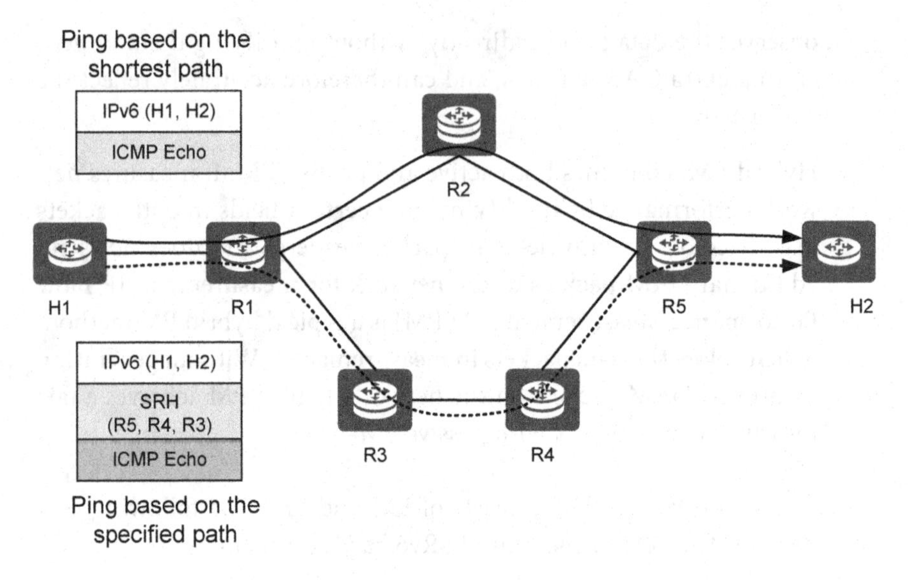

FIGURE 9.2 Classic ping and path-specific ping.

9.1.2.2 SRv6 SID Ping

Similarly, if the destination address is an SRv6 SID, ICMPv6 ping can equally be used.

For instance, when a user wants to ping SRv6 SID 2001:db8:1:1:: through the shortest path, an ICMPv6 Echo request message can be generated with the destination as 2001:db8:1:1::. When the packet arrives at a node, the node processes the SID 2001:db8:1:1::. Since the SID is a local SID instantiated by the node, the node processes the upper-layer header (the ICMPv6 header). As a result, the node responds by sending an ICMPv6 Echo reply message to the source node.

If the user wants to ping SRv6 SID 2001:db8:1:1:: through a specific path, an SRH with the SID list containing the related SIDs and 2001:db8:1:1:: can be encapsulated in the packet, where the upper-layer header is an ICMPv6 header. The packet can be forwarded to the egress node along the path specified by the SID list. When the node receives the packet, it processes the target SID. Since the SID is a local SID instantiated by the node, the node processes the upper-layer header (the ICMPv6 header). As a result, the node responds by sending an ICMPv6 Echo reply message to the source node. If the SID is not a locally instantiated one, the node discards the packet.

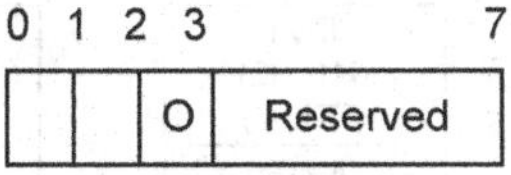

FIGURE 9.3 Format of the Flags field in the SRH.

In order to indicate OAM processing, the OAM bit (O-bit) is added to the SRH as well.

SRH.Flag.O-bit: indicates OAM operation. If SRH.Flag.O-bit is set, each segment endpoint needs to replicate the received data packet and then send the packet and corresponding timestamp to the control plane for further processing. To prevent the same packet from being repeatedly processed, the control plane does not need to respond to upper-layer IPv6 protocol operations, such as ICMP ping.

SRH.Flag.O-bit is located in the Flags field of the SRH. Figure 9.3 shows the format of the Flags field.

9.1.2.3 Classic Traceroute

Similar to ICMPv6 ping, traceroute can be implemented for common IPv6 addresses on an SRv6 network, without requiring any changes to hardware or software. Because SRv6 forwards traffic over a specified path based on SRHs, an SRH carrying a specific path can be inserted into the traceroute packet header to trace the path.

With classic traceroute, multiple probe packets (ICMP or UDP packets) are sent to the destination address under test. The TTL in the packets is used in determining the transit routers that are traversed on the way to the destination, and the value of TTL will be increased gradually from packet to packet to control the Hop Limit. Routers decrement the TTL value of packets by 1 when forwarding packets. When the TTL value reaches zero on a router, the router returns an ICMP Time Exceeded message to the source address. In this way, information about each hop on the path to the destination address is obtained. Figure 9.4 shows how classic traceroute is implemented, which is described as follows:

1. H1 sends an ICMP Echo message with the destination address being H2 and TTL being 1. As the TTL expires on R1, R1 returns an ICMP Time Exceeded message to H1. This allows H1 to obtain the address of R1 through the ICMP Time Exceeded message analysis.

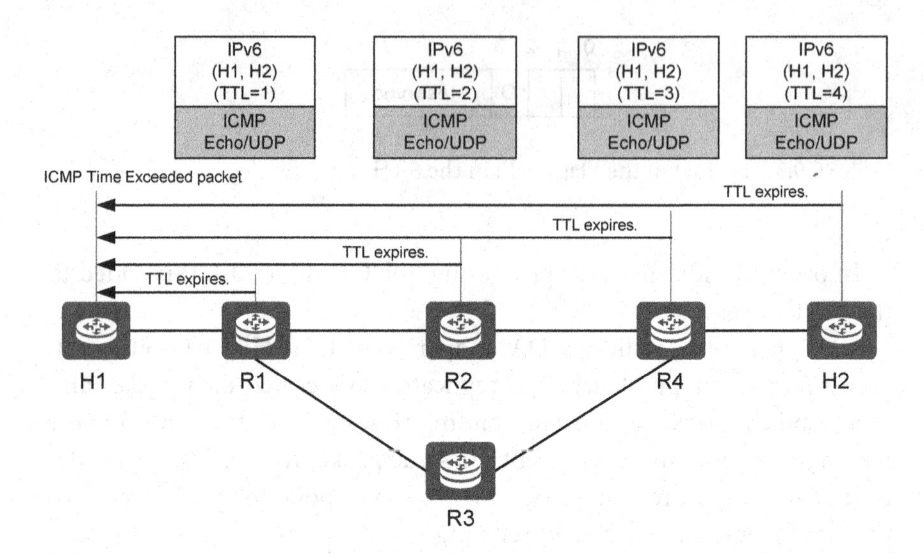

FIGURE 9.4 Traceroute implementation.

2. H1 sends an ICMP Echo message with the destination address being H2 and TTL being 2. As the TTL expires on R2, R2 returns an ICMP Time Exceeded message to H1. This allows H1 to obtain the address of R2 through the ICMP Time Exceeded message analysis.

3. H1 sends an ICMP Echo message with the destination address being H2 and TTL being 3. As the TTL expires on R4, R4 returns an ICMP Time Exceeded message to H1. This allows H1 to obtain the address of R4 through the ICMP Time Exceeded message analysis.

4. H1 sends an ICMP Echo message with the destination address being H2 and TTL being 4. After receiving the packet, H2 returns an ICMP Echo Reply message to H1. The traceroute process is completed, with the path from H1 to H2 described as H1→R1→R2→R4→H2. H1 calculates the round-trip delay from itself to each transit node based on the time when it receives the returned ICMP messages.

However, some network devices are configured to ignore or directly discard ICMP packets due to security considerations. This means that ICMP detection may not detect all devices along the path. In this case, UDP detection is a viable alternative. It can obtain a more comprehensive set of information about devices along the path because it uses UDP packets — which are more likely to pass through the firewall — rather than ICMP. UDP detection is implemented as follows:

If UDP packets are used for detection, the UDP port number must be set to a value greater than 30000. UDP detection is implemented in a similar way to ICMP detection with each transit node returning an ICMP Time Exceeded message to the source address when the corresponding TTL expires. When the packet arrives at the target node, it is sent to the transport layer for processing, and an ICMP Port Unreachable message is returned because the UDP port number is greater than 30000. Thus, traceroute through UDP detection is implemented.

On an SRv6 network, traceroute can be performed for an IPv6 address by using the existing traceroute mechanism, without requiring additional modification.

9.1.2.4 SRv6 SID Traceroute

Similar to SRv6 SID ping, traceroute for an SRv6 SID can reuse the existing IPv6 traceroute mechanisms.

For instance, when a user wants to trace SRv6 SID 2001:db8:1:1:: through the shortest path, ICMPv6 Echo request messages with gradually increasing TTL values and a destination address of 2001:db8:1:1:: can be generated. The procedure of traceroute for an SRv6 SID is identical to that of traceroute for an IPv6 address. The source node can trace the transit nodes by analyzing the received ICMPv6 Time Exceeded messages.

When the packet arrives at the target node, the node processes the SID 2001:db8:1:1::. Since the SID is a local SID instantiated by the node, the node processes the upper-layer header, that is, the ICMPv6 header or UDP header. As a result, the node responds by sending an ICMPv6 Echo reply message or an ICMP Port Unreachable message to the source node.

If the user wants to trace SRv6 SID 2001:db8:1:1:: through a specific path, an SRH with the SID list containing the related SIDs and 2001:db8:1:1:: can be encapsulated in the packet, where the upper-layer header is an ICMPv6 or UDP header. The packet can be forwarded to the egress node along the path specified by the SID list. When the node receives the packet, it processes the target SID. Since the SID is a local SID instantiated by the node, the node processes the upper-layer header and sends the related ICMP reply message or an ICMP Port Unreachable message to the source node. If the SID is not a locally instantiated one, the node discards the packet.

9.1.3 SRv6 PM

PM is another key function in OAM and includes packet loss, delay, jitter, and throughput measurement.

Many SRv6 PM methods have been proposed, such as RFC 6374-based,[8] TWAMP-based, and coloring-based PM methods, and they are currently being standardized. Both RFC 6374- and TWAMP-based PM methods implement active measurement, whereas the coloring-based PM method implements hybrid measurement. As for active measurement, this section focuses on the TWAMP-based PM method, as it is more versatile — The TWAMP-based PM method applies to various scenarios, such as SR-MPLS, SRv6, and IP/MPLS, while the RFC 6374-based PM method is applicable only to MPLS scenarios. This section will first introduce the fundamentals of TWAMP/TWAMP Light and then describe how to use TWAMP Light in SRv6.

9.1.3.1 TWAMP Fundamentals

Currently, SRv6 typically uses the Light mode of TWAMP for PM, that is, TWAMP Light.[6] TWAMP generates UDP-based test flows based on 5-tuple information (source IP address, destination IP address, source port number, destination port number, and transport protocol). The test packets in each test flow are sent through the measured path, and the response packets are sent back through the same path. TWAMP analyzes received UDP response packets to measure the performance and state of IP links.

TWAMP consists of four logical entities: Session-Sender, Session-Reflector, Control-Client, and Server. The Control-Client and Server belong to the control plane and are responsible for the negotiation (initialization), starting, and stopping of test sessions. The Session-Sender, which sends test packets, and the Session-Reflector, which responds to the packets, both belong to the data plane. Based on the deployment positions of these four logical entities, TWAMP is classified into the TWAMP Full mode (TWAMP for short) and TWAMP Light mode (TWAMP Light for short).

- TWAMP: The Session-Sender and Control-Client are integrated as one entity called the Controller, and the Session-Reflector and Server are integrated as one entity called the Responder, as shown in Figure 9.5. The Controller uses TCP-based TWAMP control packets

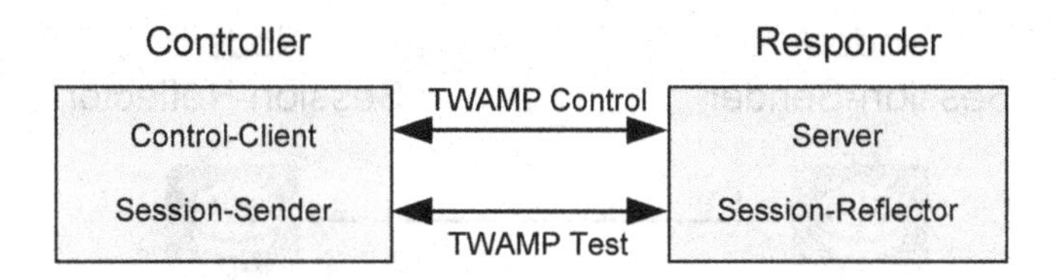

FIGURE 9.5 Architecture of TWAMP.

to establish a test session with the Responder. After the session is established, the Controller sends UDP-based TWAMP test packets to the Responder, and the Session-Reflector on the Responder responds to the test packets.

- TWAMP Light: The Session-Sender, Control-Client, and Server are integrated as one entity called the Controller, and the Session-Reflector is an independent entity called the Responder, as shown in Figure 9.6. In TWAMP Light mode, key information is directly delivered to the Controller through the Graphical User Interface (GUI). The Controller does not need to negotiate with the Responder in the control plane, and the Responder only needs to receive test packets and respond to the packets. As such, TWAMP Light is simpler than TWAMP.

Figure 9.7 uses DM as an example to illustrate the packet exchange process in TWAMP Light. Node A functions as the Session-Sender to send a test packet carrying the sending timestamp T1. Node B functions as the Session-Reflector to record the receiving timestamp T2 after receiving the packet, and then records the packet sending timestamp T3 in the response packet sent to the Session-Sender. After receiving the response packet, the Session-Sender records the receiving timestamp T4. The delays of corresponding paths from node A to node B and node B back to node A in a single period are calculated based on these four timestamps.

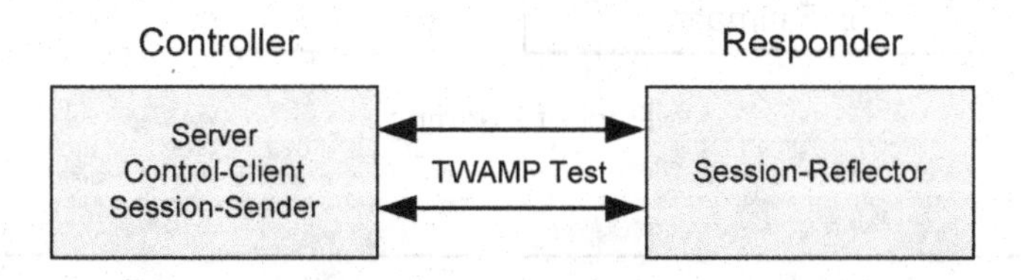

FIGURE 9.6 Architecture of TWAMP Light.

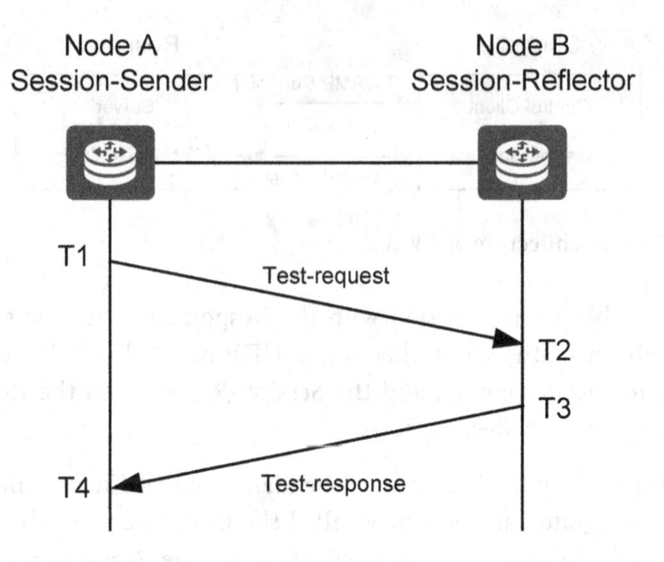

FIGURE 9.7 TWAMP Light packet exchange process.

Forward delay = T2 – T1
Backward delay = T4 – T3
Round-trip delay = (T2 – T1) + (T4 – T3)

A test packet sent by the Session-Sender carries information such as the packet sequence number and timestamp. Figure 9.8 shows the format of such a DM test packet.

Table 9.1 describes the fields of a DM test packet in TWAMP Light.

A response packet sent by the Responder carries information such as the sending timestamp of the Session-Sender and receiving and sending

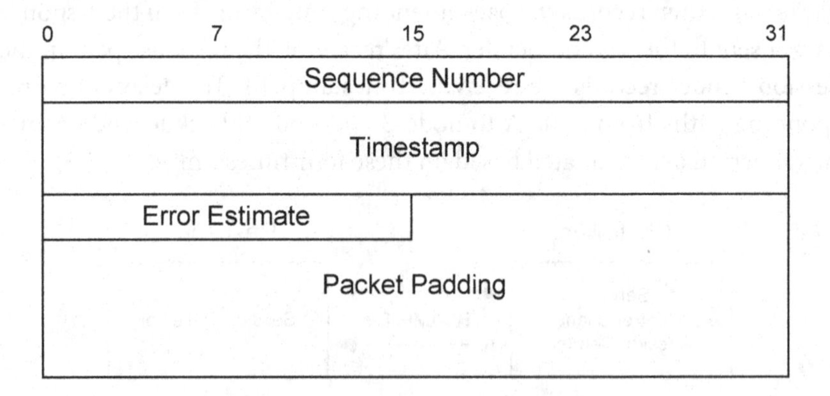

FIGURE 9.8 Format of a DM test packet in TWAMP Light.

TABLE 9.1 Fields of a DM Test Packet in TWAMP Light

Field	Length	Description
Sequence Number	32 bits	Order in which the test packet is transmitted. It starts at zero and increments by one for each subsequent packet in each test session.
Timestamp	64 bits	Time when the Session-Sender transmits the test packet.
Error Estimate	16 bits	Estimate of the error and synchronization.
Packet Padding	Variable length	Padding for alignment.

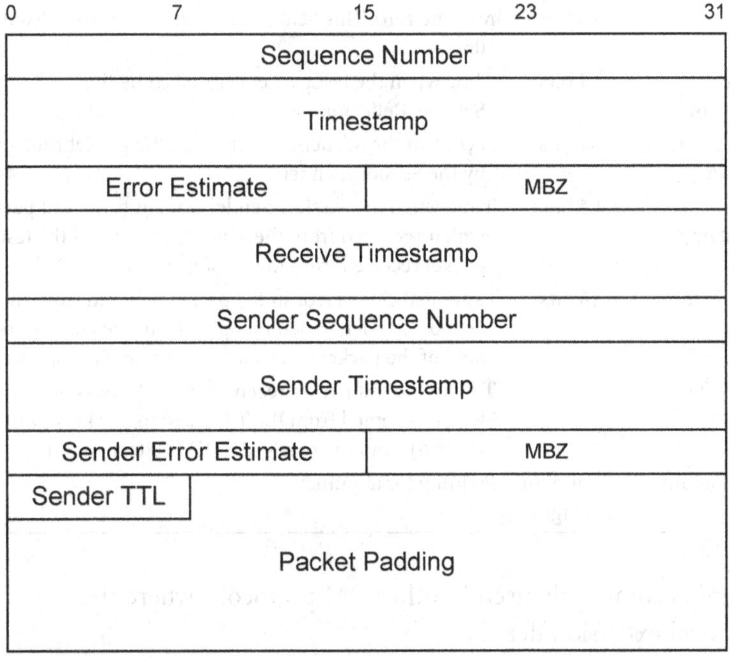

FIGURE 9.9 Format of a DM response packet in TWAMP Light.

timestamps of the Receiver. Figure 9.9 shows the format of a response packet.

Table 9.2 describes the fields of a DM response packet in TWAMP Light.

After receiving the response packet, the Session-Sender calculates the one-way delay of each direction and the total round-trip delay based on the three timestamps in the packet and the timestamp when it receives the response packet. The measured delays in a test session refer to the delays of the corresponding paths. This is the basic principle of TWAMP-based

TABLE 9.2 Fields of a DM Response Packet in TWAMP Light

Field	Length	Description
Sequence Number	32 bits	Packet sequence number. In the stateless mode, the Session-Reflector copies the value from the received TWAMP test packet's Sequence Number field; in the stateful mode, the Session-Reflector counts the transmitted TWAMP test packets. It starts at zero and increments by one for each subsequent packet in each test session.[9]
Timestamp	64 bits	Time when the Session-Reflector transmits the test packet.
Error Estimate	16 bits	Estimate of the error and synchronization.
MBZ	16 bits	Must be zero. This field is reserved. The value is fixed at all 0s.
Receive Timestamp	64 bits	Time when the test packet is received by the Session-Reflector.
Sender Sequence Number	32 bits	A copy of the Sequence Number of the packet transmitted by the Session-Sender.
Sender Timestamp	64 bits	Time when the Session-Sender transmits the test packet, which is copied from the Timestamp field of the test packet received from the Session-Sender.
Sender Error Estimate	16 bits	Estimated clock error in the packet received from the Session-Sender, which is copied from the corresponding field of the packet received from the Session-Sender.
Sender TTL	8 bits	TTL in the test packet received from the Session-Sender, which is copied from the TTL field in IPv4 (or Hop Limit in IPv6) from the received TWAMP test packet.
Packet Padding	Variable length	Padding for alignment.

DM and is commonly used in other DM protocols, where they vary only in protocol extension details.

9.1.3.2 TWAMP-Based Active SRv6 PM

TWAMP Light can be directly applied to an SRv6 network without any modification. TWAMP Light can be configured through the SDN controller to enable SRv6 PM, involving the configuration of various parameters such as the measurement protocol, destination UDP port, and measurement type. Figure 9.10 shows the configuration model.

After configuration, the SRv6 Session-Sender sends a TWAMP Light test payload (compliant with RFC 5357) encapsulated into a UDP message to the Responder. The payload of the UDP message carries the corresponding timestamps or count values. If the measured path is an SRv6

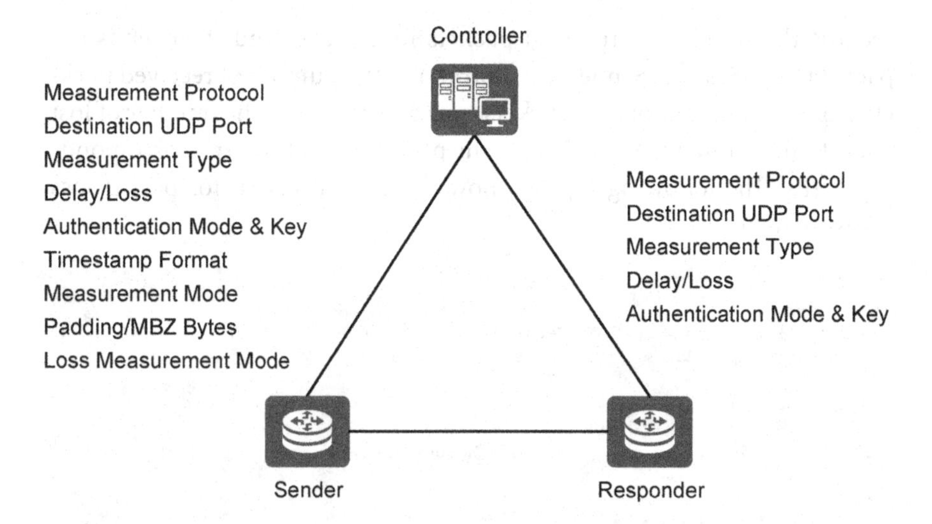

FIGURE 9.10 TWAMP Light configuration model on an SRv6 network.

BE path, the encapsulation of the TWAMP Light test packets is identical to that of the packets in IPv6, with the destination address being an SRv6 SID. If the measured path is an SRv6 TE path, an SRH is encapsulated in the test packets. Therefore, the TWAMP Light test packets can be forwarded through the measured path.

The following uses DM as an example to describe the packet format for measuring an SRv6 TE path.

```
---------------------------------------------------------------
IP Header
 Source IP Address = Session-Sender's IPv6 address
 Destination IP Address = Responder's IPv6 address
 Next Header = RH
---------------------------------------------------------------
 Routing Header (Type = 4, SRH, Next Header = UDP)
 SID[0]
 SID[1]
 …
---------------------------------------------------------------
UDP Header
 Source Port = Port selected on the Session-Sender
 Destination Port = Delay measurement port configured by users
---------------------------------------------------------------
Payload as shown in Figure 9.8
Payload as shown in Figure 9.9
---------------------------------------------------------------
```

Regarding LM, the Session-Sender sends test packets, which carry the packet count of the corresponding flow, to the Responder. The Responder records the number of received packets in a packet and then sends the packet to the Session-Sender. By subtracting the number of received packets from the number of sent packets, we can determine the number of lost packets and subsequently calculate the packet loss rate on the corresponding paths. The following figures show the packet format for packet loss measurement.

```
---------------------------------------------------------------
IP Header
 Source IP Address = Session-Sender's IPv6 address
 Destination IP Address = Responder's IPv6 address

 Next Header = RH
---------------------------------------------------------------
 Routing Header (Type = 4, SRH, Next Header = UDP)
 SID[0]
 SID[1]
 ...
---------------------------------------------------------------
UDP Header
 Source Port = Port selected on the Session-Sender
 Destination Port = Delay measurement port configured by users
---------------------------------------------------------------
Payload as shown in Figure 9.11
Payload as shown in Figure 9.12
---------------------------------------------------------------
```

Table 9.3 describes the fields of an LM test packet in TWAMP Light.

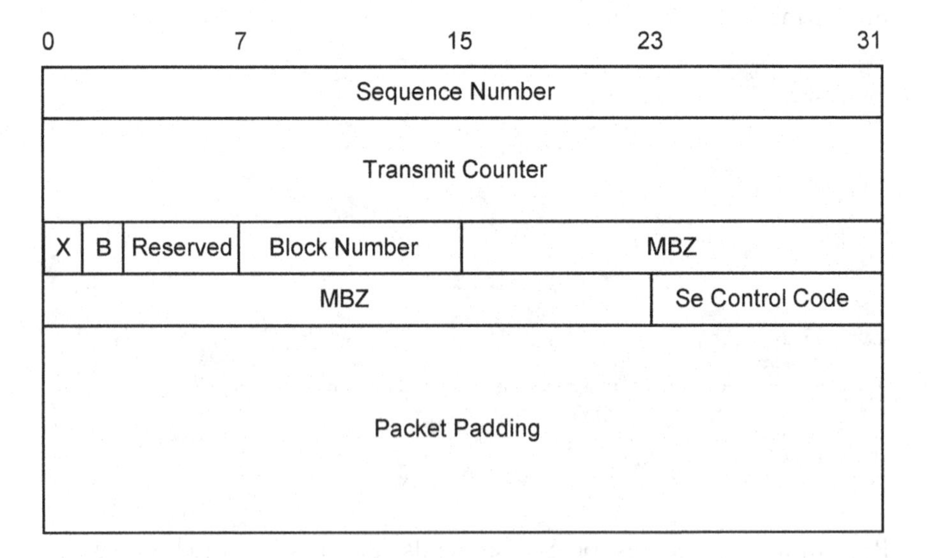

FIGURE 9.11 Format of an LM test packet.

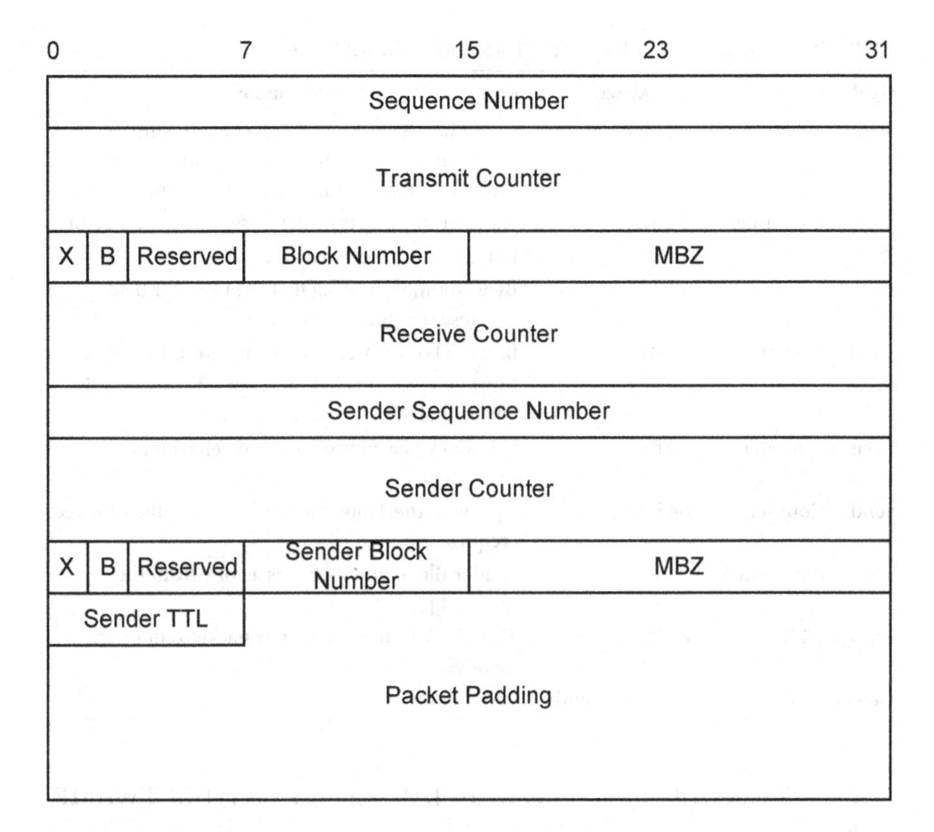

FIGURE 9.12 Format of an LM response packet.

TABLE 9.3 Fields of an LM Test Packet in TWAMP Light

Field	Length	Description
Sequence Number	32 bits	Order in which the test packet is transmitted. It starts at zero and increments by one for each subsequent packet in each test session.
Transmit Counter	64 bits	Number of packets or octets sent by the sender node in the request message.
X	1 bit	Extended counter format indicator.
B	1 bit	Byte count, indicating that the Counter fields represent octet counts.
Block Number	8 bits	Block ID of the traffic sent from the sender to the reflector.
MBZ	40 bits	Must be zero. This field is reserved. The value is fixed at all 0s.
Se Control code	8 bits	Sender control code, indicating the type of a response.
Packet Padding	Variable length	Padding for alignment.

TABLE 9.4 Fields of an LM Response Packet in TWAMP Light

Field	Length	Description
Sequence Number	32 bits	Order in which the test packet is transmitted. It starts at zero and increments by one for each subsequent packet in each test session.
Transmit Counter	64 bits	Number of packets or octets sent by the reflector.
X	1 bit	Extended counter format indicator.
B	1 bit	Byte count, indicating that the Counter fields represent octet counts.
Block Number	8 bits	Block ID of the traffic sent back from the reflector.
MBZ	40 bits	Must be zero. This field is reserved. The value is fixed at all 0s.
Receive Counter	64 bits	Number of packets or octets received on the reflector.
Sender Counter	64 bits	A copy of the Transmit Counter from the received request message.
Sender Block Nu	8 bits	Sender Block Number. It is a copy from the request packet.
Sender TTL	8 bits	TTL or Hot Limit of the test packet sent by the sender.
Packet Padding	Variable length	Padding for alignment.

Table 9.4 describes the fields of an LM response packet in TWAMP Light.

Transmit Counter carries the packet count of the corresponding flow, and Block Number records the block information of packets for coloring-based LM.[7] The Session-Sender calculates the number of lost packets and the packet loss rate based on the Transmit Counter values in the received packets and the Receiver Counter values on the Responder. For details, see the related draft.[10]

9.1.3.3 Coloring-Based Hybrid SRv6 PM

The active PM method described earlier — TWAMP Light — infers the network performance by measuring the performance of additional OAM packets. Consequently, this method cannot deliver highly accurate measurement results. For higher accuracy, the passive and hybrid PM methods are the better choices, as passive PM directly monitors the data packets and hybrid PM monitors modified data packets. In passive mode, Internet Protocol Flow Information Export (IPFIX)[11] is one of the methods that can send the real traffic to the analyzer for traffic analysis and PM, whereas in hybrid mode, we can use alternate-marking (coloring)-based IP FPM[7] or IFIT. This section focuses on the hybrid PM using IP FPM, while the next section covers IFIT.

IP FPM alternately marks packets with colors in a data flow to virtually split the flow into blocks and measures the performance of colored packets based on blocks. For example, the packet loss rate can be calculated according to the sent and received packets in a block, which is a group of packets marked with the same color (e.g., color 0). For details, see Figure 9.13.

A colored block may be set based on a given quantity of packets, for example, 1,000 consecutive packets, or it can be set based on a time period, for example, consecutive packets within 1 second. The color refers to the value of a field in a packet. Using the alternate-marking method, the accuracy of LM can be improved since it can tolerate the out-of-order packets, and it is not based on "special" packets whose loss could have a negative impact.

Take LM as an example. The coloring node colors and sends packets. It marks every 1,000 packets with a color, which alternates between 0 and 1. The statistics collection node collects statistics based on the color. For example, if the device continuously receives 998 packets whose color is 1, the number of lost packets is 2, and the packet loss rate is 0.2%.[7]

On an SRv6 network, IP FPM can color packets based on the IPv6 coloring field or SRv6 Path Segment.[12] By alternating the value of Path Segment, we can color SRv6 packets to support IP FPM-based PM.

An SRv6 Path Segment is defined to identify an SRv6 Path,[12] and it cannot be copied to a destination address for routing. Currently, SRv6 Path Segment is designed to be located at the top of the segment list, that is, the location indicated by Last Entry, and the P-flag is used to indicate whether Path Segment is present. Figure 9.14 shows the format of SRv6 Path Segment.

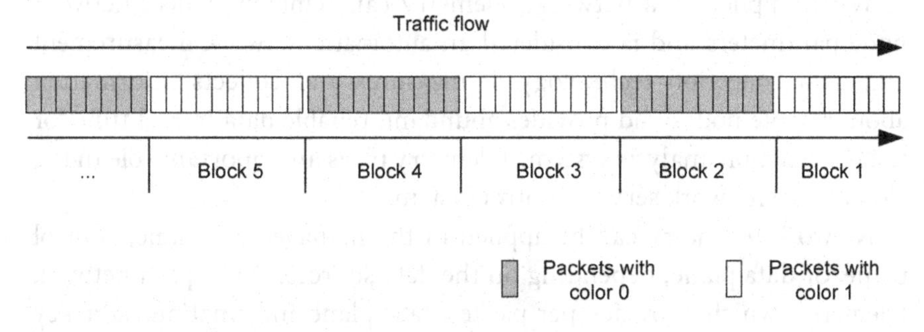

FIGURE 9.13 IP FPM implementation.

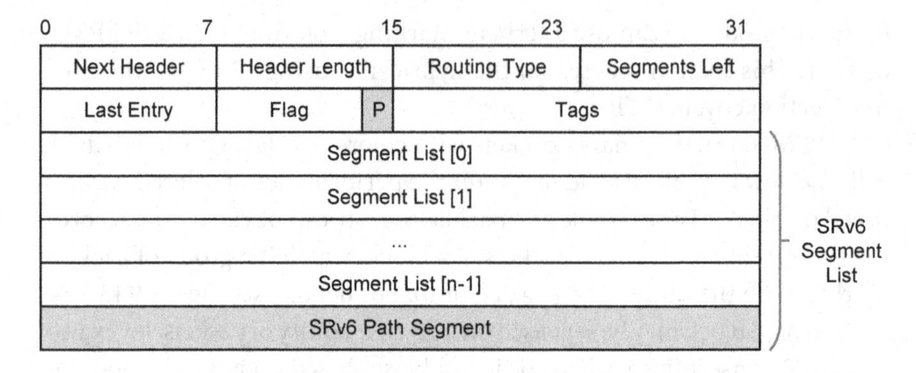

FIGURE 9.14 SRv6 Path Segment.

An SRv6 Path Segment is used for identifying an SRv6 path, which is very useful in some scenarios, such as passive delay or packet loss measurement, bidirectional path association, and 1+1 end-to-end protection.

In SRv6, different candidate paths in different SRv6 Policies may use the same SID list. Therefore, the traffic of different candidate paths will merge, resulting in inability to measure the specific candidate path. With SRv6 Path Segment, each SRv6 candidate path can be identified and measured, even when they use the same SID List.

9.2 ON-PATH NETWORK TELEMETRY

9.2.1 On-Path Network Telemetry Overview

Telemetry is a technology used to remotely obtain measurement parameters, such as satellite and sensor data in the space and geology fields. This technology is becoming increasingly important in the network field, too, as automatic network O&M becomes more prevalent.

When applied to a network, telemetry can remotely collect network node parameters and is considered an automatic network measurement and data collection technology. It measures and collects information about remote nodes and provides abundant, reliable data in real time for the information analysis system. Telemetry plays an important role in the closed-loop network service control system.

Network telemetry can be applied in the management plane, control plane, or data plane, depending on the data sources.[13] On-path network telemetry, which provides per-packet data plane information, is a key technology of data-plane telemetry.

As mentioned earlier, active PM injects additional test packets into the network and measures the delay and packet loss based on collected test packet statistics. For example, TWAMP measures OAM test packets to infer the packet loss rate on a network. Such an approach is unable to accurately reflect the real service performance.

On-path network telemetry is a hybrid measurement method and takes a different approach by adding OAM information to data packets rather than injecting additional test packets. In the packet forwarding process, OAM information is forwarded together with packets to complete measurement, leading to this process being called on-path network telemetry. In on-path network telemetry, nodes collect and process data according to the OAM information in received packets. Compared with active PM, on-path network telemetry has the following advantages:

- Measures real user traffic.
- Provides per-packet monitoring.
- Collects more data-plane information.

By using on-path network telemetry, we can obtain a more detailed set of OAM information. Such information includes:

- Packet forwarding path information, including device and inbound and outbound interface information
- Rules matched by packets on each network device that forwards the packets
- Time taken to buffer packets on each network device, accurate to within nanoseconds
- Flows with which a packet competes for a queue during the queuing process

9.2.2 On-Path Network Telemetry Modes

There are several proposals on on-path network telemetry technologies in the industry, such as IOAM,[14] Postcard-Based Telemetry (PBT),[15] and Enhanced Alternate Marking Method (EAM).[16] These technologies have two basic modes — passport and postcard — depending on how the collected data is processed, as shown in Figure 9.15.

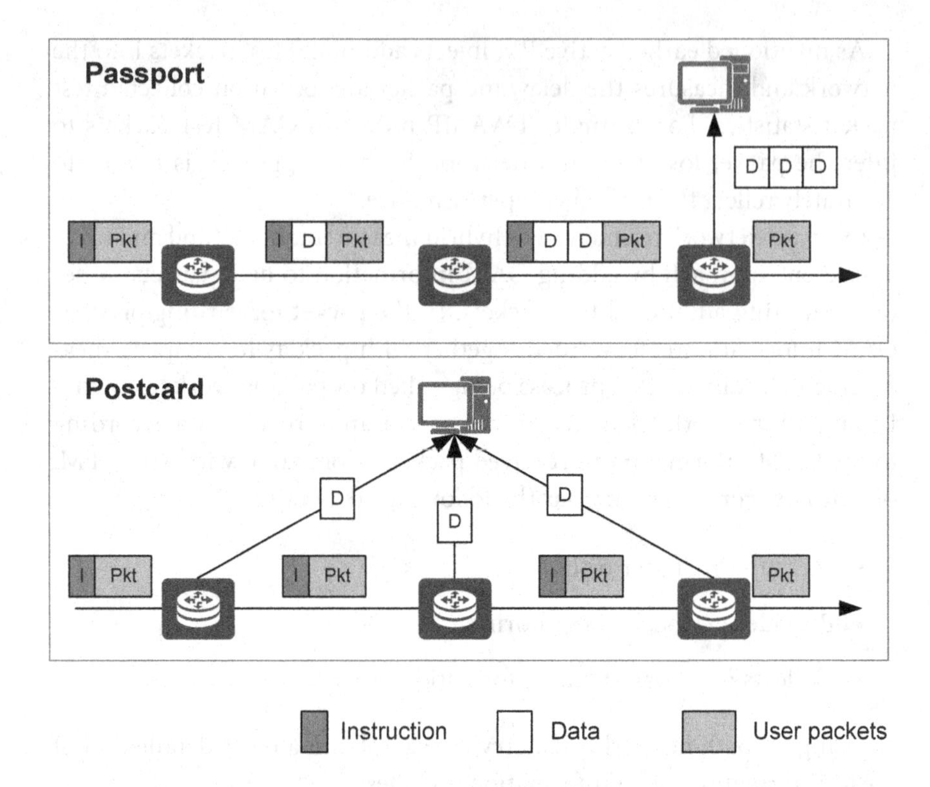

FIGURE 9.15 Comparison between the passport and postcard modes.

In passport mode, the ingress node of a measurement domain adds a Telemetry Information Header (TIH), which includes the data collection instruction, to the packets to be measured. Transit nodes collect data hop by hop according to the data collection instruction and record the data in the packets. The egress node of the measurement domain reports all the data collected along the path for processing, removes the TIH and data, and restores the data packets. Passport-based network telemetry is like a tourist whose passport receives an entry/exit stamp for each country visited.

Different from the passport mode, the postcard mode allows each node in the measurement domain to generate additional packets to carry collected data and send these packets to the collector instead of recording the collected data in the received data packets that contain instruction headers. Postcard-based network telemetry is like a tourist sending a postcard home each time he/she arrives at a scenic spot.

TABLE 9.5 Comparison between the Passport and Postcard Modes

Item	Passport Mode	Postcard Mode
Advantages	• Provides hop-by-hop data association, reducing the collector's workload. • Requires only the egress node to send data, reducing the overhead.	• Identifies the position where packet loss occurs. • Features short packet headers of fixed length. • Supports easy hardware implementation.
Disadvantages	• Unable to locate packet loss. • Increases packet header length as the number of hops increases (tracing mode).	• Requires the collector to associate packets with data generated by nodes on the path.

The two modes apply to different scenarios, and each has unique advantages and disadvantages. Table 9.5 compares these two modes.

9.2.3 IFIT Architecture and Functions

Although on-path network telemetry has many advantages, it faces multiple challenges in real-world network deployment.

- On-path network telemetry involves specifying a monitored flow object on a network device and requires monitoring resources to be allocated for corresponding operations. Such operations include inserting a data collection instruction, collecting data, or stripping an instruction and data. However, a network device can monitor only a limited number of flow objects due to its limited processing capabilities, posing a challenge to large-scale deployment of on-path network telemetry.
- Because on-path network telemetry introduces additional processing in the forwarding plane of a device, it may compromise the device's forwarding performance. The observer effect comes into play here, where the mere observation of a phenomenon inevitably changes that phenomenon. Consequently, on-path network telemetry cannot accurately reflect the state of the measured object.
- Per-packet monitoring generates a large amount of OAM data that will consume a significant amount of network bandwidth if it is all sent for analysis. Considering that a network may contain hundreds of forwarding devices, the data analyzer may be severely

compromised in terms of performance due to receiving, storing, and analyzing the massive amount of data sent from the devices.

- Intent-driven automation is the direction in which network O&M is evolving. Network virtualization, network convergence, and IP-optical convergence will impose more requirements on data acquisition, and data needs to be provided on demand in an interactive manner for data analysis applications. However, predefined datasets can provide only limited data and cannot meet flexible data requirements in the future. This means that a new method is required to implement flexible and scalable data definition and deliver the required data to applications for data analysis.

IFIT provides an architecture and solution for on-path network telemetry and supports multiple data-plane solutions. It uses technologies such as smart traffic selection, efficient data sending, and dynamic network probe, and integrates tunnel encapsulation to enable deployment of on-path network telemetry on actual networks.[17]

Figure 9.16 shows the IFIT network architecture. IFIT applications deliver monitoring and measurement tasks to network devices. Some examples of such tasks include specifying the flow objects to be measured and data to be collected, and selecting a data-plane encapsulation

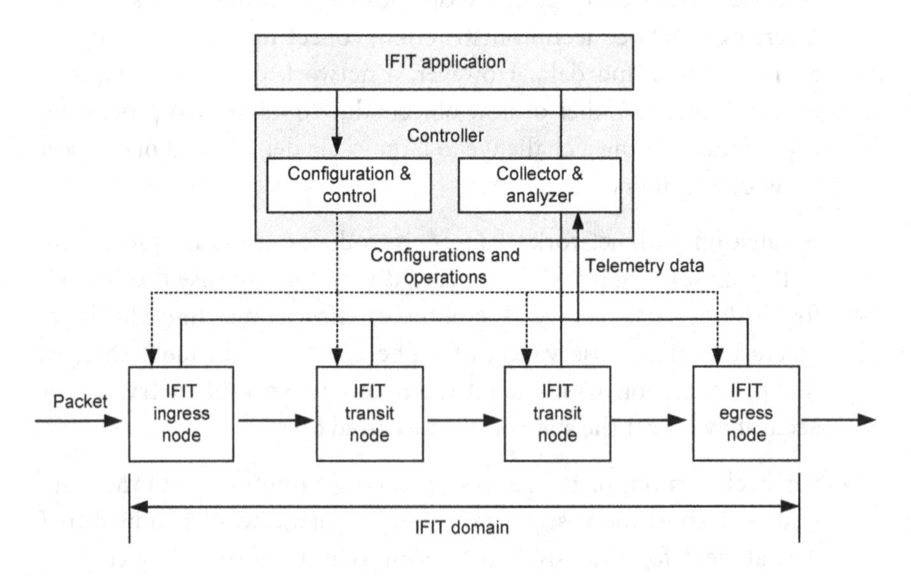

FIGURE 9.16 IFIT architecture.

mode for on-path network telemetry. When a data packet enters an IFIT domain, the ingress node adds an instruction header to the specified flow object. Based on the IFIT instructions contained in this header, the transit nodes along the path collect and send data. When the data packet leaves the IFIT domain, the egress node removes all instructions and data added to the packet for measurement.

9.2.3.1 Smart Traffic Selection

Due to the limited hardware resources, it is usually impossible to monitor all traffic on the network and collect data on a per-packet basis. This not only affects the normal forwarding of devices but also consumes a large amount of network bandwidth. A feasible workaround to this conundrum is to select some traffic for special monitoring.

The smart traffic selection technology uses a coarse-to-fine mode to trade time for space, helping users select desired traffic. Users can deploy smart traffic selection policies based on their intentions. Regardless of whether these policies are performed based on sampling or prediction, which has a certain error probability, the use of the smart traffic selection technology generally requires only a small amount of resources.

For example, to focus on monitoring the top 100 elephant flows, we can employ a smart traffic selection policy based on the Count-Min Sketch technology,[18] which uses multiple hash operations to filter data and store only the count value and thereby avoid storing flow IDs. This consumes only a very small amount of memory while also ensuring a very high level of recognition accuracy. The SDN controller generates ACL rules based on the intelligent traffic selection result and delivers the ACL rules to devices to monitor elephant flows.

9.2.3.2 Efficient Data Sending

Per-packet on-path network telemetry can capture subtle dynamic changes on a network, but the involved packets will inevitably contain a large amount of redundant information. Directly sending all the information consumes a great deal of network bandwidth and imposes an extremely heavy burden on data analysis, especially if an analyzer needs to manage tens of thousands of network nodes.

One way to achieve efficient data transmission is to use binary data transmission encoding. Currently, NETCONF-based network management information is usually encoded in Extensible Markup Language

(XML) format, which consumes a large amount of network bandwidth and is not suitable for sending flow-based on-path network telemetry measurement information. Binary encoding, such as Google Protocol Buffer (GPB),[19] effectively reduces the amount of data that needs to be transmitted.

Another way to minimize the amount of data to be sent is to use data filtering, in which a network device filters data based on certain conditions and converts the data into events to notify upper-layer applications. The following uses flow path tracing as an example. Flow-based load balancing is typically used on live networks. The path of a flow does not change easily, and a path change usually indicates an abnormality. Large amounts of duplicate path data (including information about the node and inbound and outbound interfaces of each hop) do not need to be reported and can be directly filtered out by network devices. Only information about newly discovered or changed flow paths is sent, reducing the amount of reported data.

Network devices can also cache data that does not have high real-time requirements for a certain period of time before compressing and sending it in batches. This approach reduces not only the amount of data to be sent but also the frequency at which it is sent, thereby lowering the pressure of data collection.

9.2.3.3 Dynamic Network Probe

Due to limited data plane resources, such as data storage space and instruction space, it is difficult to continuously provide complete data monitoring and sending. In addition to this, the amount of data required by applications dynamically changes. For example, a system may require only intermittent inspections while it runs normally, but requires precise real-time monitoring once a potential risk is detected. Installing and running all network telemetry functions in the data plane consumes a large number of resources, affecting data forwarding, and offers only limited benefits. To meet various service requirements by using limited resources, network telemetry functions must be loaded on demand.

Dynamic network probe is a dynamically loadable network telemetry technology that supports on-demand loading and unloading of network telemetry functions on devices, thereby reducing resource consumption in the data plane while meeting service requirements.[20] To measure the performance of a flow, for example, we can load the corresponding measurement application to the device through configuration or dynamic programming. Conversely, we can uninstall it from the device if it is not required, thereby releasing the occupied instruction space and data storage space.

The dynamic network probe technology provides sufficient flexibility and scalability for IFIT. Various functions, such as smart traffic selection, data sending, and data filtering, can be dynamically loaded to devices as policies.

9.2.4 IFIT Encapsulation Mode

Tunneling technologies are used extensively on networks, especially in multi-AS scenarios. Such technologies typically encapsulate a new tunneling protocol header outside the original packets, and only the outer encapsulation is processed during tunnel forwarding. Consequently, devices will process IFIT instructions differently depending on whether the instructions are inserted into the outer protocol header or the inner original packet.

To meet different network tunnel monitoring requirements, IFIT provides two modes: uniform mode and pipe mode.[21] O&M personnel can determine whether to monitor the network devices through which a tunnel passes based on service requirements.

In uniform mode, the ingress node copies the IFIT instruction header in a packet to the outer tunnel encapsulation. This allows IFIT instructions to be processed in the same manner on all nodes, regardless of whether the nodes are on the tunnel, thereby implementing hop-by-hop data collection. Figure 9.17 shows the uniform mode.

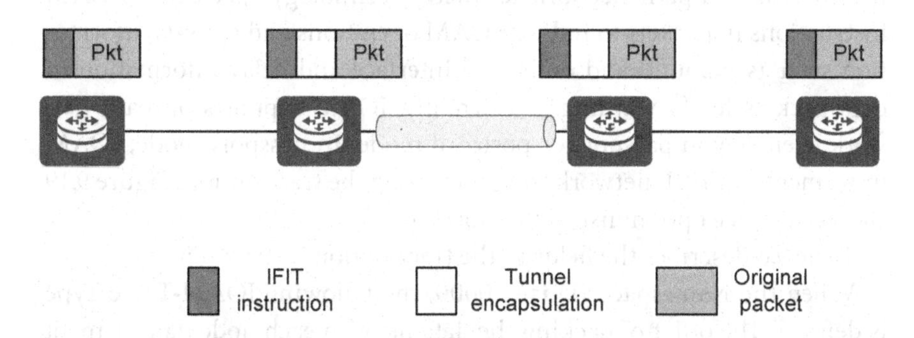

FIGURE 9.17 Uniform mode.

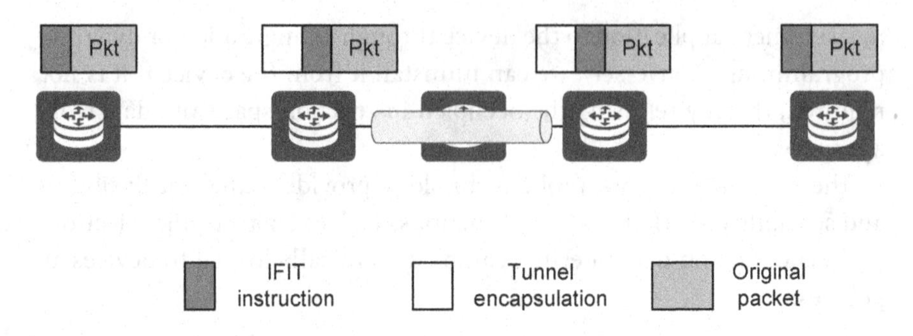

FIGURE 9.18 Pipe mode.

In pipe mode, the ingress node of a tunnel processes IFIT instructions of the received packet and collects related data. After the packet enters the tunnel, the IFIT instruction is retained in the original packet encapsulation. During packet forwarding, the transit nodes on the tunnel do not process the IFIT instruction. After the packet reaches the egress node of the tunnel, this node removes the tunnel encapsulation, continues to process the IFIT instruction in the original packet, and records the data of the entire tunnel as one node data record. From a macro perspective, the entire tunnel is considered as one hop, as shown in Figure 9.18.

9.2.5 IFIT for SRv6

SRv6 supports IFIT — both the passport and postcard modes — and provides multiple encapsulation modes (e.g., Hop-by-Hop Options header or SRH) to meet network telemetry requirements in multiple scenarios.

9.2.5.1 Passport Mode

IOAM is an on-path network telemetry technology that carries OAM instructions in packets to indicate OAM operations and records information such as inbound and outbound interface and delay information in data packets for forwarding.[14] Currently, IOAM supports on-path network telemetry in passport or postcard mode. In passport mode, IOAM implements on-path network telemetry using the trace option. Figure 9.19 shows the trace option instruction format.

Table 9.6 describes the fields in the trace option instruction.

When the Namespace-ID is 0x0000, the following IOAM-Trace-Type is defined. The order of packing the data fields in each node data element follows the bit order of the IOAM-Trace-Type field, as shown in Table 9.7.

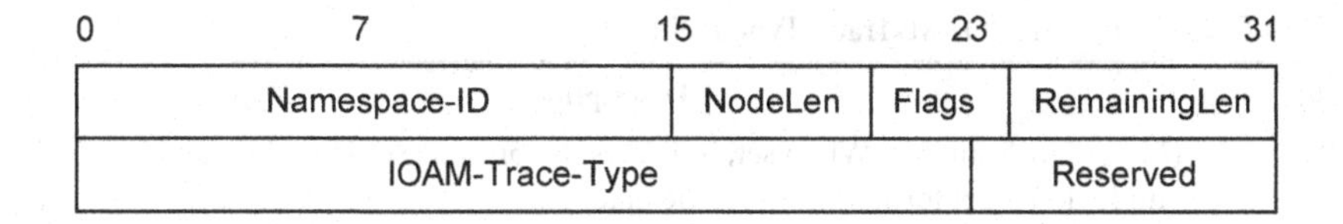

FIGURE 9.19 Trace option instruction format.

TABLE 9.6 Fields in the Trace Option Instruction

Field	Length	Description
Namespace-ID	16 bits	Indicates the IOAM namespace, which is used to distinguish the IOAM data collection types. 0x0000 is the default value of the namespace and must be identified by all devices that support IOAM.
NodeLen	5 bits	Indicates the length of the data carried in the trace option, excluding any nontransparent data without a format.
Flags	4 bits	Indicates supplementary operations in addition to data collection.
RemainingLen	7 bits	Indicates the remaining space for carrying data. When the value of this field is 0, the network node is prohibited from inserting new data.
IOAM-Trace-Type	24 bits	Indicates the type of data to be collected. Each bit indicates a type of data to be collected. Different vendors can define supported data types in a specified namespace. The data generally includes the identifier of a node, identifier of an interface for either sending or receiving a packet, time for processing a packet on a device, nontransparent data without a format, and other items.
Reserved	8 bits	Reserved field, which must be set to all 0s.

In the IOAM trace mode, a data space recording the telemetry data follows the instruction in the packet. The data space is organized based on the data collected by each node, and this collected data is sorted based on the data collection type and sequence specified in IOAM-Trace-Type, as shown in Figure 9.20.

In passport mode, the ingress node of the IOAM domain encapsulates an IOAM trace option instruction header for the monitored flow. Upon receiving the packet, the node collects data according to the type indicated in the IOAM-Trace-Type and inserts the data after the IOAM instruction header. The egress node of the IOAM domain, upon receiving the packet, sends all collected data to the collector, deletes the IOAM instruction and data from the packet, and then forwards the packet.

TABLE 9.7 Standard IOAM-Trace-Type Field

Bit	Description
0	(Most significant bit) When set, indicates the presence of Hop_Lim and node_id (short format) in the node data.
1	When set, indicates the presence of ingress_if_id and egress_if_id (short format) in the node data.
2	When set, indicates the presence of timestamp seconds in the node data.
3	When set, indicates the presence of timestamp sub-seconds in the node data.
4	When set, indicates the presence of transit delay in the node data.
5	When set, indicates the presence of IOAM-Namespace specific data (short format) in the node data.
6	When set, indicates the presence of queue depth in the node data.
7	When set, indicates the presence of the Checksum Complement node data.
8	When set, indicates the presence of Hop_Lim and node_id in a wide format in the node data.
9	When set, indicates the presence of ingress_if_id and egress_if_id in a wide format in the node data.
10	When set, indicates the presence of IOAM-Namespace specific data in a wide format in the node data.
11	When set, indicates the presence of buffer occupancy in the node data.
12–21	Undefined. An IOAM encapsulating node must set the value of each of these bits to 0.
22	When set, indicates the presence of variable length Opaque State Snapshot field.
23	Reserved: Must be set to zero upon transmission and ignored upon receipt.

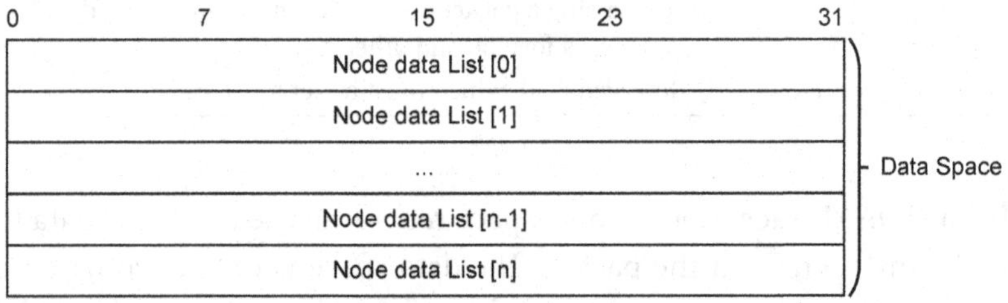

0 7 15 23 31
Node data List [0]
Node data List [1]
...
Node data List [n-1]
Node data List [n]
Data Space

FIGURE 9.20 Data collected by nodes.

9.2.5.2 Postcard Mode

The IETF draft[15] provides an option to use an instruction header for instructing data collection and sending data hop by hop to the collector. This option is called Postcard-Based Telemetry with Instruction Header (PBT-I) and is also supported in IOAM. A new IOAM Directly EXport (DEX) option is added to IOAM to implement PBT-I. Figure 9.21 shows the IOAM PBT-I format.

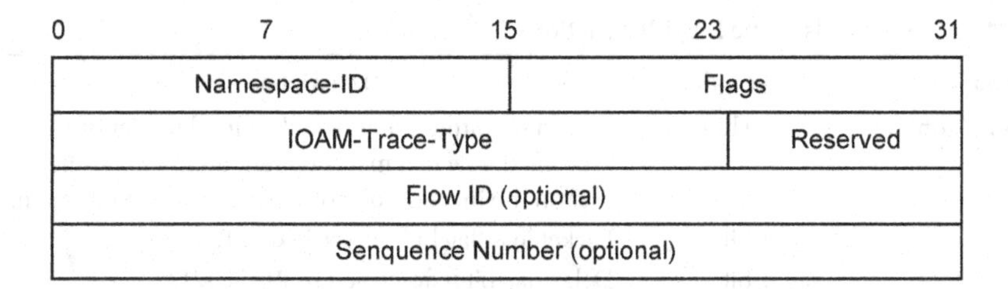

FIGURE 9.21 IOAM PBT-I format.

TABLE 9.8 Fields in the IOAM PBT-I Instruction

Field	Length	Description
Flow ID	32 bits	Uniquely identifies a monitored flow. It is generated on the ingress node in the measurement domain and encapsulated in the instruction header.
Sequence Number	32 bits	Marks the packet sequence of a monitored flow and starts from 0. It is generated on the ingress node in the measurement domain and encapsulated in the instruction header. The sequence number increments by 1 each time a packet is sent.

The Namespace-ID, Flags, IOAM-Trace-Type, and Reserved fields are the same as those described earlier in the IOAM tracing option. Table 9.8 describes the other fields.

In postcard mode, the ingress node of the IOAM domain encapsulates an IOAM DEX option instruction header for the monitored flow. Upon receiving a flow packet, a transit node collects data based on the type specified in IOAM-Trace-Type and sends the data to a configured collector. The egress node of the IOAM domain, upon receiving the packet, deletes the IOAM instruction from the packet and then forwards the packet.

IOAM DEX can provide abundant metrics from telemetry information, some of which are critical and frequently used. EAM is a simple and efficient on-path network telemetry technology that supports postcard mode and measures packet loss, delay, and jitter through a short header.[16] Figure 9.22 shows the EAM instruction format.

FIGURE 9.22 EAM instruction format.

TABLE 9.9 Fields in the EAM Instruction

Field	Length	Description
FlowMonID	20 bits	ID of a monitored flow, which is used to identify a specified flow in a measurement domain. This field is set on the ingress node of the measurement domain.
L	1 bit	Packet loss flag bit[7] described in RFC 8321.
D	1 bit	Delay flag bit[7] described in RFC 8321.
Reserved	10 bits	Reserved field, which must be set to all 0s.

Table 9.9 describes the fields in the EAM instruction.

EAM adds the preceding instruction header to packets on the ingress node of the measurement domain, allocates a FlowMonID to the monitored flow, and sets the L flag alternately to 1 or 0 (referred to as alternate coloring). A transit node identifies the monitored flow based on the FlowMonID, counts the number of packets whose L flag is 1 or 0 within a period, and uses the postcard method to send the FlowMonID, period ID, and count value within the period to the data analysis node. The data analysis node then compares the count values reported by different network devices for the same period and determines the number of lost packets and packet loss location. Using the D flag to mark delay sampling, the ingress node sets this flag to 1 for the packets whose delay needs to be monitored. Transit nodes record the timestamp for such packets and report the timestamp to the data analysis node, which can then calculate a one-way delay based on the arrival timestamps of marked packets reported by different network devices.

9.2.5.3 SRv6 IFIT Encapsulation

SRv6 provides an extensive programming space for applications in the data plane. The IFIT instruction header can be encapsulated in the IPv6 Hop-by-Hop Options header or in SRH's Optional TLV, as shown in Figure 9.23. Different encapsulation formats have different processing semantics, offering a wide array of features in SRv6 OAM.

All IPv6 forwarding nodes can process the IFIT instruction if it is encapsulated in the Hop-by-Hop Options header. This encapsulation mode allows O&M personnel to gain insight into how packets are forwarded hop by hop and facilitates fault demarcation when a network fault occurs — this is especially important in SRv6 BE or SRv6 TE loose path scenarios, where packet forwarding paths are not fixed.

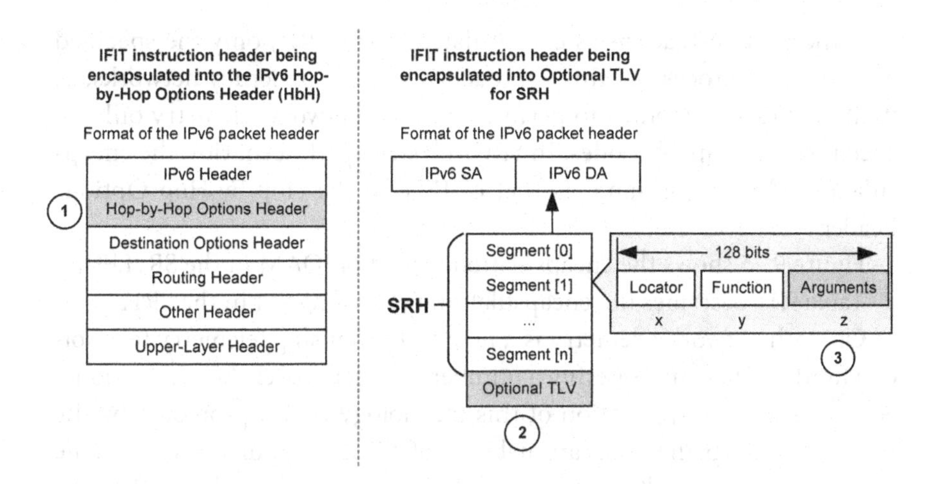

FIGURE 9.23 SRv6 IFIT encapsulation.

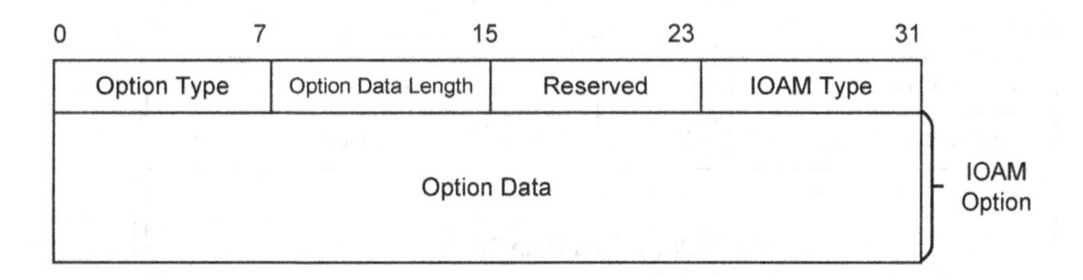

FIGURE 9.24 Reference encapsulation format of IOAM in a Hop-by-Hop Options header.

Figure 9.24 shows the encapsulation format of IOAM in a Hop-by-Hop Options header.[22]

Table 9.10 describes the IOAM fields in the Hop-by-Hop Options header.

TABLE 9.10 IOAM Fields in the Hop-by-Hop Options Header

Field	Length	Description
Option Type	8 bits	Option type. An IOAM option type needs to be defined.
Option Data Length	8 bits	Length of the option data, in bytes, indicating the length of the option header excluding the Option Type and Option Data Length.
IOAM Type	8 bits	IOAM type, corresponding to the IOAM encapsulation options, such as the IOAM tracing option, path verification option, and end-to-end option.

If the IFIT instruction is encapsulated in the SRH, only the specified endpoint can process it. This approach is suitable on legacy networks, as it allows O&M personnel to perform on-path network telemetry only on specified IFIT-capable nodes. In SRv6 TE strict path scenarios, the encapsulation effect is the same as that of IFIT in the Hop-by-Hop Options header.

Figure 9.25 shows the encapsulation format of IOAM in the SRH.[23] Table 9.11 describes the encapsulation fields of IOAM in the SRH.

On-path network telemetry is a new and promising telemetry technology for the data plane, offering a number of benefits over the conventional active OAM. The application of this technology is also promoted by the powerful data plane programmability of SRv6. Supporting data plane encapsulation of multiple on-path network telemetry technologies, the IFIT framework can work with the SDN controller to provide functions

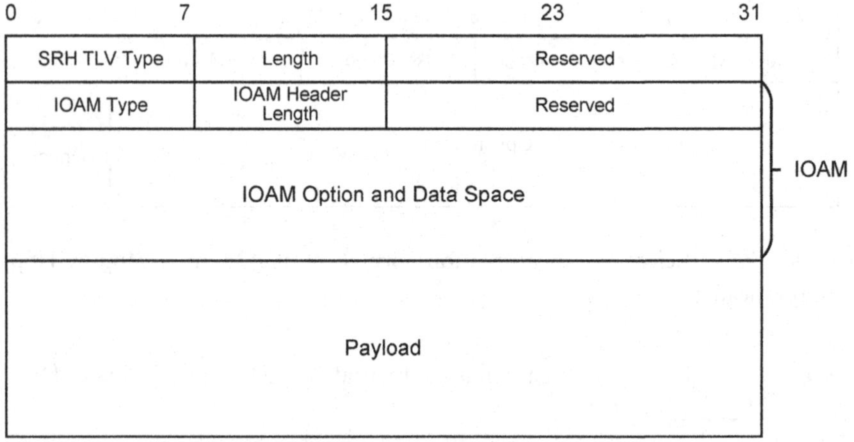

FIGURE 9.25 Encapsulation format of IOAM in the SRH.

TABLE 9.11 IOAM Encapsulation Fields in the SRH

Field	Length	Description
SRH TLV Type	8 bits	TLV type. An IOAM TLV type needs to be specified for the SRH.
IOAM Type	8 bits	IOAM type, corresponding to the IOAM encapsulation options, such as the IOAM tracing option, path verification option, and end-to-end option.
IOAM Header Length	8 bits	Length of the IOAM header, in 4-byte units.

such as smart traffic selection, efficient data sending, and dynamic network probe.

9.3 STORIES BEHIND SRv6 DESIGN

I. Path Segment and IFIT

Path segment is an important basis for SR OAM (PM). Because different SR paths can share a segment, an egress node cannot identify the SR path of a packet based on the segment. However, a path segment uniquely identifies an SR path, allowing an egress node to determine the SR path of a packet based on the path segment. The egress node can then calculate the packet loss rate and delay of the path.

Path segments can be used not only for SR performance measurement but also for bidirectional SR paths and protection switching. At the beginning, path segment was initially planned to be called Segment Routing-Transport Profile (SR-TP), the SR equivalent to Multiprotocol Label Switching-Transport Profile (MPLS-TP). However, MPLS-TP has been a source of nightmares for many IETF participants throughout its development. Due to the battle between MPLS-TP and Transfer MPLS (T-MPLS), IETF and International Telecommunication Union-Telecommunication Standardization Sector (ITU-T) have competed fiercely with each other, resulting in a lose-lose situation. Although the IETF regained its dominant position, few vendors support BFD-based MPLS-TP. To avoid reliving the past, path segment was selected as the name when the standard was promoted.

While studying IFIT, we also encountered a similar naming problem. IFIT is developed based on IOAM but IOAM alone is insufficient for IFIT research. For example, problems such as the IOAM trace option increasing the packet length, and IOAM causing devices to report a large amount of data to the analyzer, all require solutions. What name could we give to a complete IOAM solution that could be put into commercial use? At first, my colleagues were not aware of the problem, but the discussion soon became very complex and even a mess because of the lack of a concise and commonly understood definition for terminology. I therefore called a meeting to discuss the name, and because several members of the research team were losing weight to keep fit or had successfully lost weight, IFIT was quickly locked in a number of alternative acronyms. IP network O&M has always been a pain point and often received criticism in the industry. That is, although IP network products are becoming increasingly advanced, the network operation and maintenance are still very basic and inefficient. Through the research of IFIT, we hope to offer better IP O&M technologies and solutions to make the network healthier.

By unifying the terms, we can use a more common language during discussions and improve communication efficiency. Later, I took an MIT artificial intelligence course offered by the NetEase open course app. The professor mentioned the Rumpelstiltskin principle: Once you can name something, you get power over it. This resonated with me after my experience with IFIT. I defined IFIT as follows: IFIT is a data-plane telemetry solution framework that can be put into commercial use on large-scale IP networks and combines east-west on-path network telemetry technologies based on IOAM packets with north-south programmable policy for smart traffic selection and export reduction. This is probably one of the most academic definitions I have made in my years of work at Huawei.

II. Network Telemetry Framework

In my opinion, the IETF has had limited success in OAM. To a certain extent, this is related to the background of IP engineers and the working mode of IETF. They are used to installing patches and lack insight into the overall design. In the beginning, IP network basically used on-demand connectivity detection technologies such as ping and traceroute. Later, BFD technology was developed, followed by performance measurement technology being introduced to MPLS-TP. These technologies are relatively independent and make forming a unified technology system difficult. In contrast, experts with a background in telecoms are better in the field of OAM. For example, the ITU-T Y.1731 standard is defined in a unified and clear manner.

Telemetry, as an emerging technology, is very important for network O&M. I think the difficulties in IP network O&M are largely related to O&M data which has three major problems:

- Amount: The amount of data reported by network devices is insufficient.

- Speed: The data report speed of network devices is low.
- Types: The types of data reported by network devices are incomplete.

Telemetry is studied by us as a key solution to solve these problems. However, it faces challenges similar to that of OAM: Telemetry is a hot topic, but its concept and scope are not clear, and even confusing. Some people think that telemetry is equivalent to Google Remote Procedure Call (gRPC) while In-band Network Telemetry (INT) was proposed, but the reality is that gRPC and INT are totally different technologies. To clarify concepts and establish a unified technology system, we submitted the network telemetry framework draft, which divides telemetry into three layers: management-plane telemetry (collecting network management data based on gRPC/NETCONF), data-plane telemetry (collecting data-plane data based on IOAM and reporting it through mechanisms which are always UDP-based), and control-plane telemetry (reporting control protocol data based on BMP). In addition, the draft defines related terms and classifies different technologies. This draft enables us to clearly understand the telemetry framework and the differences and relationships between related technologies promoted by different IETF working groups.

IETF does not like to define use case and requirement drafts. This was closely related to the Source Packet Routing in Networking (SPRING) Working Group. SR-related protocol extensions are defined by IETF Link State Routing (LSR), Inter-Domain Routing (IDR), PCE, and other working groups. In the early stage of SR, the SPRING Working Group defined many use case and requirement drafts, but they were of limited use. Later, the IETF Routing Domain held an open meeting to discourage writing such drafts. However, the IETF operates based on many working groups and is likely to divide an overall solution into different parts and distribute them to different working groups. Consequently, people unfamiliar with the association may use a one-sided viewpoint to make an overall judgment. To solve this problem, some framework drafts become especially necessary, and this is also the driving force for us to promote the network telemetry framework draft.

III. Classification of Network Programming

In essence, IP network programming can be divided into two categories: to provide network path services, and to record network monitoring information. SRv6 segments carry instructions of network services, and an SRv6 path constructed based on segments is a set of network service instructions, including VPN service isolation, SLA guarantee for traffic engineering, and FRR reliability assurance. On-path network telemetry records network path monitoring information, which facilitates the implementation of path visualization. This is also an important reason for insisting on TLV in SRH. The SRH TLV can record variable-length on-path network telemetry information, which is hard to be supported by SRv6 segment lists.

Before the introduction of SRv6 and on-path network telemetry, IP packets are processed on a per-path basis and monitored in E2E and statistics-based mode. Thanks to the development of hardware capabilities and improvements in system processing performance, we can accurately control and monitor each packet based on technologies such as SRv6 and on-path network telemetry. This means that the basic elements (packets) on the network become more intelligent, much like ourselves as individuals. If we have accurate positioning of ourselves (measurement and monitoring) and have various ways to make a living (network services), we can live our lives to the fullest (more accurate SLA guarantee).

REFERENCES

[1] ITU-T. Operation, Administration and Maintenance (OAM) Functions and Mechanisms for Ethernet-Based Networks[EB/OL]. (2019-08-29)[2020-03-25]. G.8013/Y.1731.

[2] Mizrahi T, Sprecher N, Bellagamba E, Weingarten Y. An Overview of Operations, Administration, and Maintenance (OAM) Tools[EB/OL]. (2018-12-20)[2020-03-25]. RFC 7276.

[3] Conta A, Deering S, Gupta M. Internet Control Message Protocol (ICMPv6) for the Internet Protocol Version 6 (IPv6) Specification[EB/OL]. (2017-07-14)[2020-03-25]. RFC 4443.

[4] Katz D, Ward D. Bidirectional Forwarding Detection (BFD)[EB/OL]. (2020-01-21)[2020-03-25]. RFC 5880.

[5] Morton A. Active and Passive Metrics and Methods (with Hybrid Types In-Between)[EB/OL]. (2016-05-21)[2020-03-25]. RFC 7799.

[6] Hedayat K, Krzanowski R, Morton A, Yum K, Babiarz J. A Two-Way Active Measurement Protocol (TWAMP)[EB/OL]. (2020-01-21)[2020-03-25]. RFC 5357.

[7] Fioccola G, Capello A, Cociglio M, et al. Alternate-Marking Method for Passive and Hybrid Performance Monitoring[EB/OL]. (2018-01-29)[2020-03-25]. RFC 8321.

[8] Frost D, Bryant S. Packet Loss and Delay Measurement for MPLS Networks[EB/OL]. (2020-01-21)[2020-03-25]. RFC 6374.

[9] Mirsky G. Simple Two-way Active Measurement Protocol[EB/OL]. (2020-03-19)[2020-03-25]. draft-ietf-ippm-stamp-10.

[10] Gandhi R, Filsfils C, Voyer D, Chen M, Janssens B. TWAMP Light Extensions for Segment Routing Networks[EB/OL]. (2020-10-20)[2020-10-31]. draft-gandhi-ippm-twamp-srpm-00.

[11] Claise B, Trammell B, Aitken P. Specification of the IP Flow Information Export (IPFIX) Protocol for the Exchange of Flow Information[EB/OL]. (2020-01-21)[2020-03-25]. RFC 7011.

[12] Li C, Cheng W, Chen M, Dhody D, Li Z, Dong J, Gandhi R. Path Segment for SRv6 (Segment Routing in IPv6)[EB/OL]. (2020-03-03)[2020-03-25]. draft-li-spring-srv6-path-segment-05.

[13] Song H, Qin F, Martinez-Julia P, Ciavaglia L, Wang A. Network Telemetry Framework[EB/OL]. (2019-10-08)[2020-03-25]. draft-ietf-opsawg-ntf-02.

[14] Brockners F, Bhandari S, Pignataro C, Gredler H, Leddy J, Youell S, Mizrahi T, Mozes D, Lapukhov P, Chang R, Bernier D, Lemon J. Data Fields for In-situ OAM[EB/OL]. (2020-03-09)[2020-03-25]. draft-ietf-ippm-ioam-data-09.

[15] Song H, Zhou T, Li Z, Shin J, Lee K. Postcard-based On-Path Flow Data Telemetry[EB/OL]. (2019-11-15)[2020-03-25]. draft-song-ippm-postcard-based-telemetry-06.

[16] Zhou T, Li Z, Lee S, Cociglio M. Enhanced Alternate Marking Method [EB/OL]. (2019-10-31)[2020-03-25]. draft-zhou-ippm-enhanced-alternate-marking-04.

[17] Song H, Li Z, Zhou T, Qin F, Shin J, Jin J. In-situ Flow Information Telemetry Framework[EB/OL]. (2020-03-09)[2020-03-25]. draft-song-opsawg-ifit-framework-11.

[18] Cormode G, Muthukrishnan S. Approximating Data with the Count-Min Data Structure[EB/OL]. (2011-08-12)[2020-03-25].

[19] Google. Protocol Buffers. (2020-09-30) [2020-10-31].

[20] Song H, Gong J. Requirements for Interactive Query with Dynamic Network Probes[EB/OL]. (2017-12-21)[2020-03-25]. draft-song-opsawg-dnp4iq-01.

[21] Song H, Li Z, Zhou T, Wang Z. In-situ OAM Processing in Tunnels[EB/OL]. (2018-12-29)[2020-03-25]. draft-song-ippm-ioam-tunnel-mode-00.

[22] Bhandari S, Brockners F, Pignataro C, Gredler H, Leddy J, Youell S, Mizrahi T, Kfir A, Gafni B, Lapukhov P, Spiegel M, Krishnan S. In-situ OAM IPv6 Options[EB/OL]. (2019-09-25)[2020-03-25]. draft-ioametal-ippm-6man-ioam-ipv6-options-02.

[23] Ali Z, Gandhi R, Filsfils C, Brockners F, Kumar N, Pignataro C, Li C, Chen M, Dawra G. Segment Routing Header encapsulation for In-situ OAM Data[EB/OL]. (2019-11-03)[2020-03-25]. draft-ali-spring-ioam-srv6-02.

CHAPTER 10

SRv6 for 5G

5G HAS BEEN IN THE PIPELINE FOR MANY YEARS. During this time, it has gone from concept design to standardization and is now undergoing wide-scale commercial deployment. It's worth mentioning that the evolution of 5G networks involves not only wireless networks but also associated facilities such as mobile and fixed transport networks. Indeed, these transport networks, powered by SRv6, underpin diversified 5G service scenarios and act as a strong base for meeting the high demands of 5G services. This chapter focuses on SRv6 applications on 5G networks, including SRv6 for network slicing, Deterministic Networking (DetNet), and 5G mobile networks.

10.1 5G NETWORK EVOLUTION

4G is designed to connect everyone. 5G, on the other hand, is designed to connect everything, enabling emerging industries of an unprecedented scale and giving a new lease of life to mobile communications.

The Internet of Things (IoT) has redefined mobile communication services. No longer are connections limited to between people; rather, "things" are connecting to people, and to other "things." This is extending mobile communication technologies into more industries and fields. Building on this, 5G will further boost diverse vertical services, including mobile healthcare, Internet of Vehicles (IoV), smart home, industrial control, and environment monitoring, driving the rapid growth of various industry applications.

These services vary greatly in their characteristics, and as such pose distinct requirements on 5G networks. For instance, environment monitoring, smart home, smart farming, and smart meter reading require a vast number of device connections and frequent transmission of masses of small packets; video surveillance and mobile healthcare services require high transmission rates; IoV, smart grid, and industrial control services require millisecond-level latency and next to 100% reliability. Considering this, 5G must be highly flexible and scalable, adapting to huge quantities of device connections and diversified user requirements; only then can it penetrate more vertical services. What's more, while ensuring mobile broadband requirements are met, 5G must also meet the requirements of different industries through building flexible and dynamic networks, with vertical requirements as the focal point. This will lay the groundwork for carriers to gradually shift from selling traffic to providing on-demand, customized, and differentiated services for verticals. And this is how they will sustain growth, well into the future. Let's now look in more detail at what the industries demand from 5G networks:

- Diversified service transport

  As shown in Figure 10.1, services in the 5G era are classified into three main types[1]: Enhanced Mobile Broadband (eMBB), Ultra-Reliable Low-Latency Communication (uRLLC), and Massive Machine Type Communication (mMTC). eMBB focuses on bandwidth-hungry services, such as High-Definition (HD) video, Virtual Reality (VR), and Augmented Reality (AR); uRLLC on latency- and reliability-sensitive

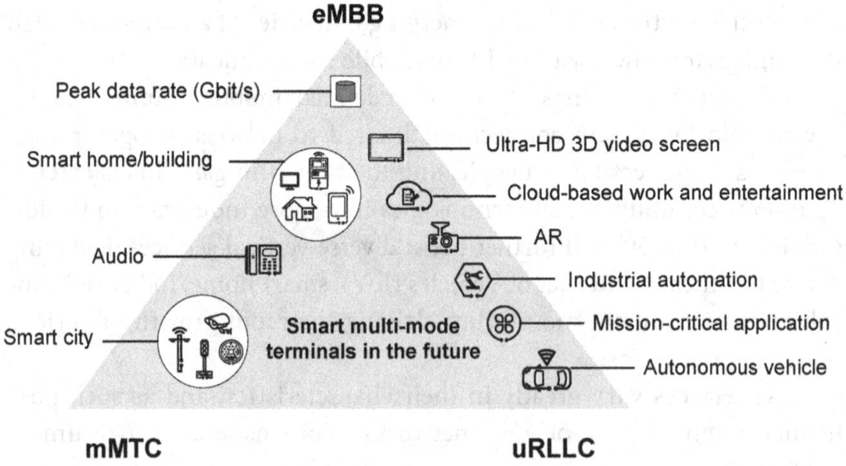

FIGURE 10.1 5G service classification.

services, such as autonomous driving, industrial control, telemedicine, and drone control; and finally mMTC on scenarios with high connection density, such as smart city and smart farming. Evidently, all three require completely different kinds of network characteristics and performance, which no single network could possibly deliver.

- High performance

  5G-oriented services usually have high requirements on multiple performance indicators. For example, VR and AR services require both high bandwidth and low latency. Terminal "users" in verticals are machines. A failure to meet their performance requirements means that the terminals lose their functionality. As such, they are more sensitive to performance than people. And fully autonomous driving of IoV will only be commercially ready once data transmission latency and reliability have reached certain levels.

- Fast provisioning

  Verticals require rapid service provisioning, yet traditional service networks typically require as long as 10–18 months to deploy a new network function, far from meeting the requirement.

- Network slicing and security isolation

  Network slicing is one of the key features of 5G. It enables a 5G network to be sliced up into logical partitions that can be allocated to different tenants over a shared network infrastructure. Industrial customers use the 5G network as slice tenants. These slices must be securely isolated. This is crucial for two reasons: both security (data and information are effectively isolated between tenants) and reliability (the network exceptions or failures of one tenant do not affect other tenants on the same network).

- Automation

  5G networks have diverse services and structures; as such, managing such networks manually is not a feasible option. Instead, managing these networks efficiently calls for automatic network management technologies, such as self-diagnosis, self-healing, automatic configuration, self-optimization, Zero Touch Provisioning (ZTP), and plug-and-play. As networks require higher-level self-management, Artificial Intelligence (AI) will start to be adopted more and more.

- New ecosystem and business models

  Once 5G networks have reached large-scale use among verticals, never-before-seen roles (infrastructure network providers, wireless network carriers, virtual network carriers, and more), business relationships, and business models will emerge in quick succession. The new roles and their business relationships will build a new telecom ecosystem for the 5G era, with carriers seeing the birth of new and diverse business relationships.

It is imperative that 5G networks, including access, core, and transport networks, evolve to meet the requirements of various service scenarios in the future, as shown in Figure 10.2.

- **Access network:** A 5G access network can provide different access architectures as required, including the Distributed Radio Access Network (D-RAN) and Cloud RAN (C-RAN). Of these, D-RAN is a RAN architecture in which both Baseband Units (BBUs) and Remote Radio Units (RRUs) are deployed on the base station side. C-RAN, in contrast, splits the BBU functionality into two functional units: Distributed Unit (DU) and Central Unit (CU). A DU is mainly responsible for physical-layer and real-time processing, whereas a CU mainly provides the non-real-time wireless upper-layer protocol stack function. A RAN can be reconstructed with CUs centrally deployed on the cloud. In this way, RAN functions can be deployed on demand — a major requirement in the 5G era.
- **Core network:** A 5G core network uses the Service-Based Architecture (SBA) for its control plane to decouple, aggregate, and servitize original control plane functions. In this way, network functions can be plug-and-play, making network functions and resources available on demand. Network slices can be constructed on top of the cloudified infrastructure to logically isolate and serve different services (eMBB, uRLLC, and mMTC) or tenants. A 5G core network also separates the control plane from the user plane, offering a simplified network structure. In addition, core network functions can be flexibly deployed in data centers at different levels based on service requirements.
- **Transport network:** Before the use of 5G, a transport network generally uses MPLS L3VPN to carry Layer 3 services and uses MPLS

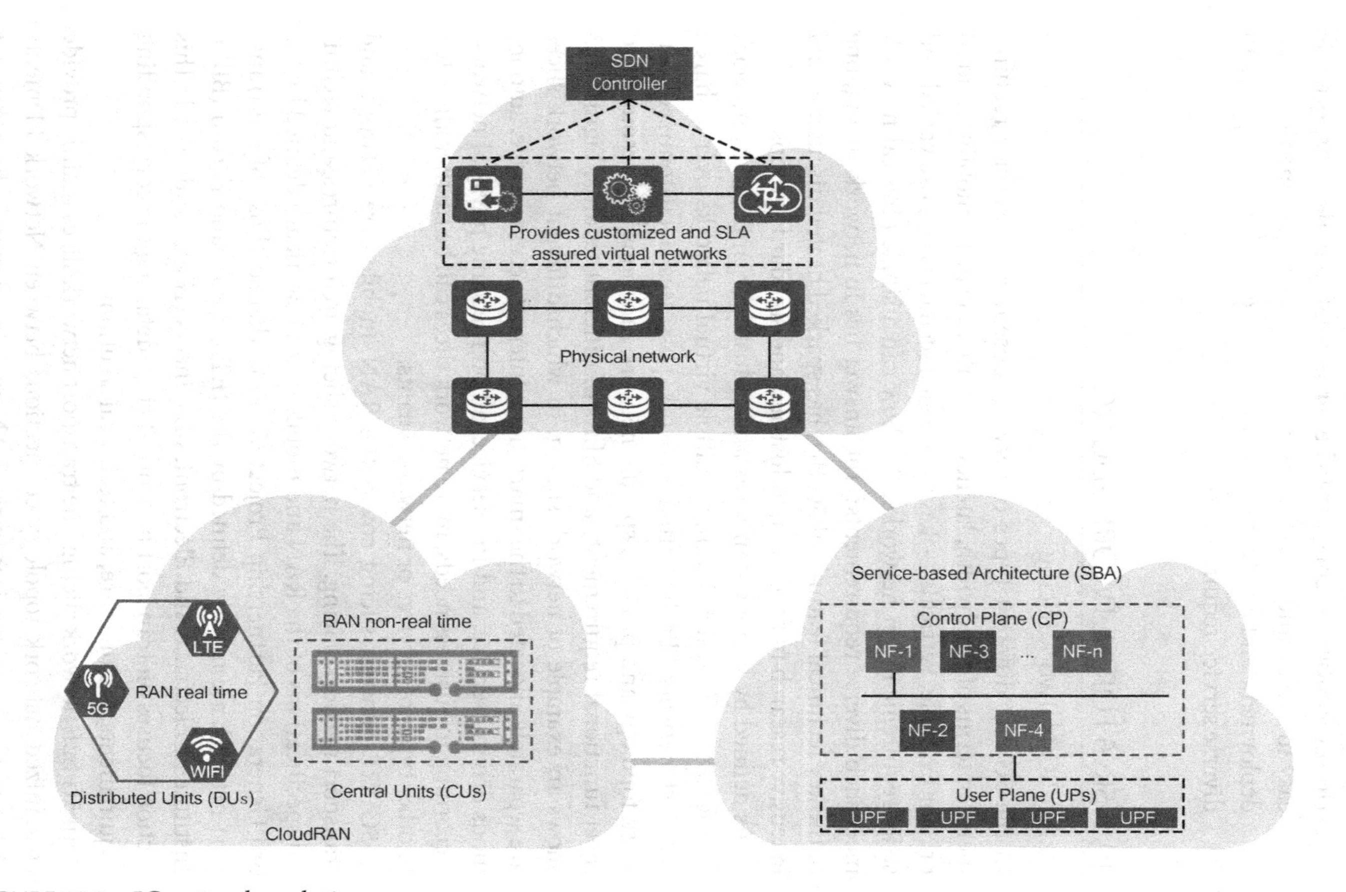

FIGURE 10.2 5G network evolution.

L2VPN VPWS or other technologies to carry Layer 2 services. However, SRv6 has emerged to better serve 5G mobile transport. It is easy to deploy and therefore meets 5G service requirements for fast deployment and automation. It can also be extended in response to diverse service requirements.

10.2 SRv6 FOR NETWORK SLICING

10.2.1 5G Network Slicing

In the 5G era, different types of services pose distinct service requirements in terms of bandwidth, latency, reliability, security, mobility, and so on. For example, verticals — key 5G service scenarios — pose varied and strict requirements on networks. But how can a single physical network meet such diverse requirements? The answer lies in network slicing, one of the key technical features of 5G that has sparked heated discussion and extensive research in academia, industries, and standards groups.

As defined by various standards and industry organizations, network slicing is a method for creating multiple virtual networks over a shared physical network. Each virtual network possesses a customized network topology and provides specific network functions and resources to meet functional requirements and SLAs of different tenants. Figure 10.3 shows an example of network slicing, in which different network slices are separately allocated to the smartphone Internet access service, autonomous driving service, and IoT service, all on the same physical network. According to business needs, more network slices can be further allocated to other types of services or network tenants.

5G E2E network slicing covers the RAN, mobile core network, and transport network slicing. The network slicing architectures and technical specifications of the RAN and mobile core network are defined by the 3rd Generation Partnership Project (3GPP), whereas those of the transport network are mainly defined by the IETF, Broadband Forum (BBF), Institute of Electrical and Electronics Engineers (IEEE), and ITU-T. This section focuses on transport network slicing, detailing the corresponding requirements, architecture, and technical solutions.

In 5G E2E network slicing, the transport network slices mainly provide customized network topology connections between Network Elements (NEs) and between services in the RAN and core network slices; but not

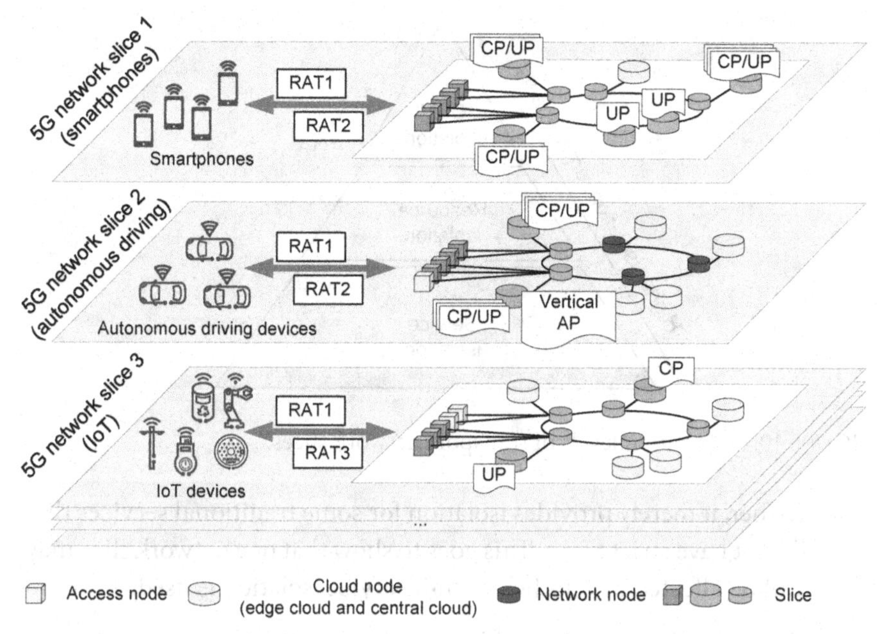

FIGURE 10.3 Example of 5G network slicing.

only this, they also provide differentiated SLA guarantee for services of different 5G E2E slices. Another thing the transport network needs to provide is open network slice management interfaces for slice lifecycle management. In this way, it can work with the RAN and core network for E2E collaborative slice management, as well as providing network slices as a new service to tenants of verticals.

As mentioned, various network slices often need to be created on the same physical network. But how can they be prevented from affecting each other throughout the service lifecycle? Isolation between network slices is the answer, and transport network slices provide three levels of isolation: service, resource, and O&M isolation, as shown in Figure 10.4.

- **Service isolation**: Service packets in one network slice will not be sent to service nodes in another network slice on the same network in any way. In other words, service connections are isolated between different network slices, making services of different tenants invisible to each other on the same network. It's important to note, however, that service isolation in and of itself cannot guarantee SLAs.

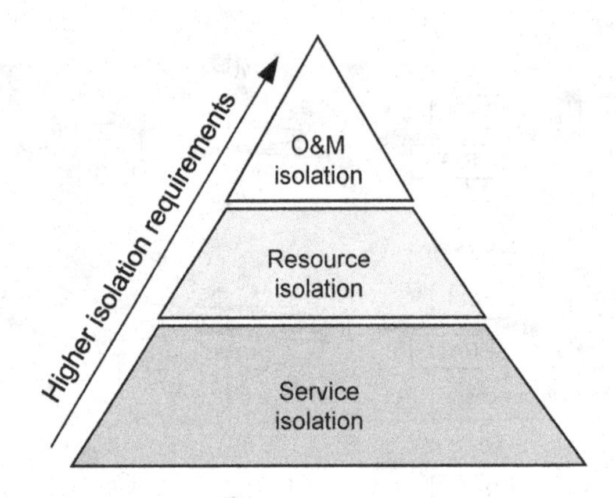

FIGURE 10.4 Isolation levels of transport network slices.

Rather, it merely provides isolation for some traditional services that do not have strict SLAs. This goes to show that one network slice may still be affected by another, even if service isolation is used.

- **Resource isolation**: Network resources are defined for exclusive use by a slice, or for sharing among multiple slices. This is paramount for 5G uRLLC services, which usually have strict SLAs and tolerate zero interference from other services. Resource isolation includes hard isolation and soft isolation, which differ in the degree of isolation.[2] Hard isolation ensures that slices are provided with exclusive network resources, preventing any interference between services. Soft isolation, on the other hand, allows each slice to use both a set of dedicated resources and the resources shared with other network slices. The benefit of soft isolation is that services are isolated to some degree, without having to sacrifice certain statistical multiplexing capabilities. By mixed use of hard and soft isolation, carriers can select the optimal combination of network slices to meet their resource requirements. This allows a single physical network to meet differentiated service SLAs.

- **O&M isolation**: In addition to service isolation and resource isolation, some tenants require independent O&M of network slices allocated by carriers, similar to using private networks. This poses higher requirements on the management plane of the network slices regarding the slice information it presents and the openness of its interfaces.

10.2.2 Transport Network Slicing Architecture

As shown in Figure 10.5, the slicing architecture of a transport network consists of three layers: network infrastructure layer, network slice instance layer, and network slice management layer. At each layer, a combination of existing and new technologies needs to be used to meet tenant requirements on transport network slicing. The following details these technologies layer by layer.

- **Network infrastructure layer**: needs to be capable of dividing the total physical network resources into multiple sets of isolated resources that are allocated to different network slices. Such resource isolation can be achieved using methods such as Flexible Ethernet (FlexE)[3] client interfaces, channelized sub-interfaces (individual channels that subdivide the total bandwidth of a physical interface), and independent queues and buffers. Depending on service requirements and capabilities of network devices, other forwarding resources on the network can also be divided and allocated to different network slices.

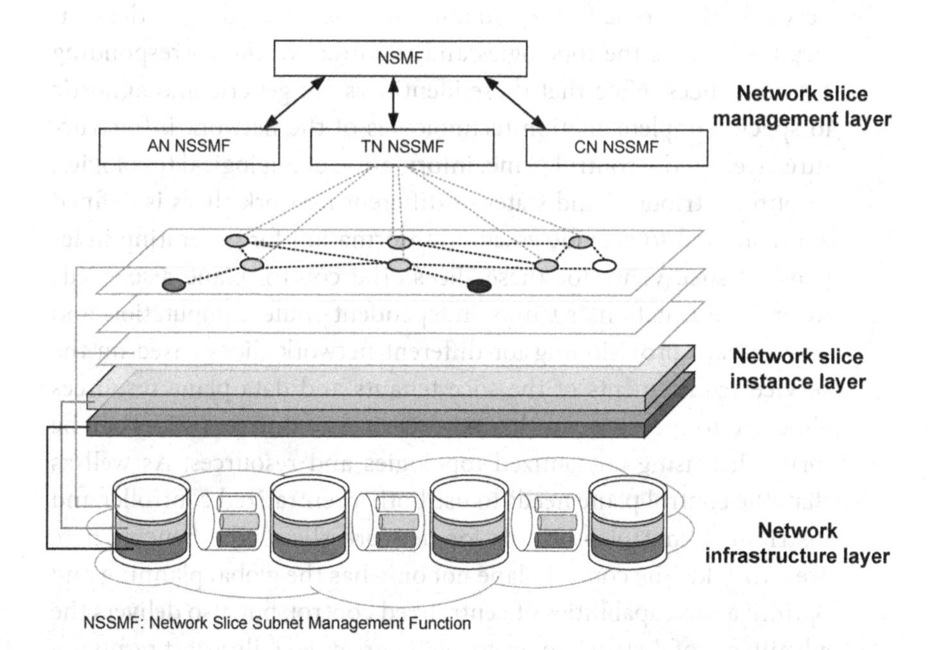

FIGURE 10.5 Transport network slicing architecture.

- **Network slice instance layer**: provides different logical network slice instances on a physical network and customized logical topology connections, and associates the logical topologies of slices with the set of network resources allocated to the slices. In this way, network slices are formed to meet specific service requirements.

  The network slice instance layer covers VPNs at the overlay layer and Virtual Transport Network (VTNs) at the underlay layer. The VPN provides logical connections for services within a network slice and can isolate services of different network slices. The VTN, on the other hand, provides customized network topologies for slice service connections and provides exclusive or partially shared network resources to meet SLAs of slice services. In this regard, a network slice instance is the integration of a VPN service as the overlay with an appropriate VTN as the underlay. As various overlay VPN technologies have been mature and widely used, this section focuses on the VTN functions only.

  VTN functions can be further broken down into data plane functions and control plane functions. In the data plane, network slice identifiers are added to data packets to instruct packets of different network slices to be forwarded and processed according to the constraints such as the topologies and resources of the corresponding network slices. Note that these identifiers are generic and agnostic to specific implementation technologies of the network infrastructure layer. In the control plane, information about logical topologies, resource attributes, and states of different network slices is defined and collected to provide essential information for generating independent slice views for these slices. The control plane also needs to provide functions such as independent route computation and service path provisioning for different network slices based on the service requirements of the slice tenants and data plane resources allocated to the slices. In this way, services of different slices can be forwarded using customized topologies and resources. As well as that, the control plane needs to use both a centralized controller and distributed control protocols for network slices. The benefits here are two-fold: The control plane not only has the global planning and optimization capabilities of centralized control, but also delivers the advantages of distributed protocols — great flexibility, fast response, high reliability, and good scalability.

- **Network slice management layer**: mainly provides slice lifecycle management functions, including planning, creation, monitoring, adjustment, and deletion. In the future, verticals will have increasingly stringent requirements on slicing. To meet these requirements, it is vital that the network slice management plane is able to support dynamic, on-demand management of the slice lifecycle. To implement 5G E2E network slicing, the management plane of the transport network slice also needs to provide open interfaces to exchange information — such as network slicing requirements, capabilities, and states — with the 5G Network Slice Management Function (NSMF), and to negotiate and interwork with RAN and mobile core network slices.

To sum up, transport network slicing involves the functions and capabilities of the network infrastructure layer, network slice instance layer, and network slice management layer. For different network slicing requirements, carriers need to select the appropriate technologies with related extensions and enhancements at each layer, thereby forming a complete network slicing solution.

The draft of VPN+ framework, which describes the transport network slicing architecture, is being standardized in the IETF and has been adopted by the working group.[2] VPN+, as its name suggests, is an enhancement to the functions, deployment modes, and business models of conventional VPN. Conventional VPN is mainly used to isolate services of different tenants but falls short in SLA guarantee and open management. VPN+ introduces resource isolation technologies and implements on-demand integration of logical service connections with the underlay network topologies and resources based on the mappings between VPN services (overlay) and VTNs (underlay). This allows VPN+ to provide differentiated SLA guarantee for different types of services. In addition, VPN+ enhances and extends management interfaces to provide more flexible and dynamically adjustable VPN services, meeting differentiated service requirements of various industries in the 5G era. The VPN+ framework also describes the layered network slicing architecture specific to the transport network and key technologies available at each layer, including existing technologies and their extensions, as well as a series of technology innovations. By leveraging the VPN+ framework and combining available technologies at different layers, network slicing can be achieved to meet diversified requirements.

10.2.3 SRv6-based Network Slicing

Among the key features of SRv6 are flexible programmability, good scalability, and great potential for E2E unified transport, and it is these features that make SRv6 one of the key technologies for 5G transport. These features can also be applied to network slicing to create network slice instances and provide an SRv6-based transport network slicing solution.

In the data plane, a data packet's SRv6 SIDs can be used to identify the network slice to which it belongs. Network devices along the packet forwarding path use the SRv6 SIDs to identify the target network slice and then forward the packet according to the rules specific to the network slice. As mentioned, unique SRv6 SIDs need to be allocated to each network slice. And another thing SRv6 SIDs can be used for is to identify network resources that network devices and links along the packet forwarding path allocate to each network slice, which is needed if the network slices have a resource isolation requirement. This ensures that packets of different network slices are forwarded and processed using only the resources dedicated to those network slices, thereby providing a reliable and deterministic SLA guarantee for services in the slices.

As shown in Figure 10.6, an SRv6 SID consists of Locator, Function, and the optional Arguments fields. Note that the Function field and the optional Arguments field are not identified or parsed by transit nodes on an SRv6 network. Instead, such nodes merely match and forward data packets based on the locators in the SIDs. To ensure processing on network slices is consistent end to end, a locator in an SRv6 SID must contain network slice identifier information. That is to say, a locator can identify both a network node and its corresponding network slice. The Function and Arguments fields can indicate the function and parameter information defined for the network slice.

Figure 10.7 shows an example of SRv6-based network slicing. In this example, a carrier receives service requirements (including service

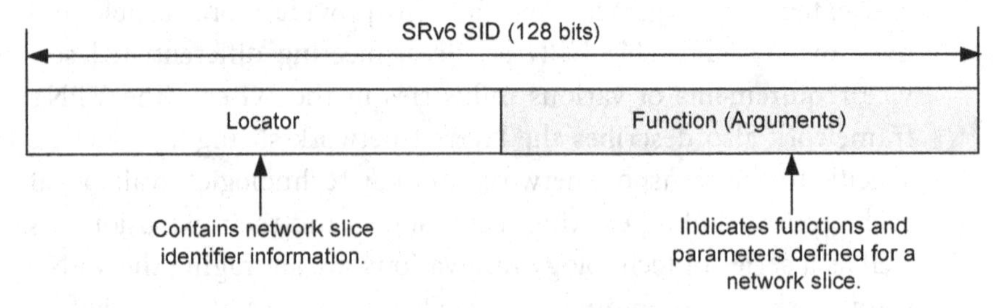

FIGURE 10.6 SRv6 SID structure.

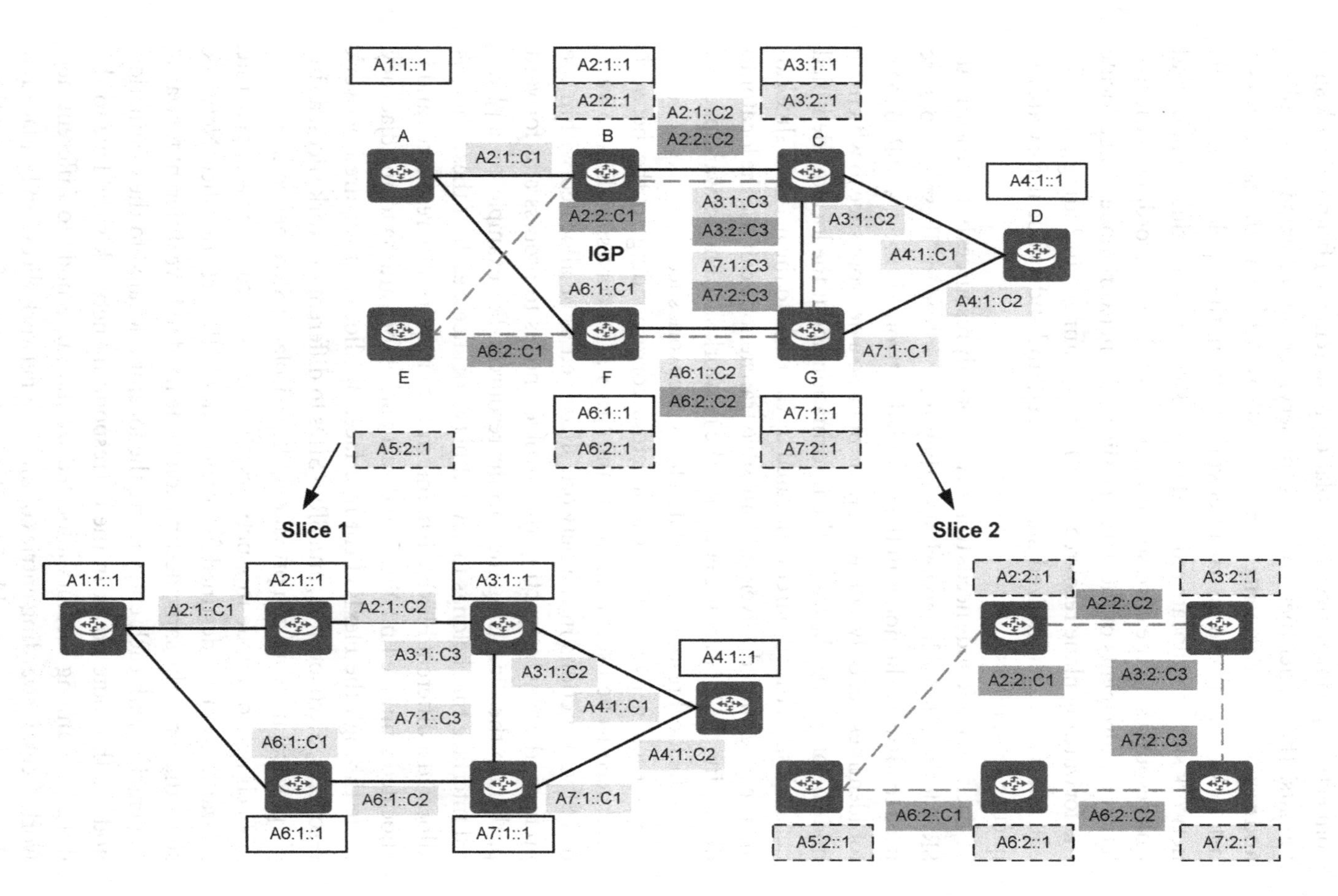

FIGURE 10.7 Example of SRv6-based network slicing.

connection, performance, and isolation requirements) from network slice tenants. The carrier then converts the service requirements into topology and resource requirements on network slices. After that, the carrier uses a network controller to instruct network nodes in the topologies of specific network slices to allocate required network resources (including node and link resources) to the corresponding network slices. According to the isolation requirements of the network slices, the network nodes can allocate the forwarding plane resources in different manners, including FlexE client interfaces, channelized sub-interfaces, and independent forwarding queues and buffers.

Each network node needs to allocate not only resources but also a unique SRv6 locator, to each network slice they are involved in. These locators are used to identify the nodes in the specified network slice. So, an SRv6 locator identifies not only a network node but also a network slice to which the node belongs. And based on the information about the locator-identified network slice, the resources allocated by the node to the network slice can also be determined. Given this, an SRv6 transit node can use the locator to determine the forwarding entry for the network slice and forward packets using the local resources allocated to the network slice.

For each network slice, the locators corresponding to the network slice need to be used as prefixes, and SRv6 SIDs corresponding to various SRv6 functions in the network slice need to be allocated within the prefixes, thereby instructing the involved nodes to process and forward packets in the slice using the specific resources. For example, for a physical interface that belongs to several network slices, each of these slices is allocated a different FlexE client interface to acquire slice resources and is allocated a unique End.X SID to direct packet forwarding to an adjacency node through the FlexE client interface. By allocating required network resources, SRv6 locators, and SRv6 SIDs to different network slices, a single physical network can provide multiple isolated SRv6 network slices.

Each network node in a network slice needs to compute and generate forwarding entries destined for other nodes in the network slice. Network slices that have the same destination node need to have different forwarding entries, with each entry using the locator allocated to the destination node as the route prefix in the corresponding network slice. Due to the difference among locators allocated by the same node to different network slices, forwarding entries of different network slices can be placed in the same forwarding table. The IPv6 prefixes in the forwarding table are matched against the SRv6 Locator field in the SIDs carried in the DA field of data packets, thereby determining the next hops and outbound interfaces of the packets. If an outbound interface allocates exclusive resources to different network slices, the sub-interfaces on the outbound interface for these slices are also determined from the forwarding table.

As shown in Figure 10.7, node B belongs to both network slices 1 and 2. The IPv6 forwarding table of node B is described in Table 10.1.

SRv6 SIDs allocated by each network node to different network slices are stored in the corresponding node's Local SID Table. Due to the fact that SRv6 SIDs allocated by a node to different network slices use the locators of these network slices as prefixes, it is possible to place SIDs of these network slices in the same Local SID Table. Each SID dictates how a packet is processed and forwarded in a specified network slice using the resources allocated to the slice. Figure 10.7 gives an example, in which the End.X SID A2:2::C2 allocated by node B instructs node B to forward the data packet to node C in network slice 2 using B's sub-interface 2 to C.

In each SRv6 network slice, the SRv6 forwarding path of a data packet can be one of two options: an explicit path computed by a network controller (centralized); or a shortest path computed by network nodes (distributed). Either path computation mode depends on constraints such as topologies and resources of network slices. In Figure 10.8, data packets are forwarded using the explicit and shortest paths in network slice 1. For a data packet to be forwarded using an SRv6 explicit path (A-F-G-D) in network slice 1, an SRH needs to be encapsulated in the packet header. The SRH carries SRv6 End.X SIDs that identify links hop by hop in the slice to explicitly specify a packet forwarding path and the sub-interface

TABLE 10.1 IPv6 Forwarding Table of Node B Outbound Interface/Sub-interface

IPv6 Prefix	Next Hop	Outbound Interface/Sub-interface
A1:1 (node A in slice 1)	Node A	B's interface to A
A3:1 (node C in slice 1)	Node C	B's sub-interface 1 to C
A3:2 (node C in slice 2)	Node C	B's sub-interface 2 to C
A4:1 (node D in slice 1)	Node C	B's sub-interface 1 to C
A5:2 (node E in slice 2)	Node E	B's interface to E
A6:1 (node F in slice 1)	Node A	B's interface to A
A6:2 (node F in slice 2)	Node E	B's interface to E
A7:1 (node G in slice 1)	Node C	B's sub-interface 1 to C
A7:2 (node G in slice 2)	Node C	B's sub-interface 2 to C

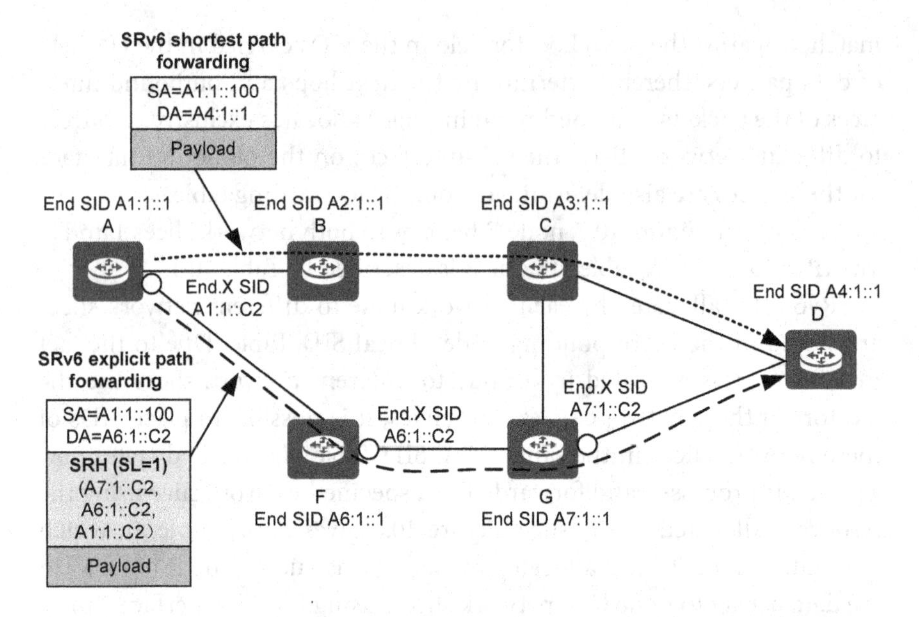

FIGURE 10.8 Data packet forwarding using the explicit and shortest paths in network slice 1.

information used by each hop. When it comes to forwarding data packets using an SRv6 shortest path, an SRH is not encapsulated in the packet header; rather, only the End SID of destination node D in the network slice is populated in the DA field of the IPv6 header. The network nodes on the path then search their forwarding tables for the destination address in the IPv6 header and forward the packet according to the matching forwarding entry. The next hop and outbound interface in the forwarding entry are computed in the network slice associated with the End SID.

As shown in Figure 10.9, SRv6 reduces the number of network protocols in the control plane as it does not require signaling protocols such as RSVP-TE for path establishment and resource reservation. Through interworking between the network slice controller and network nodes as well as cooperation among the distributed control planes of these network nodes, it is possible to deliver, distribute, and collect network slice information, and compute paths for network slices in a centralized or distributed way. The involved protocols include NETCONF/YANG, IGP, BGP-LS, BGP SR Policy, and PCEP.

- NETCONF/YANG is used by the controller to deliver network slice-related configuration information to the network nodes.

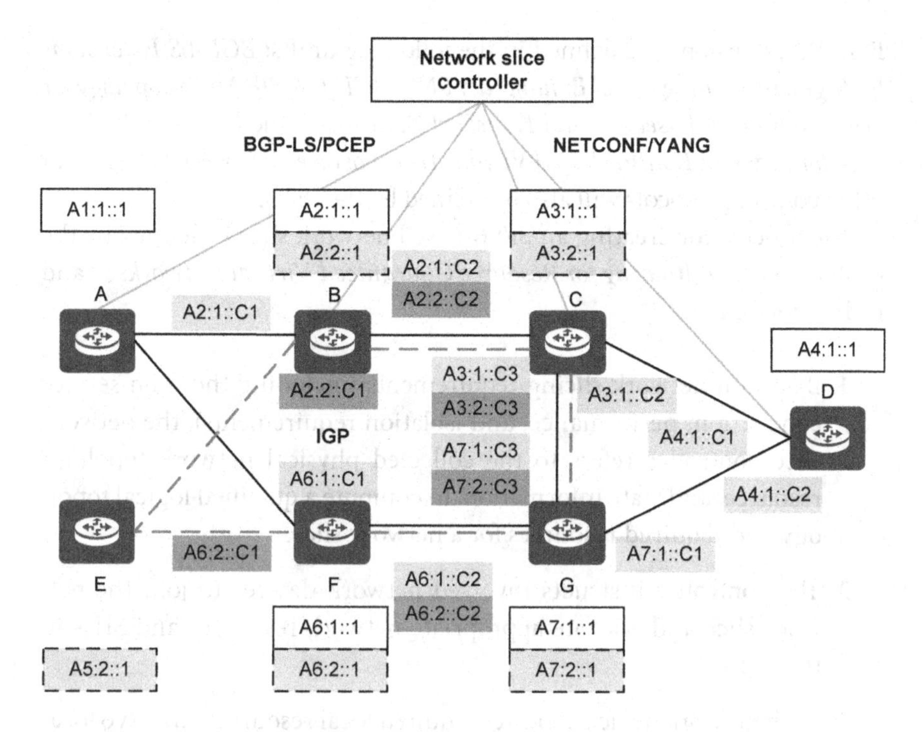

FIGURE 10.9 Control plane of network slicing.

- IGP is used to flood the network slice definitions, as well as the SIDs and the corresponding resource attributes of network slices among the network nodes.
- BGP-LS is used by the network nodes to report the flooded information to the controller.
- BGP SR Policy is used by the controller to provide the network nodes with SR paths. The controller computes these paths by taking the network slice attributes and constraints into consideration.
- PCEP is used by the controller to exchange path computation requests with the network nodes and deliver path computation results to specific network slices.

The IGP extensions for SRv6-based network slicing are defined in the following drafts: *IGP Extensions for Segment Routing based Enhanced VPN*,[4] *Using IS-IS Multi-Topology (MT) for Segment Routing based Virtual Transport Network*,[5] and *Using Flex-Algo for Segment Routing based VTN*.[6] The

BGP-LS extensions are defined in the following drafts: *BGP-LS Extensions for Segment Routing based Enhanced VPN*,[7] *BGP-LS with Multi-topology for Segment Routing based Virtual Transport Networks*,[8] and *BGP-LS with Flex-Algo for Segment Routing based Virtual Transport Networks*.[9] Extensions of other control protocols will also be defined by the IETF.

The process for creating an SRv6-based network slice is defined by the draft of *Segment Routing for Resource Guaranteed Virtual Networks*[10] and is described here:

1. Based on network slicing requirements (including those on service connection, performance, and isolation requirements), the network slice controller refers to the collected physical network topology, resource, and state information, to compute a qualified logical topology and required resources for a network slice.

2. The controller instructs involved network devices to join the network slice and allocate appropriate network resources and SIDs to the slice.

3. Each network device allocates required local resources, an SRv6 locator, and SIDs to the corresponding network slice. The locator identifies the virtual node, and the SIDs identify the virtual links and SRv6 functions of the network device in the network slice. The SRv6 locator and SIDs of the slice also specify the resources allocated by the network device to the network slice. Each network device uses the control protocol to advertise its network slice identifier to the network slice controller and other devices on the network. They also advertise SRv6 locator and SID information associated with the network slice as their data plane identifiers of virtual nodes, virtual links, and SRv6 functions in the network slice. The network devices can further advertise resource and other attribute information of the slice.

4. According to the collected network slice topology, resource, and SRv6 SID information, each network device computes paths within the network slice to generate SRv6 BE forwarding entries based on slice constraints. The network controller and network edge nodes can compute and generate SRv6 TE forwarding entries according to additional constraints specified by carriers. The network controller may also compute SRv6 BE paths based on slice constraints.

5. The services are mapped to the slice on the SRv6 network. Note that services of different tenants on the network can be mapped to exclusive or shared SRv6 network slices according to service SLAs, connection requirements, and traffic characteristics.

10.2.4 Network Slice Scalability

One question is often raised: How many network slices does a network need to provide? Currently, there is no single answer to this question. One common view is that only a few slices, around 10 or so, are required to provide coarse-grained service isolation on a network in the early stage of 5G, during which eMBB services are mainly developed. As 5G matures, the development of uRLLC services and the emergence of various vertical services will require more from network slicing. By then, hundreds or even thousands of network slices may be required. A network must therefore be able to provide more fine-grained and customized network slices. To fulfill this requirement, a network slicing solution must be capable of evolving to support more and larger slice networks, in addition to meeting the slice number and scale requirements in the early stage of 5G.

Network slice scalability mainly involves the scalability of the control and data planes, which will be detailed further in the following two sections.

10.2.4.1 Control Plane Scalability of Network Slices

The control plane of a network slice is divided into two layers: one for the VPN (overlay) and the other for the VTN (underlay). The VPN control plane distributes service connectivity and service routing information using MP-BGP. Its scalability has already been verified through wide-scale deployment, so will not be a focus here. Instead, this section focuses on the scalability of the VTN control plane.

The VTN control plane mainly distributes and collects topology and various attribute information of network slices, then computes separate routes for each network slice, and finally delivers and stores the computation results in a forwarding table. When there are a large number of network slices, there is a huge rise in the overheads involved in information distribution and route computation in the control plane. Reducing these overheads can be achieved using the following optimization methods:

1. Enable multiple network slices to exchange information over a shared control plane session. When adjacent network devices belong to different network slices, information about these slices can be distributed using the same control plane session. This method prevents the number of control plane sessions from multiplying as the number of network slices increases, thereby avoiding overheads involved in additional control plane session maintenance and information exchange. One point to note is that when network slices share a control plane session, they need to be distinguished using unique slice identifiers in control messages.

2. Decouple attributes of network slices. This method enables the control plane to separately distribute and process different types of attribute information. A network slice has various types of attributes, some of which are processed separately in the control plane. In the draft of IGP extensions for VPN+, two basic network slice attributes are proposed: topology and resource.[4]

In Figure 10.10, two different logical network topologies are defined on the physical network based on the connection requirements of network slices. Then, resources that need to be allocated to each network slice are determined based on the slice's SLAs and isolation requirements. Each network slice is a combination of its topology and resource attributes. As such, by using different resources, unique network slices can be formed over the same network topology. It is possible to decouple the topology and resource attributes of network slices to a certain extent with regard to information advertisement and computation in the control plane. For network slices with the same topology, mappings between the network slices and the topology can be specified in the control plane, in which case only one topology attribute needs to be advertised. The topology attribute can be referenced by all the network slices with the same topology, and path computation results based on the topology can also be shared by these network slices.

For example, in Figure 10.10, network slices 1, 2, and 3 share one topology, and 4, 5, and 6 share the other. To allow all the network nodes to obtain the topology information of all the six network slices, all that a carrier needs to do is to associate slices 1, 2, and 3 with topology 1, and slices 4, 5, and 6 with topology 2. The carrier then just needs to enable

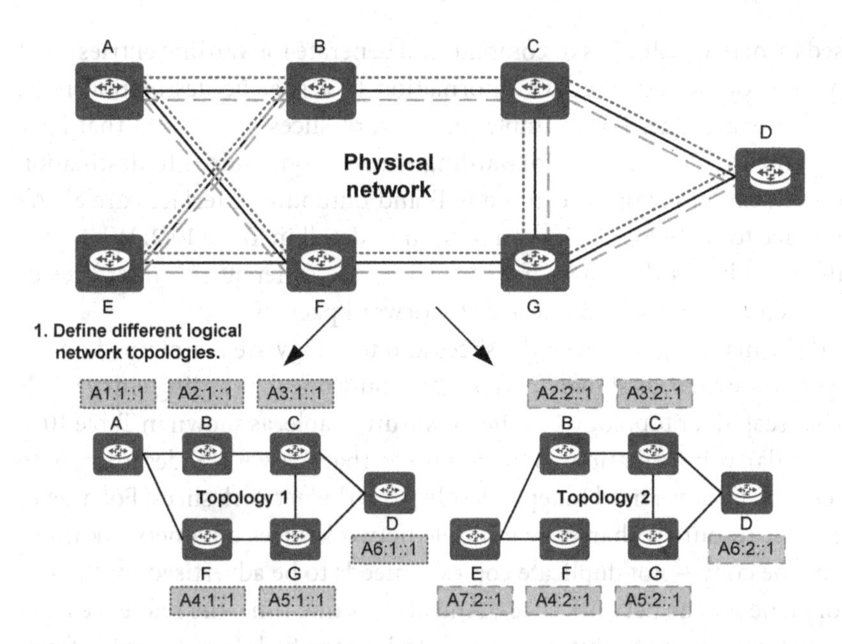

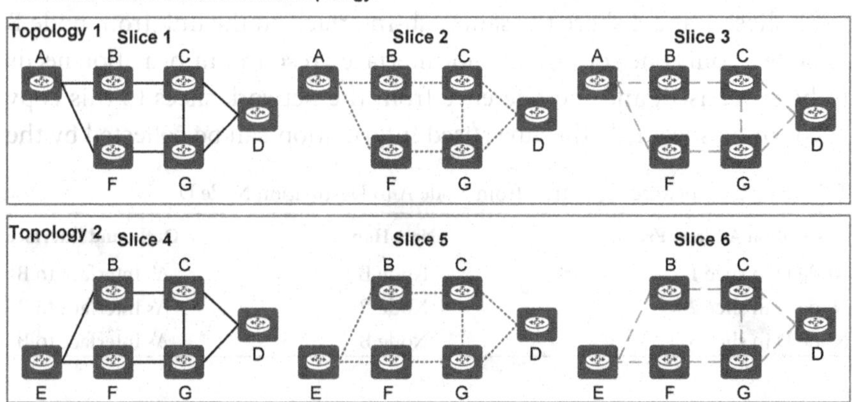

FIGURE 10.10 Two different slice network topologies defined on the physical network.

the network nodes to separately advertise the attributes of topology 1 and topology 2. Clearly, this can serve to greatly reduce the number of exchanged messages, in comparison to when the topology attribute is separately advertised for each network slice. The network slice topology information can be advertised using the Multi-Topology (MT) technology.[11] Flexible Algorithm (Flex-Algo) can also be used to describe the topological constraints for path computation.[12] The network topology information is used by network devices to compute and generate forwarding entries' next hop and outbound interface information to a specific destination node in a network slice. For example, in network slices 1, 2, and 3 that have the same topology, for the forwarding entries from node A to destination node D, the next hops are all node B and outbound interfaces are all A's interface to B. These entries can be seen in detail in Table 10.2. Where the difference lies is that these network slices use different sub-interfaces or queues on the outbound interface to forward packets.

If the mappings between the slices and topology are maintained on the involved devices, as shown in Table 10.3, only one forwarding entry needs to be created for topology 1 in the forwarding table, as shown in Table 10.4.

Similarly, information about resources that network nodes allocate to slices can be advertised independently from the network slices. For a set of network resources shared by multiple network slices on a network node, only one copy— not duplicate copies — needs to be advertised in the control plane for the network slices. And at the same time, the reference from the network slices to this copy needs to be specified. For example, if network slices 1 and 4 share the same sub-interface on the link from node B to node C, only one copy of the sub-interface's resource information needs to be advertised, and the reference from the network slices to this copy needs to be specified. The advertised information can be collected by the

TABLE 10.2 Forwarding Entries from Node A to Destination Node D

Destination Address Prefix	Next Hop	Outbound Interface
Node D in slice 1	Node B	A's interface to B
Node D in slice 2	Node B	A's interface to B
Node D in slice 3	Node B	A's interface to B

TABLE 10.3 Mappings between the Slices and Topology

Slice ID	Topology ID
Slice 1	Topology 1
Slice 2	Topology 1
Slice 3	Topology 1

TABLE 10.4 Forwarding Entry of Topology 1

Destination Address Prefix	Next Hop	Outbound Interface
Node D in topology 1	Node B	A's interface to B

TABLE 10.5 Sub-interfaces of the Outbound Interface That Forwards Packets

Next Hop	Slice ID	Outbound Interface/Sub-interface
Node B	Slice 1	A's sub-interface 1 to B
	Slice 2	A's sub-interface 2 to B
	Slice 3	A's sub-interface 3 to B

headend to generate complete information about the network slices, so as to compute constrained paths for the network slices.

When forwarding a data packet in a network slice, based on the network slice to which the packet belongs, each involved device needs to find the associated local resource, which can be a sub-interface or a queue of an outbound interface (as shown in Table 10.5).

In the IETF draft of IGP extensions for VPN+, the definition of a network slice includes a combination of a series of attributes, such as the topology attribute. The definition of the specific network slice is advertised on the network through control protocol extensions,[4] ensuring consistent network slice definition throughout the network.

In addition to network slice definitions being advertised using control protocol extensions, various network slice attributes can be separately advertised using different protocol extensions. For example, the topology attribute of a network slice can be advertised using MT or Flex-Algo,[11] and unique locators and SIDs can be allocated to the slice and topology through SRv6 control plane extensions. In terms of the resource attribute, resource information of different sub-interfaces or queues on each interface can be advertised through IGP L2 Bundle extensions.[13] At the same time, the mappings between the sub-interfaces/queues and slices can also be advertised. Following these IGP extensions, BGP-LS can be extended accordingly. This approach allows network devices or a controller in a slice to collect and combine various attribute information associated with that slice, thus obtaining complete network slice information.

10.2.4.2 Data Plane Scalability of Network Slices

In terms of the resource allocation capability of the network infrastructure layer, the number of network resources that can be sliced is somewhat limited. So, to support more network slices, the underlay network needs to improve its slicing capability by introducing new technologies that support finer-grained allocation of more resources. That said, the cost-effectiveness of network slices cannot be ignored, and a certain degree of resource sharing needs to be introduced on top of resource isolation. In other words, resources need to be allocated to network slices through a combination of soft and hard isolation. Only then can we meet differentiated SLAs and isolation requirements of different services.

In addition, network slices need to use different routing tables and forwarding entries, which leads to an increase in data plane entries each time a network slice is added. For the SRv6-based network slicing solution, allocating unique SRv6 locators and SIDs to each slice requires more locator and SID resources, bringing great difficulties to network planning.

One approach to optimizing the network slice's data plane is as follows: convert the flat structure that uses only a single data plane identifier for packet forwarding into a hierarchical structure that uses multiple data plane identifiers. In doing so, some forwarding information needs only to be maintained locally on specific network nodes and there is a drop in the number of forwarding entries that need to be maintained across the entire network. For example, a data plane slice identifier, which is independent of the routing and topology information identifiers, can be provided for the VTN layer of a network slice. When a data packet is forwarded within the network slice, a next hop is first found by matching routing and topology information in the packet, and then a sub-interface or a queue to the next hop is found in the network slice according to the slice identifier in the data packet. The introduction of multilevel identifiers demands higher flexibility and extensibility in the data plane when it comes to how the data plane encapsulates and forwards packets. Such requirements can only be met through SRv6 data plane programmability and IPv6 header extensibility. A feasible approach is to use different fields in the fixed IPv6 header and extension headers (including the SRH) of an SRv6 packet to separately carry the routing topology identifier and VTN identifier of a network slice. The involved network devices need to generate forwarding tables for the two types of data plane identifiers and forward packets by matching the different identifiers. Figure 10.11 shows how network slice topology and resource identifiers are encapsulated on an IPv6 network running SRv6.

In short, the following measures can be taken for a network to provide more larger-scale network slices to meet the requirements of the ever-growing network slice services in the 5G era:

- Improve the resource slicing and isolation capabilities of network devices.

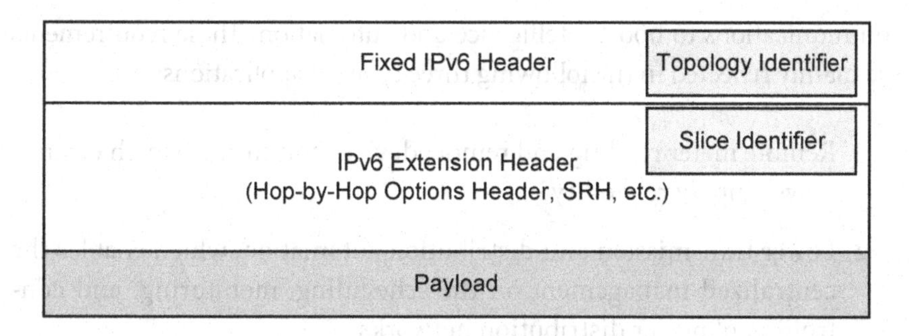

FIGURE 10.11 Network slice topology and resource identifier encapsulation.

- Combine soft and hard isolation during network slice planning.
- Decouple and optimize the functions in the network slice control plane.
- Introduce multilevel forwarding identifiers and entries in the network slice data plane.

10.3 SRv6 FOR DETNET

10.3.1 5G uRLLC

In addition to providing ultrahigh bandwidth, 5G networks can also deliver more reliable SLA guarantee for high-value services; for example, uRLLC services — one of the three major 5G application scenarios defined by 3GPP.[1] Indeed, 5G uRLLC is creating all-new business opportunities for telecom carriers. With 5G uRLLC, carriers are seeing a shift from traditional eMBB services to diverse vertical services, and from consumer-oriented services to industry-oriented services, which in turn are accelerating the maturity of industry applications, improving social production efficiency, and finally connecting all things.

It's plain to see that the vigorous development of verticals fuels the growth of 5G uRLLC services, for which smart grid, IoV, and Cloud VR are typical application scenarios. The following three sections will detail these typical application scenarios.

10.3.1.1 Smart Grid

Over recent years, the electric power industry has made great efforts in digital transformation and wider adoption of information technologies, and the resulting progress is promising. This poses further requirements for wireless communications to boost intelligence and automation. These requirements are mainly reflected in the following three types of applications:

1. Remote meter reading and remote device monitoring, which ensures power supply reliability
2. Power transmission and distribution automation, which enables the centralized management on the scheduling, monitoring, and controlling of power distribution networks
3. Energy Internet, which represents the future of smart grids

For smart grid applications, latency and reliability are paramount. To put this into context, E2E two-way latency of no more than 10 ms (not to mention extremely high reliability) is required in some scenarios, such as differential protection and precise load control. It, therefore, goes without saying that smart grids are typical uRLLC services.

10.3.1.2 IoV

IoV is a dynamic mobile communications system involving interaction between vehicles, vehicles and roads, vehicles and people, and vehicles and sensor devices. In short, it enables vehicles to communicate with public networks and provides the following functions:

- Collects information about vehicles, roads, and environments for information sharing through connections between vehicles, between vehicles and people, and between vehicles and sensor devices.
- Processes, calculates, shares, and securely releases the information collected from multiple sources on the information network platform.
- Effectively guides and monitors vehicles based on different function requirements.
- Provides professional multimedia and mobile Internet application services.

IoV has very high requirements on networks, including wide deployment, high reliability, low latency, and high bandwidth.

Currently, the industry is widely in agreement that main IoV applications served by networks may include vehicle platooning and remote/teleoperated driving. Most IoV scenarios require an E2E two-way latency of no more than 10 ms and 99.999% reliability.

10.3.1.3 Cloud VR

VR is the use of computer technologies to create a simulated three-dimensional space and provide users with simulation of various senses such as visual, auditory, and tactile senses. One thing of note is that an immersive VR experience requires shorter latency. The reason for this can be best explained as follows: In daily life, the human body mainly depends on the vestibular system to feel body movement. When the vestibular system's perception and body movement are not in sync, we feel dizzy. So, VR requires a near real-time level of synchronization: 20 ms is the level for Motion-to-Photons (MTP) latency, as is generally agreed upon in the industry. For sensitive users, the maximum MTP latency even cannot exceed 17 ms. Once the latency related to image rendering and splicing, terminal processing, image display, and other factors has been taken into account, only about 5 ms of latency is left for network processing and transmission. Therefore, VR poses higher requirements on low network latency than traditional video or audio services.

10.3.2 DetNet Fundamentals

As an integral part of 5G E2E services, the transport network must be able to meet differentiated SLA requirements of 5G services. The degree to which SLAs can be assured depends largely on the multiplexing technology used on the transport network, the most common of which are TDM and statistical multiplexing.

- TDM is a time-based multiplexing technology where a time domain is divided into recurrent time slots of a fixed length. Two or more data streams are alternately transmitted in the slots, yet they appear to be transmitted in sub-channels over one communication channel. TDM can provide a strict SLA guarantee, but its shortfalls include high deployment costs, lack of flexibility, and limited granularity for dividing minimum channels.
- Statistical multiplexing is a multiplexing technology based on statistical rules. Channel resources can be shared by any amount of traffic,

and the channel usage changes with the volume of real-time traffic. A packet switched network (IP/Ethernet) uses statistical multiplexing to forward packets. This technology features high bandwidth utilization and simple deployment but falls short when it comes to providing a strict SLA guarantee.

DetNet is a network technology that provides committed SLAs. It integrates the technical advantages of statistical multiplexing and TDM, providing TDM-like SLAs on an IP/Ethernet packet network. DetNet ensures low jitter and zero packet loss for high-value traffic during transmission and provides an expected upper bound for latency.

The IEEE 802.1 Time-Sensitive Networking (TSN) Task Group and the IETF DetNet Working Group define the TSN standard for Ethernets and the DetNet standard for Layer 3 IP/MPLS networks, respectively. This book focuses on the DetNet technology.

DetNet represents a set of technologies, including the following relatively independent technologies[14]:

- **Resource allocation:** On a statistical multiplexing network, congestion means that packets are stacked on an outbound port when the port's forwarding capability is severely exceeded, and as a result, these packets are held in the buffer for some time before being forwarded, or even discarded if the buffer overflows. Congestion is a major cause of uncertain latency and packet loss on packet switched networks. DetNet combats this by using resource reservation technologies and queue management algorithms to avoid conflicts between high-priority packets and also network congestion, and it does this while providing a guaranteed upper bound for E2E latency. Specifically, a resource reservation technology reserves outbound port resources for different traffic on transit devices, and a queue management algorithm schedules packets that may conflict with each other and allocates bandwidth based on resource reservation. Here, congestion is avoided through cooperation between the technology and algorithm. Queuing latency and packet loss on a network are consequences of packet congestion, and so avoiding congestion can effectively improve network service quality.
- **Explicit path:** To enable a network to run stably and be unaffected by network topology changes, a deterministic network needs to provide

explicit paths to constrain the routes of packets, preventing route flapping or other factors from directly affecting transmission quality. An IP network provides various protocols concerning explicit paths, including RSVP-TE and SR. Particularly, SR maintains per-flow forwarding states only on the source node, with the transit nodes only needing to forward packets based on active segments. This makes SR exceedingly scalable, a key factor in its gaining wide-scale adoption.

- **Redundancy protection:** To ensure high reliability of service transmission, redundancy protection can be used. This technology enables a service packet to be replicated and transmitted along two or more disjoint paths. The merging node then retains only the first copy of the packet that arrives. This technology is also called multi-fed selective receiving. It enables services to be switched to another path without any loss if the original path fails.

These independent technologies can be combined to form complete solutions, involving queue management algorithms and packet encapsulation design in the data plane, and resource allocation and path management in the control plane.

The following two sections describe two typical solutions: redundancy protection solution (involving explicit paths and redundancy protection) and deterministic latency solution (involving explicit paths, resource allocation, and queue management).

10.3.2.1 Redundancy Protection Fundamentals

Redundancy protection, also called multi-fed selective receiving, works by setting up two or more paths for transmitting replicated DetNet data packets at the same time. It prevents packet loss caused by the failure of a certain link or other failures. As shown in Figure 10.12, a DetNet data flow is replicated on the replication node and sent to the merging node over disjoint paths. For packets of each sequence number, the merging node forwards only the first received packet and discards others. Redundancy protection provides hitless switching against link failures and prevents packet loss in active/standby path switching of traditional path protection.

An advantage that redundancy protection has over common active/standby path switching mechanisms is that it has no switching time and can ensure hitless switching. If the active link fails in traditional path

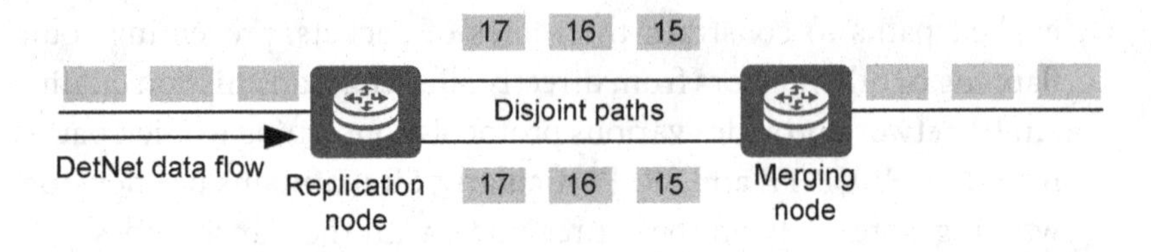

FIGURE 10.12 Redundancy protection implementation.

protection, it takes a certain period of time (about 50 ms) for services to be switched to the standby link. During the switching, packets may be lost, which leads to services being interrupted on the application side. Redundancy protection overcomes this by not distinguishing between the active and standby paths, both of which transmit data at the same time. If services on one path are interrupted, the merging node automatically receives packets from the other path. The result is that services are not interrupted on the application side.

10.3.2.2 Deterministic Latency Fundamentals

DetNet avoids any resource conflicts between DetNet traffic flows through proper resource planning. Working with a queue management algorithm, it provides deterministic latency, as shown in Figure 10.13.

An IP network uses statistical multiplexing; therefore, its packet forwarding behavior complies with the statistical pattern. As described above, if congestion occurs on the network, packets are stacked in the buffers of forwarding devices, driving up the single-hop latency of the packets. Although there is a low probability that congestion will occur on an IP network, once it occurs, the network can no longer provide reliable network quality. This is known as the long tail effect. Figure 10.13 (a) shows the latency probability distribution on an IP network.

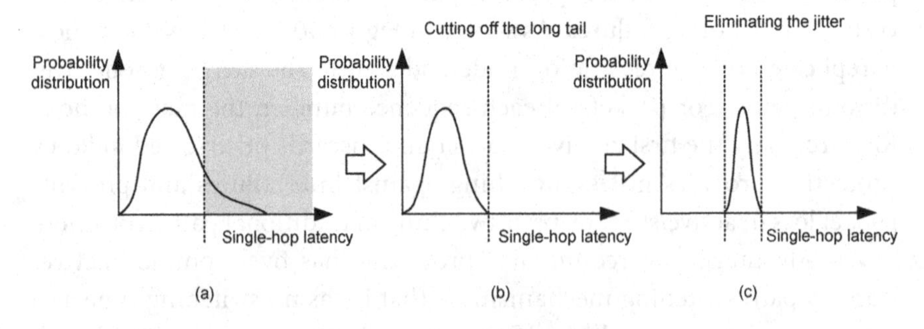

FIGURE 10.13 Probability theory for implementing deterministic latency.

Implementing deterministic latency involves two steps:

1. Cut off the long tail.

   In the outbound port queues, the DetNet traffic is set to the highest priority to eliminate the impact of non-DetNet traffic, and the controller's resource planning function is leveraged to ensure that DetNet flows do not affect each other. This cuts off the long tail caused by congestion, as shown in Figure 10.13 (b).

2. Eliminate the jitter.

   Even if the previous-hop device sends a packet at a fixed time, the time when the packet arrives at an outbound port queue of the current device is still uncertain, mainly because of the jitter caused by packet processing. The buffer of the device must be able to eliminate the jitter caused by other processes of the device, so that the packet can be forwarded within an expected time range, as shown in Figure 10.13 (c).

In the DetNet technology system, new device capabilities are also introduced. For example, IEEE 802.1 Qbv defines a new shaping mechanism—Time Aware Shaping (TAS), which can be used to eliminate the jitter. Figure 10.14 shows how TAS is implemented.

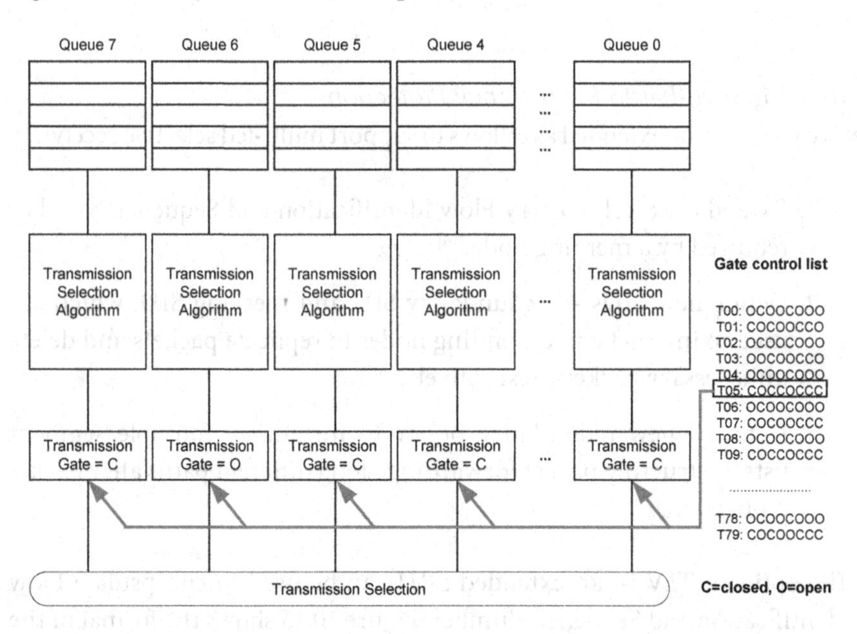

FIGURE 10.14 TAS implementation.

TAS adds a transmission gate before each queue. The Open/Closed state of the gate in each time segment (T*n*) is determined by a gate control list. TAS provides a time-aware shaping capability, which provides a hardware basis for ensuring latency. After a packet enters a queue, the transmission selection algorithm determines the scheduling rule between queues. When the queue is scheduled at a specific time, the transmission gate may be in the C (Closed) or O (Open) state at that time. The C state indicates that the shaper is closed and packets are buffered; the O state indicates that the shaper is opened and packets can be sent.

According to the mechanism shown in Figure 10.14, when a packet arrives at outbound port queues, a queue is selected according to the traffic priority. The packet is then buffered in the queue until all the packets ahead of it are forwarded. When the transmission selection algorithm determines that the queue can be scheduled and the state of the transmission gate is O at that time, the packet is forwarded.

10.3.3 SRv6-based DetNet

This section describes SRv6-based redundancy protection and deterministic latency solutions, which are used to ensure an extremely low packet loss rate and committed latency upper bound, respectively, on a deterministic network.

10.3.3.1 SRv6-Based Redundancy Protection

SRv6 needs to be extended as follows to support multi-fed selective receiving:

1. Extend the SRH to carry Flow Identification and Sequence Number required by a merging node.[15]
2. Define new SIDs — redundancy SID and merging SID, which are used to instruct corresponding nodes to replicate packets and delete unnecessary packets, respectively.[16]
3. Define a new redundancy policy to instantiate multiple segment lists, instructing packet forwarding along different paths after packet replication.[17]

The optional TLV in an extended SRH can be used to encapsulate Flow Identification and Sequence Number. Figure 10.15 shows the format of the optional TLV that carries Flow Identification.

Table 10.6 describes the fields in the optional TLV.

Figure 10.16 shows the format of the optional TLV that carries Sequence Number.

Table 10.7 describes the fields in the optional TLV.

Figure 10.17 shows the forwarding process of a DetNet flow on an SRv6 path from the ingress to the egress. Relay nodes 1 and 2 are the replication and merging nodes, respectively, involved in the multi-fed and selective receiving process.

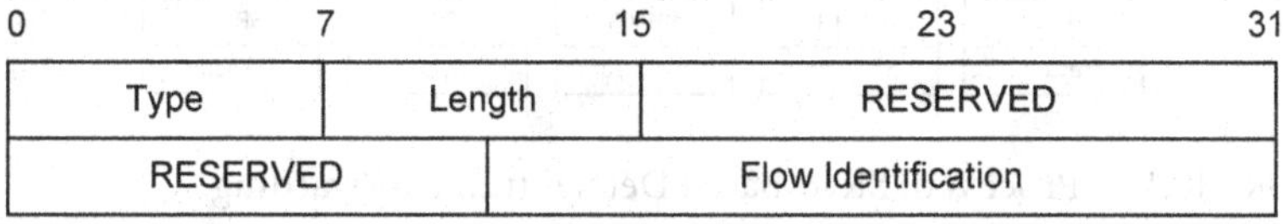

0 – 7	8 – 15	16 – 31
Type	Length	RESERVED
RESERVED	Flow Identification	

FIGURE 10.15 Optional TLV carrying Flow Identification.

TABLE 10.6 Fields in the Optional TLV That Carries Flow Identification

Field	Length	Description
Type	8 bits	Type of the TLV
Length	8 bits	Length of the TLV
RESERVED	28 bits	Reserved field, with a default value 0
Flow Identification	20 bits	Identifier of a DetNet flow

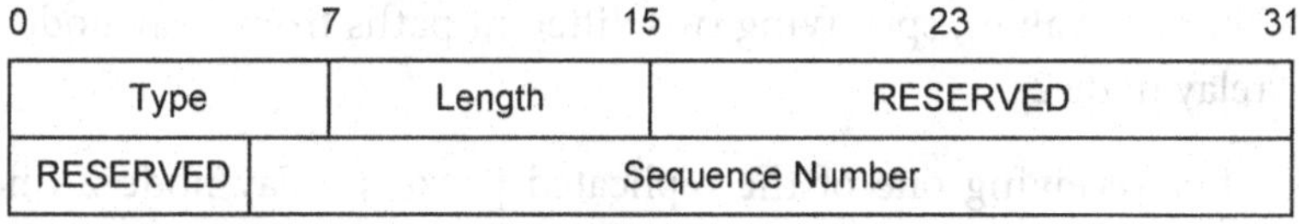

0 – 7	8 – 15	16 – 31
Type	Length	RESERVED
RESERVED	Sequence Number	

FIGURE 10.16 Optional TLV carrying Sequence Number.

TABLE 10.7 Fields in the Optional TLV That Carries Sequence Number

Field	Length	Description
Type	8 bits	Type of the TLV
Length	8 bits	Length of the TLV
RESERVED	20 bits	Reserved field, with a default value 0
Sequence Number	20 bits	Sequence number of a packet in a DetNet flow

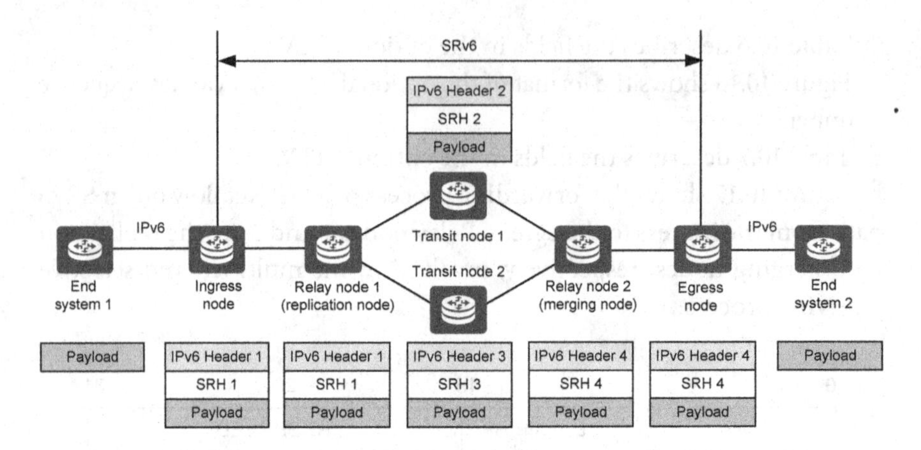

FIGURE 10.17 Process of SRv6-based DetNet traffic forwarding.

The forwarding process is as follows:

1. The ingress encapsulates an outer IPv6 header and SRH 1 into a DetNet packet, where SRH 1 contains information about the path from the ingress to relay node 1.

2. After receiving the packet, relay node 1 (replication node) replicates the DetNet packet into two and re-encapsulates an outer IPv6 header with SRH 2 and SRH 3, respectively, for each of the replicated packets. SRH 2 and SRH 3 carry unique Flow Identification and Sequence Number values, specifying two different paths from relay node 1 to relay node 2.

3. After receiving one of the replicated packets, relay node 2 (merging node) determines whether the packet is redundant according to Flow Identification and Sequence Number in the SRH of the packet. The packet received earlier is forwarded, and the one received later is considered redundant and discarded. The forwarded packet is re-encapsulated with an outer IPv6 header and SRH 4, specifying the path from the merging node to the egress.

SRv6-based redundancy protection ensures high reliability of data transmission.

10.3.3.2 SRv6-Based Deterministic Latency

Cycle Specified Queuing and Forwarding (CSQF) is an SRv6-based deterministic latency solution. It uses a controller to calculate the time at which packets should be sent out from the outbound port at each hop.

Conventional TE reserves resources based on bandwidth. The issue here is that bandwidth represents the average rate in a certain period of time, not real-time traffic changes. So, this resource reservation method can improve service quality to some extent but cannot avoid traffic bursts and packet conflicts that last a very short period of time.

The CSQF method reserves resources based on time. It uses a controller to assign the time when traffic leaves the outbound ports of involved devices. This method enables accurate traffic planning and avoids traffic congestion. A cycle — a time concept introduced in the CSQF method and derived from IEEE 802.1 Qch — refers to a time interval of a network device. The length of a cycle is the length of the associated time interval and can be a length of time such as 1 ms or 10 μs. A cycle ID is used to identify a specific time interval, with the first one starting from the time when TAS takes effect. For example, if the start time is 0 μs and the cycle length is 10 μs, cycle 10 refers to the time interval from 90 to 100 μs.

The controller needs to maintain a network's traffic state and expected period for port occupation. When receiving a new transmission request, it computes a path that meets the user's latency requirement and specifies the time (cycle ID) when the traffic leaves the outbound port on each device along the path. The cycle ID of a specific outbound port is represented by an SRv6 SID. Such SIDs are combined to form a SID list that represents a path with deterministic latency. This path is then delivered by the controller to the ingress. Devices along the path forward packets at the specified time indicated by the SIDs in the SRv6 SID list to ensure that the E2E latency meets user requirements, as shown in Figure 10.18.

In addition to resource reservation, CSQF needs to work with the queue scheduling algorithm of outbound ports to provide deterministic queue latency and absorb the possible jitter caused by packet processing latency. As shown in Figure 10.18, when a packet is transmitted from device A to device B and traverses a link with a known latency, jitter may occur during packet processing. After the packet enters a queue, TAS can be used to absorb the jitter, and CSQF can provide upper bounds for the E2E latency and jitter.

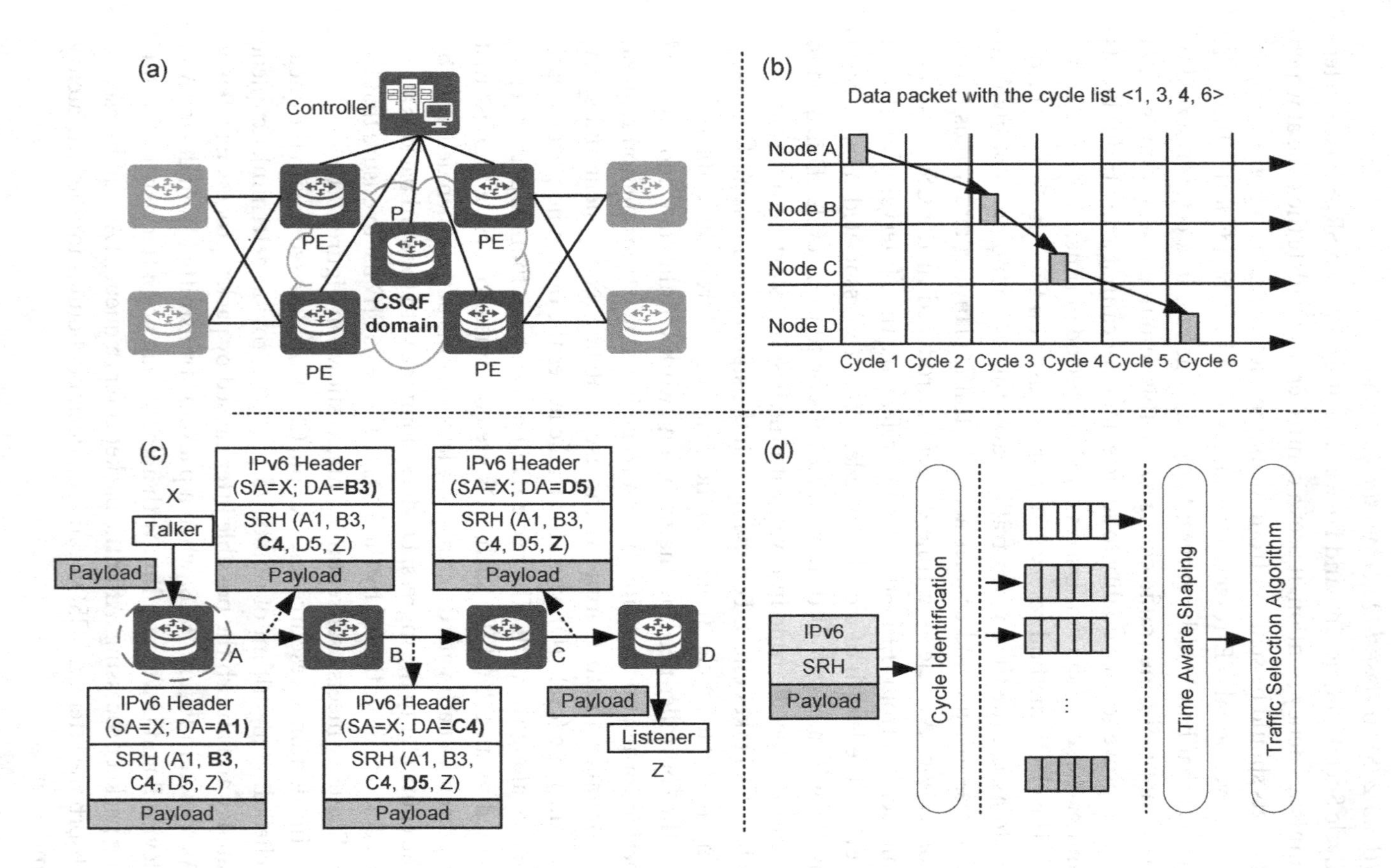

FIGURE 10.18 CSQF approach.

Figure 10.18 shows how CSQF is generally implemented, and the following describes the four parts in more detail:

1. (a): A controller collects network topology and cycle information.
2. (b): The controller plans a path and sending cycles for nodes along the path based on the model and latency requirement of the newly added traffic.
3. (c): The controller delivers SRv6 encapsulation information for the traffic, and the packets are forwarded as instructed.
4. (d): Devices use the time-aware scheduling mechanism to ensure that packets are forwarded within certain periods.

That brings this section on DetNet to a conclusion. To learn more details about DetNet, search for other relevant materials.

10.4 SRv6 FOR 5G MOBILE NETWORKS

A mobile communication network comprises three main parts: RAN, mobile core network, and transport network. If these networks are unable to be effectively coordinated, the E2E mobile communication network will fail to meet 5G services' strict SLAs (such as low latency). To effectively support new 5G services end to end, some carriers are actively deploying SRv6-based transport networks and trying to extend SRv6 from transport networks to mobile core networks. In doing so, they look to build a next-generation mobile communication network architecture featuring high scalability, reliability, flexibility, and agility to reduce Capital Expenditure (CAPEX) and Operating Expense (OPEX). Indeed, deploying SRv6 on a 5G mobile communication network and delivering E2E user plane functions based on IPv6 simplify network layers and make the network more controllable and flexible.

10.4.1 5G Mobile Network Architecture

At the end of 2016, 3GPP defined the 5G network architecture, involving an access network (R)AN and core network, as shown in Figure 10.19. In this architecture, a User Equipment (UE) accesses the 5G core network through the (R)AN (5G base station, gNB).[18]

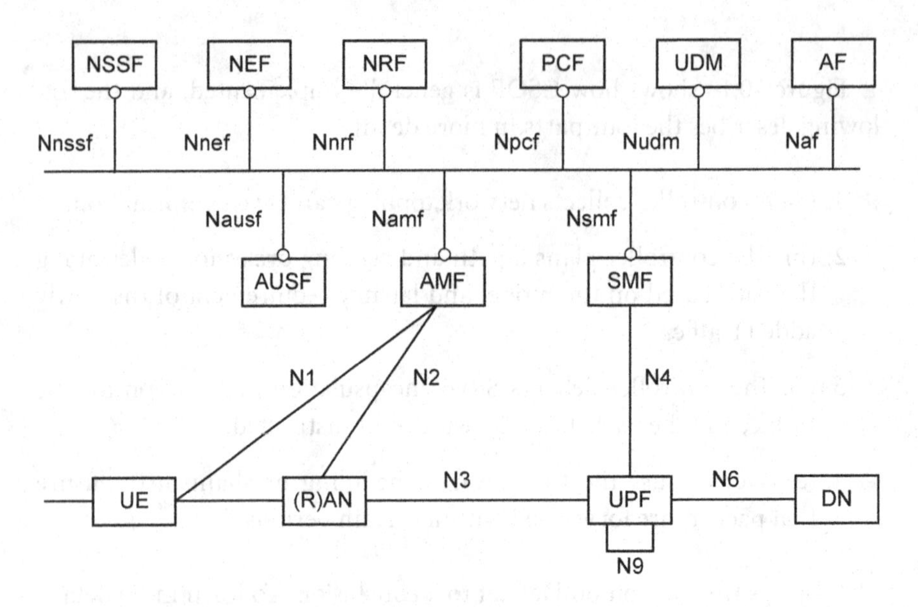

FIGURE 10.19 5G mobile network architecture.

The 5G core network consists of control and user planes, of which the latter provides only the User Plane Function (UPF) module.

The control plane uses the SBA to decouple, aggregate, and servitize the original control plane functions. As shown in Figure 10.19, the control plane provides the following function modules[18]:

- Access and Mobility Management Function (AMF)
- Session Management Function (SMF)
- Authentication Server Function (AUSF)
- Policy Control Function (PCF)

These function modules are used for the management of user mobile access, sessions, and policies, and the delivery of traffic control policies (policy enforcement) and traffic steering rules to the user plane. They can also flexibly invoke each other through RESTful APIs.

The control plane of the 5G core network is connected to the UE, (R)AN, and UPF module through standard interfaces, which are described below:

- N1: directly connects the UE and AMF.

- N2: connects the (R)AN and AMF. UE traffic can indirectly access the AMF through this interface.
- N3: connects the (R)AN and UPF over the GPRS Tunneling Protocol-User Plane (GTP-U) tunnel[19] across the transport network.
- N4: connects the UPF and SMF. This interface is used to deliver policies to the UPF, which performs traffic steering and QoS guarantee based on the policies delivered by the control plane.
- N6: connects the UPF and Data Network (DN).
- N9: connects two UPFs.

The service-oriented control plane makes network functions plug-and-play and network functions and resources deployable on demand. The separation of the control plane from the user plane simplifies the connections between the two planes, improves their extensibility, and allows core network functions to be flexibly deployed in data centers at different levels based on service requirements. For example, the user plane can be moved closer to the edge DC; that is, closer to users and application servers, and in doing so meet the requirements of low-latency services (such as autonomous driving).

10.4.2 SRv6 Deployment on a 5G Mobile Network

There are two possible modes for deploying SRv6 on a 5G mobile network.[20] Which mode to use depends on whether gNBs support SRv6, as outlined below:

Mode 1 is used when both the gNB and UPF support SRv6, as shown in Figure 10.20. In this mode, an E2E SRv6 tunnel can be established between the gNB and UPF. The gNB and UPF perform SRv6 encapsulation and decapsulation, respectively, for service packets. In the upstream direction, the gNB can function as the headend of an SRv6 tunnel, directly encapsulating received service packets into SRv6 packets and sending them through the corresponding SRv6 tunnel. It is possible to specify the transport network devices and Cloud-native Network Functions (CNFs) through which each SRv6 tunnel passes. In this example, C1 is specified as the transport network device and S1 as the CNF. After the packets reach the UPF, the UPF performs SRv6 decapsulation, after which the original packets enter the DN through the N6 interface. This E2E SRv6 tunnel establishment mode between the gNB and UPF enables the transport network to become aware of upper-layer services and their requirements, and to provide SRv6 TE explicit paths for them, thereby meeting their SLA requirements. For a uRLLC service, for example, a low-latency path from the gNB to the UPF can be provided.

Mode 2 is used when the gNB does not support SRv6 but the UPFs do, as shown in Figure 10.21. In this mode, a GTP-U tunnel can still be used between the gNB and UPF1 for the N3 interface, whereas an SRv6 tunnel is used between UPFs for the N9 interface. The GTP-U tunnel is therefore connected to the SRv6 tunnel. This means that UPF1 can function as an SRv6 gateway to perform GTP-U decapsulation and SRv6 encapsulation on service packets received through the GTP-U tunnel before sending the packets to the corresponding SRv6 tunnel. It is possible to specify the transport network device (C1) and service function (S1) that each SRv6

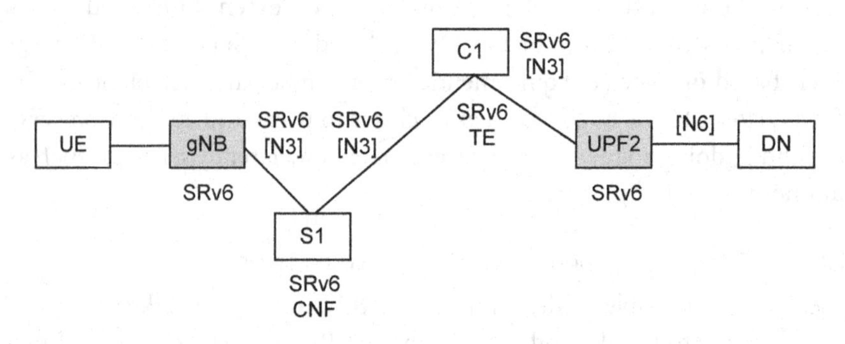

FIGURE 10.20 User plane of an SRv6-aware 5G transport network (gNB supporting SRv6).

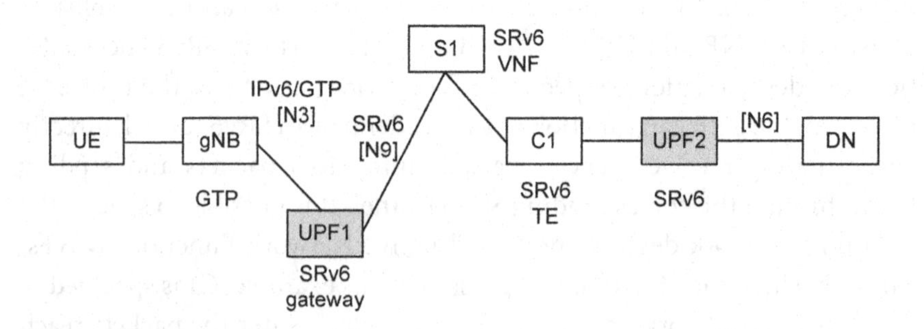

FIGURE 10.21 User plane of an SRv6-aware 5G transport network (gNB not supporting SRv6).

tunnel passes through. The service function can also be implemented through the Virtual Network Function (VNF).

This mode requires the GTP-U and SRv6 tunnels to be stitched on the first UPF, which must be able to map packet information, transfer packet parameters, and convert the packet encapsulation format between GTP-U and SRv6. Accordingly, some new SRv6 functions need to be defined, which will be described in the next section.

10.4.3 Key SRv6 Functions on a 5G Mobile Network

To make 5G networks easy to deploy and highly scalable, it is crucial to integrate 5G mobile network functions into an E2E SRv6-capable IPv6 network layer. This section describes the main SRv6 functional extensions of the mobile user plane, which are currently being standardized by the IETF.[20]

10.4.3.1 Args.Mob.Session

The Arguments (Args) field in an SRv6 SID can be used to carry Packet Data Unit (PDU) session information, such as a PDU session ID, a QoS Flow Identifier (QFI), and a Reflective QoS Indicator (RQI). On a mobile network, this field is called Args.Mob.Session[20] and can be used for UE charging and buffering. Figure 10.22 shows the format of this field.

Table 10.8 describes the fields in Args.Mob.Session.

Args.Mob.Session is usually used together with End.Map, End.DT, and End.DX. Note that End.Map is short for Endpoint Behavior with SID Mapping and is used in several scenarios. In mobility, End.Map is used on the UPFs for the PDU session anchor functionality.[20]

Multiple sessions can generally use the same aggregated SRv6 function, in which case each PDU session's parameters carried by Args.Mob.Session can help the UPF implement refined function control on a per-session basis.

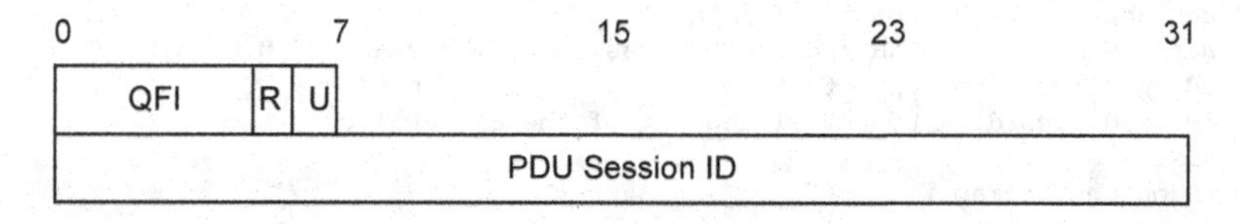

FIGURE 10.22 Encoding format of Args.Mob.Session.

TABLE 10.8 Fields in Args.Mob.Session

Field	Length	Description
QFI	6 bits	QoS flow identifier[21].
R	1 bit	Reflective QoS indicator.[18] This parameter is used to indicate the reflective QoS of the packet sent to a UE. Reflective QoS enables the UE to map uplink user services to corresponding QoS flows without the need for the QoS rules provided by the SMF.
U	1 bit	Reserved field. It is set to 0 before transmission and ignored by the receiver.
PDU Session ID	32 bits	PDU session. For a GTP-U tunnel, this field is a Tunnel Endpoint Identifier (TEID).

10.4.3.2 End.M.GTP6.D/End.M.GTP6.E

When the gNB supports IPv6/GTP but not SRv6, the SRv6 gateway (UPF1) needs to support interworking between IPv6/GTP and SRv6 tunnels. Tunnel interconnection functions that need to be supported on UPF1 include Endpoint Behavior with IPv6/GTP Decapsulation into SR Policy (End.M.GTP6.D) and Endpoint Behavior with IPv6/GTP Encapsulation for IPv6/GTP Tunnel (End.M.GTP6.E).

For uplink services, UPF1 needs to support End.M.GTP6.D to convert IPv6/GTP packets to SRv6 packets. For downlink services, UPF1 needs to support End.M.GTP6.E to convert SRv6 packets to IPv6/GTP packets.

The following uses End.M.GTP6.D as an example and assumes that information in this SID corresponds to SRv6 Policy <S1, S2, S3> and source IPv6 address A. When UPF1 receives a service packet destined for S and S is a local End.M.GTP6.D SID, the processing logic pseudocode on UPF1 is as follows:

```
1. IF NH=UDP & UDP_DST_PORT = GTP THEN
2. copy TEID to form SID S3
3. pop the IPv6, UDP and GTP headers
4. push a new IPv6 header with a SR policy in SRH <S1, S2, S3>
5. set the outer IPv6 SA to A
6. set the outer IPv6 DA to S1
7. set the outer IPv6 NH //Each PDU session type has an End.M.GTP6.D SID,
based on which, the type of the next header can be predicted.
8. forward according to the S1 segment of the SRv6 Policy
9. ELSE
10. Drop the packet
```

10.4.3.3 End.M.GTP4.E

When the gNB supports IPv4/GTP but not SRv6, the SRv6 gateway (UPF1) needs to support interworking between IPv4/GTP and SRv6 tunnels. For downlink services, UPF1 must support the Endpoint Behavior with IPv4/GTP Encapsulation for IPv4/GTP Tunnel (End.M.GTP4.E) to convert SRv6 packets to IPv4/GTP packets.

When UPF1 receives a packet destined for S and S is a local End.M.GTP4.E SID, UPF1 uses the information carried in the SID to construct an IPv4/GTP packet. In this case, the processing logic pseudocode on UPF1 is as follows:

```
1. IF (NH=SRH and SL = 0) or ENH=4 THEN
2.   store IPv6 DA in buffer S
3.   store IPv6 SA in buffer S'
4.   pop the IPv6 header and its extension headers
5.   push UDP/GTP headers with GTP TEID from S
6.   push outer IPv4 header with SA, DA from S' and S
7. ELSE
8.   Drop the packet
```

Figure 10.23 shows the encoding format of the End.M.GTP4.E SID.

Table 10.9 describes the fields in the End.M.GTP4.E SID.

10.4.3.4 End.Limit SID

End.Limit is defined to support the rate-limiting function on the mobile user plane.

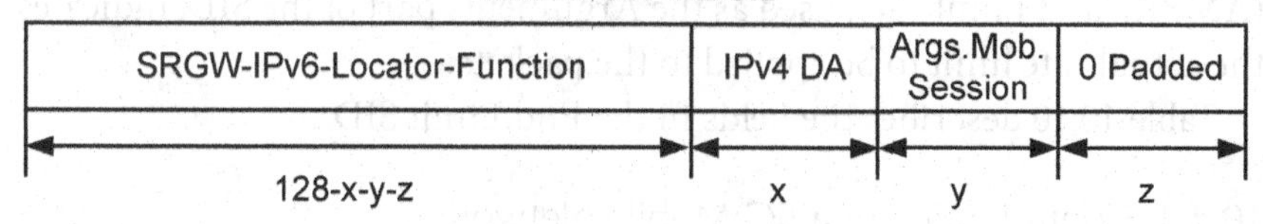

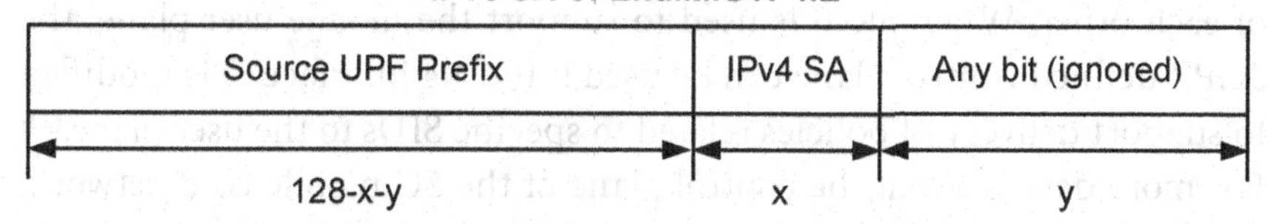

FIGURE 10.23 Encoding format of the End.M.GTP4.E SID.

TABLE 10.9 Fields in the End.M.GTP4.E SID

Field	Description
SRGW-IPv6-LOC-FUNC	Locator and Function parts of the SRv6 gateway
IPv4 DA	Destination IPv4 address
Args.Mob.Session	As discussed above
Source UPF Prefix	Source UPF prefix
IPv4 SA	Source IPv4 address
Any bit (ignored)	Used for padding

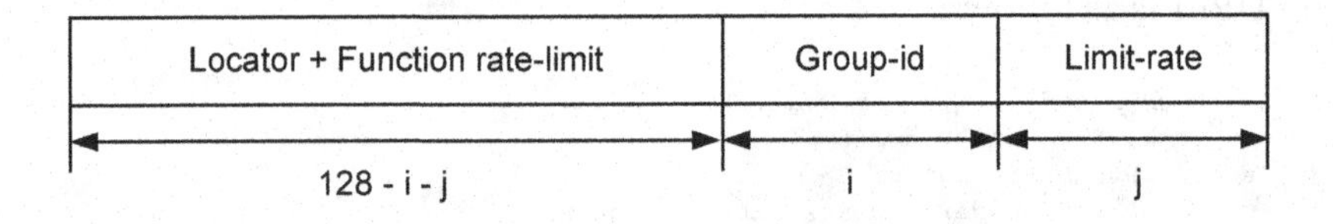

FIGURE 10.24 Encoding format of the End.Limit SID.

TABLE 10.10 Fields in the End.Limit SID

Field	Description
Locator+Function rate-limit	Locator and Function parts of the rate limiting function
Group-id	AMBR group of the packets whose rates are limited
Limit rate	Rate limiting parameter

Figure 10.24 shows the format of the End.Limit SID. Group-id specifies a group of data flows with the same Aggregate Maximum Bit Rate (AMBR), and Limit-rate, used as the Arguments part of the SID, indicates the actual rate limit to be applied to the packets.

Table 10.10 describes the fields in the End.Limit SID.

10.4.4 Control Plane of a 5G Mobile Network

The control plane and user plane of a mobile network are independent of each other. When SRv6 is used to support the mobile user plane, the 3GPP-defined control plane can be used if the N4 interface[22] is modified to support delivery of policies related to specific SIDs to the user plane.[23] For more details about the control plane of the 5G mobile core network, refer to 3GPP TS23.501.[18]

10.5 STORIES BEHIND SRv6 DESIGN

I. Network Slicing Design

1. Origin of SR Network Slicing

   In 2013, we submitted a draft about SR-based network virtualization[24] based on two considerations:

   - Unlike RSVP-TE, SR is unable to reserve bandwidth in the data plane. Segments, therefore, must be extended to indicate not only nodes and links but also resources that ensure SLAs.
   - Carriers must be capable of customizing networks to meet all kinds of user requirements. In the past, they provided VPN services on edge nodes, whereas they now have to allow customization within network domains.

   SR can also be understood from two aspects:

   - SR is commonly understood as a path service featuring high scalability. It provides different SR paths by combining different segments, such as node and link segments, to meet specific requirements of customers. This is what we call network programming.
   - SR is a technology for creating virtual networks. Node and link segments in SR can be seen as virtual nodes and links, respectively, and they can be combined to easily create a virtual network.

Although we submitted this draft a long time ago, there were few relevant customer requirements at that time, and as a result, we did not thoroughly promote the draft. It was not until about 2017, when discussion on network slicing emerged in the IETF, that this draft found its way into public view.

2. Bandwidth Access Control and Resource Reservation

   To support bandwidth guarantee, SR was designed at the early stage to deduct the bandwidth required by SR paths from the available link bandwidth during path computation by a controller. If the link bandwidth is insufficient, path computation fails. This bandwidth guarantee method falls far short of what is required. Strictly speaking, it should be called Bandwidth Admission Control (BAC), due to the fact that no bandwidth is reserved for SR paths on live networks. And because of this, it is challenging to ensure that the SR paths always obtain the expected guaranteed bandwidth. While this issue is not prominent on lightly loaded networks, once congestion occurs, service quality deteriorates sharply due to multiple services competing for limited resources. It is the lack of accurate information about all service traffic on the controller and the burstiness of IP traffic that make this issue inevitable. This in turn means that a gap will undoubtedly appear between BAC performed by the controller and the actual network state. To provide services with guaranteed bandwidth, bandwidth resource reservation needs to be provided for SR paths on a network.

Technologies that emerged later, such as Flex-Algo, can provide multiple virtual networks based on SR, compute constrained SR BE paths based on different metrics and path computation algorithms, and apply the paths to network slicing. That said, Flex-Algo was not originally designed specifically for network slicing. Although it supports multiple slices of different topologies, these slices share resources on the forwarding plane, and therefore affect each other. This means that bandwidth cannot be guaranteed. Instead, an additional mechanism needs to be introduced to associate Flex-Algo with the resources allocated to each slice.

3. IGP Scalability

   Network slicing poses great challenges to IGP scalability. Originally, IGP advertises only the topology information of the physical network. If the basic SR-based slice solution (refer to Section 10.2.3), which introduces more node SIDs, adjacency SIDs, and other SIDs, is adopted, the advertised network topology information will multiply each time a new slice is introduced, as will path computation based on virtual network topologies. So, for IP transport networks that need to support a large number of slices, the scalability of IGP becomes a bottleneck.

Zhibo Hu, one of our experts in SR research, first realized this issue, which led him to propose the concept of decoupling network slices from topologies. That is to say, multiple network slices could share the same topology, meaning that the result from only one topology calculation could serve multiple slices, ultimately reducing the IGP load and improving IGP scalability. The idea instantly wowed our research team and gained unanimous approval. Where a consensus was lacking, however, was in its implementation schemes, of which two were proposed:

Scheme 1: Bind each network slice to an IGP MT or IGP Flex-Algo, and then develop a new concept of Common Topology in IGP to advertise shared topology information and compute shared paths for involved network slices.

Scheme 2: Add the new function for IGP to advertise network slice information and use IGP MT or IGP Flex-Algo for the shared topology advertisement and path computation.

Critics of scheme 1 thought that IGP MT had been used for topology isolation, and so introducing the Common Topology would affect the existing behavior of IGP MT. Critics of scheme 2, on the other hand, thought that IGP should not be used to distribute specific service information (e.g., network slicing) since it is always used for topology information flooding and route computation. After much discussion, scheme 2 came out on top. The reason behind this is the trend that IGP is no longer protocols used only for route/path computation; rather, it is becoming protocols that support services, just like BGP, which first was only used for Internet route advertisement, before being widely used to support services such as VPN.

Another way to address the IGP scalability issue is to replace IGP with BGP. This has been achieved by some OTT providers, who have succeeded in using BGP as the underlay technology of data centers. This design exceeded many people's expectations and features good scalability. BGP, however, has its own challenges. It is more complex and requires more configurations, making automatic configuration a necessity. This is also a subject worth researching further.

II. Challenges of Extending SRv6 to the Core Network

Applying SRv6 to the mobile core network will extend SRv6 application and set a good example for unified programming of services and transport, greatly simplifying network layers and service deployment. In January 2018, the CT4 group of 3GPP started a research project focused on the application of SRv6 to the mobile core network. In September 2019, the research was completed, yet further research and the incorporation of SRv6 into the R16 standard for 5G was not approved. The main reasons were as follows:

First, 5G standards needed to be frozen immediately. The work focused on air interfaces instead of the core network, on which GTP had been determined for use a long time ago. Replacing GTP with SRv6 required re-standardization, which would slow down the process of 5G standardization.

Second, SRv6, as a possible alternative to GTP, has its own challenges. GTP has been developed for many years and so is already relatively mature. Before SRv6 is ready to replace GTP, SRv6 still needs to be improved and standardized to meet the existing requirements of GTP, which may take many years.

For these reasons, SRv6 was ultimately not accepted in 5G standards. However, the research and standardization on applying SRv6 to the core network is still going on in the Distributed Mobility Management (DMM) Working Group of the IETF. SRv6, as it continues to grow in maturity, can be further promoted in 3GPP in the future.

REFERENCES

[1] ITU-R. IMT Vision – Framework and Overall Objectives of the Future Development of IMT for 2020 and beyond[EB/OL]. (2015-09-29)[2020-03-25]. ITUR M.2083.

[2] Dong J, Bryant S, Li Z, Miyasaka T, Lee Y. A Framework for Enhanced Virtual Private Networks (VPN+) Services[EB/OL]. (2020-02-18)[2020-03-25]. draft-ietf-teas-enhanced-vpn-05.

[3] OIF. Flex Ethernet Implementation Agreement[F]. (2016-03)[2020-10-31].

[4] Dong J, Hu Z, Li Z, Bryant S. IGP Extensions for Segment Routing based Enhanced VPN[EB/OL]. (2020-03-09)[2020-03-25]. draft-dong-lsr-sr-enhanced-vpn-03.

[5] Xie C, Ma C, Dong J, et al. Using IS-IS Multi-Topology (MT) for Segment Routing based Virtual Transport Network[EB/OL]. (2020-07-130)[2020-10-31]. draft-xie-lsr-isis-sr-vtn-mt-02.
[6] Zhu Y, Dong J, Li, Z. Using Flex-Algo for Segment Routing based VTN[EB/OL]. (2020-09-11)[2020-10-31]. draft-zhu-lsr-isis-sr-vtn-flexalgo-01.
[7] Dong J, Hu Z, Li Z. BGP-LS Extensions for Segment Routing based Enhanced VPN[EB/OL]. (2020-03-09)[2020-03-25]. draft-dong-idr--bgpls-sr-enhanced-vpn-01.
[8] Xie C, Ma C, Dong J, et al. BGP-LS with Multi-topology for Segment Routing based Virtual Transport Networks[EB/OL]. (2020-07-13)[2020-10-31]. draft-xie-idr-bgpls-sr-vtn-mt-01.
[9] Zhu Y, Dong J, Li Z. BGP-LS with Flex-Algo for Segment Routing based Virtual Transport Networks[EB/OL]. (2020-03-09)[2020-10-31]. draft-zhu-idr-bgpls-sr-vtn-flexalgo-00.
[10] Dong J, Bryant S, Miyasaka T, Zhu Y, Qin F, Li Z. Segment Routing for Resource Guaranteed Virtual Networks[EB/OL]. (2020-03-09)[2020-03-25]. draft-dong-spring-sr-for-enhanced-vpn-07.
[11] Przygienda T, Shen N, Sheth N. M-ISIS: Multi Topology (MT) Routing in Intermediate System to Intermediate Systems (IS-ISs)[EB/OL]. (2015-10-14)[2020-03-25]. RFC 5120.
[12] Psenak P, Hegde S, Filsfils C, Talaulikar K, Gulko A. IGP Flexible Algorithm[EB/OL]. (2020-02-21)[2020-03-25]. draft-ietf-lsr-flex-algo-06.
[13] Ginsberg L, Bashandy A, Filsfils C, Nanduri M, Aries E. Advertising L2 Bundle Member Link Attributes in IS-IS[EB/OL]. (2019-12-06)[2020-03-25] RFC 8668.
[14] Finn N, Thubert P, Varga B, et al. Deterministic Networking Architecture [EB/OL]. (2019-10-24)[2020-03-25]. RFC 8655.
[15] Geng X, Chen M. SRH Extension for Redundancy Protection[EB/OL]. (2020-03-09)[2020-03-25]. draft-geng-spring-redundancy-protection-srh-00.
[16] Geng X, Chen M. Redundancy SID and Merging SID for Redundancy Protection[EB/OL]. (2020-03-09)[2020-03-25]. draft-geng-spring-redundancy-protection-sid-00.
[17] Geng X, Chen M. Redundancy Policy for Redundant Protection[EB/OL]. (2020-03-23)[2020-03-25]. draft-geng-spring-redundancy-policy-01.
[18] 3GPP. System architecture for the 5G System (5GS)[EB/OL]. (2019-12-22)[2020-03-25]. 3GPP TS 23.501.
[19] 3GPP. General Packet Radio System (GPRS) Tunnelling Protocol User Plane (GTPv1-U)[EB/OL]. (2019-12-20)[2020-03-25]. 3GPP TS 29.281.
[20] Matsushima S, Filsfils C, Kohno M, Camarillo P, Voyer D, Perkins C. Segment Routing IPv6 for Mobile User Plane[EB/OL]. (2019-11-04)[2020-03-25]. draft-ietf-dmm-srv6-mobile-uplane-07.
[21] 3GPP. Draft Specification for 5GS container (TS 38.415)[EB/OL]. (2019-01-08)[2020-03-25]. 3GPP TS 38.415.

[22] 3GPP. Interface between the Control Plane and the User Plane Nodes[EB/OL]. (2019-12-20)[2020-03-25]. 3GPP TS 29.244.
[23] 3GPP. Study on User-plane Protocol in 5GC[EB/OL]. (2019-09-23)[2020-03-25]. 3GPP TR 29.892.
[24] Li Z, Li M. Framework of Network Virtualization Based on MPLS Global Label[EB/OL]. (2003-10-21)[2020-03-25]. draft-li-mpls-network-virtualization-framework-00.

CHAPTER 11

SRv6 for Cloud Services

THIS CHAPTER DESCRIBES THE APPLICATION OF SRv6 IN CLOUD SERVICES. SRv6 provides network programming capabilities in edge telco cloud scenarios, and with IPv6 reachability, SRv6 connections can span multiple domains directly, simplifying service deployment across domains. In addition, SRv6 allows the explicit specification of forwarding paths by ingress, which is ideal for SFC implementation. SRv6 also provides strong support for SD-WAN. Specifically, by advertising binding SIDs corresponding to different paths, SRv6 enables devices such as Customer-Premises Equipment (CPE) to select paths based on binding SIDs.

11.1 SRv6 FOR TELCO CLOUD

11.1.1 Telco Cloud Overview

The traditional mode of telecom network construction tightly integrates software and hardware, using dedicated hardware devices purchased from equipment vendors to provide telecom services. Such devices include the Mobility Management Entity (MME), Serving Gateway (SGW), and Provisioning Gateway (PGW) for mobile data services and BNG for fixed services. With the rapid development of telecom services (including emerging services like 5G), telecom networks have to respond fast, adapt to diversified service scenarios, and support frequent service deployments. Traditional telecom devices with dedicated hardware rely heavily on equipment vendors. It often takes several months to expand network capacity or bring a new release online, driving up service rollout time and costs.

With the development and successful application of cloud-native design in the IT field and Network Functions Virtualization (NFV) technologies, virtualization and cloudification have gradually matured and evolved to provide a new level of productivity, offering a new approach to telecom network construction.

Telecom carriers hope for telecom devices with decoupled software and hardware, as is the case for IT devices. They hope to use universal or simplified hardware to improve device capabilities and forwarding throughput while lowering costs. With software decoupled from hardware, carriers can rapidly add new functions and roll out new services to quickly respond to customer requirements.

Against this backdrop, building cloud-based telecom networks (also known as telco clouds) becomes a new way of telecom network construction. Telco cloud construction refers to the NFV of telecom service nodes. NFV development generally goes through three phases, as shown in Figure 11.1.

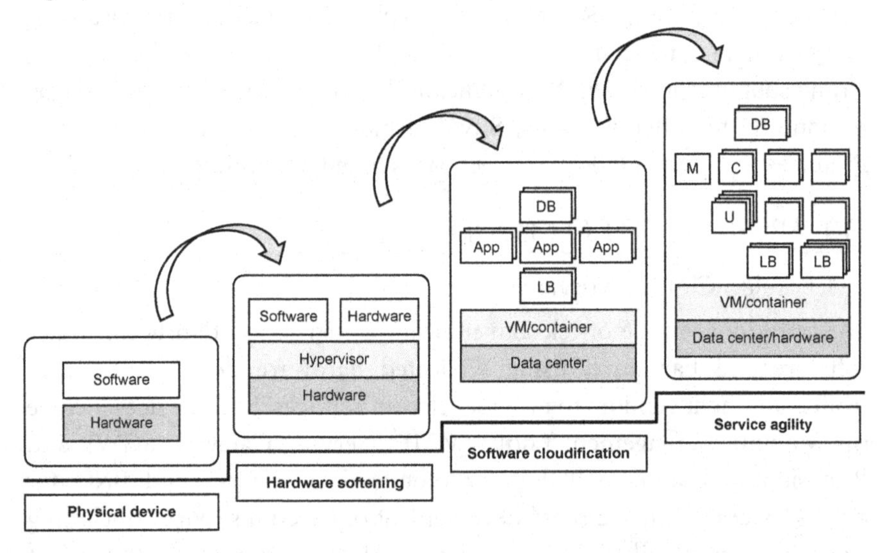

FIGURE 11.1 NFV development phases.

NOTICE

The three phases do not necessarily reflect the evolution timeline, as some telecom function entities may not need to complete all three phases.

Phase 1: hardware softening – hardware restructuring

- Software and hardware are decoupled, and functions become software-based. Capabilities are migrated from dedicated hardware to universal servers.

Phase 2: software cloudification – software restructuring

- Resources are pooled and virtualized. Compute, storage, and network resources are scheduled and orchestrated in a unified manner.
- Service and state data is separated, forming a three-layer architecture consisting of load balancers, applications, and databases.
- Key cloud capabilities, such as distributed deployment, auto-scaling, and elastic deployment, are supported.

Phase 3: service agility – service restructuring

- Software architecture is further optimized for the cloud environment, and fine-grained service deconstruction is supported, such as decoupling between the control plane and user plane as well as decoupling between the control plane and management plane.
- The network is given increased support toward cloudification and future service capabilities, such as flexible service deployment across clouds, on-demand service assembly, and service-based network slicing.

Telecom NEs are currently in different phases of NFV due to their processing characteristics. For example, CPU-intensive telco cloud VNFs mainly responsible for processing services, such as IP Multimedia Subsystems (IMSs) on the core network, are in phase 2, and telecom NEs that require high-bandwidth forwarding, such as Evolved Packet Core (EPC) nodes, UPFs, and Gateway Network Element (GNEs) on the core network, are in phase 3.

Telecom NEs are classified into the following types:

1. Host VNF: The IP address of such a VNF can be used as the source or destination address of service traffic. Other devices on the infrastructure network forward packets to the VNF by searching for a host or MAC route to the VNF.

2. Static routed VNF, such as Virtual Firewall (vFW) or Virtualized WAN Optimization Controller (vWOC): The IP address of such a VNF cannot be used as the source or destination IP address of service traffic. The infrastructure network cannot route traffic to the VNF (usually VAS VNF). Typically, an SFC is used to guide packet forwarding to such a VNF.
3. Dynamic routed VNF, such as Virtualized Evolved Packet Core (vEPC), Virtual Broadband Network Gateway (vBNG), or Cloud Service Load Balancer (CSLB): The IP address of such a VNF cannot be used as the source or destination IP address of service traffic. Instead, the user address registered within the VNF is used as the source or destination address of service traffic. As a result, a dynamic routing protocol is needed between the VNF and infrastructure network to exchange routes for packet forwarding.

There are two possible ways to use the current network architecture to support the telco cloud:

1. Using the proven IT cloud transport architecture: Public cloud and data center hosting services have been developing rapidly for years, and the existing cloud transport architecture is mature. The existing data center/cloud transport architecture can be used to make telco cloud VNFs available as new applications.

   This transport mode can well meet the requirements of host VNFs but may struggle when it comes to static and especially dynamic routed VNFs. The existing IT cloud transport network cannot meet the requirements of dynamic routed VNFs, such as dynamic routing, large routing tables, and connectivity detection.
2. Using the traditional CT transport architecture: The traditional CT transport architecture, such as MPLS VPN, supports communication between VNFs and can meet the requirements of routed VNFs. This architecture, however, faces challenges in providing automation capabilities such as cloud-based auto scaling, dynamic migration, cloud-based Virtual Machine (VM) management, and dynamic VNF onboarding and configuration.

To sum up, providing telecom functions with the traditional IT cloud transport architecture is quite difficult, and the traditional CT transport architecture cannot provide cloud-based functions. As such, a new architecture is required for telco cloud networks.

11.1.2 Telco Cloud Transport Network Architecture

Figure 11.2 shows a typical E2E telco cloud transport network architecture.

Telco clouds are generally categorized into edge telco clouds deployed in EDCs or RDCs and central telco clouds deployed in CDCs, as described in Table 11.1.

These telco cloud data centers are distributed in different locations and have different scales and capability requirements, and their transport solutions vary accordingly.

The central telco cloud involves large numbers of services and servers, requiring an independent Data Center Network (DCN). The DCN uses VXLAN as the transport technology, the carrier WAN uses MPLS/SRv6 VPN as the transport technology, and the two networks interconnect

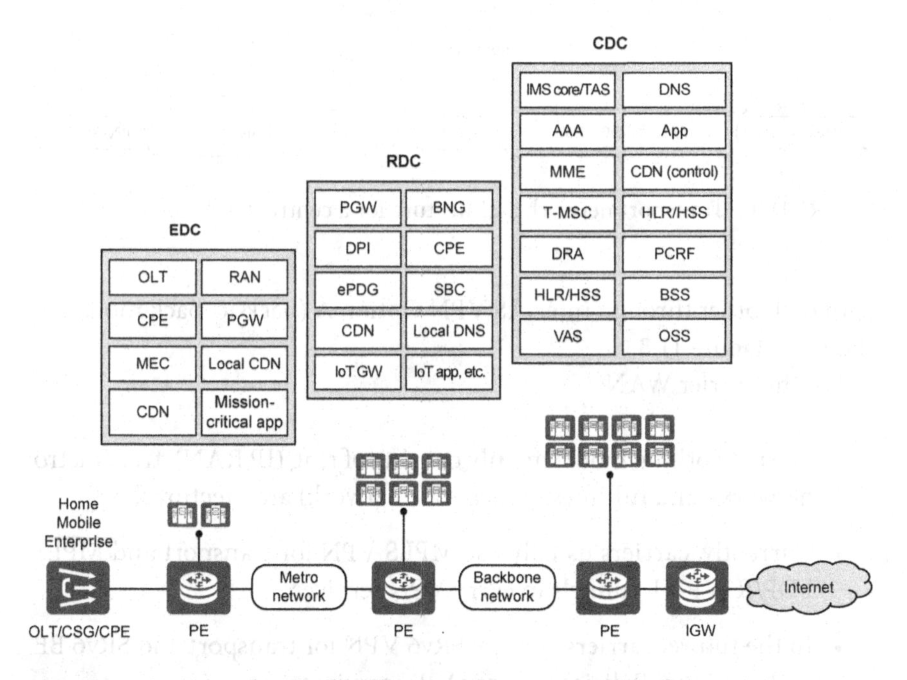

FIGURE 11.2 Typical E2E telco cloud transport network architecture.

TABLE 11.1 Telco Cloud Categorization and Characteristics

Category		Typical Service Scenario	E2E Latency (ms)	Deployment Location
Central telco cloud		4K Cloud AR/VR	<30	Municipal core equipment room or common equipment room (low latency and bandwidth requirements)
Edge telco cloud	RDC	8K Cloud VR	<10	Integrated service equipment room (high latency and bandwidth requirements)
	EDC	V2X, electric power, and other vertical industry applications	<5	Access equipment room (extremely high latency and bandwidth requirements; data transmission within a specific area required by some industries)

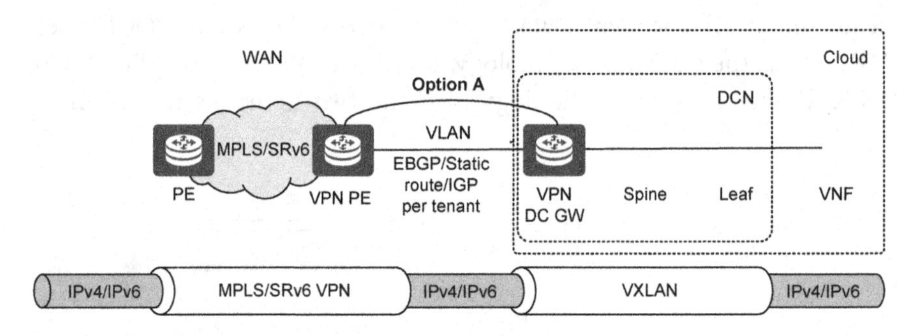

FIGURE 11.3 Transport network architecture for a central telco cloud.

with each other through inter-AS VPN Option A (back-to-back mode), as shown in Figure 11.3.

For the carrier WAN:

- The networking model mainly consists of ring (IP RAN), tree (metro network), and full-mesh (backbone network) architectures.
- Currently, carriers usually use MPLS VPN for transport and MPLS LDP/RSVP-TE tunnels to carry VPN services.
- In the future, carriers can use SRv6 VPN for transport and SRv6 BE paths or SRv6 Policies to carry VPN services.

For the telco cloud DCN:

- The networking model mainly consists of a three-layer architecture with DC gateways, spine nodes, and server leaf nodes, as shown in Figure 11.4. In the traditional DCN architecture, DC gateways are also called border leaf nodes.
- DCNs generally use VXLAN as the transport technology.

If the back-to-back transport solution for central telco cloud scenarios is used in the edge telco cloud scenarios, there are some daunting challenges, as shown in Figure 11.5.

These challenges include:

1. The edge telco cloud often carries a large amount of user-plane service traffic, which is much heavier than the control-plane service traffic carried by the central telco cloud. The MPLS/SR VPN

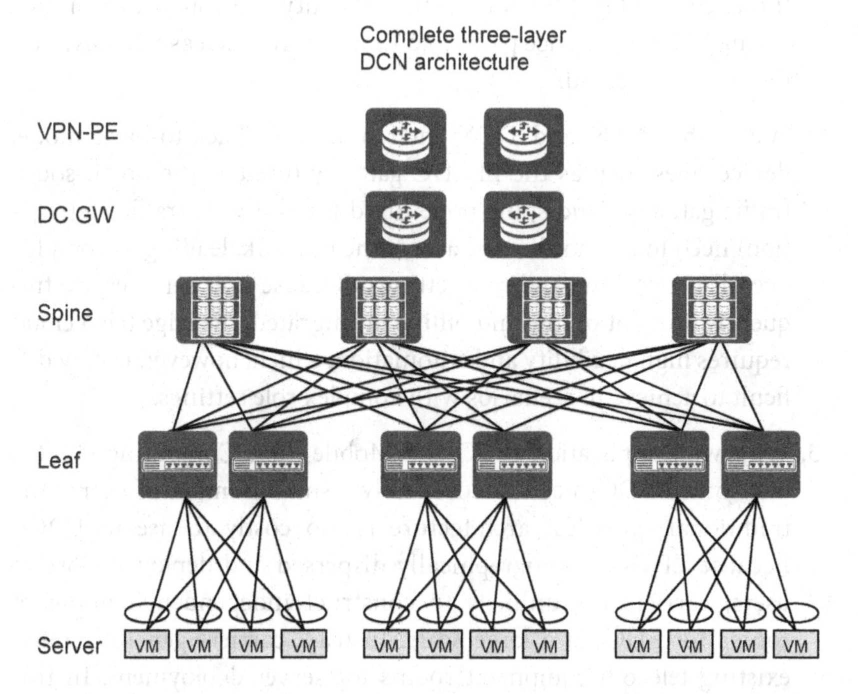

FIGURE 11.4 Three-layer DCN architecture.

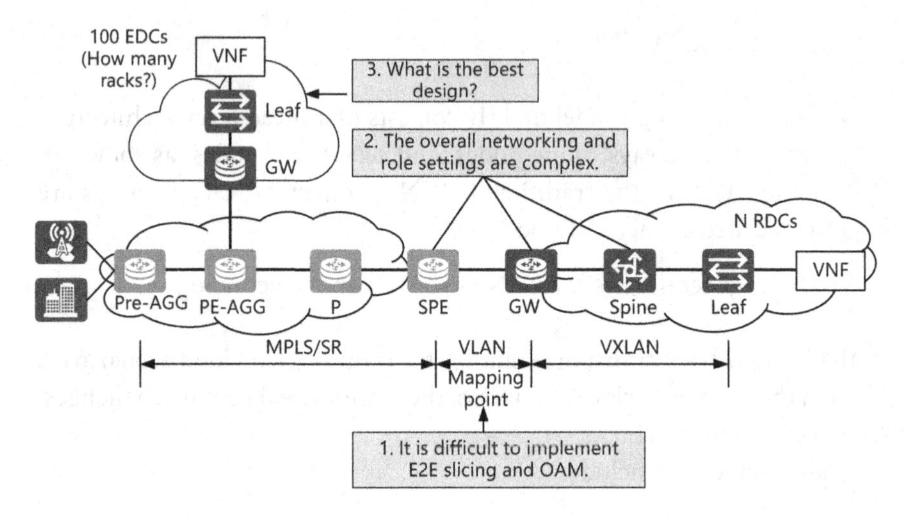

FIGURE 11.5 Transport network architecture for an edge telco cloud.

(WAN)+VXLAN (DCN) networking adopted by the traditional central telco cloud further adds to the difficulty and complexity of providing E2E service slice paths and E2E features such as E2E OAM on the edge telco cloud.

2. When the WAN and DCN interconnect in back-to-back mode, device roles such as the PE, DC gateway (used as the north-south traffic gateway), and spine node (used for east-west traffic aggregation) need to be coordinated across the network, leading to complex overall networking and role settings. Because user services are frequently brought online and offline or migrated, the edge telco cloud requires high flexibility and automation, which, however, is very difficult to achieve in scenarios with complex role settings.

3. With wide application of 5G and Mobile Edge Computing (MEC), distributed EDCs need to cover only a small number of users. The traditional spine-leaf architecture is too costly to use in EDCs. Because EDCs are geographically dispersed and deployed close to users, carriers are unlikely to construct independent equipment rooms for EDCs at a large scale. Instead, carriers generally reuse existing telecom equipment rooms for server deployment. In this case, reusing WAN PEs as DC gateways is recommended, as shown in Figure 11.6.

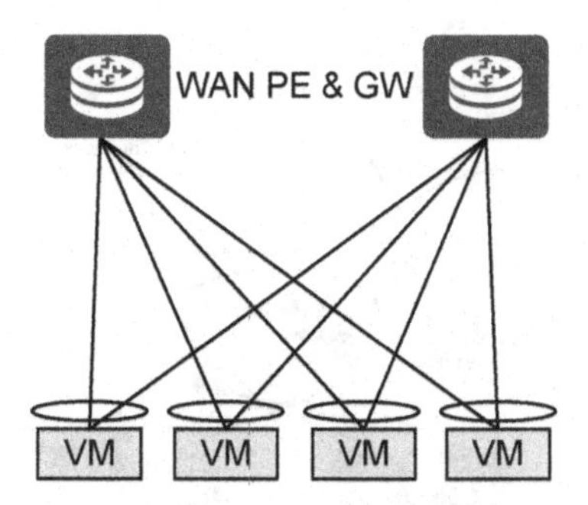

FIGURE 11.6 Reuse of WAN PEs as DC gateways to build an edge telco cloud.

When functioning as DC gateways, WAN PEs can directly connect to the servers for cloud-based applications and serve as the first hops for cloud-based services. The traditional WAN devices need to have their technology and architecture upgraded to provide cloud-based service capabilities (such as auto-scaling and load balancing).

11.1.3 Edge Telco Cloud Architecture

11.1.3.1 NAAF Physical Architecture

The general idea of the solution to issues facing the edge cloud transport network is to integrate the DCN and WAN to form a spine-leaf architecture called Network as a Fabric (NAAF).

The NAAF roles shown in Figure 11.7 are described as follows:

- Leaf node: an access node on a fabric network. WAN PEs are usually used as leaf nodes to connect various network devices to the fabric network. Depending on access devices, leaf nodes can be classified into access leaf nodes, server/service leaf nodes, and border leaf nodes.
    - Access leaf node: used for user access, such as mobile access over a base station or fixed access over an Optical Line Terminal (OLT)
    - Server/Service leaf node: used for VNF service access, including VAS, vCPE, vUPF, and BNG-UP access
    - Border leaf node: connected to one or more external networks, such as CDCs, other network domains (such as backbone networks), or other carrier networks

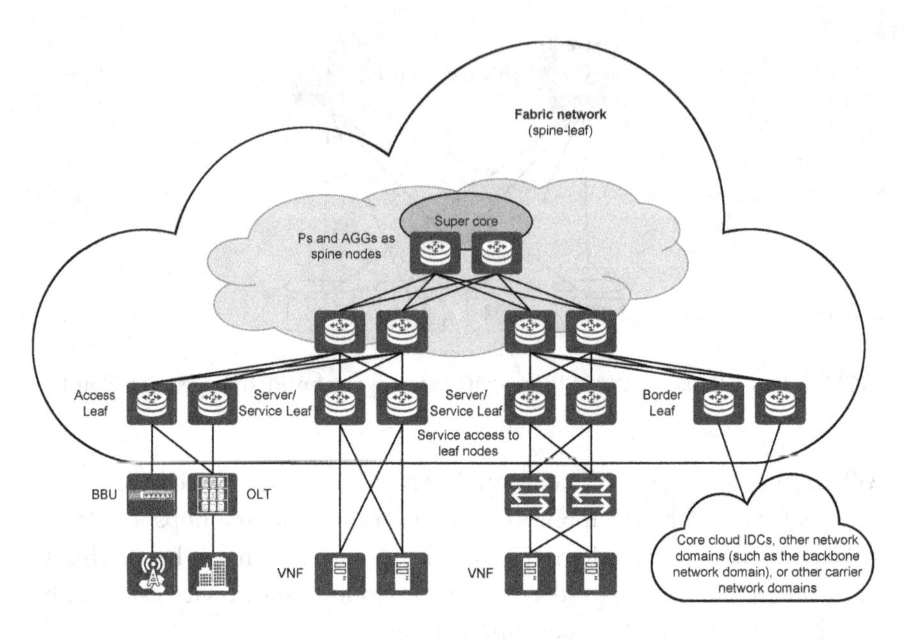

FIGURE 11.7 Spine-leaf fabric architecture.

- Spine node: a node used to forward traffic at a high speed. A spine node is usually a P device on the WAN and does not function as a service access device. Spine nodes have the following characteristics:

  - Spine nodes use high-speed interfaces to connect to leaf nodes, eliminating the need to establish full-mesh connections between all leaf node pairs. This also enables smooth service/leaf node scale-out, as capacity expansion will not affect existing services.

  - On a large network, spine nodes can be deployed hierarchically.

11.1.3.2 NAAF Transmission Protocols and Key Technologies

Because the NAAF edge cloud architecture involves both the DCN and traditional WAN, transmission protocols used on these networks need to be unified. In addition, because the edge telco cloud has many requirements for telecom connection but few for cloudification, mature telecom transmission solutions (VPN and SRv6) are more suitable for use.

NAAF uses SRv6+EVPN to support IPv4/IPv6 dual-stack and provides 5G, enterprise, MEC, and other service capabilities. Figure 11.8 shows key NAAF technologies.

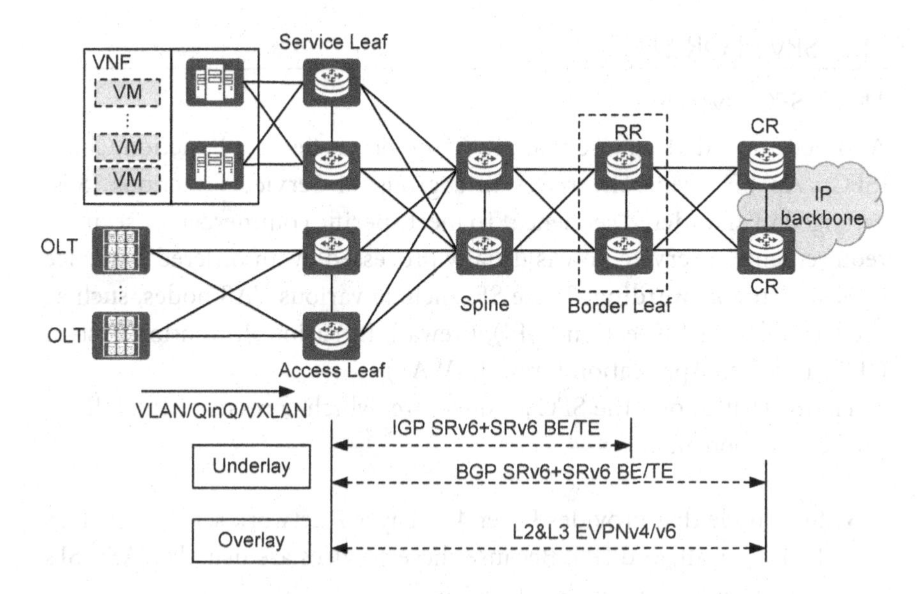

FIGURE 11.8 Key NAAF technologies.

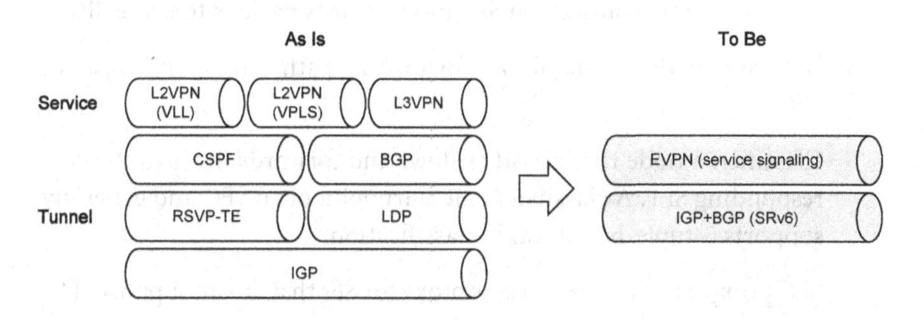

FIGURE 11.9 Technology evolution at the protocol layer using the NAAF architecture.

Figure 11.9 shows the technology evolution of the protocol stack in the NAAF architecture. The overall objective of this evolution is to simplify the architecture. At the tunnel/underlay layer, LDP and RSVP-TE are replaced by SRv6. The underlay and tunnel functions can be implemented with only IGPs and BGP, simplifying signaling protocols. Moreover, EVPN integrates L2VPN VPWS (based on LDP or MP-BGP), L2VPN VPLS (based on LDP or MP-BGP), and L3VPN (based on MP-BGP) technologies, simplifying technical complexity at the service layer.

11.2 SRv6 FOR SFC

11.2.1 SFC Overview

Another important service that SRv6 supports is Service Function Chain (SFC). An SFC generally refers to a group of Service Functions (SFs) arranged in an ordered sequence. To meet specific commercial or security requirements, a service flow is usually processed by an ordered sequence of SFs during forwarding. These SFs include various VAS nodes, such as the Deep Packet Inspection (DPI), firewall, Intrusion Prevention System (IPS), and Web Application Firewall (WAF).

Figure 11.10 shows the SFC architecture, which consists of the following key components:

- SF: a node that provides Layer 4 to Layer 7 network services such as DPI, firewall, and IPS. Because these services are usually VASs, SFs are considered equivalent to VASs.
- Service Function Forwarder (SFF): a forwarder that supports SFC. An SFF usually connects to SFs and forwards packets to these SFs.
- Service Function Path (SFP): a forwarding path corresponding to an SFC.
- Classifier: a node that classifies flows and forwards flows to the corresponding SFP. A classifier is the start point of an SFC and generally supports 5-tuple-based traffic classification.
- SFC proxy: provides SFC access proxy for SFs that do not support SFC.

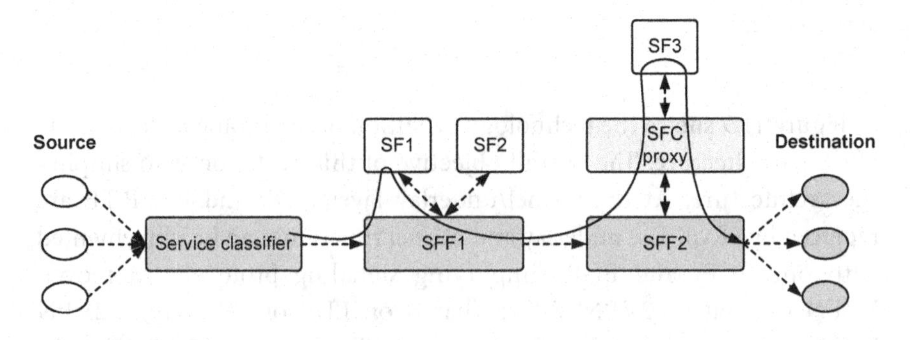

FIGURE 11.10 SFC architecture.

SFC is mainly used in scenarios where the packets are processed based on user or service policies, such as transport network Gi-LAN and multi-tenant cloud data center scenarios.

11.2.1.1 Transport Network Gi-LAN

As shown in Figure 11.11, a Gi-LAN is a LAN between the EPC and Internet. The Gi-LAN serves as an egress for mobile phones and data cards to access the Internet. The data packets of users who subscribe to different services are attached with the corresponding service tags after being processed (such as accounted) by the EPC. After entering the Gi-LAN, these packets are processed by the corresponding SFC based on these tags and then forwarded to the Internet. The Gi-LAN provides VASs such as firewall, Carrier-Grade NAT (CGN), DPI, and video acceleration.

Generally, an SFC consists of multiple VASs in a Gi-LAN. One SFC corresponds to one service package, but one service package can contain multiple SFCs. The mapping between each user and service package can be modified.

Gi-LAN SFCs are mainly used by carriers to provide customized services for OTT providers. For example, a carrier cooperates with a video content supplier to provide the video acceleration service. By redirecting user access traffic to a specific SFC in the Gi-LAN, the OTT provider delivers better video service experience to users. Subsequently, improved user experience brings new business opportunities for the OTT provider, and the carrier expands its subscriber base and revenue streams through cooperation with the OTT provider.

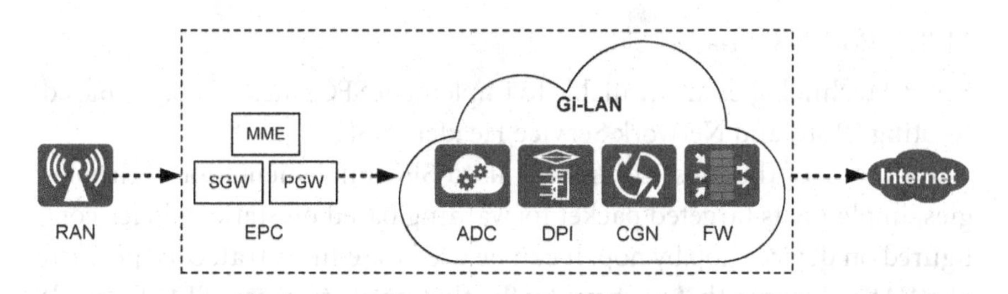

FIGURE 11.11 Gi-LAN networking.

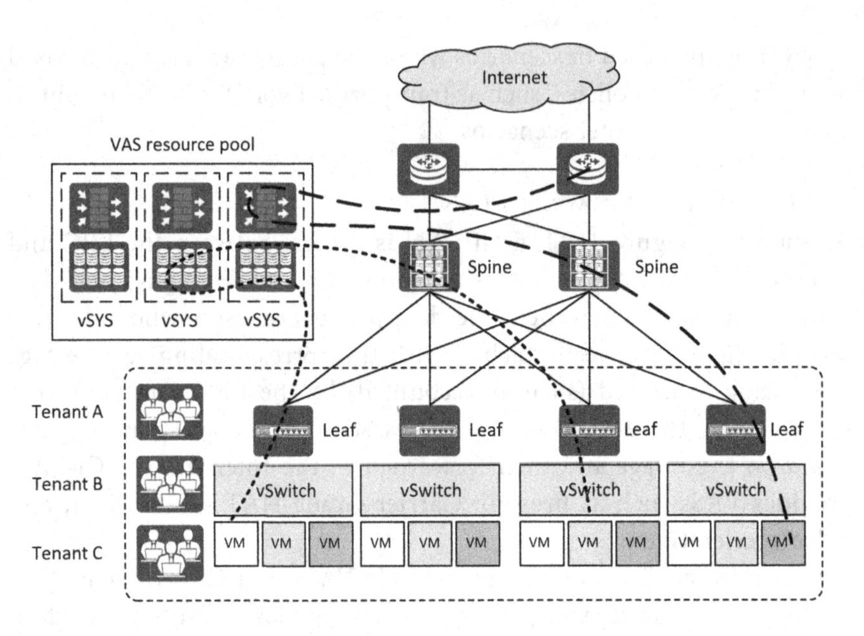

FIGURE 11.12 SFC application in a data center.

11.2.1.2 Multi-tenant Data Center

The multi-tenant cloud data center is a typical SFC application scenario, as shown in Figure 11.12. Generally, cloud computing service providers provide multiple VASs, which tenants can select on demand. For example, tenant A deploys a simple website service in a Virtual Private Cloud (VPC) of the data center and selects only the firewall and video acceleration VASs. Tenant B deploys an important service in a VPC of the data center and selects multiple security-related VASs, such as firewall, DPI, IPS, and WAF. The service traffic of the two tenants needs to pass through different SFCs in the data center as required.

11.2.2 PBR/NSH-based SFC

Several technologies are available to implement SFC, such as Policy-Based Routing (PBR) and Network Service Header (NSH).

PBR, one of the most commonly used SFC implementation technologies, implements targeted packet forwarding based on static policies configured on devices hop by hop. It can be used to redirect traffic to specified SFs/VASs. Assume that we have traffic that needs to access SF1 (firewall) and SF2 (DPI). Then we can configure policies on devices connected to SF1 and SF2 to forward packets to the specified SFs.

As networks develop, network service functions start to move toward virtualization and are no longer closely coupled with hardware devices. Instead, they are flexibly distributed in the form of software entities (NFV). As a result, SFC services change more and more frequently, especially in scenarios such as multi-tenant data centers.

In PBR-based SFC, the destination address of traffic is the actual destination address. We need to configure PBR on devices such as SFFs to forward SFC traffic over the specified links, so that the traffic can be forwarded to the specified SFs. When the location of an SF changes, PBR needs to be updated accordingly. This means that PBR is tightly coupled with the physical topology. In other words, statically configured PBR cannot support flexible SF or SFC adjustments. To summarize, this solution has the following drawbacks[1]:

1. PBR-based SFC deployment is tightly coupled with the physical topology. New service rollout depends on the physical topology and is prone to encounter problems such as limited locations and complex configurations. The service processing sequence depends on the topology and is relatively fixed. As a result, service scalability is limited and inflexible. Moreover, service adjustment is complex, slow, and costly.
2. Service transport and service-layer policies depend on the underlying transport protocol and are complex to configure.
3. Traffic classification is coarse-grained, based on either PBR or access control. In addition, re-classification lowers performance.
4. Insufficient E2E visualization makes it impossible to provide visualized OAM.
5. SFs are independent of each other and cannot transmit metadata (basic elements[2] such as application IDs used for context information exchange between classifiers and SFs and between different SFs).

To address the preceding drawbacks, the industry proposed NSH.

An NSH is a protocol packet header designed for an SFC. It carries the Service Path Identifier (SPI) and Service Index (SI) to guide packet forwarding in the SFC. Specifically, SFFs match NSHs carried in packets against local NSH forwarding entries and forward packets through the corresponding SFCs accordingly.

Moreover, NSHs can carry different types of metadata (including metadata with a fixed or variable length) to share information between SFs in scenarios requiring more complex operations. Figure 11.13 shows the NSH format.

An NSH consists of three parts: base header, service path header, and context header. Table 11.2 describes these fields.[3]

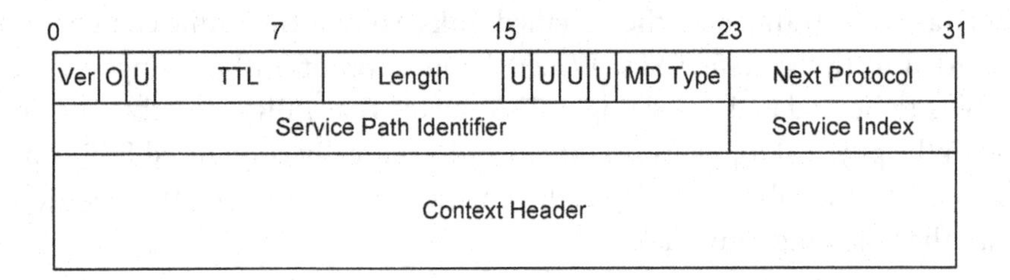

FIGURE 11.13 NSH format.

TABLE 11.2 Fields in an NSH

Category	Field	Length	Description
Base header	Ver	2 bits	Version information, which must be set to 0 in the current version.
	O	1 bit	OAM flag, which is used to indicate the OAM packet.
	U	1 bit	Unallocated field.
	TTL	6 bits	Number of SFP hops. The maximum value is 63.
	Length	6 bits	NSH length.
	MD Type	4 bits	Metadata type, which determines the format of the context header.
	Next Protocol	8 bits	Payload protocol type, which can be IPv4, IPv6, MPLS, NSH, Ethernet, etc.
Service path header	Service Path Identifier	24 bits	SFP ID, also called SPI, which uniquely identifies an SFP.
	Service Index	8 bits	Index of an SF in an SFP.
Context header	Context Header	Variable	Header used to carry metadata. The content of this field is determined by the MD Type field. • If the MD Type value is 0x1, this field contains metadata composed of four 32-bit fixed-length headers. • If the MD Type value is 0x2, this field contains metadata of variable length.

TABLE 11.3 Entries in the NSH Forwarding Table on an SFF

SPI	SI	Next Hop	Transport Encapsulation
10	255	192.168.2.1	VXLAN-GPE
10	254	172.16.100.10	GRE
10	251	172.16.100.15	GRE
40	251	172.16.100.15	GRE
50	200	01:23:45:67:89:ab	Ethernet
15	212	Null (end of path)	None

SFFs must maintain an NSH forwarding table to forward NSH packets. Table 11.3 provides an example of entries in the NSH forwarding table on an SFF.

When an NSH packet arrives at an SFF, the SFF searches for the next hop (SF) based on the SPI and SI, encapsulates the packet based on the encapsulation information of the forwarding entry, and forwards the packet to the next hop (SF). The SI value decrements by 1 each time the NSH packet is processed by an SF. After processing the packet, the SF returns the updated NSH packet to the SFF. The SFF then searches the NSH forwarding table based on the updated SI to forward the packet to the next hop (SF).

The following uses an example to describe the NSH packet forwarding process, as shown in Figure 11.14.

1. The controller computes SFC policies. In this example, SFC1 consists of SF1 and SF2.

2. The controller delivers SFC policies to the service classifier and SFFs.

    a. SFC classification policy for the service classifier: instructs the classifier to classify data flows based on 5-tuple information and add the specified NSH to matching packets.

    b. Forwarding policy for SFFs: guides SFFs to perform operations defined in the NSH.

3. After a data packet arrives at the service classifier, the classifier classifies the packet, adds an NSH to the packet, and forwards the packet to the first SFF (SFF1).

4. Upon receipt, SFF1 finds the matching SFC policy based on the SPI and SI in the NSH and forwards the packet to SF1.

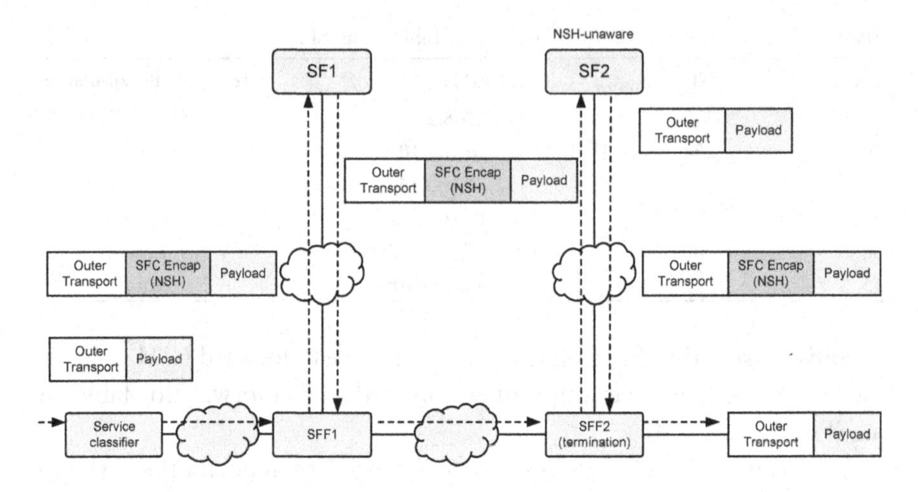

FIGURE 11.14 NSH packet forwarding.

5. After SF1 receives the packet, it processes the packet based on NSH information, updates the NSH SI, and returns the packet to SFF1.

6. SFF1 examines the SPI and SI in the updated NSH to find a matching entry in the local NSH forwarding table and forwards the packet to SFF2.

7. SFF2 forwards the packet to SF2 based on the SPI and SI in the NSH.

8. SF2 processes the packet based on NSH information, updates the SI in the NSH, and returns the packet to SFF2.

9. SFF2 searches its NSH forwarding table based on the SPI and SI in the NSH and finds that the SFC has reached an end. SFF2 then removes the NSH from the packet and forwards the packet to the next hop.

According to RFC 8300, NSH supports multiple underlying transport protocols, such as Ethernet, VXLAN-GPE, and GRE.[3] NSH only requires packets to be forwarded to SFs, regardless of the type of the underlying transport protocol. This means that NSH is unaware of the transport network.

Because NSH only requires SFs to be reachable, NSH policy deployment does not have requirements on the physical locations of SFs. When the physical location of an SF changes, the NSH forwarding policy on SFFs does not need to be changed. This solves the issue of PBR dependency on the physical network topology.

Currently, the NSH solution mainly uses Ethernet or VXLAN-GPE as the transport technology. The NSH-based SFC solution offers the following benefits:

- Provides a service plane, which is decoupled from the physical topology.
- Requires only network reachability and does not care what transport protocol is used.
- Provides visualization and OAM capabilities based on location information provided by SPIs and SIs.
- Supports both centralized and distributed control planes, fast service update, and path programming.
- Allows traffic to be classified once and forwarded multiple times or reclassified based on the SF processing result.

Although NSH has the preceding advantages, it requires each SFF to maintain the forwarding state of each SFC. This means that an NSH-based policy needs to be configured on multiple network nodes during service deployment, increasing control plane complexity. Moreover, the standardization process is still immature.[4] From the perspective of implementation, transparent NSH transport involves too many network encapsulation options, causing excessive work. In addition, different vendors also need to negotiate the NSH transport protocol in commercial deployment, incurring high interoperability costs. As a result, many carriers and OTT providers still choose the PBR solution to deploy SFCs.

11.2.3 SRv6-based SFC

As described in Chapter 2, SR supports explicit programming of data packet forwarding paths on the ingress. This capability intrinsically supports SFC. In addition, SR does not need to maintain the per-flow forwarding state on transit nodes, making service deployment much easier compared with NSH. An SR-based SFC requires SFC policies to be deployed only on the headend, which effectively simplifies SFC deployment, especially control plane deployment.

Both SR-MPLS and SRv6 can implement SFC. This section focuses on SRv6-based SFC.

Currently, the following two SRv6 SFC solutions are available to the industry:

- Stateless SRv6 SFC: This solution does not require SFFs to maintain the forwarding state of each SFC.[5]
- Stateful SRv6 SFC: This solution requires SFFs to maintain the forwarding state of each SFC. It is a combination of SRv6 and NSH.[6]

11.2.3.1 Stateless SRv6 SFC

On an SRv6 network, forwarding path programming, or explicit path programming, is implemented through SID combination. The main idea of stateless SRv6 SFC[5] is to program a path to pass through specified SFs in sequence.

The stateless SRv6 SFC solution uses a segment list to determine an SFP and the SRH TLV to carry SF-related metadata. It does not require NSH to guide packet forwarding or require SFFs to maintain the forwarding state of each SFC. Figure 11.15 shows the architecture of the stateless SRv6 SFC solution.

The stateless SRv6 SFC solution consists of the following key components:

- SRv6-aware SF: an SRv6-capable SF that can directly connect to the SFF. Such an SF needs to distribute an End.AN behavior SID corresponding to the provided service function. The End.AN behavior indicates that the SF supports SRv6 as well as a specific service function.[7]

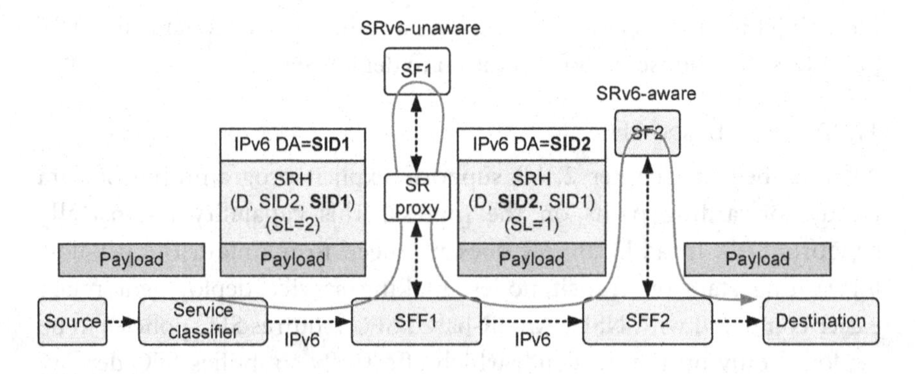

FIGURE 11.15 Stateless SRv6 SFC solution architecture.

- SRv6-unaware SF: an SRv6-incapable SF which needs an SRv6 proxy to support SRv6 SFC.
- SRv6 proxy: a proxy that forwards packets from an SRv6 network to an SRv6-unaware SF and returns packets from the SRv6-unaware SF to the SRv6 network. The SRv6 information, such as SRH, is cached at the proxy before a packet is sent to an SRv6-unaware SF, and the cached SRH is restored after the packet is received from the SRv6-unaware SF. SRv6 proxy behaviors, which vary according to proxy types, are essential to SRv6 proxy implementation. The proxy behaviors defined for SIDs used in SFC are as follows[5]:
  - End.AS: a static proxy SID advertised by an SRv6 proxy. The End.AS SID represents an instruction to perform SRv6 decapsulation and send the original packet to an SF through the corresponding physical interface or virtual interface. For example, if a packet carrying the specified VLAN ID is returned from an SF to the SRv6 proxy through a VLAN-based virtual interface, the proxy searches the corresponding cached SRv6 information according to the VLAN ID and inserts it into the packet before forwarding the packet. Because the mapping between the SRH and interface is statically configured, the End.AS SID is also called static proxy SID.
  - End.AD: a dynamic proxy SID advertised by an SRv6 proxy. Unlike the static proxy SID, the End.AD SID supports dynamic proxy. Specifically, the mapping between the SRH and interface is dynamically generated instead of statically configured.
  - End.AM: a masquerading proxy SID advertised by an SRv6 proxy. The End.AM SID represents an instruction to set the actual destination address, that is, SID [0] in the segment list, as the DA of an IPv6 packet and forward the forged packet to the SF. The SRv6 proxy replaces the DA with the next SID in the packet returned from the SF before forwarding the packet.

Figure 11.16 shows the SRv6 proxy architecture.

During SRv6 SFC deployment, nodes send service SIDs to the controller through BGP-LS for the controller to compute the segment list for SFC

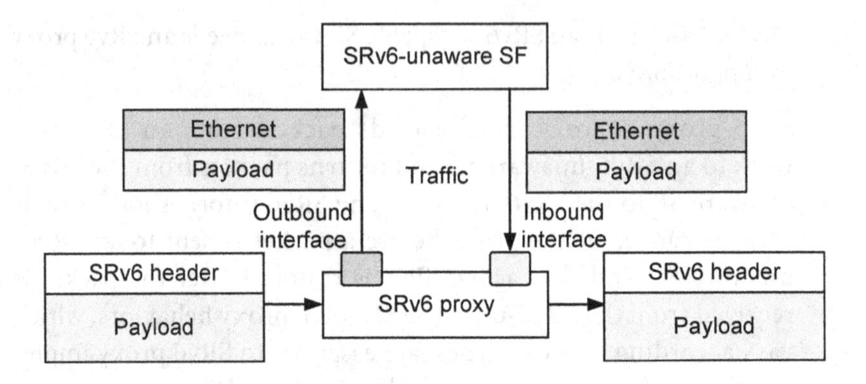

FIGURE 11.16 SRv6 proxy architecture.

forwarding. The controller then directly delivers the computed segment list to the headend.[7] The overall information reported by BGP-LS includes:

- Service SID: 128-bit IPv6 address in SRv6
- Function Identifier: behavior identifier of a service, such as static proxy, dynamic proxy, masquerading proxy, or SRv6-aware service
- Service Type: SF type, such as DPI, firewall, classifier, or LB
- Traffic Type: SFC-supported flow type, such as IPv4, IPv6, or Ethernet
- Opaque Data: other information, such as version information

For details about packet encapsulation, see the related IETF draft.[7]

The following uses the stateless SRv6 SFC solution architecture shown in Figure 11.15 as an example to describe the SRv6 SFC deployment and forwarding process.

1. The SRv6 proxy corresponding to SF1 sends End.AD SID1 to the controller through BGP-LS. The behavior defined in SID1 is to perform dynamic proxy and forward traffic to SF1. SF2, an SRv6-aware SF, directly sends End.AN SID2 to the controller through BGP-LS.
2. The controller computes the SFP based on service requirements and delivers the corresponding SRv6 Policy and traffic classification policy to the headend, which is also the service classifier.

3. When a packet arrives at the service classifier, the classifier classifies the packet according to the traffic classification policy. After finding the corresponding SRv6 Policy, the classifier encapsulates the packet with an outer IPv6 header, which contains an SRH with segment list <SID1, SID2, D>.

4. The classifier then forwards the packet to the SRv6 proxy based on SID1 in the IPv6 DA field. The SRv6 proxy processes End.AD SID1 advertised by itself, removes the outer IPv6 header and SRH, adds a VLAN ID to the packet, and sends the packet to SF1. The SRv6 proxy stores the mapping between the cached IPv6 header/SRH and VLAN ID, so that the SRv6 packet can be restored based on the VLAN ID after it is returned from the SF.

5. After the packet is returned from SF1, the SRv6 proxy first removes the VLAN ID and searches its cache mapping table for the original SRv6 information that matches the VLAN ID. The proxy then encapsulates the packet with an outer IPv6 header and SRH, updates the IPv6 DA field based on the next SID in the segment list of the SRH, and forwards the packet to the next node.

6. Upon receipt of the packet from SFF1, SFF2 forwards the packet to SF2 based on the IPv6 destination address.

7. After SF2 receives the packet and finds that SID2 carried in the DA field is the locally advertised End.AN SID, it executes DPI indicated by SID2, updates the IPv6 destination address to SID D, and forwards the packet to the next hop (SFF2).

8. SFF2 forwards the packet to the destination D.

Stateless SRv6 SFC requires only the advertisement of service SIDs and does not require SFFs to maintain the state of each SFC. The stateless SRv6 SFC solution, which is much simpler than the NSH solution, has become a new choice for SFC deployment.

However, stateless SRv6 SFC also has some drawbacks. One drawback is the relatively large packet overhead, as the number of SIDs in the segment list is large when there are a large number of SFs. Another drawback is that most SFs do not currently support SRv6 and require SRv6 proxies, increasing the difficulty of SFC deployment. With the development of

SRv6, many VAS vendors have planned to support SRv6, and this will help resolve this problem.

11.2.3.2 Stateful SRv6 SFC

The other type of SRv6-based SFC is stateful SRv6 SFC.[6] Stateful SRv6 SFC combines SRv6 and NSH technologies and is an interim solution for evolving a non-SRv6 network to an SRv6 network. It is mainly used in scenarios where SFs support NSH but not SRv6.

Stateful SRv6 SFC drafts mainly describe two solutions:

Solution 1: The service plane uses NSH to guide the packet forwarding of the entire SFC. SRv6 is only used for tunnels between SFFs.

Solution 2: The segment list carries information about the entire SFP, which passes through multiple SFFs. However, traffic is forwarded from SFFs to SFs based on the NSH forwarding table.

In essence, solution 1 still uses NSH to implement SFC but uses SRv6 as the tunnel protocol between SFFs.

Figure 11.17 shows stateful SRv6 SFC solution 1.

In stateful SRv6 SFC solution 1, the SFC forwarding process is as follows:

1. The service classifier adds an NSH to the packet based on the traffic classification policy. In the NSH, the SPI is SFP1, and the SI is 255. The classifier then searches its NSH forwarding table. After finding that the next hop of the NSH packet is SFF1 and the tunnel type is SRv6, the classifier encapsulates segment list <SID1, SID2> into the

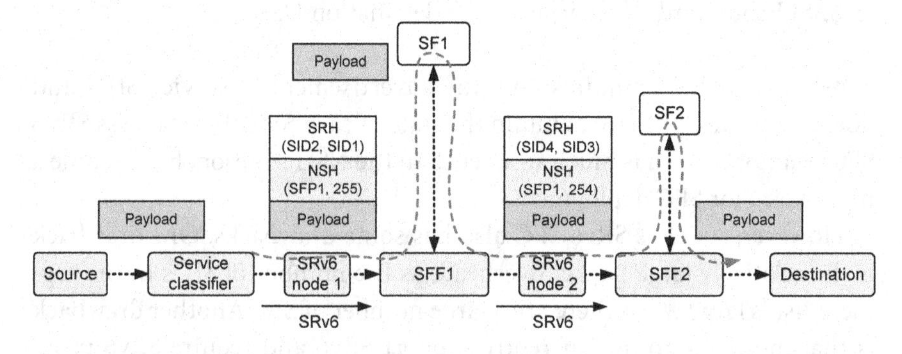

FIGURE 11.17 Stateful SRv6 SFC solution 1.

packet and forwards the packet. Here, SID1 refers to the SID advertised by a node (SRv6 node 1) on the SFP, and SID2 the SID pointing to SFF1.

2. Upon receipt of the packet, SRv6 node 1 corresponding to SID1 forwards the packet to SFF1 corresponding to SID2.

3. After the packet arrives at SFF1, SFF1 removes the SRv6 header, searches the NSH forwarding table based on the inner NSH, and forwards the packet to SF1.

4. SF1 performs the corresponding network function on the packet, decrements the SI value by 1, and returns the packet to SFF1.

5. SFF1 searches its NSH forwarding table for the next hop based on the SPI and updated SI and finds that the destination is SFF2, the underlying transport tunnel is an SRv6 tunnel, and the segment list is <SID3, SID4>. SFF1 then performs SRv6 encapsulation on the NSH packet and forwards the SRv6 packet to SRv6 node 2 (corresponding to SID3), which in turn forwards the packet to SFF2 (corresponding to SID4).

6. SFF2 removes the SRv6 header, searches its NSH forwarding table for a matching entry, and forwards the packet to SF2.

7. SF2 performs the corresponding network function on the packet, decrements the SI value by 1, and returns the packet to SFF2.

8. Upon receipt, SFF2 searches its NSH forwarding table. After finding that the next hop is none, SFF2 removes the NSH header and forwards the original packet to subsequent nodes.

From the preceding forwarding process, it can be seen that this solution is essentially an NSH solution using SRv6 tunnels, which are established between SFFs. In this solution, TE paths can be established between SFFs, but paths between SFFs and SFs are still traversed based on NSH information. Therefore, SFFs still need to maintain the state of each SFC.

Solution 2 differs from solution 1 in that the headend uses SRv6 to explicitly program the entire SFP into a segment list, instead of using NSH to connect multiple SRv6 tunnels. The similarity between the two solutions is that the path from the SFF to the SF is still determined based on NSH information. Figure 11.18 shows stateful SRv6 SFC solution 2.

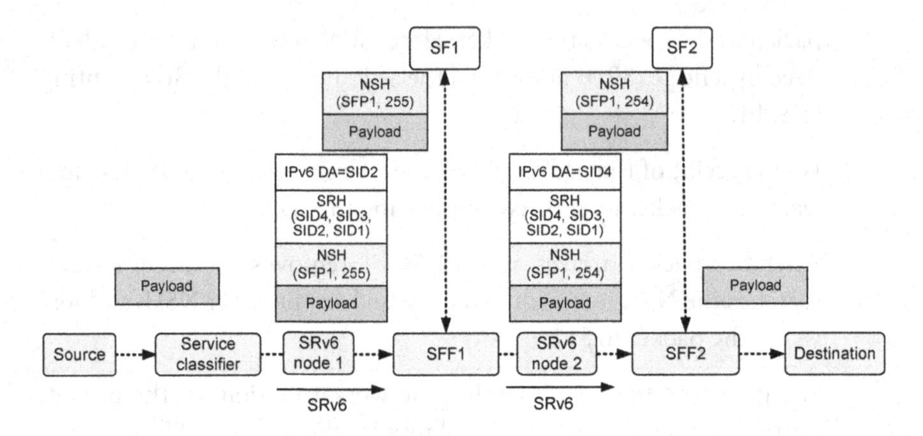

FIGURE 11.18 Stateful SRv6 SFC solution 2.

In stateful SRv6 SFC solution 2, the SFC forwarding process is as follows:

1. The service classifier adds an NSH to the packet based on the traffic classification policy. In the NSH, the SPI is SFP1, and the SI is 255. The classifier then performs SRv6 encapsulation on the NSH packet. The segment list in the SRH contains the complete SFP information, that is, <SID1, SID2, SID3, SID4>. Finally, the classifier forwards the packet to SFF1 based on the segment list.

2. Upon receipt of the packet, SRv6 node 1 corresponding to SID1 forwards the packet to SFF1 corresponding to SID2.

3. After the packet arrives at SFF1, SFF1 performs SRv6 decapsulation according to the behavior defined in SID2, caches the mapping between the SRv6 information and NSH SPI, searches the NSH forwarding table based on the SPI and SI, and forwards the packet to SF1.

4. SF1 processes the packet, decrements the SI value by 1, and returns the packet to SFF1.

5. SFF1 searches its mapping cache table for the SRv6 encapsulation information that matches the SPI and SI in the NSH, performs SRv6 encapsulation on the packet, and changes the destination IPv6 address to SID2. SFF1 then searches its IPv6 forwarding table to forward the packet to the next hop (SRv6 node 2).

6. After receiving the packet from SRv6 node 2, SFF2 removes the SRv6 packet header and caches the information based on the behavior defined in SID4, searches its NSH forwarding table for a matching entry, and forwards the packet to SF2.

7. SF2 processes the packet, decrements the SI value by 1, and returns the packet to SFF2.

8. Similarly, after the packet returns to SFF2, SFF2 searches the mapping cache table for matching SRv6 encapsulation information, performs SRv6 encapsulation, and forwards the packet.

This solution is more like the stateless SRv6 SFC solution, where an SRv6 proxy forwards packets based on the NSH forwarding entries. This solution requires a new SID, which can be called End.NSH SID, to indicate that the SRv6 packet header is followed by an NSH, so that an SRv6 node will cache the SRv6 packet header and search the NSH forwarding table to forward the packet. In the preceding example, the SIDs used by SFF1 and SFF2 are both End.NSH ones.

In general, stateless SRv6 SFC does not require SFFs to maintain the state of each SFC. If SFs support SRv6, the solution is easier to implement. Although both stateful SRv6 SFC solutions require SFFs to maintain the state of each SFC, they are compatible with the NSH-based SFC solution and do not require SFs to support SRv6. NSH+SRv6 forwarding also reduces the number of SIDs required in the segment list of the SRv6 packet header and, as a result, the packet header length. It is an optional interim SFC solution for evolution from NSH-based SFC to SRv6-based SFC.

11.3 SRv6 FOR SD-WAN

11.3.1 SD-WAN Overview

In May 2014, the Open Network User Group (ONUG) Spring Conference clearly defined a new concept — SD-WAN. As a combination of SDN and WAN, SD-WAN applies the SDN architecture and concepts to WANs and reshapes WANs with SDN.

SD-WAN can be viewed as a service made available after SDN is applied to WANs. This service connects enterprise networks, data centers, Internet apps, and cloud services across a broad geographical area, aiming to reduce WAN spending and improve network connection flexibility.

TABLE 11.4 Gartner's Definition of SD-WAN

No.	Characteristic
1	• Must support multiple connection types MPLS, Internet, LTE, etc.
2	• Can do dynamic path selection Allows for load sharing across WAN connections.
3	• Provides a simple interface for managing WAN Must support zero-touch provisioning at a branch, should be as easy as setting up a home WiFi.
4	• Support for multiple Third-Party Services Must support VPN, WAN optimization, controllers, firewalls, web gateways, and other services.

Since its release, SD-WAN has attracted wide attention. Gartner's definition of SD-WAN further summarizes its four characteristics, as listed in Table 11.4.

In traditional enterprise WAN scenarios, enterprise branches (sites) usually access the WAN over MPLS VPN private lines. As the enterprise traffic model changes, the WAN link bandwidth increases sharply, driving up enterprises' costs. This is why, in addition to MPLS VPN private lines, enterprises aim at using low-cost and high-quality links to carry a certain amount of traffic. As Internet coverage extends and network performance improves, network quality is getting significantly better. On top of that, the gap between Internet links and traditional private lines is quickly narrowing down. As a result, Internet links are becoming an alternative to carriers' expensive private lines.

According to the definition set by ONUG and Gartner, SD-WAN allows enterprise sites to be interconnected over different types of links (such as MPLS VPN, Internet, and LTE) across the WAN, as shown in Figure 11.19. Compared with the traditional WAN links for enterprises, hybrid WAN links provide more traffic transmission choices. For example, an enterprise can transfer traffic with low quality requirements from the MPLS VPN private line to the Internet for transmission. This not only lowers the spending on MPLS VPN private lines but also fully utilizes Internet resources.

In the traditional enterprise WAN environment, it might take one or more months just to connect a site to the WAN. Such a network deployment period is too long. To add on to this, service configurations need to be delivered one by one to the new site in a process that involves complex and error-prone operations. For an enterprise with large numbers of sites, service changes involve heavy maintenance workloads.

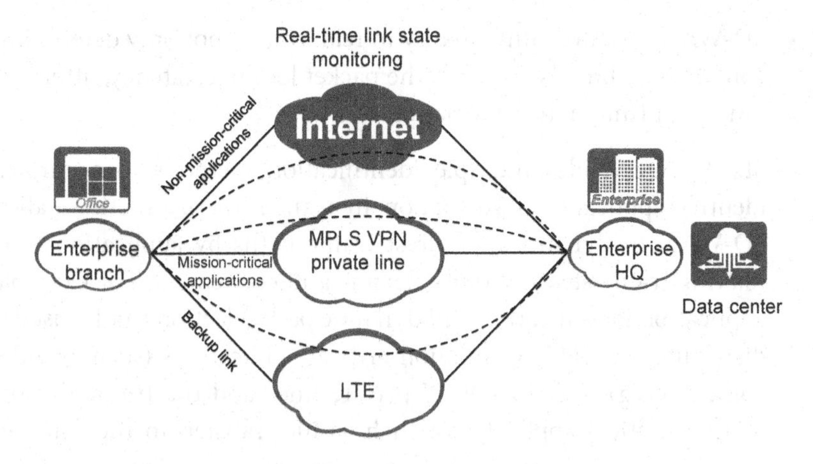

FIGURE 11.19 Hybrid WAN links.

SD-WAN solves this problem from the following perspectives:

- From the network perspective, SD-WAN introduces ZTP, among other means, to facilitate site deployment and network access. This plug-and-play deployment mode significantly lowers technical barriers, eliminating the need for professional network engineers to deploy services onsite and therefore saving labor costs. Another highlight of SD-WAN is its support for automatic setup of connections between sites through network orchestration, helping improve deployment efficiency.
- From the service perspective, SD-WAN integrates the functions of multiple hardware devices into one device through NFV, achieving fast service provisioning. The open SD-WAN architecture enables enterprise users to use VASs on demand.

Traditional enterprise WANs use routes to determine traffic paths. One drawback is that this traffic steering mode is rigid and does not support dynamic adjustment of traffic paths according to the network environment and link quality. Another drawback is that traditional enterprise WANs are unaware of applications and therefore cannot provide application-specific SLA assurance with varying link quality. When link quality deteriorates, the quality of mission-critical services cannot be guaranteed.

SD-WAN solves the preceding problems from the following aspects:

- SD-WAN monitors link quality in real time. It not only detects link connectivity but also records the packet loss rate, latency, jitter, and other real-time state information.
- SD-WAN provides multiple identification methods to accurately identify application information in traffic. These functions allow SD-WAN to dynamically adjust traffic paths by application type, making traffic steering more flexible and convenient. For example, application identification and dynamic path selection can be used to distribute the traffic of mission-critical applications (such as video conferencing) to MPLS VPN private lines and the traffic of non-mission-critical applications, such as file transfer, to the Internet. This meets the requirements of mission-critical applications on link quality and therefore delivers good application experience.

SD-WAN inherits SDN's centralized control design, which makes it possible to perform centralized management and control on the entire network. With centralized control, SD-WAN provides networkwide monitoring and can obtain site and link state and performance in real time for networkwide state visualization.

In addition to the preceding features, SD-WAN provides different O&M modes to limit users' access permissions to the bare minimum necessary to perform their work. Specifically, the access permissions can be granted by user role. To add on to that, SD-WAN integrates a wide array of fault diagnosis tools to quickly demarcate complex problems. Compared with traditional enterprise WANs, SD-WAN reduces management costs while improving O&M efficiency by providing automated and intelligent O&M capabilities.

11.3.2 SRv6-based SD-WAN

The IETF SPRING Working Group describes the SRv6 SD-WAN solution in a draft, providing a scenario where an SD-WAN controller works with an SRv6 controller to offer underlay SLA differentiation for the apps of a single carrier.[8]

11.3.2.1 Basic SRv6 SD-WAN Networking

As shown in Figure 11.20, the SD-WAN consists of sites A and Z, which are connected to the Internet through edge nodes E1 and E2, respectively.

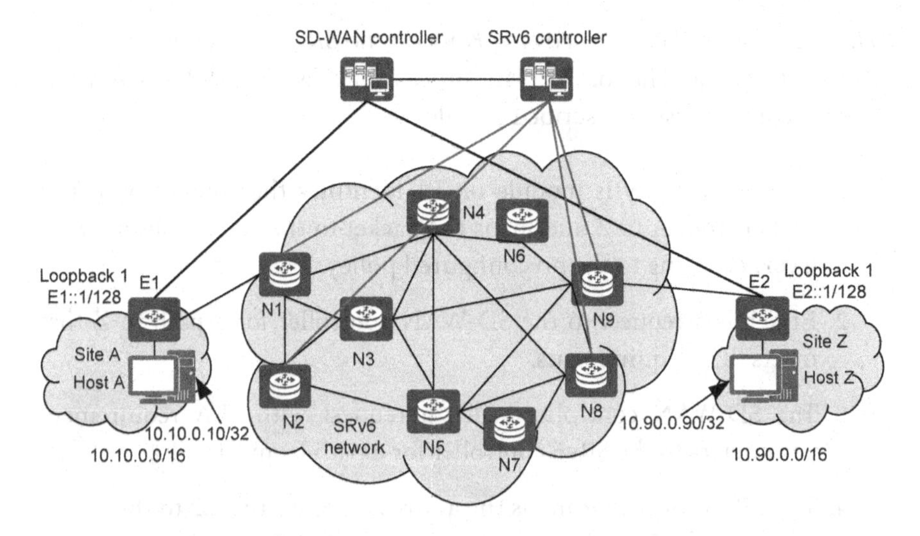

FIGURE 11.20 Basic SRv6 for SD-WAN networking.

E1 and E2 are connected over a Service Provider (SP) network to form a VPN between sites A and Z.

N1 to N9 on the SP network shown in Figure 11.20 are PEs. IS-IS SRv6 is deployed on the SP network to advertise SRv6 information, so that the network can support SRv6. Here, SRv6 SID Cj::1 refers to the End SID that has the PSP flavor on node Nj (j is a number ranging from 1 to 9).

An SRv6 controller is deployed on the SP network to compute constraint-based TE paths. The SD-WAN controller is used to manage the overlay network of the SD-WAN, and it can also communicate with the SRv6 controller to request services of various SLA levels.

E1 and E2 each have a loopback interface. The IPv6 addresses of the two loopback interfaces are E1::1/128 and E2::1/128, respectively. E1 and E2 connect to N1 and N9, respectively. The shortest path from N1 to N9 is a best-effort path computed by the IGP, and traffic from N1 to N9 travels the best-effort path by default.

Assume that the IP address of host A at site A is 10.10.0.10/32, and that of host Z at site Z is 10.90.0.90/32. E1 and E2 respectively advertise subnet addresses 10.10.0.0/16 and 10.90.0.0/16 to the SD-WAN controller over a secure channel (such as an IPsec tunnel) on the Internet. This solution applies to any traffic exchanged between sites, including IPv4, IPv6, and Layer 2 Ethernet service traffic. For convenience, we use IPv4 sites on the SD-WAN overlay network as an example in this book.

11.3.2.2 Control Plane Processing Flow on the SRv6 SD-WAN

Figure 11.21 shows the control plane processing flow on the SRv6 SD-WAN. The specific process is described as follows:

1. The traffic analysis module on E1 identifies the content of a user packet from A to Z and maps the packet to the corresponding SLA level according to the preconfigured policy.

2. E1 sends a request to the SD-WAN controller for a path to Z that meets SLA requirements.

3. The SD-WAN controller sends a request with SLA requirement parameters to the SRv6 controller for a path from E1 to E2.

4. The SRv6 controller maps the addresses of E1 and E2 to the nodes it manages (N1 and N9), computes an SRv6 Policy that meets the SLA requirements from N1 to N9, and allocates a binding SID. It then uses the BGP IPv6 SR Policy protocol extension to deliver the binding SID and the segment list of the qualified SRv6 Policy to N1.

5. N1 installs the SRv6 Policy and then sends the installation state to the SRv6 controller through BGP-LS.[9]

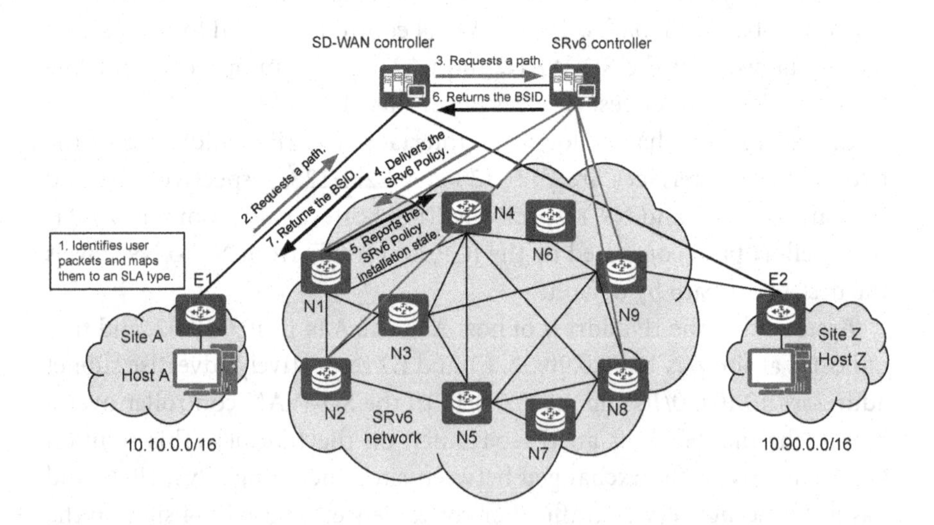

FIGURE 11.21 Control plane processing flow on the SRv6 SD-WAN.

6. After receiving a message indicating successful installation of the SRv6 Policy, the SRv6 controller returns the binding SID bound to the SRv6 Policy to the SD-WAN controller.

7. The SD-WAN controller returns the binding SID received from the SRv6 controller to E1. E1 generates an SRv6 Policy from E1 to E2. The corresponding segment list is <Binding SID, E2::1>. E1 also generates a traffic steering rule based on flow characteristics to steer target traffic to the SRv6 Policy.

11.3.2.3 SRv6 SD-WAN Data Forwarding Process

The SP network supports the creation of paths meeting different SLA requirements upon application requests. Figure 11.22 shows three different paths.

Three paths from N1 to N9 that meet different SLA requirements are used as an example.

1. The segment list of the default path (SRv6 BE) is <C9::>.

2. The segment list and binding SID of the low-latency path created upon application request are <C4::1, C6::1, C9::1> and C1::888:1, respectively.

3. The segment list and binding SID of the 100 Mbit/s path created upon application request are <C2::1, C5::1, C8::1, C9::1> and C1::888:2, respectively.

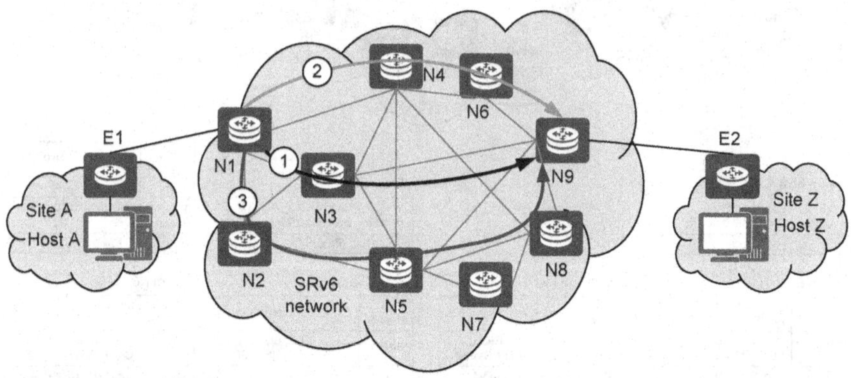

FIGURE 11.22 Creating multiple paths that meet different SLA requirements on the SP network.

After paths are created following the procedure described in Figure 11.22, traffic can be forwarded along these paths. Figure 11.23 shows the packet forwarding from host A to host Z over the low-latency path. In this example, an IPsec tunnel is deployed between E1 and E2.

1. Host A encapsulates source IP address 10.10.0.10 and destination IP address 10.90.0.90 into the IP header of an IP packet destined for host Z and forwards the packet to E1.
2. After receiving the packet, E1 processes the packet as follows:
   - Encrypts the packet and adds an ESP header to the packet.
   - Encapsulates an SRH into the packet. In the SRH, the segment list is <C1::888:1, E2::1>, SL is 1, and Next Header is ESP.
   - Finally, encapsulates an IPv6 packet header into the packet. In the IPv6 packet header, the IPv6 destination address is binding SID C1::888:1, the IPv6 source address is E1::1, and Next Header is SRH.

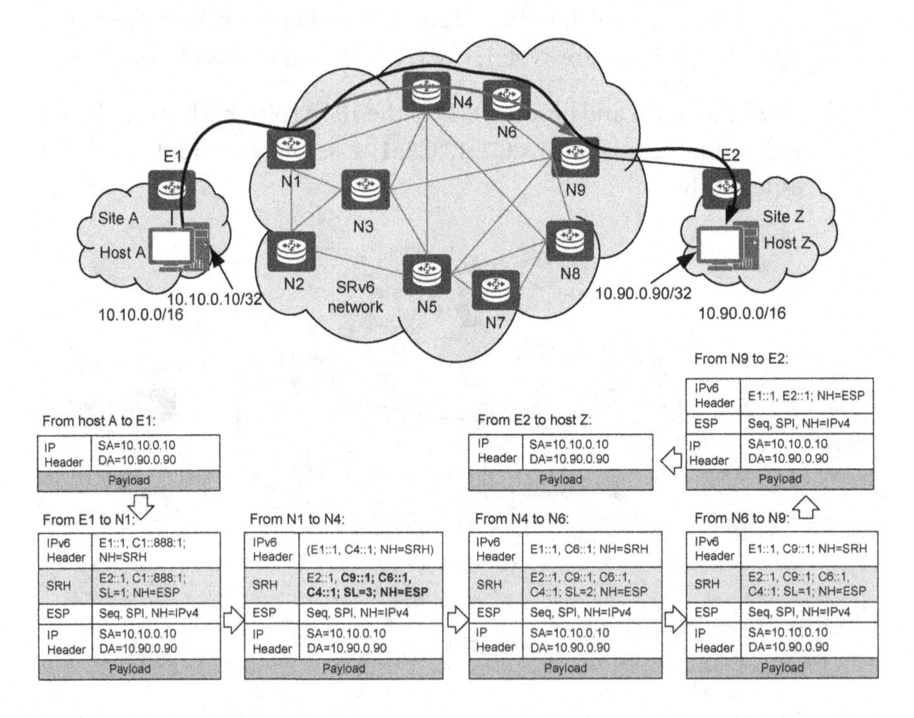

FIGURE 11.23 Packet forwarding process from host A to host Z over a low-latency path.

E1 then forwards the encapsulated packet to N1.

3. After the packet arrives at N1, N1 processes the binding SID C1::888:1 in the packet header, selects the SRv6 Policy deployed on N1 based on the binding SID, and inserts the segment list <C4::1, C6::1, C9::1> corresponding to the binding SID into the SRH of the packet. In the new SRH, the segment list is <C4::1, C6::1, C9::1, E2::1>, the SL is 3, and Next Header is ESP.

   N1 then updates the IPv6 header in the packet. In the IPv6 header, the destination IPv6 address is changed to C4::1, the source IPv6 address is still E1::1, and the Next Header is SRH.

4. Upon receipt of the packet, N4 searches the local SID table based on the destination address (C4::1) in the IPv6 header and finds that the address is a local End SID. According to the behavior defined in the End SID, N4 decrements the SL value by 1, updates the DA field in the outer IPv6 packet header with SID C6::1 indicated by the SL value, searches the IPv6 forwarding table for a matching entry, and sends the packet along the shortest path.

5. After receiving the packet, N6 searches the local SID table based on the destination address (C6::1) in the IPv6 header and finds that the address is a local End SID. According to the behavior defined in the End SID, N6 decrements the SL value by 1, updates the DA field in the outer IPv6 packet header with SID C9::1 indicated by the SL value, searches the IPv6 forwarding table for a matching entry, and sends the packet along the shortest path.

6. After the packet arrives at N9, N9 searches the local SID table based on the destination address (C9::1) in the IPv6 header and finds that the address is a local End SID. According to the behavior defined in the End SID, N9 decrements the SL value by 1 and updates the DA field in the outer IPv6 packet header with SID E2::1 indicated by the SL value. Because N9 is the penultimate segment on the path, it pops out the SRH according to PSP. N9 then searches the IPv6 routing table for a matching entry and forwards the packet to E2 along the shortest path.

7. Upon receipt of the packet, E2 finds that the destination IPv6 address of the packet is its SID E2::1. E2 then removes the IPv6 packet to

expose the ESP packet header. E2 further processes the ESP encapsulation. After the ESP authentication succeeds, E2 forwards the packet to host Z according to destination address 10.90.0.90 in the inner packet header.

11.3.2.4 Advantages of the SRv6 SD-WAN Solution

To sum up, the SRv6 SD-WAN solution offers the following benefits:

1. High scalability
   - The SP network does not maintain any per-SD-WAN-flow state in the core of the network.
   - The SP network does not perform any complex classification on Layer 4 to Layer 7 flows at the network edge.
   - The SP network is unaware of any policy change of the SD-WAN instance in terms of which flow to classify, as well as when and to which path to steer the flow.
   - The SP network statefully maintains SRv6 Policies only at the network edge and maintains hundreds of SIDs on the network, fully utilizing SR's stateless property.
2. Strong privacy protection
   - The SP network does not share any information about its infrastructure, topology, capacity, and internal SIDs.
   - The SD-WAN instance does not share any information about its traffic classification, steering policies, and service logic.
3. Flexible billing
   - The traffic sent to the binding SID can be individually accounted. The SP network and SD-WAN instance can reach an agreement on the billing methods for using the preferred path.
4. Guaranteed security
   - A binding SID (and related preferred path) can only be accessed by the specific SD-WAN instance (and site) that has subscribed to the service.

- The security solution supports any SD-WAN site connection type and allows SD-WAN sites to directly connect to various SP network edges.

11.4 STORIES BEHIND SRv6 DESIGN

The intensifying competition between DCN technologies serves as the spark igniting continuous technology evolutions. DCNs have witnessed the replacement of traditional Layer 2 networking technologies with large Layer 2 networking technologies, such as Transparent Interconnection of Lots of Links (TRILL) and Shortest Path Bridging for MAC (SPBM), and replacement of large Layer 2 networking technologies with Network Virtualization over Layer 3 (NVO3) networking technologies (such as VXLAN). What might be somewhat perplexing is the fact that MPLS, which has always been the preferred technology for various scenarios, has somehow failed to become the mainstream technology in the DCN solution, despite the fact that MPLS experts made tremendous efforts to promote MPLS in the DCN field. Put differently, MPLS has a proven track record and is widely deployed on different types of transport networks. It, therefore, goes without saying that one would expect MPLS to be the natural choice for DCNs. Along those lines, in around 2014, we held discussions with MPLS industry experts on how to promote the seamless MPLS solution to integrate data center networks with WANs. Some experts even proposed using the ARP label distribution solution to solve the problem of MPLS label distribution at the last hop.[10] Even so, these efforts failed to bring any substantive changes, and only MPLS over UDP won a certain level of support.[11,12] Ultimately, VXLAN outshone MPLS to become the de facto DCN standard. In my opinion, the reasons for this outcome can be summarized as follows:

1. Competition in the industry ecosystem: Data centers consist of compute, storage, and network devices, and the data center market is almost dominated by IT vendors, as opposed to network device vendors. For this reason, IT vendors have a bigger say in DCN construction, and quite early on, they realized the technical trend involving data center cloudification and network virtualization, and developed the VXLAN technology which consequently gained more support.

2. Weak standard binding force and low operational efficiency: A DCN generally consists of devices from a single vendor and does

not have high interoperability requirements. The NVO3 Working Group established by the IETF encountered difficulty in advancing the standardization process due to disputes between different parties and the lack of effective outputs. The IETF even once considered disbanding this working group.

3. User habits: DCNs require simple, efficient service deployment and O&M. Network management personnel generally believe that MPLS is too complex to use on DCNs. To add on to that, data centers use switches for short-distance networking and the bandwidth costs are low. In light of this, as ECMP can meet the quality requirements of most services, the need for MPLS TE has essentially been eliminated. This has also hindered the application of MPLS on DCNs.

MPLS has not been used on DCNs due to the preceding reasons. This has led to increasing divergence between IP networks, that is, there is not only the existing issue to build MPLS networks across domains, but also the interconnection problems between VXLAN and MPLS networks. That being said, the opportune emergence of SRv6 makes it possible to unify IP networks. More precisely, like VXLAN, SRv6 has the native IP attribute and can be used as a substitute for the DCN technology. We can even take a step further and use SRv6 to replace MPLS on the WAN. Taking all this into account, SRv6 can substitute traditional network technologies to better provide E2E services. This vision, however, depends on industry ecosystem development. In other words, the difficulty in applying SRv6 to DCNs lies in VXLAN ecosystem development, which is similar to the difficulty that SRv6 faces in the mobile core network caused by the GTP ecosystem. We still have quite a way to go before we solve the industry ecosystem problem and unify network protocols. Nevertheless, according to certain reports, a Japanese company has already started to deploy the SRv6-based DCN solution. It is therefore safe to say that as the industry widely recognizes and welcomes SRv6, it is entirely within the realm of possibilities that we will realize this vision one day.

REFERENCES

[1] Quinn P, Nadeau T. Problem Statement for Service Function Chaining[EB/OL]. (2018-12-20)[2020-03-25]. RFC 7498.

[2] Napper J. NSH Context Header Allocation for Broadband[EB/OL]. (2018-12-21)[2020-03-25]. draft-ietf-sfc-nsh-broadband-allocation-01.

[3] Quinn P, Elzur U, Pignataro C. Network Service Header (NSH)[EB/OL]. (2020-01-21)[2020-03-25]. RFC 8300.
[4] Farrel A, Drake J, Rosen E, Uttaro J, Jalil L. BGP Control Plane for NSH SFC[EB/OL]. (2019-12-19)[2020-03-25]. draft-ietf-bess-nsh-bgp-control-plane-13.
[5] Clad F, Xu X, Filsfils C, Bernier D, Li C, Decraene B, Ma S, Yadlapalli C, Henderickx W, Salsano S. Service Programming with Segment Routing[EB/OL]. (2019-11-04)[2020-03-25]. draft-ietf-spring-sr-service-programming-01.
[6] Guichard J, Song H, Tantsura J, Halpern J, Henderickx W, Boucadair M, Hassan S. Network Service Header (NSH) and Segment Routing Integration for Service Function Chaining (SFC)[EB/OL]. (2019-10-04)[2020-03-25]. draft-ietf-spring-nsh-sr-01.
[7] Dawra G, Filsfils C, Talaulikar K, Clad F, Bernier D, Uttaro J, Decraene B, Elmalky H, Xu X, Guichard J, Li C. BGP-LS Advertisement of Segment Routing Service Segments[EB/OL]. (2020-01-07)[2020-03-25]. draft-dawra-idr-bgp-ls-sr-service-segments-03.
[8] Dukes D. SR For SDWAN: VPN with Underlay SLA[EB/OL]. (2019-12-12)[2020-03-25]. draft-dukes-spring-sr-for-sdwan-02.
[9] Previdi S, Talaulikar K, Dong J, Chen M, Gredler H, Tantsura J. Distribution of Traffic Engineering (TE) Policies and State using BGP-LS[EB/OL]. (2019-10-14)[2020-03-25]. draft-ietf-idr-te-lsp-distribution-12.
[10] Kompella K, Balaji R, Thomas R. Label Distribution Using ARP[EB/OL]. (2020-03-07)[2020-03-25]. draft-kompella-mpls-larp-07.
[11] Xu X, Sheth N, Yong L, et al. Encapsulating MPLS in UDP[EB/OL]. (2020-01-21)[2020-03-25]. RFC 7510.
[12] Xu X, Bryant S, Farrel A, et al. MPLS Segment Routing over IP[EB/OL]. (2019-12-06)[2020-03-25]. RFC 8663.

CHAPTER 12

SRv6 Multicast/BIERv6

THIS CHAPTER INTRODUCES BIT Index Explicit Replication IPv6 Encapsulation (BIERv6), a fast-developing next-generation multicast technology. Compared with conventional multicast technologies, BIERv6 simplifies the control-plane protocols and allows the network programming by the ingress to steer multicast packets to multiple destination nodes. As such, it is suitable for both current and future large-scale network deployment. Combined with SRv6, BIERv6 allows both unicast and multicast services to be implemented based on the unified IPv6 data plane and the unified IGP/BGP control plane, thereby simplifying protocols. This is aligned with the development trend of IP networks.

12.1 OVERVIEW OF MULTICAST TECHNOLOGIES AND SERVICES

IP multicast implements P2MP real-time data transmission on IP networks and is extensively used for the services (such as IPTV) on carrier IP networks.

IP multicast protocols can be classified into multicast group member management protocols and multicast routing protocols based on their application. Multicast group member management protocols run between a host and a router, and include IGMP[1,2] and Multicast Listener Discovery (MLD).[3] Multicast routing protocols run between routers and mainly include Protocol Independent Multicast (PIM),[4] Multicast VPN (MVPN),[5] BIER,[6] etc.

This section briefly introduces basic concepts of PIM and MVPN, as well as the evolution of multicast from PIM to MVPN to BIER.

12.1.1 PIM

PIM is the most basic and widely used multicast routing protocol. In PIM, group join information of multicast receivers is collected, based on which PIM Join messages are sent to the multicast source hop by hop to establish a Multicast Distribution Tree (MDT). The goal is to forward the multicast packets sent by the multicast source to the multicast receivers along the MDT.

Figure 12.1 shows how an MDT is established. (S, G) indicates the multicast group G with the multicast source S.

The details of this process are as follows:

1. IGMP report: Multicast receivers H1 and H2 send IGMP (S, G) Report messages to Last Hop Routers (LHRs) C and D, respectively. Like routers C and D in this example, the routers that receive IGMP Report messages from multicast receivers (hosts) are also referred to as receiver PEs or receiver DRs.
2. PIM join: After receiving the IGMP (S, G) Report messages, routers C and D each search for a route to multicast source S and send a PIM (S, G) Join message to the next hop of the route, which in this instance is router B.

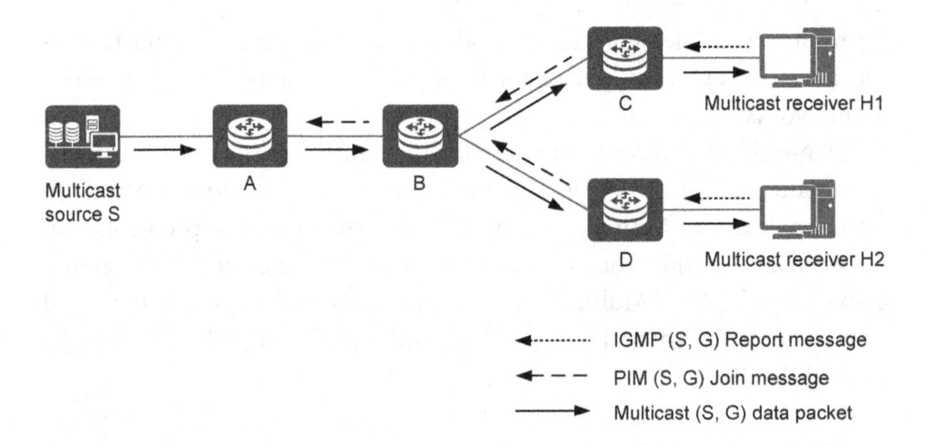

FIGURE 12.1 MDT establishment.

3. PIM join: After receiving the PIM (S, G) Join message, router B searches for a route to the multicast source address in the Join message and forwards the message upstream to router A, which is connected to the multicast source and also referred to as the First Hop Router (FHR), multicast source PE, or multicast source DR. The multicast forwarding path established between the FHR and LHR using PIM is called an MDT.

4. Multicast packet replication and sending: After receiving a multicast (S, G) data packet from the multicast source, the FHR forwards it along the MDT. The packet is forwarded along the MDT until it reaches each LHR, which then replicates the packet and sends a copy to all connected multicast receivers. In this manner, the multicast data packet is distributed from the multicast source to multicast receivers.

The multicast process in which both the multicast source and the multicast group are specified in IGMP Report and PIM Join messages is referred to as Source Specific Multicast (SSM) or PIM-SSM.[4] By way of comparison, the multicast process in which only a multicast group is specified (no multicast source specified), presented as (*, G), in IGMP Report messages or PIM Join messages is referred to as Any-Source Multicast (ASM) or PIM-SM.[4]

The process of PIM-SM is more complex than PIM-SSM. In PIM-SM, multicast receivers know the multicast group but not the multicast source address. Therefore, the IGMP Report messages sent by the multicast receivers carry the multicast group G, but no the multicast source S. This requires PIM to discover the multicast source corresponding to the multicast group through a Rendezvous Point (RP), which functions as an intermediary to advertise multicast group and source information and find the multicast source for a multicast group.

Figure 12.2 shows the process for a multicast receiver to join a multicast group and receive multicast traffic in ASM mode.

The details of this process are as follows:

1. After receiving multicast traffic, the multicast source DR (router A) notifies the RP (router B) of the (S, G) information through a Register message. The RP then parses the Register message to learn

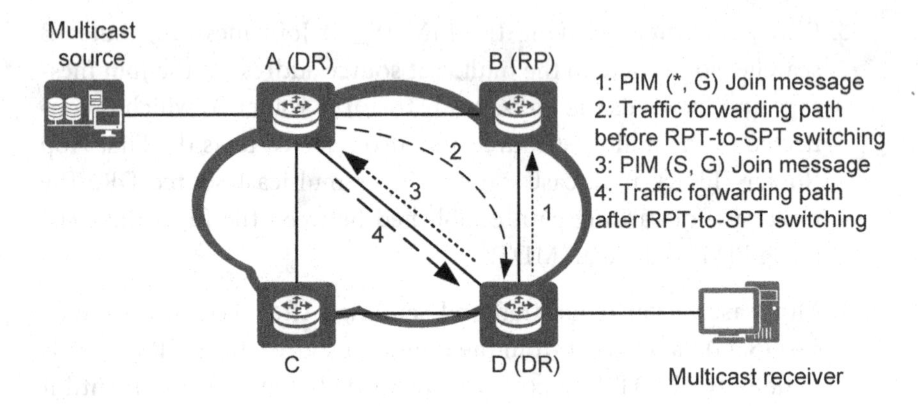

FIGURE 12.2 Multicast processing in PIM-ASM mode.

the multicast source S corresponding to the multicast group G. After receiving an IGMP (*, G) Report message from the multicast receiver, the receiver DR (router D) sends a PIM (*, G) Join message to the RP corresponding to the multicast group G. After the RP receives this message, a Rendezvous Point Tree (RPT) from the RP to the receiver DR is established.

2. Based on the received (*, G) Join message and the stored mapping between the multicast source S and multicast group G, the RP searches for a route to the multicast source S and then sends a PIM (S, G) Join message to the next hop of the route. Subsequently, the PIM (S, G) message is sent to the FHR hop by hop, and then an MDT from the FHR to the RP is established.

3. The FHR sends the received multicast data packets along the MDT to the RP, which forwards them along the RPT. The packets are then forwarded until they reach the LHR, which replicates them and sends a copy to the multicast receiver. After receiving the multicast data packets, the LHR parses them to obtain information about the multicast source S corresponding to the multicast group G. It then searches for a route to the multicast source S, and sends a PIM (S, G) Join message to the next hop of the route. After the multicast source receives this message, an SPT from the FHR to the LHR is established. It is worth pointing out that the process of obtaining the multicast source S from data packets is data-driven.

4. After the RPT and SPT are established, the LHR receives two copies of multicast data packets. In this case, the LHR performs a Reverse

Path Forwarding (RPF) check to match the interfaces that receive the multicast data packets against the RPF inbound interface (the one connected to the RP). As a result, the LHR selects the packets received through the RPT and discards those received through the SPT. Because the RPT is not the shortest path on the network, the LHR must perform RPT-to-SPT switching. This switching is initiated by the LHR changing the RPF inbound interface to the interface pointing to the multicast source S, and then sending a PIM (S, G, RPT) Prune message to the RP. After receiving this message, the RP prunes the RPT (from the RP to the LHR), after which the LHR receives only one copy of multicast data packets (through the SPT). The aim of pruning is to optimize forwarding of multicast data packets.

Considering the preceding process, the aspects where PIM-SM is more complex than PIM-SSM are expressed as follows:

1. PIM-SM involves a series of complex processes, including RPT establishment, SPT establishment, RPT-to-SPT switching, and RPT teardown.

2. PIM-SM requires each multicast group address to be configured with an RP as an intermediary, with all network nodes needing to be aware of which RP each multicast group uses.

3. PIM-SM requires multiple RPs to be deployed to serve different multicast groups for load balancing and high availability purposes.

4. PIM-SM supports the automatic RP election mechanism if manual RP configuration is not used due to the heavy workload involved. However, automatic RP election requires a BootStrap Router (BSR) to flood RP information, thus further increasing the complexity of PIM-SM.[4]

12.1.2 MVPN

With the emergence of IP/MPLS VPN services, MVPN was designed to run different multicast services concurrently while also isolating them from each other.

Early on, MVPN was designed based on PIM and IP GRE encapsulation. In the data plane, after receiving a multicast packet from a VPN interface, a PE of an MVPN encapsulated the packet with an outer IP header (using a multicast address as the destination address) and a GRE header. Following this, the PE sent the encapsulated multicast packet to other PEs of the MVPN along the MDT. This type of technology is called Rosen MVPN,[7] and GRE packet encapsulation in which a multicast address is used as the destination address in the outer IP header is called Multipoint Generic Routing Encapsulation (mGRE) encapsulation.

With the development of MPLS technologies, some carriers require a converged MPLS transport network, on which the MPLS data plane carries all services including multicast services, with the following improvements to Rosen MVPN:

1. Multicast VPN tunnels use MPLS encapsulation in the same way as unicast VPN tunnels, while also using the encapsulated MPLS label stack to implement multicast FRR.

2. Control plane protocols of MVPN are extended based on existing MPLS unicast protocols, including LDP[8] and RSVP-TE,[9] from which control plane protocols for multicast MPLS tunnels, Multipoint Extensions for Label Distribution Protocol (mLDP),[10] and RSVP-TE Point to Multipoint (P2MP)[11] are introduced. The multicast tunnels established using mLDP or RSVP-TE P2MP are referred to as P2MP Label Switched Paths (LSPs), and the multicast based on P2MP LSPs is referred to as Label Switched Multicast (LSM).

3. BGP is used to advertise information, such as PE address information, VPN multicast join information, and VPN multicast source discovery information, between MVPN sites.

This improved MVPN is referred to as Next-Generation MVPN (NG MVPN). Unless otherwise stated, MVPN in the following sections of this book refers to NG MVPN. To support NG MVPN, BGP uses the BGP MVPN address family to exchange BGP MVPN messages in NG MVPN.

Figure 12.3 shows the typical networking and working model of MVPN.

12.1.2.1 MVPN Routes

MVPN routes are advertised between PEs on an MVPN using BGP MVPN[12] and are classified into three categories based on their functions.

In the first category, routes are used to transmit VPN multicast group join information. After receiving a PIM Join or IGMP/MLD Report

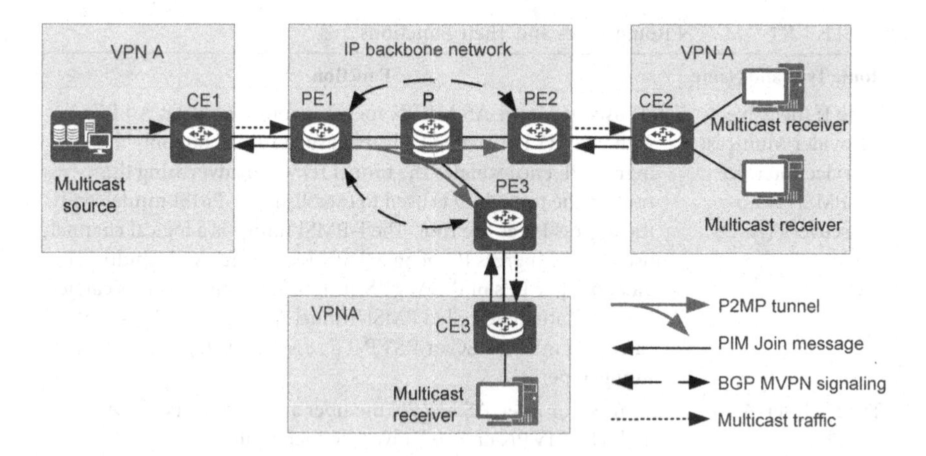

FIGURE 12.3 Typical networking and working model of MVPN.

message through a VPN interface, a leaf PE (receiver PE) sends a join request through a BGP route to the ingress PE (source PE), which then sends a PIM Join message through its VPN interface connected to the multicast source. The Type 6 and Type 7 routes listed in Table 12.1 belong to this category.

In the second category, routes are used for multicast source discovery. This involves the ingress PE sending the (S, G) information to other PEs (egress PEs) of the MVPN. After receiving multicast (*, G) join requests, each egress PE sends a BGP route carrying (S, G) information to the ingress according to the stored (S, G) information. The Type 5 route listed in Table 12.1 belongs to this category.

In the third category, routes are used to advertise IP addresses and tunnel IDs for PEs to initiate MPLS P2MP tunnel establishment. The Type 1, Type 2, Type 3, and Type 4 routes listed in Table 12.1 belong to this category. It is also worth noting that the exchange processes involved in establishing tunnels of different types may be different.

For some types of tunnels, leaf PEs are responsible for initiating tunnel establishment. For example, if an mLDP P2MP tunnel needs to be established, the ingress PE advertises a Type 1 or Type 3 route carrying the Forwarding Equivalence Class (FEC)[10] of the mLDP P2MP tunnel, but no Leaf Information Required (LIR)[12] flag. Based on the ingress PE address in the FEC, leaf PEs directly initiate an mLDP P2MP tunnel join request (through an mLDP Mapping message), without sending a Type 4 route to the ingress PE. In this process, the message is forwarded hop by hop, and

TABLE 12.1 MVPN Route Types and Their Functions

Route Type and Name	Function
Type 1: Inclusive-Provider Multicast Service Interface (I-PMSI) Auto-Discovery (A–D)	It is used for intra-AS MVPN member auto-discovery. All PEs with NG MVPN enabled advertise such routes, but only the ingress PE encapsulates the tunnel ID when advertising the routes. The tunnel ID is used to establish an I-PMSI tunnel, with the ingress PE as the root. The I-PMSI tunnel is a logical channel used by the ingress PE of an MVPN to transmit VPN multicast data to other PEs of the MVPN instance. The tunnel ID is carried in a BGP attribute called PMSI Tunnel Attribute (PTA), and it can be an mLDP FEC or RSVP-TE session ID, depending on the tunnel type.
Type 2: Inter-AS A–D	It is used for inter-AS MVPN member auto-discovery. All ASBRs with NG MVPN enabled advertise such routes.
Type 3: Selective Provider Multicast Service Interface (S-PMSI) A–D	It is used by the ingress PE to collect information about the PEs that need to receive multicast data. Based on the information, a tunnel is established. When advertising the route, the multicast ingress PE specifies (*, G) or (S, G) and adds an mLDP or RSVP-TE P2MP tunnel ID to the route. The leaf PEs that need to receive the (*, G) or (S, G) traffic then join the established tunnel, which is called a Selective PMSI (S-PMSI) tunnel. The ingress PE then sends the corresponding (*, G) or (S, G) traffic along the S-PMSI tunnel only to these leaf PEs.
Type 4: Leaf A–D	It is used by an egress PE to respond to the ingress PE's S-PMSI A–D route in which the PTA carries the LIR flag. The ingress PE sends the S-PMSI A–D route when it needs to explicitly trace egress PEs to deliver forwarding entries or initiate tunnel establishment.
Type 5: Source Active A–D	It is used by the ingress PE to advertise multicast source information to other MVPN PEs when the ingress PE is aware of a new multicast source.
Type 6: Shared Tree Join	It is advertised by a receiver PE to the ingress PE after the receiver PE receives an (*, G) join request from the user side.
Type 7: Source Tree Join	It is advertised by an egress PE to the ingress PE after the egress PE receives an (S, G) join request, or receives an (*, G) join request when the egress PE has information about the multicast source corresponding to the multicast group.

each hop determines which upstream node the message is sent to based on the IP address of the ingress PE in the FEC.

For other types of tunnels, the ingress is responsible for initiating tunnel establishment. For example, if an RSVP-TE P2MP tunnel needs to be established, the ingress PE advertises a Type 1 or Type 3 route carrying an RSVP-TE P2MP tunnel's control plane identifier (RSVP Session ID)[11] and the LIR flag. Based on this flag, each leaf PE sends a Type 4 route carrying its IP address to the ingress PE, which then initiates the establishment of a P2MP tunnel to leaf PEs.

Table 12.1 describes the seven MVPN route types and their functions.

12.1.2.2 MVPN over an mLDP P2MP Tunnel

Figure 12.4 shows MVPN signaling exchanges over an mLDP P2MP tunnel.

The details of this process are as follows:

1. The ingress PE and egress PEs advertise I-PMSI A-D routes to each other. The I-PMSI A-D routes advertised by the ingress PE carry an mLDP FEC,[10] which is used as the I-PMSI tunnel ID by egress PEs to initiate the establishment of an mLDP P2MP tunnel to the ingress PE.

2. Based on the mLDP FEC, each egress PE sends an mLDP Mapping message to the ingress PE so that an mLDP P2MP tunnel is established, with the ingress PE serving as the root and the egress PEs as leaf nodes.

3. The ingress PE usually functions as the RP of the MVPN. After receiving the data traffic from the multicast source or a PIM Register

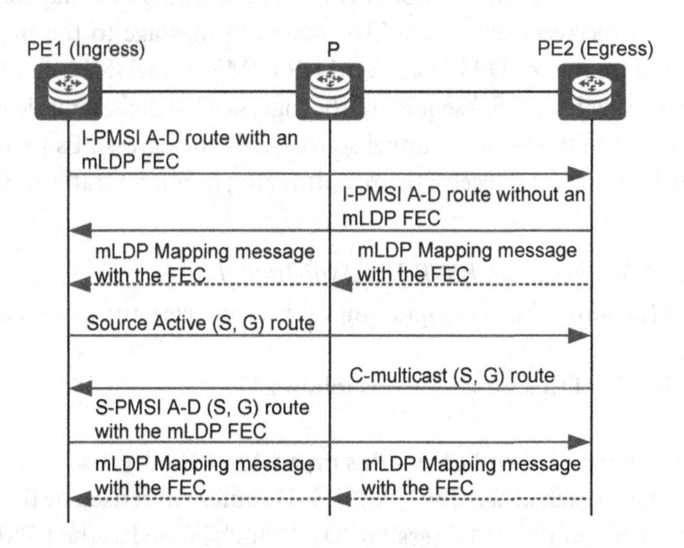

FIGURE 12.4 MVPN signaling exchanges over an mLDP P2MP tunnel.

message from the FHR, the ingress PE obtains information about the (S, G), and then advertises it to each egress PE through a Source Active (S, G) route.

4. After receiving a join request from a multicast receiver or a PIM Join message from a CE on the MVPN, an egress PE advertises a Source Tree Join (S, G) route (C-multicast routes include Share Tree Join routes and Source Tree Join routes, and the latter is used as an example in this section) to the ingress PE. If the join request received by the egress PE is of the (*, G) type, the egress PE obtains information about the multicast source corresponding to the multicast group based on the received Source Active (S, G) route and sends a Source Tree Join (S, G) route to the ingress PE. With the Source Active route, this process neither requires an RP in the MVPN nor involves RPT establishment or RPT-to-SPT switching. After receiving the Source Tree Join (S, G) route, the ingress PE sends a PIM (S, G) Join message to its upstream CE, which directs multicast traffic to the ingress PE. The ingress PE then sends the multicast traffic to each egress PE through the I-PMSI tunnel.

5. The ingress PE advertises an S-PMSI A-D (S, G) route carrying an mLDP FEC as an S-PMSI tunnel ID.

6. After receiving the S-PMSI A-D route, the egress PEs that connect to (S, G) receivers send an mLDP Mapping message to the ingress PE based on the mLDP FEC. An mLDP P2MP tunnel (S-PMSI tunnel) is then established. Subsequently, the ingress PE switches multicast (S, G) packets to the S-PMSI tunnel so that only the egress PEs that require such packets can receive them, optimizing multicast traffic replication.

12.1.2.3 MVPN over an RSVP-TE P2MP Tunnel

Figure 12.5 shows MVPN signaling exchanges over an RSVP-TE P2MP tunnel.

The details of this process are as follows:

1. The ingress PE and egress PEs on an MVPN advertise I-PMSI A–D routes to each other. The I-PMSI A–D routes advertised by the ingress PE carry an RSVP-TE session ID,[11] which is used as the I-PMSI tunnel ID by the ingress PE to initiate the establishment of an RSVP-TE P2MP tunnel to each egress PE.

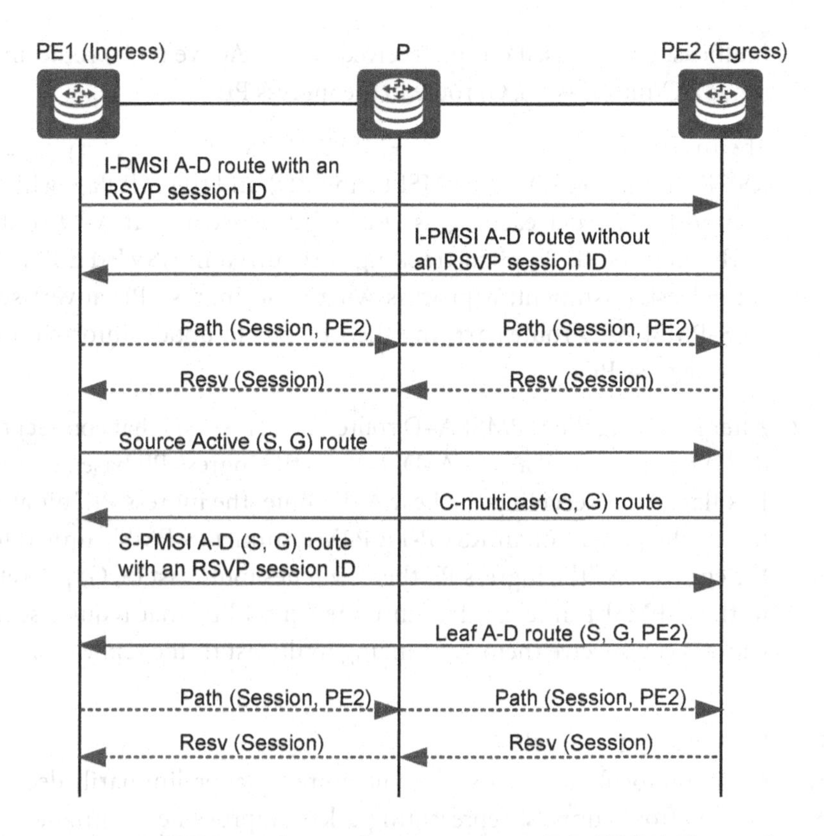

FIGURE 12.5 MVPN signaling exchanges over an RSVP-TE P2MP tunnel.

2. The ingress PE sends RSVP Path messages to the egress PEs based on the received I-PMSI A–D routes. The egress PEs respond with RSVP Resv messages, and a P2MP tunnel is then established, with the ingress PE serving as the root and the egress PEs as leaf nodes.

3. The ingress PE usually functions as the RP of the MVPN. After receiving the data traffic from the multicast source or a PIM Register message from the FHR, the ingress PE obtains information about the (S, G) and then advertises it to each egress PE through a Source Active (S, G) route.

4. After receiving a join request from a multicast receiver or a PIM Join message from a CE on the MVPN, an egress PE advertises a C-multicast (S, G) Join route to the ingress PE. If the join request that the egress PE receives from the multicast receiver is of the (*, G) type, the egress PE obtains information about the multicast source corresponding to the

multicast group based on the received Source Active (S, G) route and sends a C-multicast (S, G) route to the ingress PE.

5. The ingress PE advertises an S-PMSI A–D (S, G) route carrying an RSVP-TE session ID[11] (S-PMSI tunnel ID) and the LIR flag, which is used to instruct each egress PE to advertise a Leaf A–D route to the ingress PE. Explicit tracking is required in RSVP-TE P2MP tunnel establishment, a process where the ingress PE advertises an S-PMSI A–D route carrying the LIR flag to obtain information about egress PEs.

6. After receiving the S-PMSI A–D route, the egress PEs that connect to (S, G) receivers send a Leaf A–D route to the ingress PE based on the LIR flag. After receiving the Leaf A–D route, the ingress PE initiates the establishment of an RSVP-TE P2MP tunnel (S-PMSI tunnel) to the egress PEs. The ingress PE then switches multicast (S, G) packets to the S-PMSI tunnel so that only the egress PEs that require such packets can receive them, optimizing multicast traffic replication.

12.1.2.4 Advantages of MVPN

The MVPN protocol establishes an architecture[5] that preliminarily decouples services from tunnels, representing a key improvement to multicast technologies. The technical advantages of the MVPN protocol are as follows:

1. MVPN allows multicast data packets to be encapsulated with outer tunnel information to support multiple multicast VPN services, which are isolated from each other and Internet access services.

2. BGP MVPN signaling is used to support multiple types of tunnels. Based on the MPLS data plane, MVPN extends the existing LDP and RSVP-TE protocols to form P2MP tunnels, thereby implementing converged MPLS transport.

3. MVPN supports I-PMSI and S-PMSI tunnels. It allows multiple multicast flows of the same VPN to be carried over one I-PMSI tunnel or each multicast flow to be carried over one S-PMSI tunnel.

4. Multicast join requests can be directly sent from a leaf PE to the ingress PE through BGP MVPN signaling, which is more efficient and less drawn-out.

12.1.2.5 Disadvantages of MVPN

MVPN defines a multicast service architecture in which the multicast services of PE sites are decoupled from the tunnel type and signaling of the IP backbone network. On a converged MPLS transport network in particular, MVPN supports MPLS encapsulation as unicast VPN does, uses the signaling extended based on the unicast LDP and RSVP-TE protocols, and leverages FRR in MPLS. That being said, as some problems emerge in MPLS and networks evolve toward SR and SRv6, the disadvantages of MPLS P2MP-based MVPN are becoming more and more prominent. These shortcomings are as follows:

1. An MVPN uses one I-PMSI tunnel to carry multiple multicast flows of a VPN, wasting traffic bandwidth resources.
2. Even though MVPN can prevent traffic bandwidth waste by using one S-PMSI tunnel to transmit each multicast flow, a dedicated P2MP tunnel needs to be established for each multicast flow using mLDP or RSVP-TE. Consequently, as the number of multicast flows increases, there is a need for more P2MP tunnels. In turn, this increases the network costs and increases the service recovery time if a link failure occurs.
3. Both I-PMSI and S-PMSI tunnels need to use signaling (such as mLDP or RSVP-TE) to establish P2MP tunnels on the network. This is not ideal as the signaling for tunnel establishment is complex, and BGP MVPN needs to interact with the protocol responsible for tunnel establishment. After a tunnel ID is obtained and carried in a BGP MVPN message, BGP MVPN then instructs mLDP or RSVP-TE to establish a tunnel.
4. The MDT establishment through mLDP is complex. Specifically, mLDP on the ingress PE needs to generate a tunnel ID (FEC) and advertise the FEC to each egress PE through BGP. Then, mLDP on each egress PE searches for a route based on the root IP address in the FEC and sends an mLDP Mapping message to the upstream node. The message is then forwarded hop by hop to the ingress PE to establish a P2MP MDT.
5. P2MP MDTs established through RSVP-TE are maintained based on the RSVP soft state.[13] Because RSVP-TE periodically sends Path

and Resv signaling,[14] this gives rise to issues of high costs and low scalability specifications.

6. Multicast service maintenance is complex, requiring operators to be familiar with not only BGP MVPN signaling but also mLDP and RSVP-TE. They must also be proficient in locating failures hop by hop on the network. On transit nodes, operators can view the MPLS labels of P2MP tunnels and tunnel IDs, such as mLDP FECs or RSVP session IDs. As the mapping between the label or tunnel ID and the specific (S, G) is indirect and complex, repeated table lookup is required.

12.1.3 BIER

Regardless of whether PIM is used to transmit non-VPN multicast services or mLDP/RSVP-TE is used to transmit MVPN services, an MDT needs to be established explicitly for each multicast flow, and transit nodes (such as Ps) must also be service-aware.

Figure 12.6 shows the transmission of IPTV services in multicast mode.

An MDT needs to be established explicitly for each multicast flow. In view of this, MDTs need to be established on demand upon requests of IPTV services, which involves steps 2, 3, and 4 in Figure 12.6.

If a link fails (at point 5 in Figure 12.6), MDTs must be reestablished through Join message transmission, involved in step 6 in Figure 12.6. For example, if there are 1,000 multicast flows, 1,000 MDTs must be reestablished.

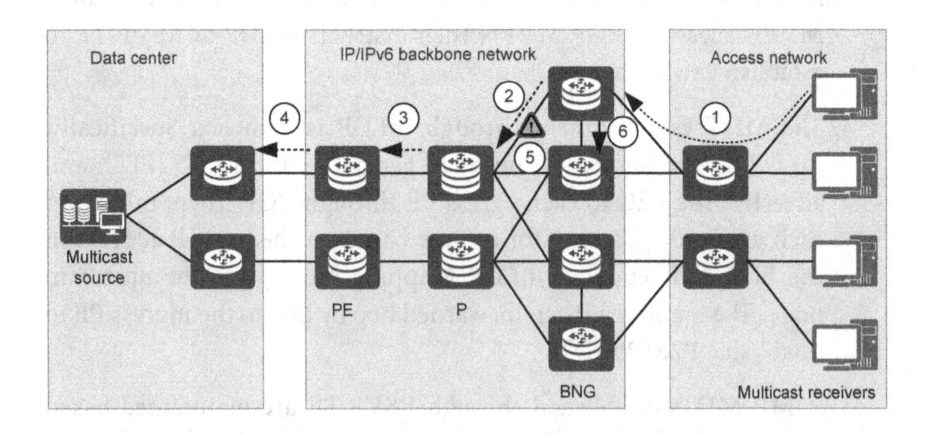

FIGURE 12.6 Transmission of IPTV services in multicast mode.

Although MVPN establishes a protocol architecture that supports multiple multicast services, if mLDP or RSVP-TE is used to establish MDTs explicitly for these services, the number of multicast flows supported is limited because an MDT needs to be established for each multicast flow, and transport and services are not thoroughly decoupled. As a result, carriers cannot provide a large number of multicast MVPN services for enterprises.

BIER overcomes this problem, as it does not depend on protocols that require explicit MDT establishment, such as PIM, mLDP, and RSVP-TE. On top of that, it meets technical requirements, such as network simplification as well as transport and service decoupling. The IETF set up the BIER Working Group in 2014 and has released the following standards:

- RFC 8279[6] — Multicast Using Bit Index Explicit Replication (BIER)
- RFC 8296[15] — Encapsulation for Bit Index Explicit Replication (BIER) in MPLS and Non-MPLS Networks
- RFC 8401[16] — Bit Index Explicit Replication (BIER) Support via IS-IS
- RFC 8444[17] — OSPFv2 Extensions for Bit Index Explicit Replication (BIER)
- RFC 8556[18] — Multicast VPN Using Bit Index Explicit Replication (BIER)
- RFC 8534[19] — Explicit Tracking with Wildcard Routes in Multicast VPN

The BIER architecture and BIER encapsulation standards are applicable both to MPLS and non-MPLS networks, whereas other standards focus on MPLS encapsulation and some basic mechanisms are also applicable to non-MPLS encapsulation. The BIER multicast technology described in 12.2 BIER Multicast Technology uses BIER-MPLS encapsulation as an example.

In 2018, the BIER Working Group incorporated into its charter the BIER standards development for native IPv6 based on RFC 8279, RFC 8296, and the proposals of participants in the working group, and embarked on BIERv6 standardization. Following this, the BIER Working

Group adopted the BIERv6 requirement draft,[20] and the following drafts were submitted: *Encapsulation for BIER in Non-MPLS IPv6 Networks,*[21] *BIER IPv6 Encapsulation (BIERv6) Support via IS-IS,*[22] *Use of BIER IPv6 Encapsulation (BIERv6) for Multicast VPN in IPv6 networks,*[23] and *Inter-Domain Multicast Deployment using BIERv6.*[24] These drafts describe how the BIER multicast architecture and IPv6 encapsulation are used to carry multicast services on native IPv6 or SRv6 networks. These multicast services include MVPN, non-VPN multicast which is also referred to as Global Table Multicast (GTM), and inter-AS multicast services. The BIERv6-encapsulated multicast technology is described in Section 12.3 BIERv6 Multicast Technology.

12.1.3.1 BIER Working Group Anecdotes

With a nice ring to it, BIER sounds like beer. As it turns out, IETF participants are easy going and they enjoy beer, which makes the existence of a "beer" working group even more fitting. Key members of the BIER Working Group had commemorative T-shirts made, with a large beer cup in front, which is very impressive.

The fundamentals of BIER and SR are fundamentally similar. Specifically, multicast forwarding is programmed on the ingress, and transit nodes do not maintain MDT states, which improves scalability. Unlike SR, which has the existing MPLS label stack mechanism and IPv6 routing extension header mechanism to depend on, BIER poses new challenges to forwarding. Against this backdrop, when the BIER Working Group was established, it was identified as an EXPERIMENTAL one, as were the related standards drafts. Nowadays, with the mature technologies and standards achieved through several years of development, the BIER Working Group and standards have officially been formalized.

12.2 BIER MULTICAST TECHNOLOGY

BIER provides a new multicast data forwarding architecture. In this section, we describe the BIER multicast technology, including the basic concepts, hierarchical architecture, forwarding fundamentals, data encapsulation, control plane, and BIER-based multicast services.

12.2.1 Basic Concepts of BIER

Before describing the fundamentals of BIER, we will first introduce its basic concepts, as listed in Table 12.2.

TABLE 12.2 Basic Concepts of BIER

Concept	Definition
BIER domain	Network domain that supports BIER forwarding. Each BIER domain must contain at least one Sub-Domain (SD): the default sub-domain 0. BIER also supports multiple SDs in IGP multi-topology or inter-area deployment scenarios. For example, sub-domain 0 and sub-domain 1 can be configured on each Bit-Forwarding Router (BFR), with sub-domain 0 using the default topology and sub-domain 1 using the multicast topology.
BFR	Router that supports BIER forwarding. In a BIER domain, it is called a Bit-Forwarding Ingress Router (BFIR) when it functions as an ingress router, and a Bit-Forwarding Egress Router (BFER) when it functions as an egress router.
BFR-prefix	IP address of a BFR. Using the BFR's loopback interface address as the BFR-prefix is recommended. Each BFR needs to be configured with a BFR-prefix for each sub-domain to which it belongs. If a BFR belongs to multiple sub-domains, it can use the same or different BFR-prefixes in these sub-domains.
BFR-ID	ID of an edge router (BFIR or BFER) in a BIER sub-domain. It is an integer ranging from 1 to 65535 and must be configured for each BFIR and BFER, but not for transit BFRs. It is recommended that the BFR-IDs in the BIER domain be densely distributed. For example, if the number of BFERs in a sub-domain is less than 256, it is recommended that BFR-IDs in the range of 1–256 be allocated to the BFERs; if there are more than 256 but less than 512 BFERs, it is recommended that BFR-IDs in the range of 1–512 be allocated to the BFERs.
BitString	Binary bit string, indicating a set of destination nodes of BIER packets. After receiving a multicast data packet from outside a BIER domain, the BFIR encapsulates the packet as a BIER packet by adding a BIER header and forwards it in the BIER domain. After receiving the packet, the BFER decapsulates the BIER header and sends the packet to the corresponding CEs or multicast receivers. Multicast packet forwarding in a BIER domain is based on the BitString field in the BIER header. Each bit in the BitString indicates a BFR-ID. If the value of a bit is 1, the packet needs to be sent to the BFER whose BFR-ID corresponds to the bit. This is unnecessary if the value of the bit is 0. Because each bit in the BitString represents a BFER, the number of BFERs described by the BitString is subject to the length of the BitString. For example, if the length of a BitString is 256 bits, the BitString can describe a maximum of 256 BFERs.
Set ID	A Set ID (SI) identifies a group of BFERs. BIER encapsulation contains not only a BitString but also an SI that indicates the set to which the BFERs are allocated. SIs are used to divide BIER-IDs into different ranges to support larger-scale network addressing. Assuming that a 256-bit BitString is used, nodes whose BFR-IDs range from 1 to 256 belong to a set 0 (SI=0), and those whose BFR-IDs range from 257 to 512 belong to a set 1 (SI=1). For the BIER packet to be sent to set 0, the BFR-IDs represented by the bits from right to left in the BitString are 1–256, and they are 257–512 for the BIER packet to be sent to set 1. We can deploy larger-scale multicast networks by employing SIs.

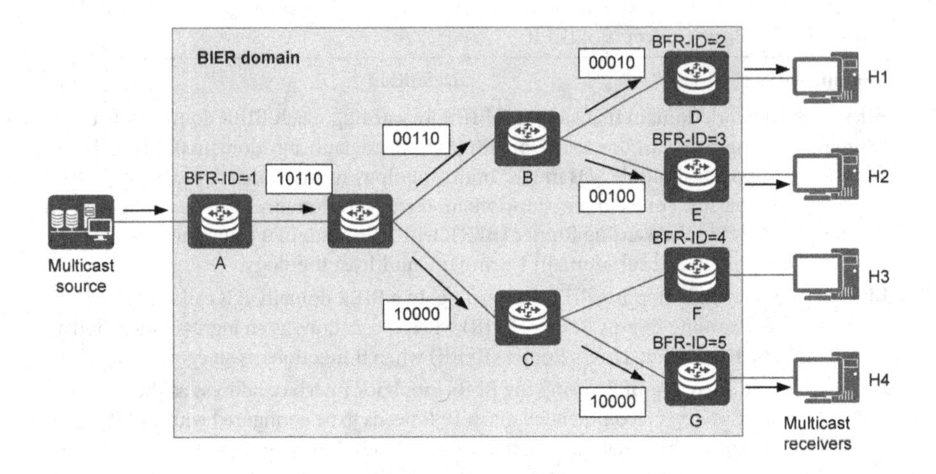

FIGURE 12.7 BIER multicast fundamentals.

A simple loop-free topology (as shown in Figure 12.7) is used as an example to describe the basic concepts of BIER. As shown in Figure 12.7, the BFR-IDs of nodes A, D, E, F, and G are set to 1, 2, 3, 4, and 5, respectively.

As shown in Figure 12.7, multicast packets enter the BIER domain through ingress A, and one copy of each packet needs to be sent to nodes D, E, and G. As such, ingress A encapsulates the packet with a BIER header, in which bits 2, 3, and 5 in the BitString are set to 1. The rightmost eight bits in the BitString (5-bit BitStrings are used in Figure 12.7 for short) are 00010110, in which bits 2, 3, and 5 from right to left are 1. This indicates that one copy of the packet needs to be sent to the nodes (D, E, and G) whose BFR-IDs are 2, 3, and 5, respectively.

12.2.2 BIER Architecture

BIER architecture consists of three layers: the routing underlay, BIER layer, and multicast flow overlay.[6]

12.2.2.1 Routing Underlay

The routing underlay is responsible for determining the next-hop BFR to each BFER. As shown in Figure 12.7, BFR A determines the next-hop BFR and the corresponding outbound interface for each BFR-prefix. Generally, the routing underlay protocol is an IGP, such as IS-IS or OSPF. On some networks, such as large-scale data center networks, BGP may be used as the routing underlay protocol.

12.2.2.2 BIER Layer

The BIER layer is responsible for transmitting multicast data packets in the BIER domain and provides the following functions:

- Each BFR advertises BIER information, including BFR-prefixes, sub-domains, the BFR-IDs of nodes in each sub-domain, and the BitString length used by nodes in each sub-domain.
- The BFIR encapsulates each multicast data packet with a BIER header.
- BFRs update the BIER header of each received BIER packet and forward the packets.
- Each BFER decapsulates received BIER packets and distributes them to the multicast service processing module.

The BIER layer consists of the control plane and data plane.

The control plane of the BIER layer uses routing underlay protocols for BIER information advertisement. We can consider that the IGP extension required by BIER belongs to the BIER layer, and that the basic IGP functions and mechanisms belong to the routing underlay.

The data plane of the BIER layer processes BIER data packets and serves the multicast flow overlay. Specifically, the control plane of the multicast flow overlay on the BFIR determines to which BFERs multicast data packets are to be sent, and the data plane of the BIER layer encapsulates the BIER header for multicast data packets. The BIER layer distributes packets to the multicast flow overlay for processing, which determines the VPN or GTM instance to which the packets belong, decapsulates the packets, and forwards them according to the inner multicast packets.

12.2.2.3 Multicast Flow Overlay

The multicast flow overlay is responsible for processing each multicast data packet. Specifically, the BFIR determines the BFERs to which the multicast traffic received from outside the BIER domain is to be sent. After receiving a BIER packet, each BFER determines the VPN or GTM instance to which the packet belongs, replicates the packet, and forwards packet copies according to the VPN or GTM instance and the inner multicast packet.

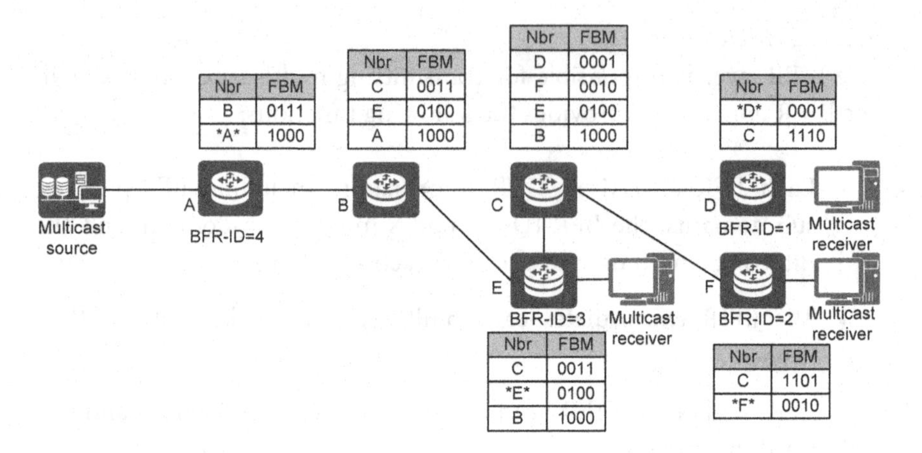

FIGURE 12.8 BIER forwarding.

12.2.3 BIER Forwarding Fundamentals

This section uses Figure 12.8 as an example to describe BIER forwarding fundamentals, including BFR-ID configuration, Bit Index Forwarding Table (BIFT) as well as its establishment, and BIFT-based forwarding.

12.2.3.1 BFR-ID Configuration

As shown in Figure 12.8, there are six BIER devices. As edge nodes, A, D, E, and F need to be configured with valid BFR-IDs. In this example, their BFR-IDs are 4, 1, 3, and 2, respectively.

12.2.3.2 BIFT Introduction

A BIFT is used to guide packet forwarding. Nodes A–F establish their BIFTs based on the information advertised by the control plane. Each BIFT entry includes a BFR Neighbor (represented by **Nbr**) and Forwarding Bit Mask (FBM). The FBM indicates the set of BFERs that are reachable through the BFR neighbor.

The FBM is also represented by a BitString, with the same length as the one used for packet forwarding. For example, if the BitString Length (BSL) used for packet forwarding is 256 bits, the FBM in the BIFT is also 256 bits long. A bitwise AND operation is performed between the BitString in the packet and the FBM in the BIFT during packet forwarding. Figure 12.8 shows a simplified depiction of BIFTs in use. The following explains the BIFT entries for two nodes in this figure.

Node A has two BIFT entries:

- In the entry where the neighbor is B, the FBM is 0111, indicating that the BFR-IDs of the BFERs reachable through node B are 1, 2, and 3.
- In the entry where the neighbor is A, the FBM is 1000. The neighbor A is surrounded with asterisks (*), indicating that the neighbor is node A itself.

Node B has three BIFT entries:

- In the entry where the neighbor is C, the FBM is 0011.
- In the entry where the neighbor is E, the FBM is 0100.
- In the entry where the neighbor is A, the FBM is 1000.

Note: In Figure 12.8, 4-bit FBMs are used for illustration. The actual length of each FBM is at least 64 bits.

12.2.3.3 BIFT Establishment

Node B is used as an example to describe BIFT establishment.

First, node B receives the BIER information flooded by other nodes through an IGP, including each BFR's BFR-prefix, sub-domain, and BFR-ID in the sub-domain, as well as the maximum SI and label block corresponding to each <SD, BSL>. If the BFR has not been allocated a BFR-ID, the BFR-ID is set to 0, an invalid ID. In this example, the BSL is 4, the maximum SI is 0, and the BFR-IDs range from 1 to 4.

Then, node B generates the forwarding information for each node with a valid BFR-ID.

- The next hop for packets to reach the node whose BFR-ID is 4 is node A.
- The next hop for packets to reach the node whose BFR-ID is 3 is node E.
- The next hop for packets to reach the node whose BFR-ID is 1 or 2 is node C.

Finally, node B establishes its BIFT which includes three neighbors and corresponding FBMs.

Similarly, nodes A, C, D, E, and F establish their own BIFTs. The BIFT of each edge node contains a special entry with the neighbor as itself, and the bit with value 1 in the FBM indicates the edge node itself. If the result of the bitwise AND operation between the BitString in a packet and the FBM of the special entry is not 0, the edge node removes the BIER header and performs forwarding according to the inner IP (multicast) packet.

Note: In this example, node A functioning as the ingress is configured with a valid BFR-ID and therefore can generate a special entry pointing to itself. Node A can also function as an egress to receive BIER packets sent by nodes D, E, and F. Similarly, nodes D, E, and F functioning as egresses are configured with valid BFR-IDs, and therefore can also generate such special entries.

12.2.3.4 BIFT Lookup and Packet Forwarding

When receiving a BIER packet, node B performs a bitwise AND operation between the BitString in the packet and the FBM in each entry in the BIFT. If the result of the operation is not 0, node B replicates the packet, updates the BitString in the copy to the result of the operation, and sends the copy to the corresponding neighbor. If the result is 0, node B does not replicate the packet for the neighbor.

A typical characteristic of BIER forwarding is that the BitString is updated to the result of the bitwise AND operation before a packet is forwarded to the next hop. This is to prevent duplicate packets from being sent when a ring topology exists on the network.

For example, in Figure 12.8, nodes B, C, and E form a ring topology. If node A sends a packet with the BitString 0111 to node B, and node B sends copies of the packet to nodes C and E without updating the BitString to the result of the bitwise AND operation between the BitString and the FBM, the BitString in the packet received by node C or E remains as 0111. In this case, node C also sends a copy of the packet to node E, meaning that node E receives two duplicate packets. Similarly, after receiving the packet from node B, node E also sends a copy of the packet to node C, which consequently also receives two duplicate packets. To make matters worse, after node E receives the copy from node C, node E sends it back to node C, which may cause a traffic storm.

The bitwise AND operation prevents this problem. When receiving a BitString of 0111, node B performs a bitwise AND operation between the

BitString and the FBM in each entry in the BIFT, and updates the BitString to 0011 to be sent to node C, and to BitString 0100 to be sent to node E. With different BitString values, nodes C and E will process received packets differently. Specifically, after receiving the packet, node C will send a copy only to nodes D and F, not to node E; and node E will decapsulate the packet and send the multicast packet to the multicast receivers, without sending a copy to node C.

12.2.4 BIER Data Plane

To implement BIER forwarding in the data plane, RFC 8296 defines the format of the BIER header, as shown in Figure 12.9.[15]

The preceding BIER header format applies to both MPLS encapsulation and non-MPLS encapsulation. The formats are the same for the two encapsulation types, but the meanings and usage of some fields are different.

Table 12.3 describes the fields in the BIER header in MPLS or BIER-MPLS encapsulation.

RFC 8296 provides an example of non-MPLS BIER encapsulation (Ethernet-based BIER encapsulation), in which the Ethernet type 0xAB37 is used to identify the BIER header following the Ethernet header.

The format of the BIER header in non-MPLS encapsulation is the same as that in BIER-MPLS encapsulation. However, the meanings and usage of some fields, mainly BIFT-ID, TC, and S, are slightly different. In non-MPLS BIER encapsulation, the BIFT-ID field is not an MPLS label but a common integer. The TC and S fields are meaningless in non-MPLS BIER. The TC field must be set to 0, and the S field must be set to 1. TTL has the same meaning as that in BIER-MPLS encapsulation and is used to prevent packet storms when a loop occurs *transiently* due to a network topology change.

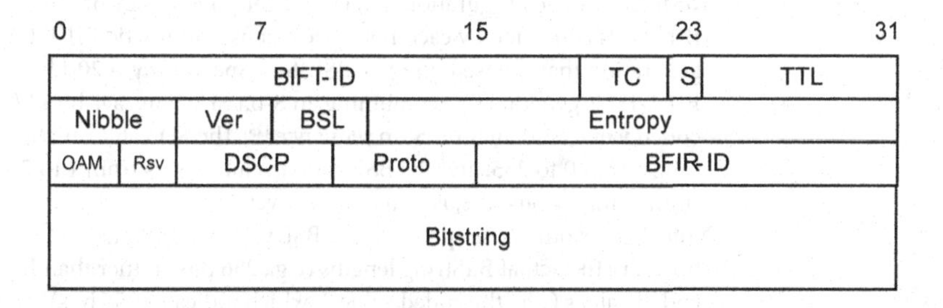

FIGURE 12.9 BIER header format.

TABLE 12.3 Fields in the BIER Header

Field	Length	Description
BIFT-ID	20 bits	In BIER-MPLS encapsulation, the BIFT-ID field is an MPLS label (also called a BIER label). The BIFT-ID corresponds to a triplet of <SD, BSL, SI>. The BIFT-ID field can be used to obtain a unique <SD, BSL, SI> triplet. This field has the following functions: 1. The BSL is used to obtain the length of the BitString in the BIER header, so that the length of the entire BIER header can be obtained (12 bytes+length of the BitString). 2. The BSL and SI together determine the BFR-ID range represented by the BitString, such as from 1 to 256 or from 257 to 512. In BIER-MPLS encapsulation, BIER labels are allocated by each BFR, and labels of multiple SIs under each <SD, BSL> tuple must belong to the same label block. For example, if a router is configured with only <SD=0, BSL=256> and the maximum SI is set to 3, a label block with four contiguous labels must be allocated to <SD=0, BSL=256>. The start label of the label block is L1, and the last label is L1+3. The mapping between each label and <SD, BSL, SI> is as follows: 1. L1 corresponds to <SD=0, BSL=256, SI=0>. 2. L1+1 corresponds to <SD=0, BSL=256, and SI=1>. 3. L1+2 corresponds to <SD=0, BSL=256, and SI=2>. 4. L1+3 corresponds to <SD=0, BSL=256, and SI=3>. The labels allocated by routers to the same <SD, BSL, SI> triplet may be the same or different. For example, the following allocation is possible: • For node 1, the labels corresponding to <SD=0, BSL=256, SI=0/1/2/3> are 100/101/102/103. • For node 2, the labels corresponding to <SD=0, BSL=256, SI=0/1/2/3> are 200/201/202/203. • For node 3, the labels corresponding to <SD=0, BSL=256, SI=0/1/2/3> are 300/301/302/303. • For node 4, the labels corresponding to <SD=0, BSL=256, SI=0/1/2/3> are 400/401/402/403. To simplify the configuration and ensure the consistency of the BIFT-IDs generated by each node, we can use automatic BIFT-ID generation that is based on <SD, BSL, SI>. Specifically, a 20-bit BIFT-ID is generated by combining an 8-bit SD value, a 4-bit BSL coded value, and an 8-bit SI in sequence.[25] The SD value and SI range from 0 to 255, and the BSL coded value ranges from 1 to 7 (indicating 64 bits to 4,096 bits, respectively).[15] Note: Unless otherwise specified, the BSL values mentioned in this book are the actual BitString lengths (e.g., 256 bits) rather than BSL coded values (e.g., the coded value 3, which indicates 256 bits).

(*Continued*)

TABLE 12.3 (*Continued*) Fields in the BIER Header

Field	Length	Description
TC	3 bits	This field is used for traffic classification.
S	1 bit	This field is the stack bottom flag, the value of which is 1 in every BIER header, indicating that the MPLS label is the bottom label in the entire label stack.
TTL	8 bits	This field specifies the TTL value.
Nibble	4 bits	This field has a fixed value of 0101 and is used to distinguish from IPv4 and IPv6 headers carried in MPLS packets. In MPLS encapsulation and forwarding, sometimes the IPv4 header (first nibble value 0100) or IPv6 header (first nibble value 0110) following the label stack needs to be checked. The nibble value 0101 can prevent the BIER header from being misinterpreted as an IPv4 or IPv6 header.
Ver	4 bits	This field specifies the version of the BIER header. The current version is 0.
BSL	4 bits	This field specifies the BitString coded value, which ranges from 1 to 7, indicating 64 bits to 4,096 bits. For details, see RFC 8296. Note: The BSL field is used by the packet analyzer to determine the BitString length of the BIER header. The forwarding plane does not rely on this field during packet forwarding but obtains the corresponding BSL based on the BIFT-ID field.
Entropy	20 bits	This field specifies an entropy value that can be used for load-balancing purposes. Different multicast data flows can use different entropy values so that the flows are forwarded on different paths for load balancing. Packets of a multicast data flow use the same entropy value so that they pass through the same path.
OAM	2 bits	This field is used to support functions such as Performance Measurement.
Rsv	2 bits	This field is reserved.
DSCP	6 bits	Currently, this field is not used in BIER with MPLS encapsulation.
Proto	6 bits	This field specifies the payload type and is used in MVPN over BIER scenarios. Value 2 indicates an MPLS packet with an upstream label. Note: Multicast flows are transmitted in P2MP mode. The ingress PE may allocate a unique label (used to identify an MVPN instance) and advertise it to egress PEs through the control plane, and the data packet destined for the egress PEs carries the label. For the egress PEs, the label is allocated by the ingress PE rather than by themselves. This label is referred to as an upstream label.
BFIR-ID	16 bits	This field specifies the BFR-ID of a BFIR. When encapsulating the BIER header into a received multicast data packet, a BFIR must set this field to its BFR-ID.
BitString	Variable	This field specifies the set of destination nodes of a BIER packet.

In non-MPLS BIER encapsulation, each node needs to allocate a BIFT-ID to each <SD, BSL, SI> triplet. The BIFT-ID must be unique in the BIER domain and can be configured manually.

12.2.5 BIER Control Plane

In the control plane, BIER uses an IGP to flood BIER information, based on which each BFR establishes its BIFT. RFC 8401 defines the IS-IS extension required by BIER, and the extension only supports BIER-MPLS encapsulation.[16] RFC 8444 defines the OSPFv2 extension required by BIER, and the extension also only supports BIER-MPLS encapsulation.[17] This section uses IS-IS as an example to describe the format and fundamentals of BIER-MPLS encapsulation on an IPv4 network. For details about OSPFv2 extension, see RFC 8444.

BIER requires the use of the Wide cost style, which is currently the most common cost style used by IS-IS. In Wide cost style, IS-IS uses a TLV with a Type of 135 or 136 to advertise IPv4 address prefixes. The TLV with the Type 135 is the Extended IS Reachability TLV,[26] and its format for IPv4 is shown in Figure 12.10.

Table 12.4 describes the fields in the Extended IS Reachability TLV (IPv4).

BIER information is carried through a sub-TLV of TLV 135. Figure 12.11 shows the format of the sub-TLV.

Table 12.5 describes the fields in this sub-TLV.

The preceding BIER sub-TLV may contain one or more sub-sub-TLVs. Figure 12.12 shows the format of a BIER MPLS Encapsulation sub-sub-TLV.

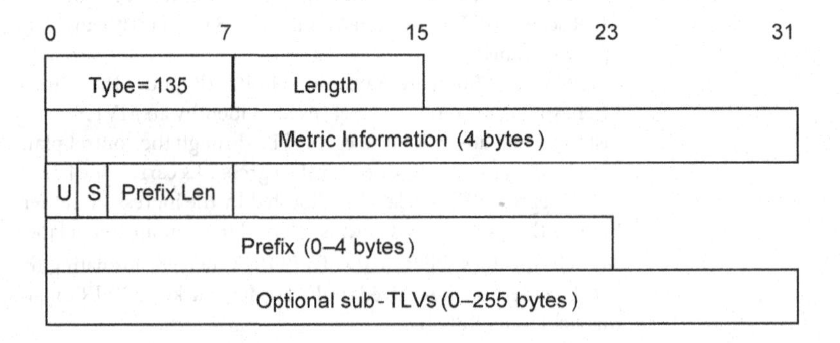

FIGURE 12.10 Extended IS Reachability TLV (IPv4).

TABLE 12.4 Fields in the Extended IS Reachability TLV (IPv4)

Field	Length	Description
Type	8 bits	TLV type. The value 135 indicates the Extended IS Reachability TLV.
Length	8 bits	TLV length.
Metric Information	32 bits	Metric information carried by the TLV.
U	1 bit	Up/Down bit.
S	1 bit	Whether a sub-TLV is present.
Prefix Len	6 bits	The value of this field is 32 when BIER information is carried. The prefixes with the 32-bit mask are also called host route prefixes.
Prefix	0–4 bytes	The length is determined by the Prefix Len field. When BIER information is carried, the length of this field is 4 bytes, and the content is a complete IP address (BFR-prefix).
Optional sub-TLVs	Variable	This field is optional. Whether a sub-TLV is present is determined by the S field.

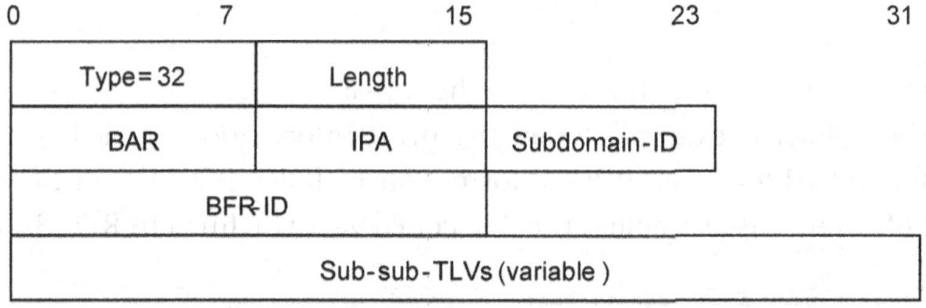

0 7 15 23 31

Type=32 | Length
BAR | IPA | Subdomain-ID
BFR-ID
Sub-sub-TLVs (variable)

FIGURE 12.11 Sub-TLV used to carry BIER information.

TABLE 12.5 Fields in the Sub-TLV Used to Carry BIER Information

Field	Length	Description
Type	8 bits	The value is 32, indicating that this sub-TLV carries BIER information.
Length	8 bits	Length of the sub-TLV.
BAR	8 bits	BIER algorithm. Currently, the value can only be 0, indicating the default algorithm.
IPA	8 bits	IGP algorithm. Currently, the value can only be 0, indicating the default algorithm.
Subdomain-ID	8 bits	BIER sub-domain ID.
BFR-ID	16 bits	BFR-ID of the node in the sub-domain. If no BFR-ID is configured, 0 (invalid BFR-ID) is carried in packets.
Sub-sub-TLVs	Variable	This field is optional. Whether a sub-sub-TLV is present is determined by the Length field.

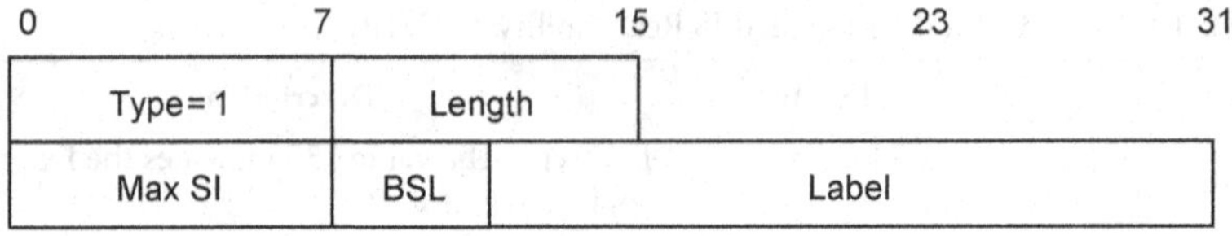

0 – 7	8 – 15	16 – 23	24 – 31
Type=1	Length		
Max SI	BSL	Label	

FIGURE 12.12 Sub-sub-TLV format.

TABLE 12.6 Fields in the Sub-sub-TLV

Field	Length	Description
Type	8 bits	The value is 1, indicating BIER-MPLS encapsulation information.
Length	8 bits	The value is 4, indicating that 4 bytes follow the Length field.
Max SI	8 bits	Maximum SI in a specific <Sub-domain, BSL>.
BSL	4 bits	BSL coded value. Coded values 1, 2, 3, 4, 5, 6, and 7 indicate 64 bits, 128 bits, 256 bits, 512 bits, 1,024 bits, 2,048 bits, and 4,096 bits, respectively.[15]
Label	20 bits	Start label value of the label block in a specified <Sub-domain, BSL>.

Table 12.6 describes the fields in the sub-sub-TLV.

The following example shows the BIER information carried in IS-IS prefix 10.1.1.10/32. The BIER information includes BSL <3> in SD <1>. The BSL 3 in this example is the BSL coded value (defined in RFC 8296).

```
<HUAWEI> display isis lsdb verbose

                        Database information for ISIS(1)
                        --------------------------------

                          Level-1 Link State Database

LSPID                   Seq Num Checksum HoldTime      Length ATT/P/OL
-------------------------------------------------------------------------------
0000.0000.0001.00-00*   0x00000070 0xb273    732           326     0/0/0
 SOURCE        0000.0000.0001.00
 NLPID        IPV4
 NLPID        IPV6
 AREA ADDR 49.0001
 INTF ADDR 10.1.1.10
 INTF ADDR 10.1.1.20
 INTF ADDR V6 2001:DB8::192:168:12:10
 INTF ADDR V6 2001:DB8::10:1:1:10
 Topology      Standard
+NBR ID        0000.0000.0001.01   COST: 10
+IP-Extended  10.1.1.10         255.255.255.255 COST: 0
 Bier-SD       1         BAR: 0 IPA: 0 BFR-id: 20
    Encapsulation Type MPLS Max SI: 0      BS Len: 3 Label: 331776
 Extended Reach Attr Flag: X:0 R:0 N:1
```

12.2.6 MVPN over BIER

After BIFTs are established, the network can transmit BIER multicast services. Typical multicast services include the MVPN service and GTM IPTV service.

- The MVPN service is the isolated multicast service deployed with L3VPN. Only the sites in the same VPN can communicate with each other, whereas Internet users cannot access VPN sites.
- The GTM IPTV service is generally one of the integrated broadband services that carriers provide (and operate) to home users. These services do not require L3VPN to be configured.

The following uses MVPN as an example to describe how BIER-MPLS is used to carry multicast services.

MVPN over BIER complies with the MVPN framework[5] and is processed at the multicast flow overlay. MVPN services consider BIER as a P2MP tunnel. This tunnel is not explicitly established using signaling. Instead, after the BIFT is established using an IGP, the ingress explicitly tracks the egresses of each multicast flow, combines the BFR-IDs of these egresses into a BitString, and encapsulates it into the BIER header.

Figure 12.13 is a diagram of explicit tracking.

As shown in Figure 12.13, node A is a multicast source PE with a BFR-ID of 4, and nodes D and F are multicast receiver PEs with the BFR-IDs being 1 and 2, respectively. The multicast source PE sends (S1, G1) information to the receiver PEs through a BGP-MVPN Type 5 route (Source Active A–D route).

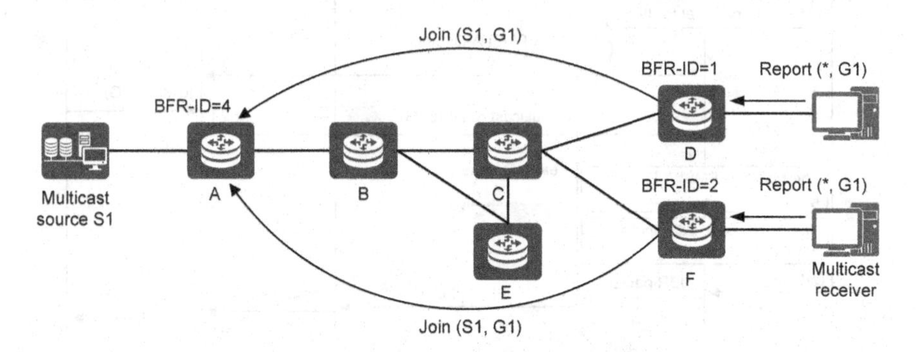

FIGURE 12.13 Explicit tracking.

After receiving an (*, G1) Report message, node D generates the (S1, G1) entry according to the multicast source and group information stored locally. After finding that the next hop of the route to S1 is the multicast source PE, node D sends a join request to the multicast source PE through a BGP-MVPN Type 7 route (C-multicast route) and a Type 4 route (Leaf A–D route). The Leaf A–D route contains (S1, G1) and BFR-ID information.

Similarly, once node F receives an (*, G1) Report message, it sends a join request to the multicast source PE through a C-multicast route and a Leaf A–D route. The Leaf A–D route contains (S1, G1) and BFR-ID information.

The multicast source PE, after receiving the C-multicast routes from the receiver PEs, establishes the mappings between (S1, G1) and BFR-ID 1 and between (S1, G1) and BFR-ID 2. This is the result of explicit tracking. After receiving an (S1, G1) multicast data packet from multicast source S1, the multicast source PE encapsulates the packet with a BIER header, in which the bits corresponding to BFR-ID 1 and BFR-ID 2 in the BitString are set to 1.

Figure 12.14 shows the basic MVPN over BIER procedure.

The basic MVPN over BIER procedure is as follows:

1. The S-PMSI tunnel type for an MVPN instance is set to BIER on the ingress PE. The ingress PE advertises an S-PMSI A–D route with (*, *) information to all egress PEs, indicating that the ingress PE

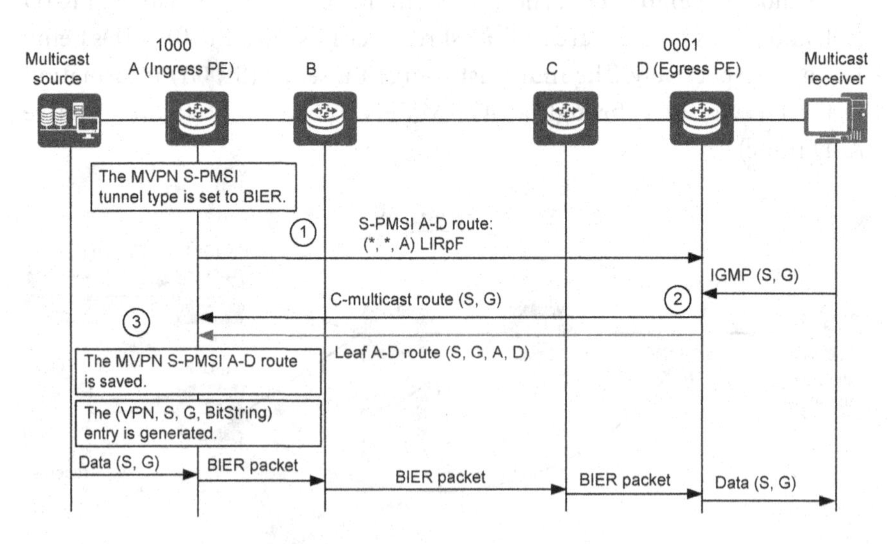

FIGURE 12.14 Basic MVPN over BIER procedure.

tracks all join information, regardless of the multicast source or group.

2. When an egress PE receives a join request (e.g., an IGMP Report message in Figure 12.14) from the VPN side downstream, the egress PE converts the (VPN, S, G) information into a BGP message and sends it through a BGP-MVPN Type 7 route (C-multicast route) and a Type 4 route (Leaf A–D route) to the ingress PE. The C-multicast route contains (S, G) information, but no egress PE address information. The Leaf A–D route contains not only (S, G) information, but also egress PE address, BIER sub-domain, and BFR-ID information.

3. Based on the Leaf A–D route, the ingress PE establishes the mappings between the (S, G) and the set of egress PEs, and uses a BitString to represent the egress PEs. Then, the ingress PE can send traffic to the egress PEs.

As shown in Figure 12.14, each egress PE sends the join request directly to the ingress PE, regardless of the number of hops in between. Transit nodes remain unaware of this (S, G) join process, which is a service process performed by the multicast flow overlay, independent of the process for establishing a BIFT through IGP flooding, and takes place only when a multicast service is involved, for example, when a user requests a multicast flow. The process for establishing a BIFT through IGP flooding paves the way for network transport and is performed by the routing underlay. As long as the network is ready, the BIFT is established, and subsequent service processes do not affect network transport.

BIER encapsulation is crucial to decoupling of transport and services. After tracking the receiver information of a multicast flow and encapsulating the BitString in the BIER header, the multicast flow overlay hands subsequent processing over to the BIER layer. The BIER layer replicates and forwards packets according to the destination node (BFER) set represented by the BitString. In BIER forwarding, a set of destination addresses is combined into a BitString, and the packet is sent to each destination accordingly. This is similar to unicast IP forwarding, where a unicast destination IP address is encapsulated into the packet, and the packet is sent to the destination accordingly. Figure 12.15 shows a BIER-MPLS-based MVPN forwarding process.

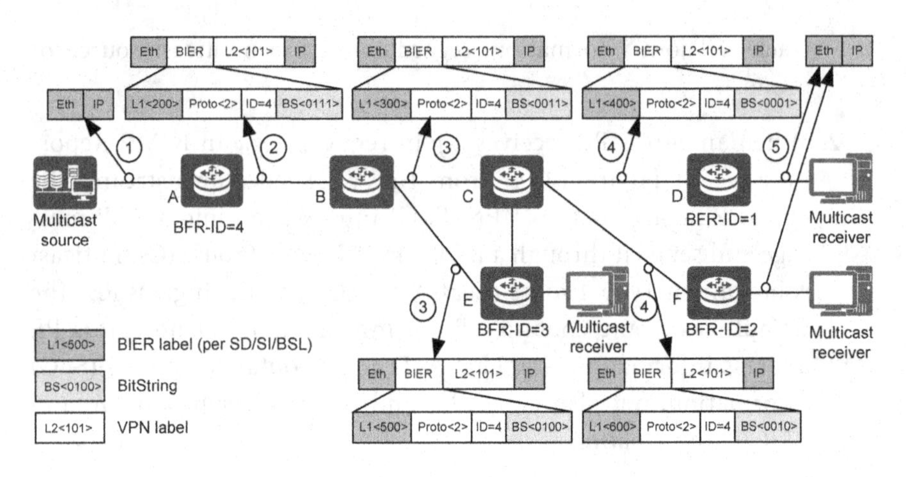

FIGURE 12.15 BIER-MPLS-based MVPN forwarding process.

The detailed procedure of BIER-MPLS-based MVPN forwarding is as follows:

1. After receiving a multicast packet, node A (BFIR) encapsulates a BIER header and a VPN label L2<101> and sends the packet to node B. In the BIER header:
    - The label is the BIER label 200 allocated by node B.
    - The value of the Proto field is 2, indicating that a VPN label exists between the BIER header and the inner IP multicast packet.
    - The BFIR-ID is 4, which is the BFR-ID of node A.
    - The BitString is 0111, indicating that the packet is to be sent to nodes D, E, and F.
2. After receiving the BIER packet, node B replicates the packet according to the BitString and its BIFT, and sends a copy (with BIER label 300 and BitString 0011) to node C, and a copy (with BIER label 500 and BitString 0100) to node E.
3. After receiving the BIER packet, node C replicates the packet according to the BitString and its BIFT, and sends a copy (with BIER label 400 and BitString 0001) to node D and a copy (with BIER label 600 and BitString 0010) to node F. Note that the BIER labels allocated by nodes C, D, E, and F are 300, 400, 500, and 600, respectively.

4. After receiving the BIER packet, node E checks the BitString and its BIFT and finds that the bit corresponding to its BFR-ID is 1. Then node E performs multicast overlay processing on the packet. Specifically, it determines the VPN instance to which the packet belongs according to the Proto field in the BIER header, as well as the following VPN label. Next, it identifies the outbound interface used for packet forwarding according to the inner IP multicast packet. Finally, it sends the inner IP multicast packet to the multicast receiver.
5. Nodes D and F perform the processing similar to that on node E.

12.2.7 Characteristics of BIER

The preceding sections have described BIER's basic concepts, architecture, forwarding fundamentals, encapsulation format in the data plane, IGP extension in the control plane, and MVPN over BIER, all of which have been standardized in RFCs. BIER-MPLS is currently the most common format for encapsulation since IGP extension[16,17] required for BIFT establishment, and procedures for MVPN over BIER as well as GTM over BIER[18] are only defined based on MPLS encapsulation. The BIER-MPLS encapsulation format is mainly applicable to MPLS and SR-MPLS networks.

In addition to the BIER-MPLS encapsulation format, RFC 8296 defines the Ethernet-based BIER (BIER-ETH) encapsulation format. The BIER-ETH encapsulation format uses the Ethernet frame type 0xAB37 to indicate that the Ethernet header is followed by the BIER header, which precedes the IP packet payload. This form of encapsulation can be referred to "2.5-layer" encapsulation.

BIER simplifies deployment, as it does not depend on protocols that require explicit MDT establishment, such as PIM, mLDP, and RSVP-TE. In addition, with BIER, transit nodes (such as Ps) do not need to establish an MDT for each multicast flow if they do not have multicast services, greatly reducing costs.

BIER-ETH encapsulation depends on the Ethernet link layer and is mainly targeted at data center networks. It is not applicable to carrier networks, where BIER-MPLS encapsulation is mainly used. However, BIER-MPLS faces the following challenges during deployment:

1. BIER-MPLS deployment is complex. It depends on MPLS and is applicable to MPLS networks. Considering the trend that unicast SR-MPLS is replacing LDP and RSVP-TE, BIER mainly applies

to SR-MPLS networks. On a network where a node does not support BIER forwarding, an SR-MPLS unicast tunnel has to be used. Although an LDP tunnel can also be used, it means that LDP has to be deployed along with BIER. Such a solution is technically feasible, but it is only transitional and complex.

2. Some existing multicast services, such as IPTV services, are deployed on non-MPLS networks or MPLS-capable networks without MPLS multicast used to carry the IPTV services. Deploying BIER-MPLS on such networks is difficult in terms of management and maintenance, as without MPLS, BIER deployment requires the upgrade of all devices on the network.

3. Even on a network where MPLS MVPN has been deployed, deploying BIER-MPLS faces extra challenges, especially in inter-AS multicast deployment (which is common). For example, an IPTV multicast source server might be connected to a PE on a carrier's IP backbone network, and IPTV receivers are connected to BNGs on metro networks belonging to an AS different from that of the IP backbone network. In this case, mLDP signaling packets can be sent from the BNGs hop by hop to the PE on the IP backbone network to establish an MDT. However, it is difficult to configure BIER-MPLS to advertise BIER information and establish BIFTs across ASs.

4. On IPv6 networks, the focus in the industry has been shifted from SR-MPLS (based on the MPLS data plane) to SRv6 (based on the IPv6 data plane). It is imperative to devise a solution to implement MPLS-independent native IPv6 BIER encapsulation, forwarding, and deployment with BIER architecture and encapsulation.

12.3 BIERv6 MULTICAST TECHNOLOGY

In this section, we describe the BIERv6 multicast technology, which adopts non-MPLS BIER encapsulation, covering the BIERv6 proposal and design, fundamentals, and IGP extensions, BIERv6-based multicast service, and characteristics of BIERv6.

12.3.1 Proposal and Design of BIERv6

BIER was initially proposed and designed based on the MPLS data plane, which was fitting for that phase of technical development. BIER also inherited some limitations of MPLS, including difficulties in inter-AS deployment and extensibility for new features. With the introduction and rapid development of SRv6, the industry gradually began to focus on network programming based on the IPv6 data plane. As such, a solution was badly needed to provide the BIER multicast architecture based on the IPv6 data plane. BIERv6 (which is based on native IPv6) was proposed within this context, aiming to achieve the following objectives:

- Allow both unicast and multicast services to be transmitted in the IPv6 data plane to further simplify protocols, and avoid the need to allocate, manage, and maintain MPLS labels.
- Take advantage of IPv6 unicast route reachability to allow services to be transmitted across the network where some nodes may not support BIER, and facilitate inter-AS deployment.
- Reap the benefits of mechanisms such as the IPv6 extension header to support the evolution and addition of new features.

Based on the IPv6 data plane specification and the design ideas of SRv6 network programming, the following key designs were made for BIERv6:

- An IPv6 source address in an IPv6 packet header is used to identify the source of each BIER packet, and an IPv6 extension header is used to carry the BIER header information, in which the BitString identifies a set of destination nodes of each packet. In this way, each BIER packet can be replicated and forwarded according to the IPv6 header and IPv6 extension header.
- A new type of SID, End.BIER, was defined for BIERv6 to support forwarding based on the IPv6 extension header. It can be used as an IPv6 destination address and instruct the forwarding plane to process the BIER header in the IPv6 extension header. It can also fully utilize IPv6 unicast route reachability to allow services to be transmitted across the network where some nodes do not support BIER, and facilitate inter-AS deployment.
- MVPN instances need to be identified by identifiers in packets to support MVPN services. BIERv6 directly uses the IPv6 source address

in each packet to identify an MVPN instance, avoiding additional identifiers, such as MPLS labels.

- In BIERv6 forwarding, the source IPv6 address remains unchanged, whereas the IPv6 DA field is updated based on the destination address determined by the BitString in the BIER header. In this way, the entire forwarding process is presented as native IPv6-based source routing multicast. This design not only allows BIERv6 to directly inherit the existing features related to the IPv6 extension header, such as IPv6 packet fragmentation and reassembly as well as IPsec-based encryption and authentication, but also lays a foundation for supporting new features for multicast, such as network slicing and In-situ Network Telemetry, based on IPv6 extension headers in the future.

12.3.2 BIERv6 Fundamentals

BIERv6 is based on the IPv6 extension header mechanism. Figure 12.16 shows the fundamentals of BIERv6 multicast from the data plane perspective.

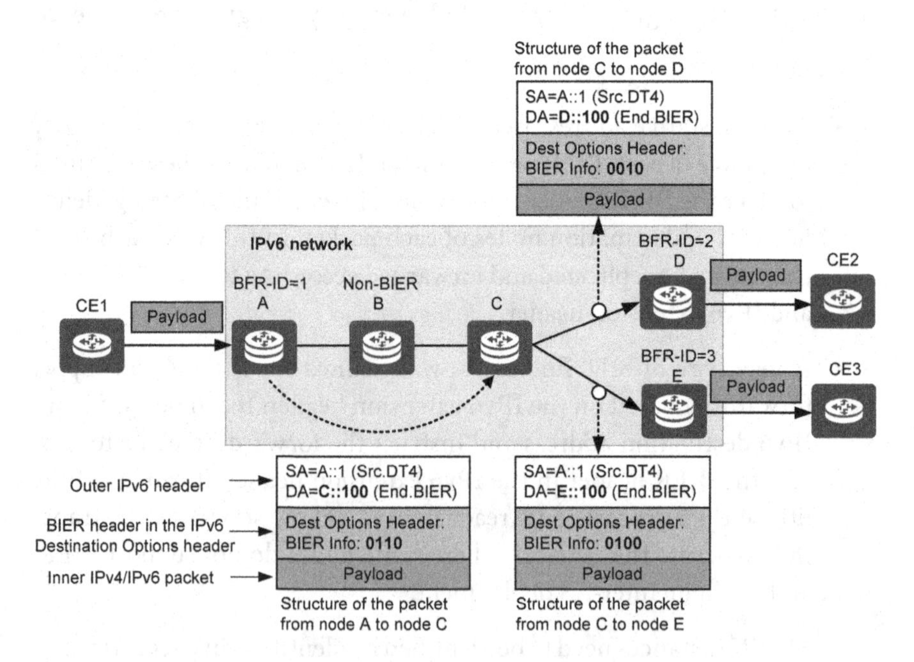

FIGURE 12.16 BIERv6 fundamentals.

After receiving a multicast packet, node A encapsulates an outer IPv6 header and a Destination Options header which carries the BIER header information containing the BitString.

According to the BIER header and BitString, node A sends the packet to the unicast address C::100 of node C.

According to the BIER header and BitString, node C replicates the packet and sends a copy to the unicast addresses D::100 and E::100 of nodes D and E, respectively.

Throughout the entire packet forwarding process, unicast IP addresses are used. Node B does not support BIERv6 forwarding, but it supports IPv6. In this case, it forwards BIERv6 packets as ordinary IPv6 packets between nodes A and C, without requiring any additional configuration or processing. As shown in Figure 12.17, the format of the BIER header is retained and the BIER header information is encapsulated in the Destination Options header, which is an existing IPv6 extension header recommended by IPv6 specification.[27]

BIERv6 data packet forwarding uses special unicast IPv6 addresses called End.BIER addresses to process BIER packets. After an End.BIER address is configured for a BFR, a forwarding entry with the address as the destination address and a 128-bit mask is generated in the FIB, and this address is marked as an End.BIER address in the forwarding entry.

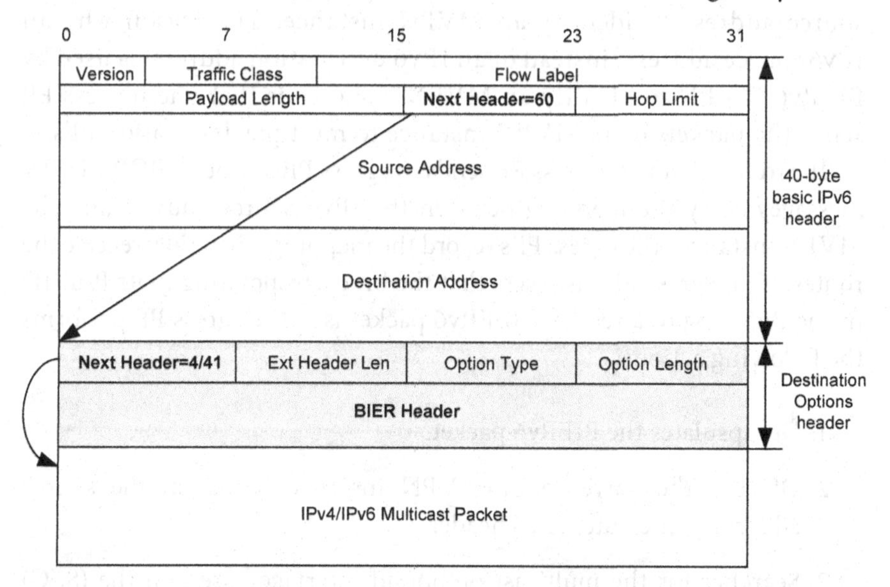

FIGURE 12.17 Format of an IPv6 packet encapsulated with a BIER header.

The End.BIER address and other information required for the encapsulation of BIERv6 packets (such as the BIFT-ID, BIER sub-domain, and BFR-ID) are flooded in a sub-TLV of BFR-prefix information through an IGP. Nodes in IGP areas establish their BIFTs according to BFR-prefixes, BFR-IDs, BIFT-IDs, and End.BIER addresses. In these entries, neighbor information contains the End.BIER address corresponding to each neighbor.

When a node receives an IPv6 packet, it searches its FIB for an entry based on the destination address of the packet. If the node determines that the address is an End.BIER address, it performs the End.BIER-related action by processing the BIER header in the IPv6 extension header.

The End.BIER address is used to guide BIER packet processing, and BIER forwarding depends on the BIER header, in which the BIFT-ID field determines the <SD, BSL, SI> to which a packet belongs, and the BitString field determines the set and BFERs to which the packet is to be sent. If an ordinary IPv6 node that does not support BIERv6 forwarding exists between two BFRs, this node only needs to forward the packet according to the IPv6 destination address in the packet.

To support MVPN over BIERv6, identifiers are required to differentiate multiple VPNs. Like SRv6 VPN, which uses an IPv6 destination address to identify a VPN instance, BIERv6 MVPN uses an IPv6 source address to identify an MVPN instance. The reason why an IPv6 source address (instead of an IPv6 destination address) is used by BIERv6 MVPN to identify an MVPN instance is that one ingress PE sends the packets of an MVPN instance to multiple destination PEs.

The routes that the ingress PE sends to egress PEs through BGP-MVPN messages carry the mapping between the IPv6 source address and the MVPN instance. The egress PEs record the mapping when they receive the routes. If an egress PE discovers that the bit corresponding to its BFR-ID in the BitString of a received BIERv6 packet is 1, the egress PE performs the following actions:

1. Decapsulates the BIERv6 packet.
2. Obtains the corresponding VPN instance based on the source address in the outer IPv6 header.
3. Searches for the multicast outbound interface based on the (S, G) information in the inner multicast packet.

4. Replicates the inner multicast packet.
5. Sends copies to receiver CEs through outbound interfaces.

In BIERv6-based MVPN services, the source IPv6 address used to identify an MVPN instance is called a Source Address for Decapsulation and MFIB Table lookup (Src.DT) address. An MVPN can be enabled with an IPv4 address family (IPv4 MVPN or MVPN), an IPv6 address family (IPv6 MVPN or MVPN6), or both. Two different IPv6 source addresses (Src.DT4 and Src.DT6 addresses) can be allocated to an MVPN instance IPv4 address family and an MVPN instance IPv6 address family, respectively, or one IPv6 source address (Src.DT46 address) can be allocated to the IPv4 and IPv6 address families of an MVPN instance.

12.3.3 BIERv6 Control Plane

Like the BIER control plane, the BIERv6 control plane can be implemented through IGP extensions. Based on the existing IGP for BIER, the IGP extension for BIERv6 uses a sub-sub-TLV to carry BIERv6 encapsulation information and another sub-sub-TLV to carry End.BIER address information. Note that among the current IGPs, only IS-IS extensions have been defined for BIERv6.[22]

In IS-IS extension for BIERv6, BIER information may be carried in a sub-TLV (Type=32) of a TLV with Type being 236 or 237. Figure 12.18 shows the format of the Extended IS Reachability TLV (IPv6), which is a TLV with Type being 236.

Table 12.7 describes the fields in the Extended IS Reachability TLV (IPv6).

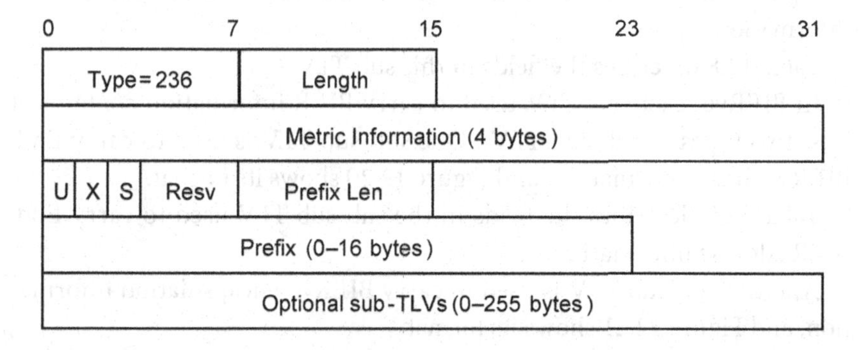

FIGURE 12.18 Extended IS Reachability TLV (IPv6).

TABLE 12.7 Fields in the Extended IS Reachability TLV (IPv6)

Field	Length	Description
Type	8 bits	TLV type. Value 236 indicates the Extended IS Reachability TLV (IPv6).
Length	8 bits	Length of the TLV.
Metric Information	32 bits	Metric information of the IPv6 prefix.
U	1 bit	Up/Down bit.
X	1 bit	External bit.
S	1 bit	Whether a sub-TLV is present.
Resv	4 bits	This field is reserved.
Prefix Len	8 bits	The value ranges from 0 to 128 and is 128 when BIER information is carried.
Prefix	0–128 bits	The length is determined by Prefix Len. When BIER information is carried, the length is 128 bits, and the content is a complete IPv6 address (BFR-prefix).
Optional sub-TLVs	Variable	This field is optional and used to carry BIER information.

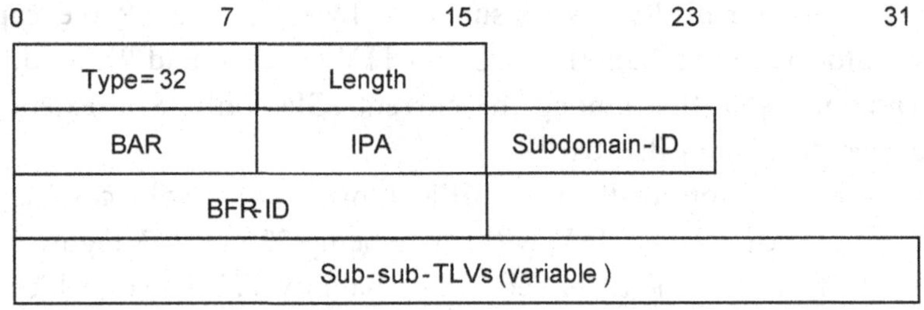

FIGURE 12.19 Sub-TLV used to carry BIER information.

Figure 12.19 shows the format of the sub-TLV used to carry BIER information.

Table 12.8 describes the fields in this sub-TLV.

In BIERv6, each sub-TLV used to carry BIER information contains at least two types of sub-sub-TLVs. One sub-sub-TLV is used to carry End. BIER address information, and Figure 12.20 shows its format.

Table 12.9 describes the fields in the sub-sub-TLV used to carry End. BIER address information.

Another sub-sub-TLV is used to carry BIERv6 encapsulation information, and Figure 12.21 shows its format.

TABLE 12.8 Fields in the Sub-TLV Used to Carry BIER Information

Field	Length	Description
Type	8 bits	The value is 32, indicating that this sub-TLV carries BIER information.
Length	8 bits	Length of the sub-TLV.
BAR	8 bits	BIER algorithm. Currently, the value can only be 0, indicating the default algorithm.
IPA	8 bits	IGP algorithm. Currently, the value can only be 0, indicating the default algorithm.
Subdomain-ID	8 bits	BIER sub-domain ID.
BFR-ID	16 bits	BFR-ID of the node in the sub-domain. If no BFR-ID is configured, 0 (invalid BFR-ID) is carried in packets.
Sub-sub-TLVs	Variable	This field is optional and determined by the Length field.

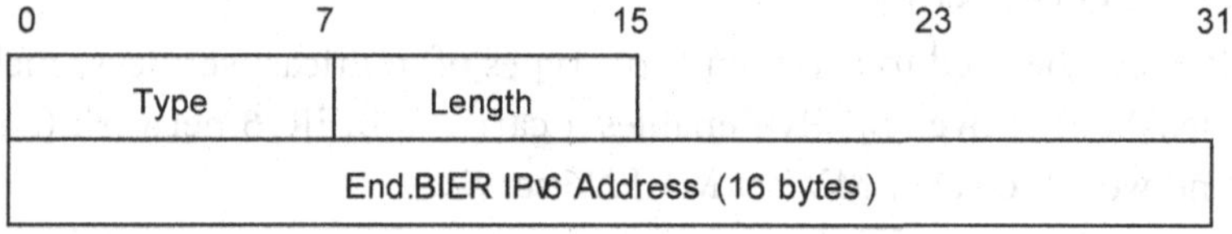

FIGURE 12.20 Sub-sub-TLV used to carry End.BIER address information.

TABLE 12.9 Fields in the Sub-sub-TLV Used to Carry End.BIER Address Information

Field	Length	Description
Type	8 bits	Type of the sub-sub-TLV.
Length	8 bits	Length of the sub-sub-TLV. The value is 16.
End.BIER IPv6 Address	128 bits	End.BIER address.

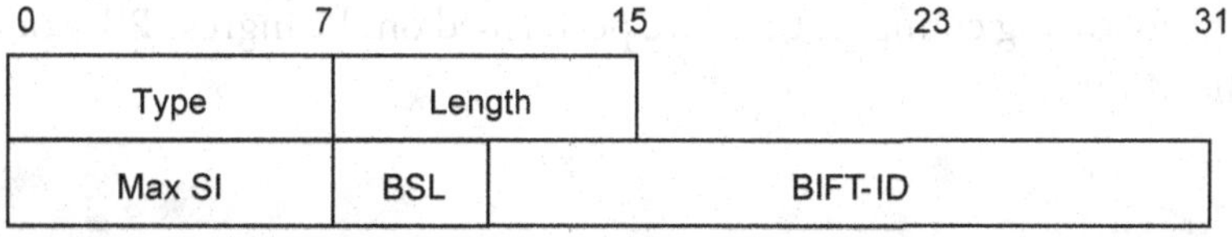

FIGURE 12.21 Format of the sub-sub-TLV used to carry BIERv6 encapsulation information.

TABLE 12.10 Fields in the Sub-sub-TLV Used to Carry BIERv6 Encapsulation Information

Field	Length	Description
Type	8 bits	Type of the sub-sub-TLV.
Length	8 bits	Length of the sub-sub-TLV.
Max SI	8 bits	Maximum SI in a specific <Sub-domain, BSL>.
BSL	4 bits	BSL coded value. For example, coded values 1, 2, and 3 indicate 64 bits, 128 bits, and 256 bits, respectively.
BIFT-ID	20 bits	Start BIFT ID in a specified <sub-domain, BSL>. For example, if Max SI is set to 3, the value of this field indicates the start BIFT ID of four contiguous BIFT IDs.

Table 12.10 describes the fields in the sub-sub-TLV used to carry BIERv6 encapsulation information.

12.3.4 MVPN over BIERv6

BIERv6 can be used to carry multiple types of multicast services, such as MVPN. MVPN over BIERv6 enables a carrier's BIERv6 network (transport network) to carry MVPN/MVPN6 services.

- MVPN over BIERv6 allows IPv4 multicast VPN services to be carried over an IPv6 network. The multicast service system, including the Set-Top Box (STB) and IPTV headend system, runs IPv4 multicast, but the transport network runs IPv6.
- MVPN6 over BIERv6 allows IPv6 multicast VPN services to be carried over an IPv6 network. Both the multicast service system and the transport network run IPv6.

The following part takes the networking example in Figure 12.16 to describe how to configure MVPN over BIERv6.

The following configurations are performed on the ingress PE (node A) of the MVPN:

```
#
interface loopback0
 ipv6 enable
 ipv6 address A::1 128
#
bier
 ipv6-block as1 2001:DB8:A1:: 96 static 32 //Configure an IPv6 address block
for BIERv6.
```

```
 sub-domain 6 ipv6
  bfr-prefix interface loopback0 //Configure a BFR-prefix.
  end-bier ipv6-block as1 opcode ::100 //Configure an End.BIER address.
  encapsulation ipv6 bsl 256 max-si 0 //Configure BIERv6 encapsulation.
#
ip vpn-instance vpn1
 ipv4-family
  multicast routing-enable
  mvpn
   sender-enable
   ipv6-underlay //Configure MVPN to use an IPv6 network as the underlay
network.
   src-dt4 locator as1 opcode ::2 //Configure an Src.DT4 address.
   spmsi-tunnel
    group wildcard source wildcard bier sub-domain 6
#
```

The following configurations are performed on egress PEs (nodes D/E) of the MVPN:

```
#
interface loopback0
 ipv6 enable
 ipv6 address D::1 128
#
bier
 ipv6-block as1 2001:DB8:D1:: 96 static 32 //Configure an IPv6 address block
for BIERv6.
 sub-domain 6 ipv6
  bfr-prefix interface loopback0 //Configure a BFR-prefix.
  end-bier ipv6-block as1 opcode ::200 //Configure an End.BIER address.
  encapsulation ipv6 bsl 256 max-si 0 //Configure BIERv6 encapsulation.
#
ip vpn-instance vpn1
 ipv4-family
  multicast routing-enable
  mvpn
   ipv6-underlay //Configure MVPN to use an IPv6 network as the underlay
network.
#
```

The preceding configurations assume that BIERv6 is deployed to carry multicast services on a network where SRv6 is not deployed. For example, on an IPv4/IPv6 dual-stack network, unicast reachability between MVPN sites can be implemented through IPv4 or IPv6 GRE tunnels, whereas multicast services between the MVPN sites are encapsulated and forwarded using BIERv6.

If SRv6 and SRv6 VPN have been deployed on the network, the End. BIER and Src.DT4 addresses required by MVPN services can be directly allocated from the locator address space of SRv6. It is not necessary to plan or allocate an extra address block for BIERv6. Assuming that a locator named **as1** has been configured for SRv6 on the ingress (PE A), the addresses in the locator address space can be used as End.BIER and Src.DT4 addresses. The following example shows the configurations performed on PE A.

```
#
segment-routing ipv6
 locator as1 ipv6-prefix 2001:DB8:A1:: 96 static 32 //Configure a locator
(named as1).
#
interface loopback0
 ipv6 enable
 ipv6 address A::1 128
#
bier
 sub domain 6 ipv6
  bfr-prefix interface loopback0 //Configure a BFR-prefix.
  end-bier locator as1 opcode ::100 //Configure an End.BIER address.
  encapsulation ipv6 bsl 256 max-si 0 //Configure BIERv6 encapsulation.
#
ip vpn-instance vpn1
 ipv4-family
  multicast routing-enable
  mvpn
   sender-enable
   ipv6-underlay //Configure MVPN to use an IPv6 network as the underlay
network.
   src-dt4 locator as1 opcode ::2 //Configure an Src.DT4 address.
   spmsi-tunnel
    group wildcard source wildcard bier sub-domain 6
#
```

Regardless of which configuration method is chosen, MVPN over BIERv6 and MVPN over BIER share similar fundamentals, except that the IPv6 source (Src.DT4) address identifying an MVPN instance required by the former is carried in the BGP Prefix-SID attribute of I-PMSI or S-PMSI A–D routes,[23] as shown in Figure 12.22.

12.3.5 GTM over BIERv6

In GTM over BIERv6, no VPN information needs to be transmitted. For this reason, the preceding MVPN signaling process is omitted, and the ingress PE only needs to configure a set of egress PEs to which each multicast flow needs to be sent and "pushes" the traffic to the egress PEs accordingly. Note that we also need to configure the IPv6 source address representing the GTM instance on the ingress PE to be encapsulated in the IPv6 SA field for BIERv6 packets.

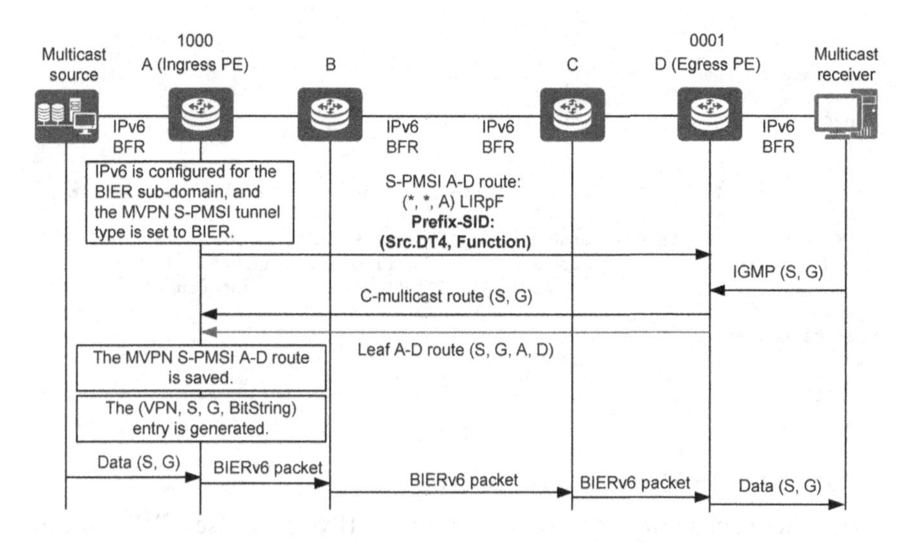

FIGURE 12.22 MVPN over BIERv6 service processing.

Using the networking example in Figure 12.16, the following configurations are performed on the ingress PE (node A):

```
#
interface loopback0
 ipv6 enable
 ipv6 address A::1 128
#
bier
 ipv6-block as1 2001:DB8:A1:: 96 static 32 //Configure an IPv6 address block.
  sub-domain 6 ipv6
   bfr-prefix interface loopback0 //Configure a BFR-prefix.
   end-bier ipv6-block as1 opcode ::100 //Configure an End.BIER address.
   encapsulation ipv6 bsl 256 max-si 0 //Configure BIERv6 encapsulation.
#
multicast-bier
 ipv6 source-address A::1 imposition
 static-imposition //Specify the BFR-IDs of egresses for which traffic needs
to be replicated.
  group 232.0.0.1 source 10.1.1.10 bier sub-domain 6 bsl 256 BFR-ID 1 to 80
  group 232.0.0.2 source 10.1.1.10 bier sub-domain 6 bsl 256 BFR-ID 1 to 80
#
```

The IPv6 source address representing the GTM instance is configured on egress PEs D and E for them to determine that the received BIERv6 packets belong to the GTM instance.

```
#
interface loopback0
 ipv6 enable
 ipv6 address D::1 128
#
bier
 ipv6-block as1 2001:DB8:D1:: 96 static 32 //Configure an IPv6 address block.
 sub-domain 6 ipv6
  bfr-prefix interface loopback0 //Configure a BFR-prefix.
  end-bier ipv6-block as1 opcode ::200 //Configure an End.BIER address.
  encapsulation ipv6 bsl 256 max-si 0 //Configure BIERv6 encapsulation.
#
multicast-bier
 ipv6 source-address A::1 disposition
#
```

12.3.6 Inter-AS Multicast over BIERv6

With the reachability and configurability of IPv6 addresses, BIERv6 can support inter-AS deployment easily. On the network shown in Figure 12.23, AS 65001 is connected to AS 65002 and AS 65003, and an IGP floods BIER or BIERv6 information only within each AS. To implement inter-AS multicast service transmission, we can configure the next hops to BFRs with BFR-IDs 1–64 in AS 65002, and to the BFRs with BFR-IDs 275–320 in AS 65003 on nodes A and B, respectively.

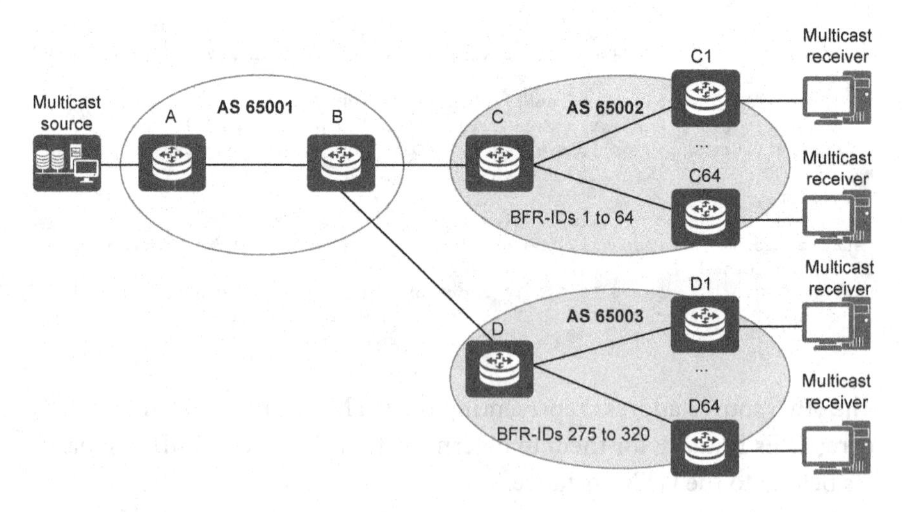

FIGURE 12.23 Inter-AS multicast over BIERv6.

For example, the following configurations are performed on node A:

```
#
bier
 ipv6-block as1 2001:DB8:A:: 96 static 32 //Configure an IPv6 address block.
  sub-domain 6 ipv6
   bfr-prefix interface loopback0
   end-bier ipv6-block as1 opcode ::300 //Configure an End.BIER address.
   static-bift //Configure the End.BIER addresses of the next hops to
involved BFRs.
    nexthop end-bier 2001:DB1:B::300 BFR-ID 1 to 64
    nexthop end-bier 2001:DB1:B::300 BFR-ID 275 to 320
#
```

The following configurations are performed on node B:

```
#
bier
 ipv6-block as1 2001:DB8:B:: 96 static 32 //Configure an IPv6 address block.
  sub-domain 6 ipv6
   bfr-prefix interface loopback0
   end-bier ipv6-block as1 opcode ::300 //Configure an End.BIER address.
   static-bift //Configure the End.BIER addresses of the next hops to
involved BFRs.
    nexthop end-bier 2001:DB1:C::300 BFR-ID 1 to 64
    nexthop end-bier 2001:DB1:D::300 BFR-ID 275 to 320
#
```

Except nodes A and B, the other nodes in AS 65001 do not need to be configured with BIERv6; instead, they only need to support IPv6 unicast forwarding, owing to the fact that they do not need to process BIER information or perform BIER forwarding. After receiving a multicast data packet from the multicast source, node A searches its BIFT and forwards the packet to node B. The nodes between nodes A and B only need to forward the packet to node B according to the packet's IPv6 unicast address. After receiving the packet, node B replicates it according to its BIFT and sends a copy each to nodes C and D. After nodes C and D receive the packet, they perform packet replication and forwarding in their respective ASs.

If a failure occurs on the link between nodes A and B, between nodes B and C, or between nodes B and D, the corresponding unicast routes change, triggering an update in the BIER forwarding path. Through this process, multicast traffic is automatically restored.

We can see that inter-AS multicast can be easily implemented by BIERv6 based on the IPv6 reachability. MVPN over BIERv6 and GTM over BIERv6 also support the inter-AS deployment like this.

12.3.7 Characteristics of BIERv6

BIERv6 combines the advantages of both native IPv6 and BIER. It encapsulates a standard BIER header into an IPv6 extension header and uses a unicast IPv6 address of the End.BIER address type as the destination address in the IPv6 header. Upon receiving a BIERv6 packet, each node searches its FIB for an entry according to the destination address in the packet, obtains the End.BIER instruction, processes the BIER header in the IPv6 extension header, and performs BIER forwarding. Based on the reachability of IPv6 unicast addresses, BIERv6 can implement hop-by-hop intra-AS or inter-AS multicast packet replication and forwarding even if some nodes on the network do not support BIERv6.

12.4 OTHER SRv6 MULTICAST TECHNOLOGIES

BIER/BIERv6 provides a brand-new multicast architecture, adapts to the development of SR/SRv6 networks, and sets the direction in which multicast technologies will develop. Like SR/SRv6, BIER/BIERv6 requires network devices to have certain hardware forwarding capabilities, and as such, routers or switches on the network must be both programmable and capable of evolving.

If some devices are non-programmable or incapable of evolving, we can use SR Replication or Tree SID,[28] a multicast transition solution based on SR/SRv6, without the need to deploy mLDP or RSVP-TE.

The Tree SID solution can use MPLS or IPv6 encapsulation, with the former being the focus of the current proposal and discussion. Tree SID based on MPLS does not require a device to be programmable or capable of evolving. Instead, it leverages the existing capability of MPLS P2MP forwarding.

By comparison, Tree SID based on IPv6 needs significant forwarding capability improvement and therefore is not suitable as a transition solution.

We describe the MPLS-based Tree SID solution and its characteristics based on the network topology shown in Figure 12.24.

12.4.1 P2MP Multicast Forwarding Path Establishment Using the Tree SID Solution

Based on the global SID concept in SR-MPLS, the Tree SID solution requires all devices on the network to be configured with the same SID

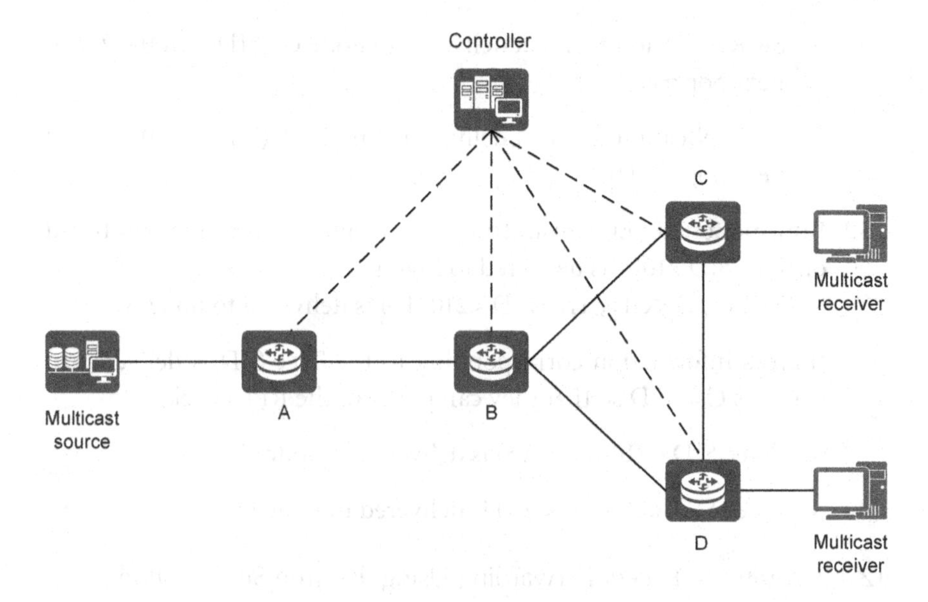

FIGURE 12.24 Topology of the MPLS-based Tree SID solution.

block. For example, nodes A, B, C, and D are each configured with a SID block ranging from 21000 to 22000 as the Tree SID block.

On the network shown in Figure 12.24, the process of establishing a P2MP multicast forwarding path using the MPLS-based Tree SID solution is as follows:

1. The controller delivers an SR Replication segment, which is identified by the tuple (SID, P2MP next hop), to each node. The combination of the SR Replication segments of the nodes from the ingress to egresses indicates the multicast forwarding path that can be identified with a Tree SID.

   For example, if SID 21001 is used to deliver the multicast tree, according to which node A sends packets to node B and node B sends copies of the packets to nodes C and D, the Tree SID is 21001.

   - SR Replication segment delivered to node A: (SID=21001, P2MP next hop =<B>)

   - SR Replication segment delivered to node B: (SID=21001, P2MP next hop=<C, D>)

- SR Replication segment delivered to node C: (SID=21001, P2MP next hop=NULL)
- SR Replication segment delivered to node D: (SID=21001, P2MP next hop=NULL)

2. Information about transmitting one or more multicast flows based on Tree SID 21001 is delivered to ingress A.
   (S <S1>, G <G1>, Tree SID <21001>) is delivered to node A.
3. Ingress information corresponding to the Tree SID is delivered to egresses C and D so that they can perform the RPF check.
   - (Tree SID <21001>, <A>) is delivered to node C.
   - (Tree SID <21001>, <A>) is delivered to node D.

12.4.2 Multicast Packet Forwarding Using the Tree SID Solution

On the network shown in Figure 12.24, the process of forwarding a multicast packet using the MPLS-based Tree SID solution is as follows:

1. After receiving a multicast data packet (S1, G1) from the multicast source, node A encapsulates the packet with SID 21001 according to its forwarding table and sends the packet to node B according to the SID and the corresponding forwarding entry.
2. After receiving the packet, node B replicates the packet and sends a copy to each of nodes C and D according to the SID and corresponding forwarding entry.
3. After receiving the packet, nodes C and D decapsulate it according to the SID and the forwarding entries, and also use the ingress information of the forwarding entries to perform the RPF check. If the packet passes the RPF check, they forward it through corresponding outbound interfaces.

12.4.3 Characteristics of the Tree SID Solution

If the controller detects a failure on a link (e.g., between nodes B and D), it delivers a new forwarding path to each corresponding node. By way of illustration, the controller sends addition signaling of (SID=21001, P2MP next hop=<D>) to node C and deletion signaling of (SID=21001, P2MP next hop=<D>) to node B.

That said, if the controller successfully delivers the addition signaling to node C but fails to deliver the deletion signaling to node B due to an exception (such as a failure or delay in data transmission between the controller and node B), and the link between nodes B and D recovers, node D will receive two copies of multicast traffic: one from link B→C→D and the other from link B→D.

From this information, we can see that the Tree SID solution depends on a controller for multicast forwarding path and service delivery. Multicast forwarding path delivery includes delivering the P2MP global SID and the corresponding P2MP next hops to each node so that a P2MP LSP from the ingress through transit nodes to egresses is established. The P2MP global SID is the incoming SID and the outgoing SID of the P2MP LSP on each node, and also the service SID representing the entire P2MP path. As for multicast service delivery, it includes delivering the mapping between (S, G) and the global SID to the ingress, as well as the ingress information of the P2MP LSP corresponding to the global SID to each egress for RPF check.

Although the Tree SID solution does not depend on PIM, mLDP, or RSVP-TE for MDT establishment, it does depend on the controller for centralized management and control. In other words, if a link or node fails, the controller has to deliver a new multicast forwarding path. This triggers convergence, which may take a long time. The maintenance methods also vary greatly from those of the existing multicast routing protocols, such as PIM.

In conclusion, we can use Tree SID as a transition technology on networks where devices are not programmable or not capable of evolving to support BIER. It goes without saying that PIM can still be used to carry multicast services on such networks.

12.5 STORIES BEHIND SRv6 DESIGN

Multicast protocols are developing along the following three paths:

- A unicast protocol is used to transmit states of multicast groups. The typical multicast protocols that fall into this category include the Distance Vector Multicast Routing Protocol (DVMRP)[29] based on Routing Information Protocol (RIP) and the Multicast Open Shortest Path First (MOSPF)[30] based on OSPF.
- An explicit multicast path is established based on multicast join information. The typical multicast protocols that can be placed

into this category include Core Based Tree,[31] PIM-SM,[4] and PIM-DM.[32]

- Multicast is implemented based on source routing. Specifically, the source node encapsulates a set of destination nodes into a packet before sending the packet. The typical multicast protocols covered by this category include IPv4 Option for Sender Directed Multi-Destination Delivery[33] and Explicit Multicast (Xcast).[34]

Later some variants of multicast protocols were developed. TRILL, which transmits states of multicast groups based on IS-IS, belongs to the first category.[35] On the other hand, the MPLS multicast technology uses mLDP or P2MP RSVP-TE signaling to explicitly establish a P2MP MDT and falls under the second category. Finally, BIER and BIERv6 encapsulate a BitString that represents the set of destination nodes in the packet and belong to the third category.

The Internet Architecture Board (IAB) is of the opinion that the second path features the best network scalability (e.g., in inter-AS deployment) and therefore chose it as the multicast development direction in RFC 2902. That being said, the IAB also acknowledged that this path failed to meet the scalability requirements of some other items (such as the number of multicast groups or sessions).[36] With the deployment of multicast services, the industry has reached a consensus on the fact that the scalability of multi-session for multicast is particularly important. It is under this circumstance that the BIER/BIERv6 technology is developed.

REFERENCES

[1] Fenner W. Internet Group Management Protocol, Version 2[EB/OL]. (2015-05-05)[2020-03-25]. RFC 2236.

[2] Cain B, Deering S, Kouvelas I, Fenner B, Thyagarajan A. Internet Group Management Protocol, Version 3[EB/OL]. (2020-01-21)[2020-03-25]. RFC 3376.

[3] Deering S, Fenner W, Haberman B. Multicast Listener Discovery (MLD) for IPv6[EB/OL]. (2013-03-02)[2020-03-25]. RFC 2710.

[4] Fenner B, Handley M, Holbrook H, Kouvelas I, Parekh R, Zhang Z, Zheng L. Protocol Independent Multicast - Sparse Mode (PIM-SM): Protocol Specification (Revised)[EB/OL]. (2020-01-21)[2020-03-25]. RFC 7761.

[5] Rosen E, Aggarwal R. Multicast in MPLS/BGP IP VPNs[EB/OL]. (2015-10-14)[2020-03-25]. RFC 6513.

[6] Wijnands IJ, Rosen E, Dolganow A, Przygienda T, Aldrin S. Multicast Using Bit Index Explicit Replication (BIER)[EB/OL]. (2018-06-05)[2020-03-25]. RFC 8279.

[7] Rosen E, Cai Y, Wijnands IJ. Cisco Systems' Solution for Multicast in BGP/MPLS IP VPNs[EB/OL]. (2015-10-14)[2020-03-25]. RFC 6037.

[8] Andersson L, Minei I, Thomas B. LDP Specification[EB/OL]. (2020-01-21)[2020-03-25]. RFC 5036.

[9] Awduche D, Berger L, Gan D, Li T, Srinivasan V, Swallow G. RSVP-TE: Extensions to RSVP for LSP Tunnels[EB/OL]. (2020-01-21)[2020-03-25]. RFC 3209.

[10] Wijnands IJ, Ed, Minei I, Ed, Kompella K, Thomas B. Label Distribution Protocol Extensions for Point-to-Multipoint and Multipoint-to-Multipoint Label Switched Paths[EB/OL]. (2015-10-14)[2020-03-25]. RFC 6388.

[11] Aggarwal R, Ed, Papadimitriou D, Yasukawa S. Extensions to Resource Reservation Protocol - Traffic Engineering (RSVP-TE) for Point-to-Multipoint TE Label Switched Paths (LSPs)[EB/OL]. (2020-01-21)[2020-03-25]. RFC 4875.

[12] Aggarwal R, Rosen E, Morin T, Rekhter Y. BGP Encodings and Procedures for Multicast in MPLS/BGP IP VPNs[EB/OL]. (2015-10-14)[2020-03-25]. RFC 6514.

[13] Awduche D, Chiu A, Elwalid A, Widjaja I, Xiao X. Overview and Principles of Internet Traffic Engineering[EB/OL]. (2020-01-21)[2020-03-25]. RFC 3272.

[14] Awduche D, Berger L, Gan D, Li T, Srinivasan V, Swallow G. RSVP-TE: Extensions to RSVP for LSP Tunnels[EB/OL]. (2020-01-21)[2020-03-25]. RFC 3209.

[15] Wijnands IJ, Rosen E, Dolganow A, Tantsura J, Aldrin S, Meilik I. Encapsulation for Bit Index Explicit Replication (BIER) in MPLS and Non-MPLS Networks[EB/OL]. (2019-02-22)[2020-03-25]. RFC 8296.

[16] Ginsberg L, Przygienda T, Aldrin S, Zhang Z. Bit Index Explicit Replication (BIER) Support via IS-IS[EB/OL]. (2018-06-07)[2020-03-25]. RFC 8401.

[17] Psenak P, Kumar N, Wijnands IJ, Dolganow A, Przygienda T, Zhang J, Aldrin S. OSPFv2 Extensions for Bit Index Explicit Replication (BIER)[EB/OL]. (2018-12-19)[2020-03-25]. RFC 8444.

[18] Rosen E, Sivakumar M, Przygienda T, Aldrin S, Dolganow A. Multicast VPN Using Bit Index Explicit Replication (BIER)[EB/OL]. (2019-04-08)[2020-03-25]. RFC 8556.

[19] Dolganow A, Kotalwar J, Rosen E, Zhang Z. Explicit Tracking with Wildcard Routes in Multicast VPN[EB/OL]. (2019-02-19)[2020-03-25]. RFC 8534.

[20] Mcbride M, Xie J, Dhanaraj S, Asati R. BIER IPv6 Requirements[EB/OL]. (2020-01-15)[2020-03-25]. draft-ietf-bier-ipv6-requirements-04.

[21] Xie J, Geng L, Mcbride M, Dhanaraj S, Yan G, Xia Y. Encapsulation for BIER in Non-MPLS IPv6 Networks[EB/OL]. (2020-03-09)[2020-03-25]. draft-xie-bier-ipv6-encapsulation-06.

[22] Xie J, Wang A, Yan G, Dhanaraj S. BIER IPv6 Encapsulation (BIERv6) Support via IS-IS[EB/OL]. (2020-01-13)[2020-03-25]. draft-xie-bier-ipv6-isis-extension-01.
[23] Xie J, Mcbride M, Dhanaraj S, Geng L. Use of BIER IPv6 Encapsulation (BIERv6) for Multicast VPN in IPv6 networks[EB/OL]. (2020-01-13)[2020-03-25]. draft-xie-bier-ipv6-mvpn-02.
[24] Geng L, Xie J, Mcbride M, Yan G. Inter-Domain Multicast Deployment using BIERv6[EB/OL]. (2020-01-13)[2020-03-25]. draft-geng-bier-ipv6-inter-domain-01.
[25] Wijnands I, Xu X, Bidgoli H. An Optional Encoding of the BIFT-id Field in the non-MPLS BIER Encapsulation[EB/OL]. (2020-02-10)[2020-03-25]. draft-ietf-bier-non-mpls-bift-encoding-02.
[26] Li T, Smit H. IS-IS Extensions for Traffic Engineering[EB/OL]. (2015-10-14)[2020-03-25]. RFC 5305.
[27] Deering S, Hinden R. Internet Protocol Version 6 (IPv6) Specification[EB/OL]. (2020-02-04)[2020-03-25]. RFC 8200.
[28] Voyer D, Filsfils C, Parekh R, Bidgoli H, Zhang Z. SR Replication Segment for Multi-point Service Delivery[EB/OL]. (2019-11-27)[2020-03-25]. draft-voyer-spring-sr-replication-segment-02.
[29] Waitzman D, Partridge C, Deering S. Distance Vector Multicast Routing Protocol[EB/OL]. (2013-03-02)[2020-03-25]. RFC 1075.
[30] Moy J. MOSPF: Analysis and Experience[EB/OL]. (2013-03-02)[2020-03-25]. RFC 1585.
[31] Ballaridie A. Core Based Trees (CBT version 2) Multicast Routing[EB/OL]. (2013-03-02)[2020-03-25]. RFC 2189.
[32] Adams A, Nicholas J, Siadak W. Protocol Independent Multicast - Dense Mode (PIM-DM): Protocol Specification (Revised)[EB/OL]. (2020-01-21)[2020-03-25]. RFC 3973.
[33] Graff G. IPv4 Option for Sender Directed Multi-Destination Delivery[EB/OL]. (2013-03-02)[2020-03-25]. RFC 1770.
[34] Boivie R, Feldman N, Imai Y, et al. Explicit Multicast (Xcast) Concepts and Options[EB/OL]. (2020-01-21)[2020-03-25]. RFC 5058.
[35] Touch J, Perlman R. Transparent Interconnection of Lots of Links (TRILL): Problem and Applicability Statement[EB/OL]. (2015-10-14)[2020-03-25]. RFC 5556.
[36] Deering S, Hares S, Perkins C, et al. Overview of the 1998 IAB Routing Workshop[EB/OL]. (2015-11-11)[2020-03-25]. RFC 2902.

IV

Summary and Future Developments

CHAPTER 13

SRv6 Industry and Future

SRv6 HAS COME A LONG WAY SINCE 2017, sparking a new wave of innovation around IPv6. This chapter describes the progress of the SRv6 industry, as well as the future innovation work including SRv6 extension header compression and Application-aware IPv6 Networking (APN6). In addition, this chapter describes the possible roadmap for the IPv6+ era.

13.1 SRv6 INDUSTRY PROGRESS

This section describes the progress of SRv6 standards, product implementations, interoperability tests, commercial deployments, and industry activities.

13.1.1 SRv6 Standardization Progress

Most SRv6 standardization work is being done in the IETF SPRING Working Group. In addition, the SRH standardization is being done in the 6MAN Working Group. The standardization of protocol extensions of IGP, BGP, PCEP, and VPN is being done in the LSR Working Group, IDR Working Group, PCE Working Group, and BGP Enabled ServiceS (BESS) Working Group, respectively.

SRv6 standardization work can be divided into two parts:

1. Basic SRv6 features: As listed in Table 13.1, basic SRv6 features include the SRv6 network programming framework, SRH encapsulation format, and protocol extensions of IGP, BGP/VPN, BGP-LS,

TABLE 13.1 SRv6 Standards

Domain	Subject	Document
Architecture/ use case	SRv6 Network Programming	draft-ietf-spring-srv6-network-programming[1]
SRH	IPv6 Segment Routing Header (SRH)	RFC 8754[2]
IGP	IS-IS Extensions for SRv6	draft-ietf-lsr-isis-srv6-extensions[3]
	OSPFv3 Extensions for SRv6	draft-ietf-lsr-ospfv3-srv6-extensions[4]
VPN	SRv6 VPN	draft-ietf-bess-srv6-services[5]
SDN interface	BGP-LS for SRv6	draft-ietf-idr-bgpls-srv6-ext[6]
	PCEP for SRv6	draft-ietf-pce-segment-routing-ipv6[7]

and PCEP, which can provide the fundamental network services including VPN, TE, and FRR. Currently, all drafts of basic SRv6 features have been adopted by the working groups, and RFC 8754 has defined the fundamental SRH encapsulation. This means SRv6 standardization work enters a new stage of maturity.

2. New SRv6 services for 5G and cloud: These new services include network slicing, DetNet, OAM, IOAM, SFC, SD-WAN, and SRv6 multicast/BIERv6. These services propose new requirements on network programming and require new information to be encapsulated in the forwarding plane. SRv6 can satisfy these requirements well by leveraging its unique advantage in network programming. The urgency of these service requirements is different, as is the level of corresponding standardization. In general, standardization for SRv6 OAM, IOAM, and SFC has been progressing rapidly, and multiple IETF working group drafts have been released. Standardization on network slicing is also active and the draft of VPN+ framework has been adopted by the working group. The concept of using SRv6 SIDs to indicate forwarding resources in the forwarding plane has been widely recognized as the way to guarantee service requirements.

As the series of SRv6 drafts are adopted by working groups, they become mature enough to support the deployment of services such as SRv6 BE, TE, VPN, and FRR. New services will continue to be deployed as standards mature.

13.1.2 SRv6-Related Products

Currently, major device, tester, and chip vendors have supported SRv6:

- Major device vendors
    a. Huawei: all series of router products
    b. Cisco: series of ASR and NCS products
- Tester vendors: Sprint and Ixia
- Chip vendors: HiSilicon and Broadcom, which have released commercial chips. The chips have been verified on major devices.

In addition, most major open-source platforms support SRv6, such as Linux Kernel, Linux srext module, and FD.io VPP. Some open-source tools, such as Wireshark, Tcpdump, Iptables, Nftables, and Snort, can also process IPv6 packets containing SRHs.

13.1.3 SRv6 Interoperability Test

Multiple rounds of interoperability tests have been conducted in the industry to ensure the interconnection of SRv6 devices provided by different vendors. This enables carriers to deploy SRv6 networks with devices from multiple vendors.

13.1.3.1 SRv6 Interoperability Test by the EANTC

So far, the European Advanced Networking Test Center (EANTC) has conducted two successful SRv6 interoperability tests, one in 2018 and the other in 2019.

The results of the interoperability test conducted in March 2019 were presented at the MPLS+SDN+NFV World Congress held in April 2019. The test verified the SRv6 implementation on five types of devices and the interoperability of SRv6 (including SRH processing) in the following scenarios:

1. IPv4 traffic over SRv6 L3VPN
2. IPv6 traffic over SRv6 L3VPN
3. SRH-based TI-LFA FRR for link protection
4. OAM process (ping and traceroute)

In the interoperability test scenarios, bidirectional traffic was sent between the ingress and egress PEs, and SRv6 encapsulation (such as H.Encaps) and decapsulation (such as End.DT4 and End.DT6) were tested. It was verified that SRv6 traffic could pass through SRv6-incapable P nodes as long as they support IPv6.

13.1.3.2 SRv6 Interoperability Test by China's Expert Committee of Promoting Large-Scale IPv6 Deployment

In November 2019, China's Expert Committee of Promoting Large-Scale IPv6 Deployment successfully organized an SRv6 interoperability test, verifying the functions of SRv6 VPN, SRv6 Policy, SRv6 SFC, etc. The test covers the following scenarios:

1. Services based on SRv6 BE, including:
    a. L3VPN over SRv6 BE (including basic functions, TI-LFA FRR, and OAM)
    b. EVPN VPWS over SRv6 BE
    c. EVPN VPLS over SRv6 BE
2. Services based on SRv6 Policy, including:
    a. L3VPN over SRv6 Policy
    b. Path optimization based on SRv6 Policy
    c. SFCs based on SRv6 Policy:
        - SFC1: defense against TCP SYN packet attacks
        - SFC2: IP flow access control and traffic monitoring
        - SFC3: web user access control and content audit

The success of the multi-vendor interoperability tests indicates that SRv6 can be deployed on commercial networks. In fact, SRv6 is being deployed quickly.

13.1.4 Commercial SRv6 Deployment

To date, SRv6 has been deployed on networks all over the world, such as SoftBank Japan, China Telecom, LINE Japan, Iliad Italy, China Unicom, MTN Uganda, NOIA Network, and CERNET2 networks.[8]

Among these deployments, the rapid development of SRv6 in China is worthy of attention. China has promoted large-scale IPv6 deployment since 2017, and after nearly 2 years of construction, almost all networks of carriers and enterprises can fully support IPv6, providing a solid foundation for large-scale SRv6 deployment. By the end of 2019, commercial deployment and field trial projects had been carried out on more than ten networks, including China Telecom, China Unicom, China Mobile, Bank of China, and CERNET2 networks. The SRv6 advantages, including simple large-scale cross-domain networking, easy incremental deployment, and fast service provisioning, have been fully verified in these deployments. Moreover, China Academy of Information and Communications Technology (CAICT) organized China Telecom, China Unicom, and Huawei to summarize the experience in SRv6 deployment, and has shared it throughout the industry.[9]

13.1.5 SRv6 Industrial Activities

Multiple SRv6 industrial activities have been held to further consolidate the industry and promote SRv6 innovations.

- In April 2019, the first SRv6 industry roundtable was held at the MPLS+SDN+NFV World Congress in Paris, France.
- In June 2019, China's Expert Committee of Promoting Large-Scale IPv6 Deployment followed up by hosting the first SRv6 industry salon in Beijing, China.
- In November 2019, China's Expert Committee of Promoting Large-Scale IPv6 Deployment approved the establishment of the IPv6+ Innovation Working Group. The working objectives are to strengthen the systematic innovation of IPv6 for the next-generation Internet based on the achievements of China's large-scale IPv6 deployment and to integrate IPv6 technology industry chains (including enterprises, universities, research institutions, application service providers) to proactively verify and demonstrate new technologies of IPv6+ (including SRv6, VPN+, DetNet, BIERv6, SFC, and OAM) for the possible directions of network routing protocols, management automation, intelligence, and security, continuously improving IPv6 technology standards and enhancing China's international competitiveness in the IPv6 field.

- In December 2019, China's Expert Committee of Promoting Large-Scale IPv6 Deployment hosted the second SRv6 industry salon, with more than 110 experts attending. Wu Hequan, an academician of Chinese Academy of Engineering (CAE) and the director of the Expert Committee, gave a keynote speech. Experts discussed the innovative technologies of SRv6 and IPv6+ as well as how to promote SRv6 technology and industry, and jointly released *SRv6 Technology and Industry White Paper* and *SRv6 Interoperability Test Report.*

These industrial activities have played an active role in promoting SRv6 innovation and deployment.

The SRv6 industry is developing rapidly, and great progress has been made in product implementation, interoperability test, and commercial deployment. However, during commercial deployment, there is a voice of concern about the size of an SRv6 header.

13.2 SRv6 EXTENSION HEADER COMPRESSION

13.2.1 Impact of the SRv6 Extension Header Length

In the encapsulation mode, the header overhead increases after an SRv6 ingress encapsulates an outer IPv6 header and an SRH into a packet before forwarding it. The SRH length increases further if there are a large number of SRv6 SIDs, which may cause the following issues:

1. Payload efficiency reduction: The header added in SRv6 acts as a transmission overhead. When there are a large number of SIDs in the SRH, the header length increases and the payload proportion decreases. This results in low payload efficiency.

2. Hardware forwarding performance reduction: As the number of SIDs increases, the SID stack depth in SRv6 packets may exceed the depth read by hardware each time. As a result, the hardware must read a second time, causing forwarding performance to reduce.

3. Packet forwarding restricted by the Maximum Transmission Unit (MTU): After the SRv6 header is added, the size of packets may exceed the MTU, causing packet fragmentation or loss.

13.2.2 SRv6 Extension Header Compression Solutions

To minimize the impact of the SRv6 header overhead, the industry proposed multiple possible solutions, including Generalized SRv6 Network Programming (G-SRv6), Micro Segment (uSID), and Segment Routing Mapped To IPv6 (SRm6).

The principles of G-SRv6 and uSID are similar. Both G-SRv6 and uSID compress the SRv6 extension header by removing redundant information from segment lists. They differ in SID encoding formats and the modes for updating the SID in an IPv6 destination address.

In an SRv6 domain, SIDs are allocated from an address block and therefore have the same prefix. If a complete segment list contains multiple SIDs, multiple redundant prefixes are carried. To reduce the SRH length and compress the header, the redundant information can be removed from the segment list so that the segment list carries only the different parts, such as the Node ID and Function ID in a SID. These parts are called Compressed SIDs (C-SIDs).

G-SRv6 is implemented by reducing redundancy and replacing the different parts of the SRv6 SID. In G-SRv6, only the first SID in a compressed SRv6 segment list carries complete 128-bit information, including the Common Prefix, C-SID, and possible Arguments/Padding. For the remaining SIDs, only the C-SIDs need to be encoded into the segment list. Therefore, when an IPv6 destination address is updated, only the following C-SID in the segment list needs to be copied to this address to form a complete SRv6 SID.

uSID is implemented by reducing redundancy and shifting bits to the left to update the SRv6 SID. In the uSID solution, a 128-bit space stores one Common Prefix and multiple C-SIDs, thereby reducing the number of times that the Common Prefix is stored in the SRH. The 128-bit data is copied to the IPv6 destination address. When the active SID needs to be replaced with the next C-SID, all bits after the Common Prefix are shifted to the left by the length of a C-SID.

SRm6 is implemented by mapping short IDs to IPv6 interface addresses. It establishes mappings between short IDs (such as 16- or 32-bit IDs) and IPv6 interface addresses, and stores the mapping table on all devices on a network. SRm6 changes a 128-bit IPv6 address stored in the SRH to a 16- or 32-bit ID, reducing the header overhead. When the next SID needs to be loaded, a device maps the next ID to an IPv6 address and then updates it to the IPv6 DA field.

13.2.2.1 G-SRv6

G-SRv6 is an SRv6-compatible compression solution proposed in draft-cl-spring-generalized-srv6-for-cmpr.[10–12] G-SRv6 not only supports SRv6 compression to reduce the overhead of SRv6 headers but also supports smooth transition from SRv6 to G-SRv6 as G-SRv6 SIDs can also be programmed in the same SRH with common SRv6 SIDs.

G-SRv6 involves SRv6 compression and compatibility with SRv6.

- In SRv6 compression, redundant information in the segment list is removed from the SRH based on the regularity of the SRv6 SID format and allocation. So the SRH carries only different parts of SRv6 SIDs, reducing the header overhead.
- In compatibility with SRv6, a flavor is introduced to indicate the format of the next SID, processing SIDs of different lengths and in turn supporting hybrid programming and processing of different types of SIDs.

1. C-SID definition

   SIDs in an SRv6 domain are allocated from a SID address block and therefore have a Common Prefix. If the SID in the destination address of the IPv6 header carries the Common Prefix, the SIDs in the SRH only need to carry the different parts. As such, when the address is updated, only the different parts are replaced to restore the original SID. In G-SRv6, the different parts are called the Compressed SID (C-SID). Figure 13.1 shows the relationship between a complete SID and a C-SID.

   As shown in Figure 13.1, multiple SRv6 SIDs can have the same prefix, which can also be referred to as the Locator Block. The C-SID consists of the locator's Node ID and the following Function ID.

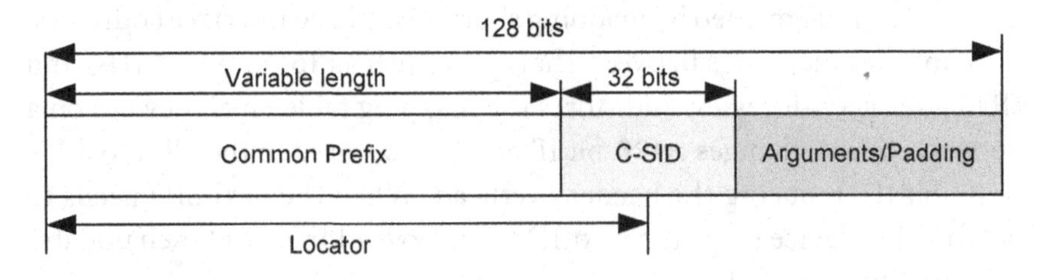

FIGURE 13.1 Relationship between a complete SID and a C-SID.

In an actual network design, the total length of the Common Prefix and C-SID may be less than 128 bits. In this case, the remaining bits can be padded with 0s.

2. G-SID definition

To be compatible with the SRH, C-SIDs need to be arranged to align with 128 bits in the SRH. That is, four 32-bit C-SIDs or multiple C-SIDs of other lengths need to be arranged in a 128-bit space. If the length is less than 128 bits, 0s are used to pad it to 128 bits. The G-SRv6 solution defines the Generalized SRv6 SID (G-SRv6 SID), Generalized SID (G-SID) for short, to indicate the 128-bit unit, which is different from the common 128-bit SRv6 SID. The G-SID is a 128-bit value and can contain a common SRv6 SID or multiple C-SIDs. A G-SID that contains multiple C-SIDs is called the Compression G-SID. Figure 13.2 shows the possible formats of a G-SID with 32-bit C-SIDs.

A Compression G-SID is also 128 bits long, the same as an existing SRv6 SID, and can be used to encapsulate multiple C-SIDs.

In addition, the G-SID can also be extended to encapsulate information of MPLS labels traversing an MPLS domain, or information of the IPv4 tunnel traversing an IPv4 domain. It can be considered as a generalization of an existing SRv6 SID and is therefore called a G-SID.

Network programming based on the G-SID is called Generalized SRv6 Network Programming (G-SRv6).

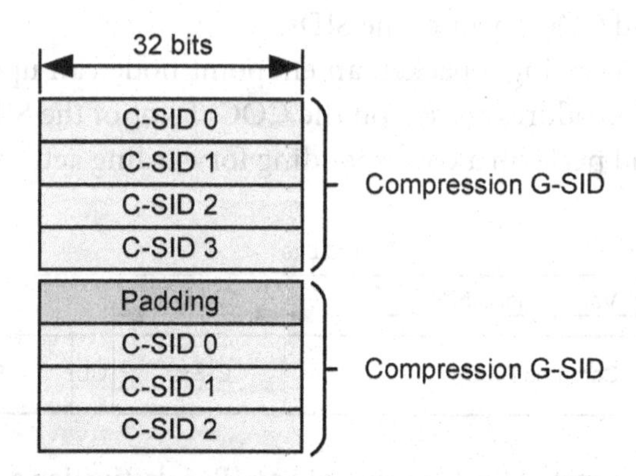

FIGURE 13.2 Format of a 128-bit Compression G-SID.

3. G-SRH encapsulation

To support hybrid encoding of a C-SID and an SRv6 SID in an SRH, a C-SID Left (CL) pointer is added to the Arguments of a complete SID. This pointer is located at the least significant bit of the Arguments (excluding the Padding) to identify the location of a C-SID in a G-SID. When the length of a C-SID is 32 bits, the CL pointer is located in the two least significant bits of the Arguments. In this case, if the total length of the Common Prefix, C-SID, and Arguments is 128 bits, the CL pointer is located in the two least significant bits of the 128 bits. Figure 13.3 shows the format of the CL pointer in an IPv6 destination address with the Padding field.

G-SRv6 extends only the format of a SID that can be carried in an SRH and does not change the definitions of the original fields in the standard SRH.[12] Such an SRH can carry SIDs of multiple formats and is therefore called the Generalized SRH (G-SRH). Figure 13.4 shows the G-SRH format.

To satisfy the programming requirements of the controller or ingress, a C flag needs to be added to identify whether the SID can be compressed when a SID is advertised. Such a SID is called the Compressible SRv6 SID.

During segment list processing, a Continuation of Compression (COC) flavor needs to be introduced to express the need of updating the next 32-bit C-SID to the DA field. By default, a SID that does not carry a COC flavor indicates that the length of the next SID is 128 bits.

The controller or ingress can encode the segment list, based on the C flag and COC flavor of the SIDs.

After receiving a packet, an endpoint node can update the IPv6 destination address based on the COC flavor of the SID and the CL value, and perform a corresponding forwarding action based on the

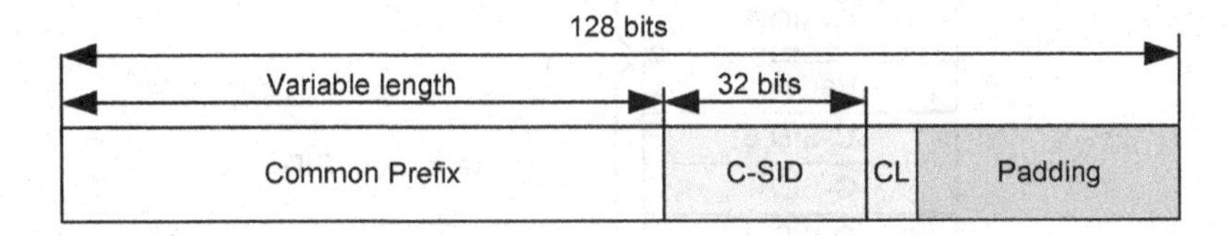

FIGURE 13.3 Format of the CL pointer in an IPv6 destination address with the Padding field.

Next Header	Hdr Ext Len	Routing Type	Segments Left
Last Entry	Flags	Tag	
Generalized Segment List [0] (128 bits)			
...			
Generalized Segment List [n] (128 bits)			
Optional TLV objects (variable)			

FIGURE 13.4 G-SRH format.

behavior of the SID, for example, to search a FIB table to forward the packet. The following pseudocode shows an example of such processing:

```
if IPv6 DA hits LOCAL SID table
   if LOCAL_SID.COC == TRUE
      if CL!= 0              //Reads the next C-SID from the current G-SID.
       CL--;
      else                   //Reads the first C-SID from the next G-SID.
         SL--;
         CL = 3;
      DA[CP..CP+31] = SRH[SL][CL]; //Updates the C-SID in the IPv6
      destination address.
      Forward the packets.     //Forwards packets.
   else
      Perform SRv6 Processing. //If no COC flavor is available, SRv6
processing is performed.
```

4. G-SRv6 mechanism

G-SRv6 supports the compression of an SRH and the hybrid programming of SRv6 SIDs and C-SIDs on an SRv6 network.

Figure 13.5 shows the forwarding process of the G-SRv6 solution when packets traverse SRv6 compression and non-compression domains.

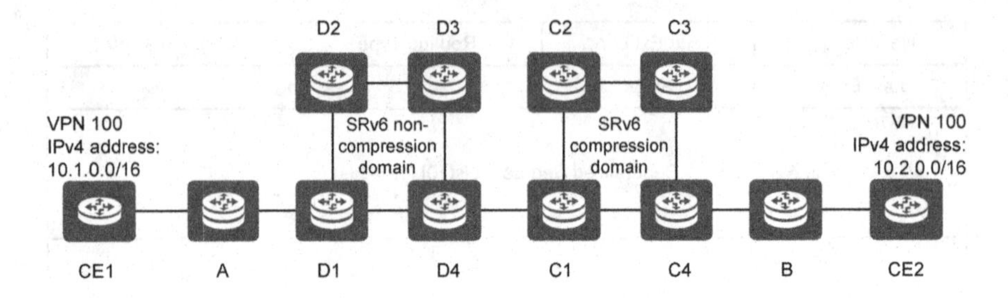

FIGURE 13.5 Forwarding process of the G-SRv6 solution.

Assume that SIDs and IP addresses are allocated as follows:

- The IP addresses of CE1 and CE2 are X (10.1.0.1) and Y (10.2.0.1), respectively.
- The SRv6 tunnel originating from node A uses the source address A::1.
- Node B allocates an SRv6 End.DT4 SID of B::100 to VPN 100.
- An End SID of D::*k*:1 is allocated to node D*k* in the SRv6 non-compression domain.
- An End SID of C::*k*:1:0:CL is allocated to node C*k* in the SRv6 compression domain, and the Common Prefix is C::/64. The 32-bit C-SID *k*:1 carries the COC flavor, whereas the C-SID *k*:2 does not. The C-SID is followed by a 32-bit Arguments, and the CL pointer is located in the two least significant bits of the Arguments.

k specifies the ID of a node in the topology.

Assume that the optimal path from CE1 to CE2 is A→D1→D2→D3→D4→C1→C2→C3→C4→B. When the G-SRv6 solution is used, the G-SRv6 path can be divided into the following sub-paths: D1→D2→D3→D4 in the SRv6 non-compression domain and C1→C2→C3→C4 in the SRv6 compression domain.

The sub-path in the SRv6 non-compression domain can be represented as a list formed by four SRv6 G-SIDs (D::1:1, D::2:1, D::3:1, D::4:1), and the sub-path in the SRv6 compression domain can be represented as a list formed by one SRv6 G-SID and one Compression G-SID (C::1:1:0:0, <2:1, 3:1, 4:2, Padding>).

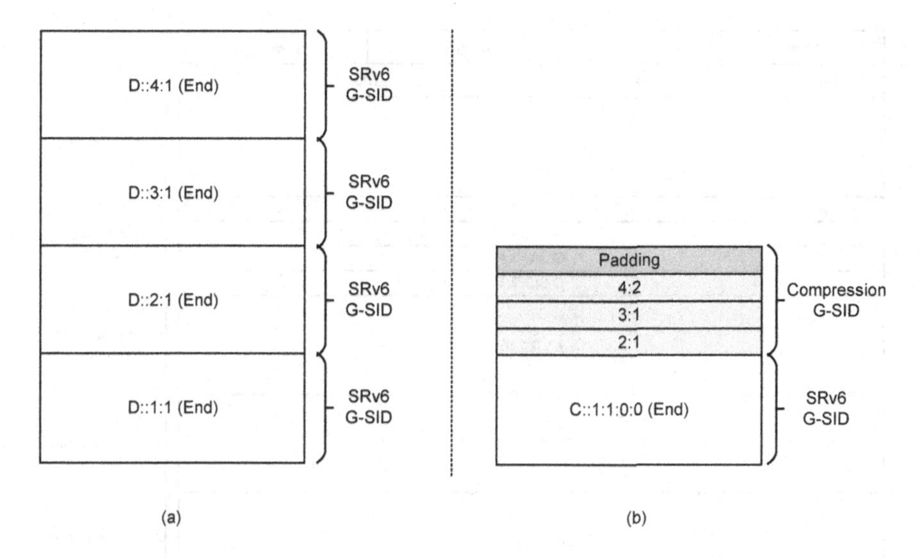

FIGURE 13.6 Packet formats corresponding to the two sub-paths.

Figure 13.6 (a) and Figure 13.6 (b) show the packet formats corresponding to the two sub-paths.

Figure 13.7 shows the complete G-SRH.

When CE1 sends a packet (X, Y) to CE2:

1. After receiving the packet, node A encapsulates it as (A::1, D::1:1) (B::100, <Padding, 4:2, 3:1, 2:1>, C::1:1:0:0, D::4:1, D::3:1, D::2:1, D::1:1) (X, Y) using G-SRv6, and the SL value in the G-SRH is 6.

2. Node A searches the IPv6 routing table and forwards the packet to node D1 based on the destination address D::1:1.

3. Node D1 determines that the destination address is a local SRv6 End SID, performs the End behavior to update the packet, and then searches the routing table to forward the packet to node D2. The packet sent by node D1 is (A::1, D::2:1) (B::100, <Padding, 4:2, 3:1, 2:1>, C::1:1:0:0, D::4:1, D::3:1, D::2:1, D::1:1) (X, Y), and the SL value in the G-SRH is 5.

4. Nodes D2, D3, and D4 perform the same action upon receipt of the packet. The packet sent by node D4 to node C1 is (A::1, C::1:1:0:0) (B::100, <Padding, 4:2, 3:1, 2:1>, C::1:1:0:0, D::4:1, D::3:1, D::2:1, D::1:1)

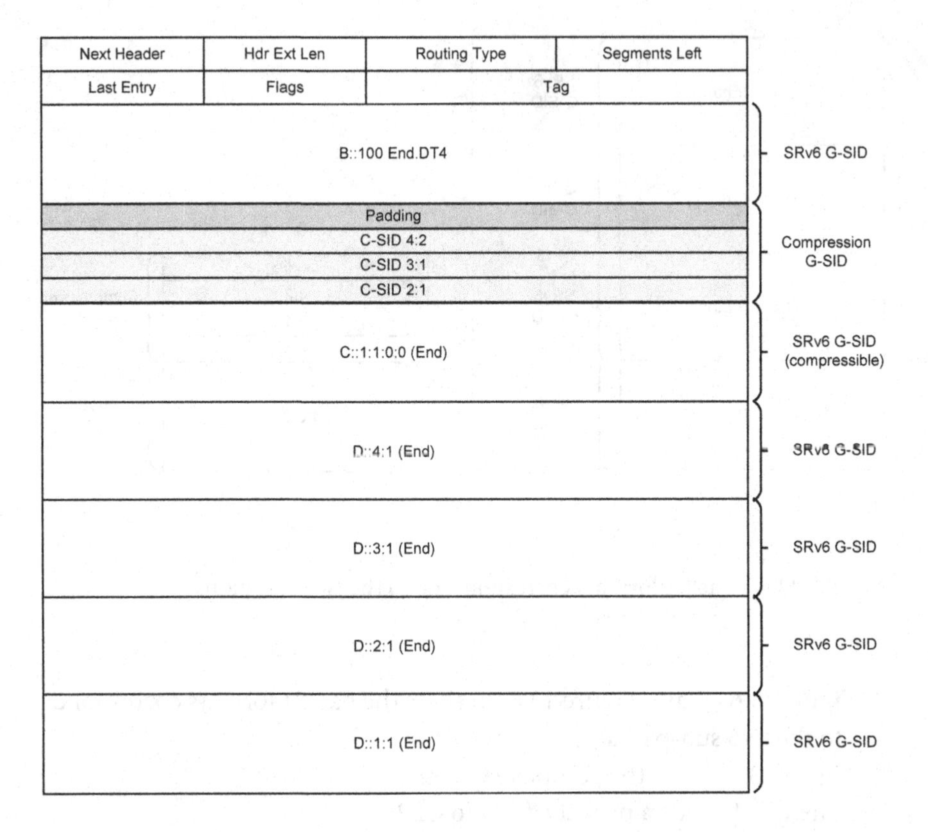

FIGURE 13.7 Complete G-SRH.

(X, Y), the SL value in the G-SRH is 2, and the CL value in the destination address is 0.

5. After receiving the packet, node C1 determines that the destination address C::1:1:0:0 is the local SRv6 End SID that carries the COC flavor. In this case, because the CL value is 0, node C1 decrements the SL value and assigns the value 3 to CL. Node C1 then reads SRH[SL=1][CL=3] based on SL and CL to obtain the C-SID 2:1 and replaces the C-SID in the corresponding position of the IPv6 destination address. The new IPv6 destination address is C::2:1:0:3. Node C1 searches the routing table based on this address and forwards the packet to node C2. The packet sent by node C1 is (A::1, C::2:1:0:3) (B::100, <Padding, 4:2, 3:1, 2:1>, C::1:1:0:0, D::4:1, D::3:1, D::2:1, D::1:1) (X, Y), the SL value in the G-SRH is 1, and the CL value in the destination address is 3.

6. After receiving the packet, node C2 determines that the destination address is the local SRv6 End SID that carries the COC flavor. In this case, because the CL value is greater than 0, node C2 decrements the CL value to 2. Node C2 then obtains the C-SID 3:1 based on SL and CL, updates the C-SID to a corresponding position of the IPv6 destination address to obtain the new address C::3:1:0:2, and forwards the packet to node C3 by searching the forwarding table.

7. The process of node C3 is similar to that of node C2. The packet sent to node C4 is (A::1, C::4:2:0:1) (B::100, <Padding, 4:2, 3:1, 2:1>, C::1:1:0:0, D::4:1, D::3:1, D::2:1, D::1:1) (X, Y), the SL value in the G-SRH is 1, and the CL value in the destination address is 1.

8. After receiving the packet, node C4 determines that the destination address is a local compressible SRv6 End SID. In this case, because C::4:2:0:1 does not carry the COC flavor, node C4 decrements the SL value. Node C4 then updates the next 128-bit G-SID B::100 to the IPv6 destination address and forwards the packet to node B by searching the forwarding table. The packet sent to node B is (A::1, B::100) (B::100, <Padding, 4:2, 3:1, 2:1>, C::1:1:0:0, D::4:1, D::3:1, D::2:1, D::1:1) (X, Y), and the SL value in the G-SRH is 0.

   If C::4:2:0:1 carries the PSP flavor, node C4 removes the G-SRH from the packet. In this case, the packet sent to node B is (A::1, B::100) (X, Y).

9. After receiving the packet, node B determines that the destination address is a local SRv6 End.DT4 SID, performs IPv6 decapsulation to obtain the packet (X, Y), searches the routing table of VPN 100, and finally forwards the packet to CE2.

13.2.2.2 uSID

Another SRv6 compression solution is uSID.[13] In this solution, the same Locator Block is adopted by the SIDs in the SRH to form redundant information in the SRH. Multiple SIDs can share one Locator Block to reduce the need to carry redundant information. In contrast to G-SRv6, the uSID solution combines one Locator Block with Locator Nodes from multiple SIDs and stores them in a 128-bit space, allowing the space to carry information of multiple SIDs and ultimately shortening the SRH. Note, however, that each 128-bit space in the SRH needs to carry a Locator Block.

The uSID solution defines a partition method for the 128-bit SID space and a new endpoint behavior.

1. uSID definition: A uSID is also a C-SID, which is called the Micro SID.

2. uSID encapsulation: The uSID draft proposes the uSID Carrier to carry uSIDs. The uSID Carrier is a 128-bit space that can carry one Locator Block and multiple uSIDs.

   As shown in Figure 13.8, the uSID Carrier carries one uSID block and multiple uSIDs. If the Carrier is not filled entirely with uSIDs, it will be padded with one or more End-Of-Carriers (EOCs) with a value of 0. Each uSID is a fixed-length ID (typically 16/32 bits). Multiple uSIDs are arranged from left to right according to the SRv6 endpoint sequence. uSIDs have three roles according to their location:

   - Active uSID: the first uSID after the uSID Block. It identifies the uSID that is being used for IPv6 forwarding or SRv6 endpoint behavior execution.
   - Next uSID: the first uSID after the Active uSID.
   - Last uSID: the rightmost uSID in the uSID sequence.

   The uSID Block and the Active uSID form a SID. If the uSID Block is C:0::/32 and the uSID 0100:1 is configured for node 1, the complete SID is C:0:0100:1::/64 (an IPv6 prefix), which needs to be configured in the Local SID Table of node 1 and associated with a behavior on node 1. Node 1 also needs to advertise the route prefix to other nodes through a routing protocol.

   The uSID solution also defines a new endpoint behavior: uN. The uN behavior is associated with a uSID, and the action to be performed is "shifting and then forwarding based on table lookup," as shown in Figure 13.9.

uSID Block	Active uSID	Next uSID	uSID	uSID	Last uSID	EOC <0000>

FIGURE 13.8 uSID Carrier format.

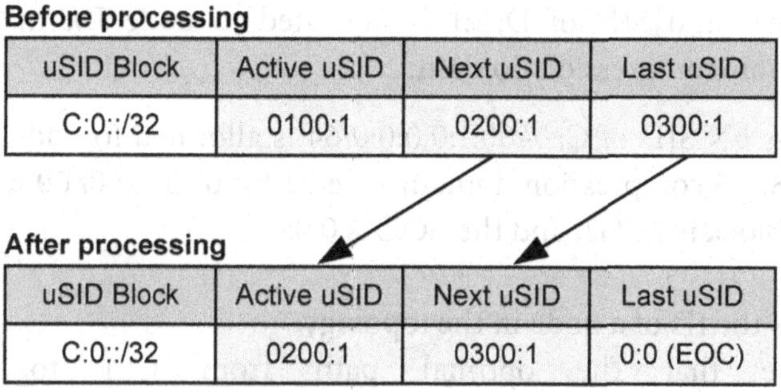

Before processing

uSID Block	Active uSID	Next uSID	Last uSID
C:0::/32	0100:1	0200:1	0300:1

After processing

uSID Block	Active uSID	Next uSID	Last uSID
C:0::/32	0200:1	0300:1	0:0 (EOC)

FIGURE 13.9 uSID behavior's operations.

Specifically, the 128-bit SID in the IPv6 DA field is processed as follows:

- If the Next uSID is not 0, the involved node shifts the Next uSID and all following uSIDs to the left and adds 0s to the end of the 128 bits. In this case, the Next uSID becomes the Active uSID, and packets are forwarded based on the new destination address.

- If the Next uSID is 0, it means that all the uSIDs in the current 128 bits have been processed. In this case, the End behavior is performed to update the next SID in the SRH to the IPv6 DA field and then packets are forwarded based on table lookup.

The order of multiple uSIDs specifies the endpoint nodes through which a packet passes. One uSID is removed each time the packet passes through an endpoint node.

3. uSID mechanism

The following describes the forwarding process of the uSID solution using the topology of G-SRv6 (see Figure 13.5).

Assume that SIDs and IP addresses are allocated as follows:

- The IP addresses of CE1 and CE2 are X (10.1.0.1) and Y (10.2.0.1), respectively.

- The SRv6 tunnel originating from node A uses the source address A::1.

- Node B allocates an SRv6 End.DT4 SID of B::100 to VPN 100.

- An End SID of D::*k*:1 is allocated to node D*k* in the SRv6 non-compression domain.
- A uN SID of C::0*k*00:1:0:0:0:0/64 is allocated to node C*k* in the SRv6 compression domain. The 32-bit uSID is 0*k*00:1, the uSID Block is C::/32, and the EOC is 0:0.

k specifies the ID of a node in the topology.

Assume that the optimal path from CE1 to CE2 is A→D1→D2→D3→D4→C1→C2→C3→C4→B. The path can be divided into the sub-path D1→D2→D3→D4 in the SRv6 non-compression domain and the sub-path C1→C2→C3→C4 in the SRv6 compression domain.

The sub-path in the SRv6 non-compression domain can be represented as a list of four SRv6 SIDs (D::1:1, D::2:1, D::3:1, D::4:1), and the sub-path in the SRv6 compression domain can be represented as a list of two uSID Carriers (C::0100:1:0200:1:0300:1, C::0400:1:0:0:0:0). Figure 13.10 (a) and Figure 13.10 (b) show the packet formats corresponding to the two sub-paths, respectively.

Figure 13.11 shows the complete uSID list.

When CE1 sends a packet (X, Y) to CE2:

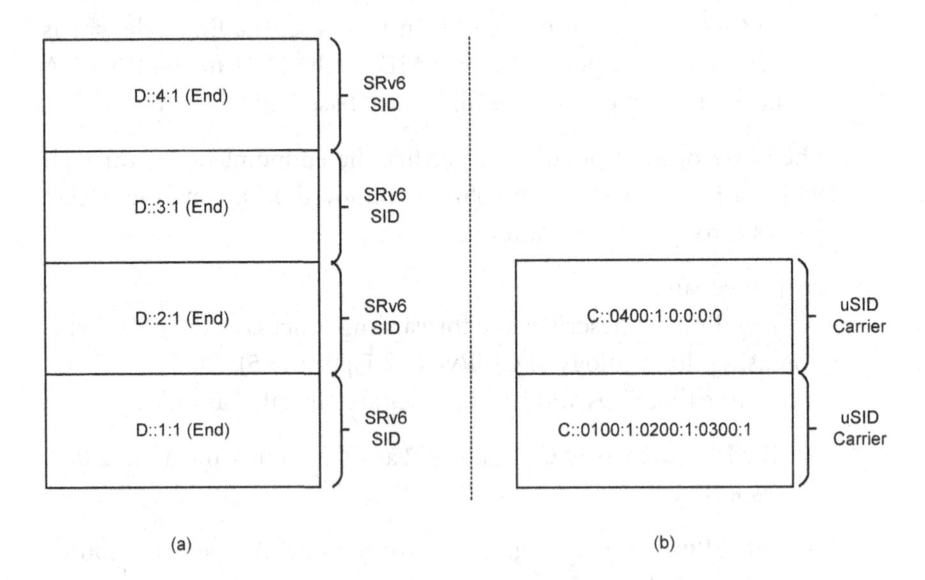

FIGURE 13.10 Packet formats corresponding to the two sub-paths.

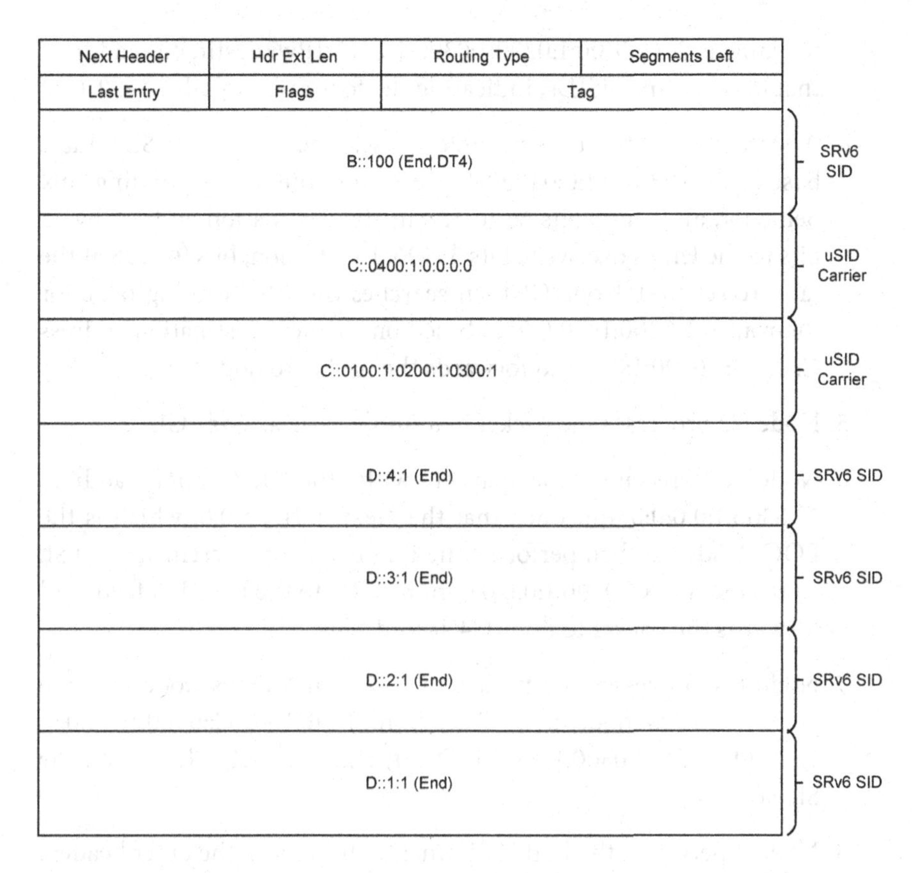

FIGURE 13.11 Complete uSID list in the SRH.

1. The packet (X, Y) sent by CE1 to CE2 arrives at node A.

   Node A encapsulates the packet as (A::1, D::1:1) (B::100, C::0400:1:0:0:0:0, C::0100:1:0200:1:0300:1, D::4:1, D::3:1, D::2:1, D::1:1) (X, Y), and the SL value is 6.

2. Node A searches the IPv6 routing table based on D::1:1 and forwards the packet to node D1.

3. Nodes D1, D2, D3, and D4 process the packet in a similar way. Specifically, they perform the End behavior and forward the packet by searching the IPv6 routing table. The packet forwarded to node C1 is (A::1, C::0100:1:0200:1:0300:1) (B::100, C::0400:1:0:0:0:0, C::0100:1:0200:1:0300:1, D::4:1, D::3:1, D::2:1, D::1:1) (X, Y), and the

SL value is 2. C::0100:1:0200:1:0300:1 is an IPv6 destination address that carries three uSIDs, indicating the forwarding path C1-C2-C3.

4. After receiving the packet, node C1 searches the Local SID Table based on C::0100:1:0:0:0:0/64, performs the corresponding uN behavior, and shifts bits 65 to 128 in the destination address by 32 bits to the left to overwrite bits 33–96. In addition, bits 97–128 at the tail are set to 0. Node C1 then searches the IPv6 routing table for the route C::0200:1:0:0:0:0/64 based on the new destination address C::0200:1:0300:1:0:0, and forwards the packet to node C2.

5. Node C2 processes the packet in a similar way as node C1.

6. Node C3 receives the packet with the destination address C::0300:1:0:0:0:0 and finds that the Next uSID is 0:0, which is the EOC. Node C3 then performs the End behavior, decrements the SL value, updates C::0400:1:0:0:0:0 in the SRH to the IPv6 DA field, and forwards the packet to node C4 accordingly.

7. Node C4 processes the packet in a similar way as node C3. The packet sent to node B is (A::1, B::100) (B::100, C::0400:1:0:0:0:0, C::0100:1:0200:1:0300:1, D::4:1, D::3:1, D::2:1, D::1:1) (X, Y), and the SL value is 0.

8. Node B performs the End.DT4 function to remove the outer headers of the packet to obtain (X, Y). Node B then searches the routing table of VPN 100 based on the inner destination address Y and forwards the packet to CE2.

The preceding example shows that uSIDs can be carried by multiple uSID Carriers when the uSID list is long.

13.2.2.3 SRm6

SRm6 is a new source routing mechanism proposed to resolve the SRv6 extension header compression issue.[14] SRm6 is incompatible with SRv6 and SRHs.

SRm6's Routing Header compression aims to use a 16- or 32-bit SID to replace a 128-bit IPv6 address and establishes a mapping between them. In this way, the IPv6 address sequence is converted into the SID sequence, shortening the Routing Header.

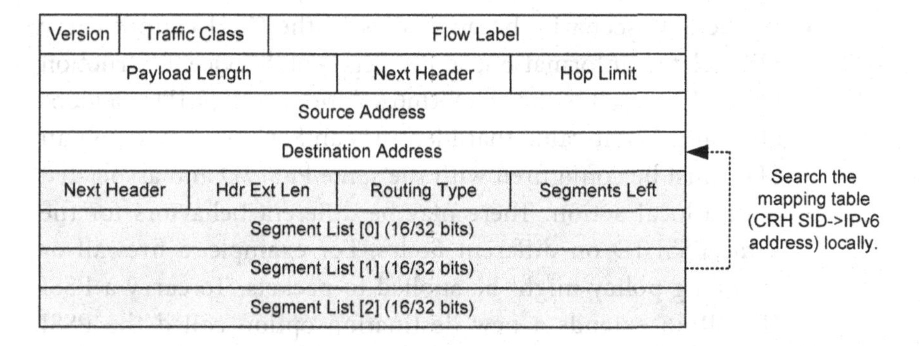

FIGURE 13.12 Format and usage of the SRm6 Routing Header.

1. Routing Header for SRm6

As shown in Figure 13.12, the SRm6 solution defines a new IPv6 Routing Header called Compressed Routing Header (CRH). Similar to the SRH, the CRH also contains a segment list, but their SIDs are different. A CRH SID is a 16- or 32-bit value and is classified as either the Adjacency or Node SID. A Node SID is mapped to a local interface address, and an Adjacency SID is mapped to a local interface address and the interface ID. SID mappings are stored on devices and advertised using routing protocols.

When a path needs to be specified for a packet, the ingress encapsulates an outer IPv6 header and CRH into the packet. The CRH carries the SIDs corresponding to the endpoint nodes. When an endpoint node receives the packet, it decrements the SL value to locate the SID of the next node, searches the SID mapping table for the next-hop IPv6 interface address, updates the address to the IPv6 DA field, searches the routing table based on the LPM rule, and then forwards the packet.

By mapping SIDs to IPv6 addresses, SRm6 implements explicit path encoding with a low header overhead without changing the original semantics of IPv6 addresses.

2. Destination Options header for SRm6

Because in SRm6 SIDs can only be mapped to interface addresses, SRm6 uses a new solution for some application scenarios (such as SFC and VPN) that are supported by associating SIDs with endpoint behaviors in SRv6. The following focuses on the SRm6 solution for SFC and VPN scenarios.

- In the SFC scenario, because SIDs in the CRH cannot carry SFC-related information, a Per-Segment Service Instruction (PSSI) ID is defined in the SRm6 solution. A PSSI ID is a globally valid 32-bit value that identifies an SFC. Each node on an SFC must be configured with the same PSSI ID and associated with a local action. There may be different behaviors for the same PSSI ID on different nodes. For example, a firewall or sampling policy might be applied to packets. To carry a PSSI ID, SRm6 extends a new destination option called the PSSI Option TLV. The Destination Options header can occur twice within a packet, once before the Routing Header and once after it. The one that occurs before the Routing Header is processed by each node specified in the Routing Header, whereas that after a Routing Header is processed only by the last node. The PSSI Option TLV can be used in the two types of Destination Options header.
- For the VPN scenario, a Per-Path Service Instruction (PPSI) ID is defined in the SRm6 solution. To carry a PPSI ID, SRm6 extends a new destination option called the PPSI Option TLV. The PPSI ID is a locally valid 32-bit value that identifies a VPN and is configured on the VPN decapsulation node. Currently, SRm6 defines two types of PPSIs. One is to decapsulate packets and forward them through a specific interface, and the other is to decapsulate packets and forward them based on a specific routing table. Because VPN information is processed only by the last node, the PPSI Option TLV is carried only by the Destination Options header after the CRH.

13.2.3 Comparison between SRv6 Extension Header Compression Solutions

Table 13.2 compares the G-SRv6, uSID, and SRm6 solutions, each of which has its own advantages and disadvantages.

When a 16-bit C-SID/uSID/CRH SID contains both Node and Function IDs, the numbers of nodes and functions are limited. As a result, a 16-bit C-SID/uSID/CRH SID is not applicable to most networks. This book uses a 32-bit C-SID/uSID/CRH SID as a reference for comparison.

In conclusion, G-SRv6 has some advantages over the other solutions.

TABLE 13.2 Comparison between SRv6 Extension Header Compression Solutions

Item	G-SRv6	uSID	SRm6
Compatibility with SRv6 and SRHs	Compatible	Compatible	Incompatible
Address planning	1. SIDs need to use the same Locator Block. 2. Existing SRv6 address blocks can be reused. 3. The prefix length is flexible. Short prefixes are not required, and address blocks are easy to obtain.	1. SIDs need to use the same Locator Block. 2. Additional Locator Blocks need to be planned. 3. Short prefixes are required. Address blocks corresponding to short prefixes are large and difficult to obtain.	Existing IPv6 interface addresses are used.

(Continued)

TABLE 13.2 (*Continued*) Comparison between SRv6 Extension Header Compression Solutions

Item	G-SRv6	uSID	SRm6
Control plane	G-SRv6 requires an enhancement to SRv6 protocol extensions. The C flag and COC flavor need to be advertised together with SID information.	uSID requires an enhancement to SRv6 protocol extensions. The information such as the uN behavior needs to be advertised together with SID information.	SRm6 requires new protocol extensions to be introduced to advertise mappings between SIDs and IPv6 interface addresses. In addition, different protocol extensions need to be defined for different application scenarios (such as SFC and VPN) to advertise PSSI/PPSI IDs and their bindings with local actions, resulting in high complexity.
	1. New protocol extensions need to be introduced in the control plane. 2. SID entries need to be added. 3. SIDs with multiple behaviors can be compressed.	1. New protocol extensions need to be introduced in the control plane. 2. SID and routing entries need to be added. 3. Currently, only SIDs with limited behaviors such as End SIDs can be compressed.	1. New, complex control-plane protocol extensions need to be introduced. 2. Mappings from SIDs to IPv6 interface addresses need to be introduced. In addition, other new mappings may be introduced for different application scenarios (such as SFC and VPN).
Forwarding-plane protocol extension	G-SRv6 does not change the SRH format but adds processing logic for the COC flavor.	uSID does not change the SRH format but adds uN series behaviors.	SRm6 defines three types of ID addressing space, mapping methods, and a new Routing Header. In addition, other types of IPv6 extension headers need to be used to implement SRv6 functions, and new forwarding-plane protocol extensions must be introduced.

(*Continued*)

TABLE 13.2 (*Continued*) Comparison between SRv6 Extension Header Compression Solutions

Item	G-SRv6	uSID	SRm6
Forwarding entry extension	G-SRv6 requires corresponding SID entries to be added for C-SIDs.	uSID requires corresponding SID and routing entries to be added.	SRm6 requires a mapping table for SIDs and IPv6 interface addresses to be added. In addition, a mapping table for PSSI/PPSI IDs and local actions needs to be added for some application scenarios (such as SFC and VPN).
Routing information	Simple. No route needs to be added when C-SIDs and common SRv6 SIDs share a locator.	Complex. A route is maintained for each uSID.	Complex. A mapping table for SIDs and IPv6 interface addresses needs to be stored on a device. In addition, PSSI and PPSI IDs need to be planned and maintained.
Address space utilization	High utilization.	Low utilization. A large number of available addresses are wasted.	SID, PSSI, and PPSI ID space are extra and independently planned, and do not affect IPv6 address space.
Forwarding performance	C-SID replacement has little impact on forwarding performance.	uSID shifting has little impact on forwarding performance.	Multiple extension headers need to be read and written, affecting forwarding performance.
Compression efficiency	High: The length of a segment list can be reduced to 25% of the original length.	Medium: The length of a segment list can be reduced to 33.3% of the original length.	High: The length of a CRH segment list can be reduced to 25% of an SRv6 segment list's length. However, in specific application scenarios (such as SFC and VPN), other IPv6 extension headers need to be used, which brings new overhead and reduces the overall compression efficiency.

13.2.4 Future of SRv6 Extension Header Compression

Although we may run into challenges with SRv6 headers when implementing SRv6 in our applications, the impact of these challenges is manageable according to current major scenarios.

1. Path MTU: In a network model, edge links have small MTU values, whereas links inside the network have large MTU values. For the link connecting the network edge to users, the MTU is usually set to 1,500 bytes by referring to the traditional Ethernet mechanism. For links inside the current IP WAN and data center network, the MTU can usually be set to a large value, for example, 9,000 bytes, which is also called jumbo frame length. Therefore, after SRv6 headers are encapsulated in user packets, it is unlikely that fragmentation or packet loss will occur due to Path Maximum Transmission Unit (PMTU) exceeding the limit.

2. Forwarding performance: The number of SIDs used in SRv6 packets is limited in SRv6 application scenarios (such as SRv6 VPN and SRv6 loose TE). Existing programmable hardware processors can already meet forwarding performance requirements. In addition, some vendors have claimed that they can implement line-rate forwarding of SRv6 packets carrying up to ten SIDs. This, together with the development of hardware, will satisfy current requirements and continue to mitigate the impact of SRv6 headers on forwarding performance over time.

3. Payload: According to statistics from some networks, the average length of packets on an IP WAN is 700–900 bytes.[15] In current scenarios, if an SRv6 header containing three SIDs is added to these packets, it will create an overhead proportion of about 10%, which does not significantly affect the network.

Typically, SRHs have limited impact in current SRv6 scenarios. While the IETF continues to focus on SRv6 standardization, the IETF SPRING Working Group has formed a design team to discuss SRv6 extension header compression solutions.

As SRv6 functions and applications grow, so will the impact of SRv6 extension headers. Therefore, SRv6 extension header compression is still an important direction for research. Considering the rapid development of the SRv6 industry and the deployment of many commercial cases, it is recommended that a future optimization solution be compatible with the existing SRH, so as to minimize evolution costs.

13.3 APN6

Application-aware IPv6 Networking (APN6) is another innovative framework triggered by SRv6. APN6 uses IPv6 extension headers that provide multidimensional programming space for carrying application information in packets, enabling networks to be aware of the applications to which received packets belong and their requirements. APN6 provides fine-granularity SLA guarantee and effectively optimizes the utilization of network resources. It enables support for new network services and brings changes to network architectures.

13.3.1 Why APN6

13.3.1.1 Current Carriers' Pain Points and Requirements

With the E2E design concept of Internet, applications and networks have traditionally been decoupled. This kind of design throws up a range of issues: The application layer is developing continuously, and various new applications are emerging, especially with the advent of the 5G era. However, networks are still considered to be only pipes that bear upper-layer applications, without being aware of application information. Such an application-unaware network pipe can neither help network service providers profit from increasing network traffic, nor enable them to implement fine-granularity operations, as the network cannot provide differentiated services based on the diverse requirements of applications. Furthermore, the network does not have the application information of service traffic and therefore usually maintains a light load to guarantee QoS, underutilizing network resources.

In 2013, Google published the paper *B4: Experience with a Globally-Deployed Software Defined WAN* at the ACM Special Interest Group on Data Communication (SIGCOMM). This paper is the first publication of Google's experience in designing and deploying the WAN that connects its data centers, claiming that WAN connection utilization could reach approximately 100%.[16]

Conventional inter-WAN data center network devices uniformly transport data packets of different applications, without differentiating them, which Google believes is the main cause of low bandwidth utilization.

As such, Google analyzed and identified the data transmitted on its inter-WAN data center network devices, and categorized the data into user data, remote storage access data, and large-scale state synchronization data. User data represents the lowest traffic volume on B4 but is most sensitive to latency and has the highest priority. Therefore, the availability and persistence of user data must be guaranteed. In contrast, large-scale state synchronization data represents the largest traffic volume but is not sensitive to latency and has the lowest priority.

The fundamental idea of Google's B4 network involves guaranteeing low latency for high-priority traffic and allowing large-volume but low-priority traffic to automatically adjust the transmission rate based on the high-priority traffic states. Specifically, the network distributes large-volume but low-priority traffic to different idle timeslots, allowing the traffic to fill in idle pipes while also effectively improving link utilization.

The B4 network resolves the problem to utilize application information in network traffic engineering. As the network is aware of applications' traffic information, it can significantly improve network resource utilization through effective traffic scheduling. By referring to Google's idea of how to adjust network resources based on the requirements and traffic characteristics of applications, we aim to break the isolation between the traditional IP transport networks and applications, as well as enable the network to effectively be aware of application information. In this way, it is possible to implement fine-granularity operations for carried services and improve network resource utilization while also guaranteeing SLAs.

13.3.1.2 Challenges Faced by Traditional Differentiated Service Solutions

Across the industry, a considerable amount of effort has been put into enabling networks to be aware of applications and differentiate traffic. Generally, a network device depends on 5-tuple information or DPI to be aware of applications. Those conventional methods have some shortcomings and challenges.

1. 5-tuple-based ACL/PBR: Although 5-tuple information is widely used in ACL/PBR, this information is insufficient for fine-granularity service processing, and can only be used as indirect application-related information to infer real applications to which the packets belong, negatively affecting the forwarding performance of service packets.

2. DPI: It can be used to learn more application information by inspecting packets in depth. However, this further compromises the forwarding performance and brings security challenges.

3. Methods based on orchestrators and SDN controllers: With the emergence of SDN, orchestrators are introduced to orchestrate networks and applications. In a typical SDN architecture, an orchestrator imports application requirements, and a controller is responsible for network control. The controller can be aware of the application's requirement on the network through the interface connected to the orchestrator, and then controls and manages network resources and traffic based on the requirement. The actual development of SDN-based architecture has encountered some challenges. Figure 13.13 shows an SDN-based architecture, in which many interfaces are used (including interfaces between the orchestrator and controllers, between the orchestrator and applications, and between controllers and network devices). The presence of various interfaces leads to a lengthy coordination period, making this architecture unsuitable for quick service deployment based on the requirements of critical applications. In addition, it is difficult to standardize those interfaces, hindering network interconnection and openness.

13.3.1.3 APN6 Solution

In fact, an IP transport network can be aware of applications and steer traffic by using some traditional methods. In the draft we submitted to the

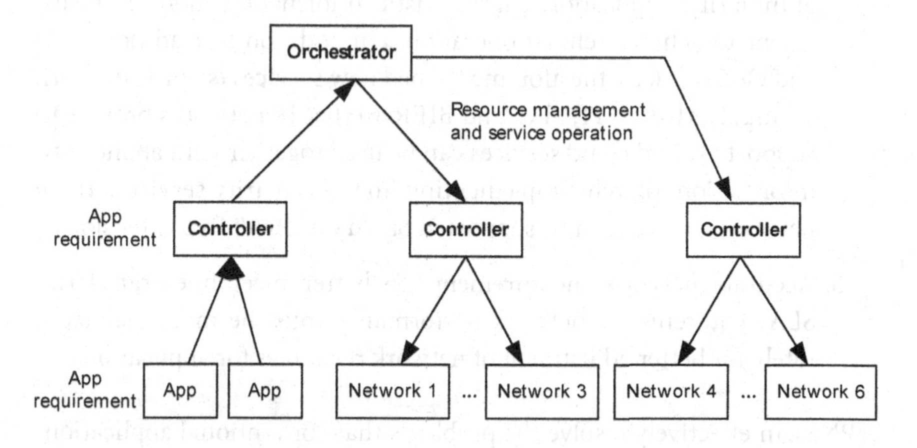

FIGURE 13.13 SDN-based architecture.

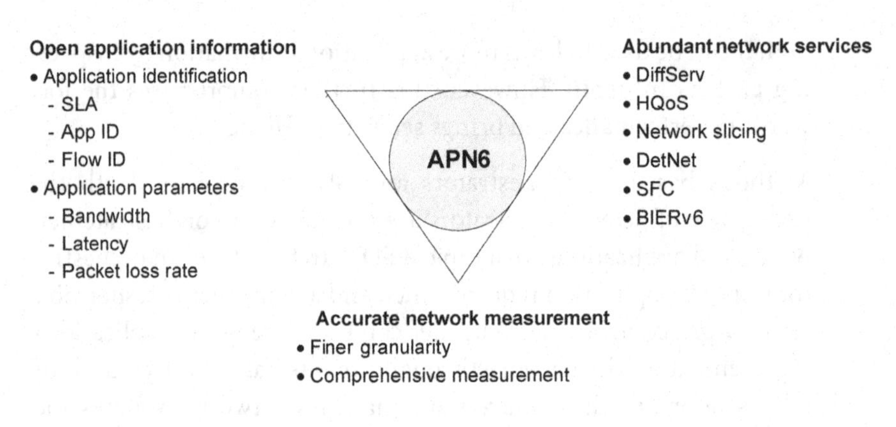

FIGURE 13.14 Key APN6 elements.

IETF, we analyzed the problems with these traditional methods, proposed the APN6 concept, and defined three key APN6 elements,[17] as shown in Figure 13.14.

1. Open application information: APN6 uses IPv6 to carry application characteristic information, including application-aware identification information and service requirement information. Application information can be further extended as required. Carrying application information is not enforced; instead, it is optional.

2. Abundant network services: The network side needs to provide abundant services to implement fine-granularity operations, as fine-granularity application characteristic information alone is insufficient to achieve refined operations. In addition to traditional TE and QoS services, the aforementioned new services (such as network slicing, DetNet, IFIT, SFC, and BIERv6) that IP networks provide to support 5G and cloud services can be used together with application information, thereby implementing fine-granularity services. These network services can be supported based on IPv6/SRv6 extensions.

3. Accurate network measurement: To better meet fine-granularity SLA requirements, network performance must be measured accurately for better adjustment of network resources for applications.

APN6 can effectively resolve the problems that conventional application-aware networking solutions encounter. It uses IPv6 extension headers to carry application characteristic information (including application-aware identification information and service requirement information) for packets. A network can quickly and effectively be aware of an application and its requirements upon receiving such a packet, enabling it to provide fine-granularity network resource scheduling and guarantee the SLA for the application.

The advantages of using IPv6 to implement APN6 include:

1. Simplicity: Application information is encapsulated in IPv6 packets and can be forwarded based on IPv6 reachability.
2. Seamless convergence: Seamless convergence can be easily implemented between the network and applications as they are both based on IPv6.
3. Great extensibility: The programming space provided by IPv6 encapsulation can be used to carry abundant application information.
4. Good compatibility: Network upgrade and service provisioning can be performed on demand. If a network node does not identify the application information in a packet, the packet is forwarded as an IPv6 packet, implementing backward compatibility.
5. Low dependency: Application information conveyance and service provisioning are based on the forwarding plane of devices, which is different from the management and control method that involves a significant number of interfaces in the SDN architecture.
6. Quick response: Flow-driven and direct response from devices is supported because the response is based on the forwarding plane.

In the IETF draft,[17] we also defined the possible application scenarios of APN6, including application-aware network slicing, DetNet, SFC, and network measurement.

13.3.2 APN6 Framework

The APN6 framework is defined in the IETF draft.[18] As shown in Figure 13.15, the APN6's components include applications, application-aware network edge devices, application-aware-process SRv6 headends, application-aware-process midpoints, and application-aware-process

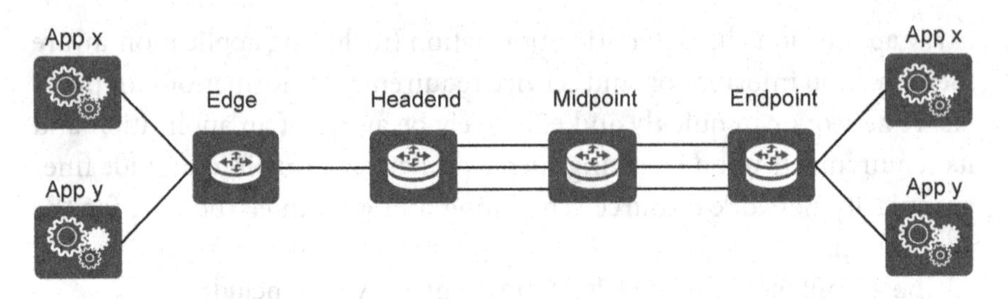

FIGURE 13.15 Application-aware IPv6 network.

endpoints. There are both network-side and application-side solutions. The difference between the two is the start point for generating and carrying application information, namely, a network edge device in the network-side solution, and an application in the application-side solution.

In the application-side solution, application-aware information is carried in IPv6 extension headers to convey the SLA level, application ID, user ID, flow ID, and specific network requirement parameters, such as bandwidth, latency, jitter, and packet loss rate.[19] APN6-enabled IPv6 on the transmission path can process the application information. The application determines whether to add its information in packets, whereas the network only serves as a medium for carrying and transmitting the information.

The application information is processed on the SRv6 headend and can be used to find a path that meets the application's requirements. If there are none, a controller can be triggered to establish a new SRv6 path. Packets can also be processed by midpoints based on the carried application information.

A network is aware of the application and its requirements in a received packet, and can then adjust resources and provide services based on the requirements. This flow-driven mechanism can effectively address the prolonged interaction between multiple controllers that has otherwise been a challenge to deal with.

In the application-side solution, an application's identification information and requirements are added to IPv6 extension headers (such as the Hop-by-Hop Options header and Destination Options header) of an IPv6 packet. Implementing the solution is simple but requires a wide range of upgrades on host operating systems and applications. To avoid illegitimate occupation of network resources, the solution must be able to prevent applications from sending arbitrary network requirements without consent from carriers. Therefore, the key to implementing this solution is effective access control. However, in the network-side solution, application information is written and encapsulated by a network edge device, which is a boundary device close to the application side, without requiring upgrades or extensions on the application side. In addition, the network-side solution controls application information using a network management and control system, which has certain defense capabilities against the security risks that may arise when application information is carried and managed.

13.3.3 APN6 Framework Requirements

APN6 requirements are defined in the IETF draft.[18] In APN6, application characteristic information is encapsulated into IPv6 packet headers (such as IPv6/SRv6 packet headers and their extension headers). After a network receives a packet carrying such information, it uses the information to determine the services to be provided, traffic forwarding action, and SLA level to be guaranteed. This section describes the requirements for supporting the APN6 framework, including requirements for carrying application characteristic information, requirements for processing application-aware information, and security requirements.

13.3.3.1 Requirements for Carrying Application Characteristic Information

Application characteristic information includes application-aware identification information and service requirement information.

The application-aware identification information can include:

- SLA level: SLA requirement level of an application, such as gold, silver, or bronze. In some scenarios, SLA levels can be identified using colors (such as red and green).
- Application ID
- User ID
- Flow ID: ID of a flow or session

Different combinations of these IDs can differentiate traffic and provide fine-granularity SLA guarantee.

The service requirement information can include an application's requirements on:

- Bandwidth
- Latency
- Jitter
- Packet loss rate

Different combinations of these parameters can indicate the network service requirements of applications in more detail. Together with the application-aware identification information, the service requirement information can be used to map the SRv6 tunnel/Policy and QoS queue that meet specific service requirements. If no appropriate SRv6 tunnel/Policy and QoS queue are matched based on the service requirement information, a new SRv6 tunnel/Policy and QoS policy will be established.

The requirements for APN6 to carry application characteristic information include:

1. Carrying an application ID in the application-aware identification information is mandatory. Application IDs identify the application to which a packet belongs.
2. Carrying an SLA level in the application-aware identification information is recommended.
3. Carrying a user ID and flow ID in the application-aware identification information is optional.
4. The service requirement information is optional.
5. If necessary, all nodes on a path must be able to process application-aware information.
6. The application-aware information can be directly generated by an application, or by an application-aware edge device through DPI or based on a locally configured policy.
7. The application-aware information must be complete if it is directly replicated and added to IPv6 packets by an application-aware edge device.

13.3.3.2 Requirements for Processing Application-Aware Information

The application-aware headend and midpoint match application IDs and/or service requirements with network resources (tunnels, SR Policies, or queues).

I. Application-Aware SLA Guarantee

To attract end-users and provide them with better Quality of Experience (QoE), networks must be able to provide fine-granularity and even application-level SLA guarantee.[17] Application-aware SLA guarantee has the following requirements:

1. The application-aware headend must be able to steer traffic to matching SRv6 tunnels/Policies.

2. The application-aware headend must be able to trigger the establishment of SRv6 tunnels/Policies that meet applications' SLA requirements.

3. The application-aware headend and midpoint must be able to steer traffic to matching QoS queues.

4. The application-aware headend and midpoint must be able to trigger the establishment of matching QoS queues.

II. Application-Aware Network Slicing

The network slicing function divides the control or data plane of the network infrastructure into multiple isolated network slices, which can run concurrently and be allocated exclusive device and link resources. Application-aware network slicing has the following requirements:

1. The application-aware headend must be able to search the forwarding table based on application information and steer traffic to corresponding network slices.

2. The application-aware midpoint must be able to search the forwarding table based on application information and allow a flow to use the resources of the network slice to which the flow belongs.

III. Application-Aware DetNet

DetNet traffic is transmitted together with best-effort traffic. Each node through which DetNet traffic passes must provide guaranteed bandwidth, upper-bounded latency, and some features related to the data which is sensitive to transmission time. Application-aware DetNet has the following requirements:

1. The application-aware headend must be able to search the forwarding table based on application information and steer traffic to appropriate transmission paths.
2. The application-aware headend must be able to search the forwarding table based on application information and establish appropriate transmission paths on demand for DetNet traffic.
3. The application-aware midpoint must be able to use application information to allow DetNet traffic to utilize the resources of a transmission path that meets service requirements.
4. The application-aware midpoint must be able to use application information to reserve resources for DetNet traffic along the transmission path.

IV. Application-Aware SFC

E2E service distribution usually involves multiple service functions, including traditional network service functions (such as firewall and DPI) and new application-related functions. These service functions can be implemented on physical or virtual devices, and SFC can be applied to fixed networks, mobile networks, and data center networks. Application-aware SFC has the following requirements:

1. The application-aware headend must be able to steer traffic to proper SFCs based on application information.
2. The application-aware headend must be able to process the application information carried in packets.

V. Application-Aware Network Measurement

Network measurement can be used to locate silent faults and predict QoE satisfaction, making real-time SLA awareness and proactive OAM possible. Application-aware network measurement has the following requirements:

1. The application-aware nodes must be able to drive IOAM based on the application IDs in the application information.

2. With application information, network measurement results can be reported based on application IDs, and whether applications' service requirements are met can be verified.

13.3.3.3 Security Requirements

APN6 has the following security requirements:

1. The security mechanism defined for APN6 must enable operators to prevent applications from sending arbitrary application-aware information without consent.

2. The security mechanism defined for APN6 must prevent applications from requesting unauthorized services.

13.3.4 Future of APN6

In March 2019, APN6 was first proposed at the IETF 104 meeting held in Prague, Czech Republic. In July 2019, the APN6 side meeting was held at the IETF 105 meeting in Montreal, Canada, attracting over 50 industry experts and leading to widespread recognition of APN6's value. Currently, APN6 research and standardization are being steadily promoted.

APN6 is the solution to the "pipe" problem that carriers urgently need to address. It makes application-based fine-granularity operations possible. APN6 complies with the development trend of cloud-network convergence and has become an important enabler for further convergence of services and transport networks. To a certain extent, APN6 breaks the Internet's hierarchical decoupling design and changes the network architecture. Taking security and privacy challenges into account, APN6 will be first deployed on the network side, allowing applications to be detected and processed within the trusted network scope of carriers. Research will be conducted on problems that may arise when application information is encapsulated at the host side.

13.4 FROM SRv6 TO IPv6+

Even after over 20 years of development, IPv6 has not yet been widely deployed. The story may change after the advent of SRv6. However, with the development of new services, SRv6 is no longer the only technology affecting IPv6's future. Specifically, the data plane supports not only SRv6 SRHs but also extensions of other types of IPv6 extension headers.

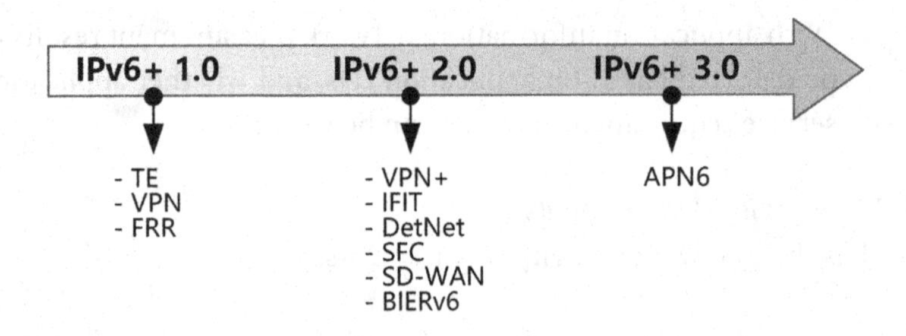

FIGURE 13.16 Three phases of IPv6+ development.

We defined IPv6 enhanced innovation as IPv6+, which was expected to support new services through network programming, revealing the characteristics of the new era of IP networks better. Considering the priorities of requirements on network services and the maturity of technical standards, we defined three phases of IPv6+ development, as shown in Figure 13.16.

1. IPv6+ 1.0: mainly provides basic SRv6 features, including TE, VPN, and FRR. These features are imperative to the success of MPLS and need to be inherited by SRv6. SRv6 simplifies network service deployment by making use of these features while leveraging its own strengths.

2. IPv6+ 2.0: focuses on new services oriented toward 5G and cloud. These new services require new SRv6 SRH extensions or extensions of other IPv6 extension headers. Possible new services include but are not limited to VPN+ (network slicing), IFIT, DetNet, SFC, SD-WAN, and BIERv6.

3. IPv6+3.0: focuses on APN6. With further convergence of the cloud and network, more information needs to be exchanged between them, and IPv6 is undoubtedly the most advantageous medium. This will bring important changes to the network architecture.

With the development of 5G and cloud, SRv6 is ushering in a new era of IPv6 applications. The IPv6+ roadmap facilitates systematic network evolution, and new research topics as well as solutions may emerge during IPv6+ development. Without question, the IPv6+ era has just begun.

13.5 STORIES BEHIND SRv6 DESIGN

13.5.1 Design of SRv6 Compression: From C-SRH to G-SRH

The competition between SRv6 compression solutions is far more intense than I imagined.

From my point of view, SRv6 compression is just an optimization of SRv6 and unable to introduce new value-added services. Moreover, it may incur high costs, while the challenges caused by the stack depth of SRv6 SRH can be overcome with the fast development of network hardware. However, people in the industry paid close attention to this, making SRv6 compression a hot topic. Fortunately, my colleague Cheng Li persisted with the SRv6 compression research and completed the C-SRH solution[20] in the early of 2019. C-SRH is much simpler for SRv6 compression than I expected, since it simply removes the redundant information of SIDs based on the regularity of SRv6 SID allocation. As we all know, address planning on a live network must comply with certain rules, that is, addresses cannot be randomly allocated. Specifically, SRv6 SIDs can be allocated from an address block, and the SIDs will share the Common Prefix. This results in redundant information in the segment list encoded in an SRH, and such information can be removed to achieve the purpose of SRv6 compression. In this way, the solution is far simpler than the possible solutions that try to provide a general mechanism for the rare scenarios with SIDs being randomly allocated.

Although C-SRH provides a simple solution to compress SRHs, it requires all SIDs in an SRH to be C-SIDs. In other words, C-SRH cannot encode 128-bit SRv6 SIDs and C-SIDs in a single SRH, and so all the nodes along a path need to support compression if SRv6 compression is applied. However, as SRv6 has been deployed on many networks, smooth upgrade from SRv6 to compressed SRv6 must be considered. To this end, myself, Cheng Li, and other colleagues continued to work on the hybrid programming of SRv6 SIDs and C-SIDs in a single SRH. We designed a new solution very soon after. In the new solution, a 128-bit SID space in the segment list of an SRH can contain an SRv6 SID or multiple C-SIDs. This means that the semantic of SIDs in an SRH is different from that of the existing SRv6 SID. In other words, the 128-bit SID space can contain multiple types of SIDs instead of only SRv6 SIDs, and its semantic is beyond the scope of SRv6 SID and becomes generalized. Inspired by Generalized MPLS (G-MPLS), we named the solution Generalized SRv6 (G-SRv6). The corresponding encapsulation is called G-SRH.

Casting my mind back to the design process from C-SRH to G-SRH, I think two aspects are very important: (1) Considering the engineering design of the live network will help us to simplify the solution and avoid overdesigning. (2) The compatibility design prevents networkwide upgrade issues and ensures smooth network evolution from an IPv6 network to an SRv6 network and to a compressible/generalized SRv6 network. The loose requirement of G-SRv6 on the address planning of the live network also contributes much to this aspect. During communication with vendors and carriers, these advantages of G-SRv6 are also well received and highly appreciated.

13.5.2 Design Principle: Internet and Limited Domains

SRv6 causes a conflict between the Internet-oriented design and limited domain-oriented design in the IETF. The IETF first adopted the Internet-oriented design. The Internet follows the end-to-end design principle: It allows users to generate various types of information on their intelligent terminals, such as computers and mobile phones while the network only transmits information in a best-effort manner, without memorizing or controlling the information. Such an architecture effectively simplifies network functions, as it leaves the greatest extent of complexity for information processing and control to terminals. Meanwhile, to minimize network interference on user traffic, terminals and applications protect privacy through encryption. On a traditional telecom network such as a telephone or ATM network, QoS assurance is an important infrastructure service. These functions were later implemented through IP-based MPLS, meaning that IP technologies replaced traditional telecom technologies. MPLS is generally applied to limited network domains, such as IP backbone networks, metro networks, and mobile transport networks. Applications do not require MPLS on the host side; instead, they require IP end to end.

IETF protocols were Internet-oriented at the beginning, but later MPLS-related protocols were applied to limited domains. As such, the Internet-oriented and MPLS-related protocols coexist harmoniously. As defined by the IETF, IPv6 and MPLS are independent as they fall within the Internet Area and Routing Area, respectively. However, a cross-area situation is occurring with the development of SRv6. SRv6 supports functions similar to MPLS and can be used in limited domains, but it is extended based on IPv6, potentially impacting IPv6 applications on the

Internet. Colleagues from the Routing Area believe that many SRv6 extensions are irrelevant to the Internet, and that SRv6's impact on the Internet does not need to be considered during SRv6 design. However, colleagues working for the traditional IPv6 technologies insist that SRv6 design must comply with the IPv6 design principles as there is no definite boundary between the Internet and limited domains. For example, SRv6 supports the Insert mode, meaning that a new SRH can be directly inserted during FRR. However, RFC 8200 for IPv6 allows only one RH to be added to an IPv6 packet and does not allow intermediate network nodes to modify packets. As such, the SRv6 Insert mode entails potential risks, and these issues have also led to debates between the two parties.

APN6 further aroused debates about the design concepts of limited domains and Internet. According to the Internet design philosophy, packets of users/applications should be encrypted when being transmitted end to end, preventing the network from being aware of packet information and limiting the network's interference on packets. Essentially, the network serves as a simple pipe. APN6 is designed for applications in limited domains and requires packets to carry application information to implement fine-granularity traffic scheduling and improve network value as well as network resource utilization. APN6 caters for carriers that do not aim to provide only pipes but does not support the design idea of colleagues who adhere to the Internet design philosophy. As a result, APN6 inevitably causes new conflicts. During the development of IPv6+, the conflict between different design principles may occur continuously.

13.5.3 Design Principle: The Greatest Truths Are the Simplest

One of the principles behind the design of the Internet is "The greatest truths are the simplest." Regarding the driving force for the development of IP network technologies, I consider the following objectives as the most pertinent:

1. Providing more IP-based services to create more values for customers
2. Simplifying IP network O&M to reduce laborious O&M workloads for network engineers

If we look further into possible technologies based on the above objectives, we can develop advanced technologies.

The essence of SRv6 is the definition and combination of segments, bringing flexible network services. This seemingly simple technology implements network programming; however, it faces many challenges due to conflicts with the traditional IPv6 design principle. Simply put, APN6 combines the network and various applications through the definition and transmission of application IDs. In addition, it not only extends network boundaries but also breaks them down, including the boundary between the network and applications, and the boundary between different layers of networks. APN6 also faces enormous challenges. But thanks to new network development opportunities, such as cloud-network convergence and computing-network convergence, it is more likely to become an effective technology to empower network development. Currently, APN6 has being won wide support across the industry. We will continuously put effort into developing it further.

REFERENCES

[1] Filsfils C, Camarillo P, Leddy J, Voyer D, Matsushima S, Li Z. SRv6 Network Programming[EB/OL]. (2019-12-05)[2020-03-25]. draft-ietf-spring-srv6-network-programming-05.

[2] Filsfils C, Dukes D, Previdi S, Leddy J, Matsushima S, Voyer D. IPv6 Segment Routing Header (SRH)[EB/OL]. (2020-03-14)[2020-03-25]. RFC 8754.

[3] Psenak P, Filsfils C, Bashandy A, Decraene B, Hu Z. IS-IS Extension to Support Segment Routing over IPv6 Dataplane[EB/OL]. (2019-10-04)[2020-03-25]. draft-ietf-lsr-isis-srv6-extensions-03.

[4] Li Z, Hu Z, Cheng D, Talaulikar K, Psenak P. OSPFv3 Extensions for SRv6[EB/OL]. (2020-02-12)[2020-03-25]. draft-ietf-lsr-ospfv3-srv6-extensions-00.

[5] Dawra G, Filsfils C, Raszuk R, Decraene B, Zhuang S, Rabadan J. SRv6 BGP Based Overlay Services[EB/OL]. (2019-11-04)[2020-03-25]. draft-ietf-bess-srv6-services-01.

[6] Dawra G, Filsfils C, Talaulikar K, Chen M, Bernier D, Decraene B. BGP Link State Extensions for SRv6[EB/OL]. (2019-07-07)[2020–03–25]. draft-ietf-idr-bgpls-srv6-ext-01.

[7] Negi M, Li C, Sivabalan S, Kaladharan P, Zhu Y. PCEP Extensions for Segment Routing leveraging the IPv6 data plane[EB/OL]. (2019-10-09)[2020-03-25]. draft-ietf-pce-segment-routing-ipv6-03.

[8] Matsushima S, Filsfils C, Ali Z, Li Z. SRv6 Implementation and Deployment Status [EB/OL]. (2020-03-09)[2020-03-25]. draft-matsushima-spring-srv6-deployment-status-06.

[9] Tian H, Zhao F, Xie C, Li T, Ma J, Peng S, Li Z, Xiao Y. SRv6 Deployment Consideration [EB/OL]. (2019-11-04)[2020-03-25]. draft-tian-spring-srv6-deployment-consideration-00.

[10] Cheng W, Li Z, Li C, et al. Generalized SRv6 Network Programming for SRv6 Compression[EB/OL]. (2020-05-19)[2020-05-29]. draft-cl-spring-generalized-srv6-for-cmpr-01.
[11] Li Z, Li C, Xie C, Lee K, Tian H, Zhao F, Guichard J, Li C, Peng S. Compressed SRv6 Network Programming[EB/OL]. (2020-02-25)[2020-03-25]. draft-li-spring-compressed-srv6-np-02.
[12] Li Z, Li C, Cheng W, Xie C, Li C, Tian H, Zhao F. Generalized Segment Routing Header[EB/OL]. (2020-02-11)[2020-03-25]. draft-lc-6man-generalized-srh-00.
[13] Filsfils C, Camarillo P, Cai D, Voyer D, Meilik I, Patel K, Henderrickx W, Jonnalagadda P, Melman D. Network Programming extension: SRv6 uSID instruction[EB/OL]. (2020-02-25)[2020-03-25]. draft-filsfils-spring-net-pgm-extension-srv6-usid-04.
[14] Bonica R, Hedge S, Kamite Y, Alston A, Henriques D, Jalil L, Halpern J, Linkova J, Chen C. Segment Routing Mapped To IPv6 (SRm6)[EB/OL]. (2019-11-19)[2020-01-22]. draft-bonica-spring-sr-mapped-six-00.
[15] AMS-IX. Frame Size Distribution[EB/OL]. (2020-01-22)[2020-01-22]. AMS-IX Statistics/sFlow Statistics.
[16] Jain S, Kumar A, Mandal S, Joon O, Leon P, Singh A, Venkata S, Jim W, Zhou J, Zhu M, Zolla J, Holzle U, Stuart S, Vahdat A. B4: Experience with a Globally-Deployed Software Defined WAN[EB/OL]. (2013-08-16)[2020-03-25]. ACM SIGCOMM Computer Communication Review, 2013.
[17] Li Z, Peng S, Voyer D, Xie C, Liu P, Liu C, Ebisawa K, Ueno Y, Previdi S, Guichard J. Problem Statement and Use Cases of Application-aware IPv6 Networking (APN6)[EB/OL]. (2019-11-03)[2020-03-25]. draft-li-apn6-problem-statement-usecases-00.
[18] Peng S, Voyer D, Xie C, Liu P, Liu C, Ebisawa K, Previdi S, Guichard J. Application-aware IPv6 Networking (APN6) Framework[EB/OL]. (2019-11-03)[2020-03-25]. draft-li-apn6-framework-00.
[19] Li Z, Peng S, Li C, et al. Application-aware IPv6 Networking (APN6) Encapsulation[EB/OL]. (2020-07-04)[2020-10-31]. draft-li-6man-app-aware-ipv6-network-02.
[20] Li Z, Li C, Xie C, et al. Compressed SRv6 Network Programming[EB/OL]. (2020-02-25)[2020-03-25]. draft-li-spring-compressed-srv6-np-02.

Postface: SRv6 Path

14.1 SRv6 PATH: SR-MPLS DEBATES

SR was first proposed in the industry in March 2013. Prior to that, MPLS technology had reached maturity after more than 10 years of development. At that time, I believed that the following were the major pain points of MPLS:

- Inability to deploy LDP FRR: Despite LDP having good scalability, FRR is unable to provide 100% network coverage, meaning that it was not possible to deploy LDP FRR on the live network. At that time, it seemed that this problem could be solved by combining the Maximally Redundant Trees (MRT) algorithm with LDP multi-topology.
- Poor scalability of RSVP-TE: The poor scalability arising from the soft state mechanism of RSVP-TE prevented MPLS TE from being widely deployed. However, this problem did not seem to be serious at that time. To start with, carrier IP networks had low requirements on the number of MPLS TE tunnels. Rather, they preferred to use technologies with higher scalability (such as IGP/LDP) to carry services, ensure service quality by increasing bandwidth, and provide high reliability through fast convergence. The other reason is that the distributed system architecture could be used to overcome the poor scalability of RSVP-TE.
- Immature MPLS multicast: MPLS multicast developed relatively late, meaning its reliability and inter-domain solutions required further improvement.

When SR was first proposed, it was difficult for industry practitioners, including myself, to accept SR immediately because it brought great changes to traditional MPLS. Back then, I believed that the problems of traditional MPLS that SR could solve could be fully solved by developing existing technologies. What's more, SR adopts brand new concepts. While it can solve some problems of traditional MPLS, it comes at a high cost, as it introduces new problems (e.g., challenges to hardware performance brought by processing of deeper label stacks and lack of support for multicast). I summarized the results of the comparative analysis between SR and LDP/RSVP-TE, and submitted a draft to the IETF.[1]

It's worth noting that I was not the only one with doubts about SR — it was also opposed by many experts in the MPLS field. Even experts studying IGP-based label distribution only expected to solve certain problems based on the MPLS architecture; they never planned to completely reconstruct MPLS.

Despite these doubts, we conducted in-depth research on SR.

1. Solutions to improve SR: Based on our analysis of SR's problems and deep insight into possible future development directions, we quickly started research to develop related technologies and standards, including:

   - MPLS global label[2]: Some use cases mentioned in the draft had been further developed, including MPLS source label[3] and MVPN/EVPN aggregation label.[4]

   - SR-based virtual network and bandwidth-guaranteed SR[5]: This work later became the foundation of network slicing.

   - SR SFC[6]

   - SR over UDP,[7] and more

   The research was carried out as early as 2013 and 2014. Later, China Mobile and Huawei jointly proposed SR Path Segment solutions, which laid the foundation for SR performance measurement.

2. Centralized MPLS TE solution: We submitted the draft *The Use Cases for Path Computation Element (PCE) as a Central Controller (PCECC).*[8] In contrast to SR, the initial solution of PCECC was to establish MPLS TE LSPs (instead of SR paths) through centralized

label distribution by the PCE. This solves poor scalability of RSVP-TE while achieving centralized traffic optimization.

In 2013 and 2014, during the early stages of SR development, the industry focused discussions on SR-MPLS. Although SRv6 was mentioned in the SR architecture, few people put much thought into it. Compared with SR-MPLS, SRv6 seemed far-fetched.

14.1.1 SRv6 Path: SDN Transition

The development of SR is closely intertwined with SDN. When it comes to SDN, the industry held contrasting views as to which path it should take. The first path is revolution, represented by OpenFlow, which aims to replace existing protocols and implement fully centralized SDN through the full separation of forwarding and control. The second path is transition, which aims to solve the problems of distributed IP networks through centralized control based on existing protocols, including NETCONF/YANG, BGP, and PCEP.

At the 88th IETF meeting in November 2013, we submitted more than 30 drafts based on our research in the SDN field and our vision on the future development of IP technologies. I made more than 15 presentations at that meeting for almost all important working groups in the Routing Area, comprehensively explaining our views on SDN transition.

1. MPLS+ on the forwarding plane: Global labels are introduced to allow MPLS labels to carry more information. Flexible combinations of these MPLS labels can meet more customized requirements of network services.

2. Control plane: We clearly outlined the southbound protocol frameworks and protocol extension requirements for SDN transition, such as Central Controlled Interior Gateway Protocol (IGPCC),[9] Central Controlled Border Gateway Protocol (BGPCC),[10] and PCECC.[11] Later, we submitted our proposal for the hierarchical controller architecture and corresponding southbound protocol extensions.[12]

3. Management plane: We proposed the construction of an open network environment based on NETCONF/YANG, including the standardization of controllers' southbound and northbound models and the tools and mechanisms required for building an open programmable environment.

These proposals truly caught the attention of those in attendance and triggered many heated debates. The distributed model is an important design concept of the Internet for which IETF works. This is why some IP experts have a natural aversion to centralized control. Even though we advocated a hybrid "distributed+centralized" network architecture over the fully centralized SDN controller architecture based on OpenFlow, some experts still felt "upset" about us strongly promoting centralized control in the IETF. However, as SDN continued to develop in the industry, these concepts gained more support and helped to promote the development of the SDN transitional path combined with other important work including stateful PCE and SR.

SR is aligned with the trend of SDN transition. This allowed it to gain wide recognition throughout the industry, which in turn promoted its rapid development. In addition, our PCECC solution achieved integration with SR, with the PCE centrally distributing SR labels to replace IGP extensions for label distribution. Furthermore, the PCE can be used to deliver SR paths, making the SR protocols simpler and more centralized. In 2015, 2 years after PCECC was proposed, when we presented at the Traffic Engineering Architecture and Signaling (TEAS) Working Group, attendees were asked to vote on whether to accept PCECC work. Almost everyone on site raised their hands in support. Dhruv, our presenter, said that this was the most successful presentation he had made in the IETF.

In addition to PCECC, we also proposed the innovative solution of MPLS path programming,[13] which abstracts SR label combinations into path programming, which is further divided into two types: transport-oriented path programming and service-oriented path programming. At that time, SR was mainly used to solve the problem of reachability, and so was defined as transport-oriented path programming. In contrast, the combination of MPLS global labels — used to indicate various network services, including VPN, load balancing, OAM, QoS, and security — allows us the flexibility to define a set of network services for a specific service flow. We categorize this type as service-oriented MPLS path programming, which can provide more value-added network services and offer more benefits to carriers.

During the development of the SDN transitional path, a whole host of technologies were integrated. Later on, SR Policy supported centralized allocation of binding SIDs, similar to SID allocation by the PCECC. In addition, SRv6 was renamed SRv6 Network Programming, and integrated service and transport programming was proposed, echoing MPLS path programming. Indeed, the development of any technology requires concerted efforts from the entire industry. Eventually, SR has become the de facto standard for SDN.

14.1.2 SRv6 Path: 2017

2013 is seen by many as the beginning of SRv6, but I see it differently. When SRv6 was first proposed, it was merely intended for inserting IPv6 addresses of nodes and links into SRHs to steer traffic. In March 2017, the *SRv6 Network Programming* draft was submitted to the IETF, upgrading SRv6 to SRv6 Network Programming. In a narrow sense, SRv6 Network Programming differs from SRv6 in that it divides a 128-bit SRv6 SID into fields including Locator and Function. In this way, it actually integrates routing and MPLS capabilities. This greatly enhances network programming capabilities of SRv6 SRHs and better supports new services. For these reasons, I believe 2017 is actually the year that marks the beginning of SRv6.

In a way, SRv6 can also be considered as the continuation of SDN transition. An important idea put forward by OpenFlow is the separation of the control and forwarding planes to enable the controller to perform forwarding plane programming through standard interfaces. However, this idea is too difficult to fully implement. In contrast, SRv6 Network Programming implements the forwarding plane programming in a way that is more compatible with existing IP networks, with far stronger capabilities than IPv4 and MPLS. From 2013 to 2016, the IETF focused more on SDN transition on the control plane. Following the submission of *SRv6 Network Programming* in 2017, its focus shifted to SDN transition on the forwarding plane.

Similar to when SR was released in 2013, we noticed the changes in SRv6 right away and began to conduct in-depth analysis and research. However, because the technology had only just been released, there was a lot of heated internal discussion. For more than half a year or so, Huawei experts had been discussing the selection of technical solutions including SRv6, SR-MPLS6, and SR-MPLS over UDP6. Of these, SR-MPLS over UDP was the first to be ruled out, as we believed it to be only a transition technology, and the combination of SR-MPLS and IPv6 will cause its deployment complex on the mobile transport networks. SR-MPLS6 was considered a mature technology at that time and was preferred by many.

However, because our customers were resolute in their preference for SRv6, we were left with little choice in the matter. In a way, we were forced to choose SRv6. So, to accelerate the maturity of SRv6, around October 2017 we submitted four drafts related to extensions of protocols (including OSPFv3, BGP-LS, PCEP, and BGP) for SRv6. These drafts, along with the existing drafts (including SRv6 SRH, SRv6 Network Programming, SRv6 IS-IS, and SRv6 VPN), constitute a relatively complete set of drafts about basic SRv6 features. We met industry experts at the IETF meeting held in Singapore in November 2017 to discuss the SRv6 standards roadmap and promotion priorities. At that time, we also comprehensively discussed how to improve the technical solutions proposed in these drafts. The IETF meeting held in Singapore was the 100th IETF meeting, making it particularly memorable. It also has a special meaning to us: It was the first time for us to work so closely with industry experts in key technical fields, including experts from equipment vendors, to jointly promote the development of the industry.

14.1.3 SRv6 Path: 2018

2018 was a crucial year for improving SRv6 standards and solutions. At each IETF meeting, we discussed SRv6 solutions and standards with industry experts, including those from equipment vendors. In addition, we conducted joint SRv6 design activities with some technical experts from carriers who, each time they came to China for 2 or 3 days, met with Huawei engineers to discuss the problems in their SRv6 network design as well as possible solutions. These activities played an important role in accelerating the improvement in SRv6 technologies and standards.

In September 2018, under the coordination of a third party, I led a Huawei development and test team, alongside that of another equipment vendor, to undertake interoperability testing for SRv6 in the lab of a Japanese company. Over 5 days, we performed a series of inter-op test cases, covering SRv6 L3VPN, SRv6 IS-IS, SRv6 BGP, SRv6 OAM, and SRv6 TI-LFA FRR. At times, these test processes felt like a roller-coaster ride. Sometimes, Huawei's devices encountered problems, so I would keep the other vendor's engineers entertained by going out for a meal, and then going on a tour of a garden. Other times, the other vendor's devices encountered problems, and I would do the same thing with Huawei's engineers. It felt as though I was visiting the garden once a day. Finally, on Friday afternoon, we completed the most complex SRv6 TI-LFA FRR interoperability test, with all scenarios passing the interoperability test. We could hardly contain our excitement. When posing for a group photo in the lab, we all instinctively gave a thumbs up. Then, on the spur of the moment, I taught the experts the Chinese hand gesture for the number six, which is interpreted as that "everything will go well smoothly" in Chinese, wishing that would be the case with SRv6. This ultimately became a standard gesture for SRv6 group photos. After the group photos, I invited everyone to a cafe for coffee, where we chatted away, celebrating the success of the interoperability test. On our way back, we passed by the Nihonbashi bridge, where I invited peer experts to take a photo in front of the bridge's plaque. Nihonbashi is the starting point of many expressways in Japan, meaning the mileage of these expressways is calculated from here. Similarly, the success of our first SRv6 third-party interoperability test with the other equipment vendor is also an important milestone for the development of SRv6. For this reason, Nihonbashi holds a deeper meaning for us.

In the early stages of SRv6 development, we focused on delivering basic features such as SRv6 TE, VPN, and FRR. However, a question always lingered in my mind: Since these features are already supported by SR-MPLS, why do we need to implement them again with SRv6? In other words, what are the unique values and significance of SRv6?

During our exploration from 2017 to 2018, my thinking on this issue went deeper and deeper thanks to the technical research carried out by our team in various emerging fields — including SRv6 Path Segment/OAM, IOAM/IFIT (on-path network telemetry), VPN+ (network slicing), and BIERv6 — as well as extensive technical exchanges with customers and technical experts in the industry. One or two days before the SRv6 Industry Forum was held in early 2019, the answer finally popped into my head. I could barely hide my excitement, and wrote a diary entry, which I then shared on my WeChat Moments:

> With thoughts bouncing around my head, I went for a long walk by myself, along the streets of Beijing, struck with an overwhelming sense of déjà vu. Over the past 20 years, I have been very fortunate to have experienced almost the entire MPLS era. I witnessed IP expanding originally from the Internet to IP backbone, metro, and mobile transport networks, fully unleashing the value of MPLS and fulfilling what I believe is the goal of the All IP 1.0 era. However, there are two bottlenecks: one is the interconnection

between isolated MPLS network islands; and the other is the network-only issue resulting from the decoupling of transport networks from applications. SRv6 can help overcome these bottlenecks by leveraging compatibility with IPv6's reachability. This attribute enables SRv6 to easily connect network domains, and SRv6 can also make use of MPLS-compatible TE capabilities to provide a wealth of options for network paths. In addition, this attribute allows SRv6 to bridge the gap between networks and applications, and IPv6 extension headers or SRHs can break the limitations of IPv4 and MPLS encapsulation, and bring application information to networks. This way, network traffic and applications can be associated to provide more room for innovations, improving the value of transport networks. In this sense, SRv6 is fundamentally different from SR-MPLS. Driven by application requirements such as 5G, cloud, and IoT, SRv6 has done its utmost to usher in the All IP 2.0 era, that is, All IPv6. This sense of the times made me excited again. Moreover, I felt extremely lucky that I am still here in this new era.

14.1.4 SRv6 Path: 2019

We eventually won the trust of customers thanks to our persistence with SRv6. However, our joint technical innovation project came to an abrupt stop due to an unexpected incident. As we lamented on this incident, things, thankfully, took a turn for the better as our joint innovation with carriers in the China region was coming along nicely. This was reminiscent of a Chinese poem, which roughly translates to "After passing through endless mountains and rivers that leave doubt as to whether there is a path out, suddenly one encounters the shade of a willow, bright flowers, and a lovely village." In Huawei Data Communication Product Line, Mr. Hu (Hu Kewen) actively promoted NetCity joint innovation activities with customers to accelerate the deployment of innovative solutions and improve the competitiveness of Huawei products and solutions through quick trials. Undoubtedly, China region is the best test field. There, the IP networks of carriers already support IPv6, facilitating SRv6 deployments. At the end of 2018, Huawei discussed with China Telecom Sichuan the possibility of deploying SRv6 on their live network. After lengthy discussions, the customer realized that SRv6 VPN could be easily deployed using their existing IPv6 network infrastructure, eliminating the need for the complex cross-departmental coordination that was required when they deployed MPLS VPN. At the beginning of 2019, China Telecom Sichuan deployed SRv6 VPN-based video services, making it the industry's first SRv6 commercial site.

Thanks to Huawei's efforts, 12 SRv6 sites were deployed within 1 year in the China region, taking full advantage of China's IPv6 network infrastructure. But it's not only in the China region where SRv6 is gaining ground: SRv6 has also been successively deployed by carriers in other countries and regions. Among these SRv6 deployments, I was particularly moved by two: The first used SRv6 TI-LFA FRR technology to improve the reliability of some networks that had poor physical infrastructure and were prone to failures. The other used SRv6 to quickly provision services in some war-stricken countries or regions. I was really happy to see that the networks in these less developed countries and regions have achieved rapid development through SRv6 technologies. As all parties strive for the newest technologies, competition inevitably arises, such competition can at times be unpleasant. However, these activities are fundamental to the development of the industry, benefiting numerous IP practitioners. Moreover, the development of these infrastructure technologies can benefit more people. Every time I think about this, I feel bright and warm inside, and become determined to put everything I have into embracing challenges and solving problems.

Along with the successful commercial deployment of SRv6, SRv6 standards continued to mature. In 2019, the *IPv6 Segment Routing Header (SRH)* draft passed Last Call and Internet Engineering Steering Group (IESG) review, and was released as RFC 8754 in early 2020. Drafts on basic SRv6 features have all been adopted by corresponding IETF working groups. The maturity of standards has entered a new stage. SRv6 industry activities are also getting into full swing, gaining more support from industry stakeholders. With the wide deployment of SRv6 in China and increasing innovation and standardization work of Chinese experts in the IETF, China's Expert Committee of Promoting Large-Scale IPv6 Deployment set up the IPv6+ Innovation Working Group. This marked the change of China's IPv6 work from accelerating deployment to leading innovation. The innovations we are promoting in the IETF are also incorporated into China's national strategy. At the end of 2019, I shared my SRv6 exploration during 2019 on WeChat's Moments:

SRv6 has made great progress over the past year. While witnessing its development, what impressed me the most about SRv6 is the sense of the times. SRv6 and the development of 5G/cloud services complement each other. 5G/cloud provides a great chance for SRv6 to succeed. As service requirements grow, technical innovation extends beyond SRv6. This means that the new encapsulation is no longer merely based on SRv6 SRHs, but instead is extended to other IPv6 extension headers. We define this as IPv6+ to better convey its characteristics of the times, and also define three possible stages of its development as follows:

1. IPv6+ 1.0: basic SRv6 features, including VPN, TE, and FRR.
2. IPv6+ 2.0: new network services oriented to 5G and cloud, including VPN+ (network slicing), IFIT (on-path network telemetry), DetNet, SFC, SD-WAN, BIERv6, etc.
3. IPv6+ 3.0: APN6, which will bring significant changes to the network architecture.

The opportunities brought by such an era also lay the foundation for us to solve many problems. The development of the IP industry, the development of IPv6 in China, and growing IP forces may all benefit from the development of SRv6. During my extensive communication with customers, I have fully felt the ambitions of emerging IP forces that took the lead in deploying SRv6, such as China, Japan, and some countries in the Middle East. Unlike the huge impact brought by SDN and the resulting chaos, division, disputes, and attacks, development of the IP industry is more orderly and constructive with SRv6. At the SRv6 Industry Salon, IPv6, SDN, and MPLS experts cooperated with each other, making us feel warm and welcome. Moreover, the might of these combined forces filled us with hope for the future.

14.2 SUMMARY

The development of SRv6 actually goes hand in hand with the rise of Huawei's IP protocol innovation. Throughout this process, I can strongly feel the growth of our company and country. After more than 20 years of development, our product and market strengths are finally able to fully support our innovation and standardization work. Chinese government began to promote large-scale IPv6 deployment at the end of 2017. After nearly 2 years of efforts, carriers' IP networks are able to fully support IPv6. At the same time, SRv6 brings many opportunities for IPv6 innovation, allowing us to quickly carry out innovation tests and deployment based on the IPv6-ready infrastructure. I feel so lucky to be part of this, but this luck is the result of our hard work and unremitting efforts. We would not have achieved any of this if we gave up each time we encountered a setback or if we failed to learn from any of their lessons. Looking ahead, we still have a long way to go, so we need to remain focused and keep pushing on.

REFERENCES

[1] Li Z. Comparison between Segment Routing and LDP/RSVP-TE[EB/OL]. (2016-03-13)[2020-03-25]. draft-li-spring-compare-sr-ldp-rsvpte-01.

[2] Li Z, Zhao Q, Yang T. Usecases of MPLS Global Label[EB/OL]. (2013-07-11)[2020-03-25]. draft-li-mpls-global-label-usecases-00.

[3] Bryant S, Chen M, Li Z, et al. Synonymous Flow Label Framework[EB/OL]. (2018-12-12)[2020-03-25]. draft-ietf-mpls-sfl-framework-04.

[4] Zhang Z, Rosen E, Lin W, et al. MVPN/EVPN Tunnel Aggregation with Common Labels[EB/OL]. (2019-10-24)[2020-03-25]. draft-ietf-bess-mvpn-evpn-aggregation-label-03.

[5] Li Z, Li M. Framework of Network Virtualization Based on MPLS Global Label[EB/OL]. (2013-10-21)[2020-03-25]. draft-li-mpls-network-virtualization-framework-00.

[6] Xu X, Bryant S, Assarpour H. Service Chaining using Unified Source Routing Instructions[EB/OL]. (2017-06-29)[2020-03-25]. draft-xu-mpls-service-chaining-03.

[7] Xu X, Bryant S, Farrel A, et al. MPLS Segment Routing over IP[EB/OL]. (2019-12-30)[2020-03-25]. RFC 8663.

[8] Zhao Q, Li Z, Khasanov B, et al. The Use Cases for Path Computation Element (PCE) as a Central Controller (PCECC)[EB/OL]. (2020-03-08)[2020-03-25]. draft-ietf-teas-pcecc-use-cases-05.

[9] Li Z, Chen H, Yan G. An Architecture of Central Controlled Interior Gateway Protocol (IGP)[EB/OL]. (2013-10-21)[2020-03-25]. draft-li-rtgwg-cc-igp-arch-00.

[10] Li Z, Chen M, Zhuang S. An Architecture of Central Controlled Border Gateway Protocol (BGP)[EB/OL]. (2013-10-20)[2020-03-25]. draft-li-idr-cc-bgp-arch-00.

[11] Farrel A, Zhao Q, Li R, et al. An Architecture for Use of PCE and the PCE Communication Protocol (PCEP) in a Network with Central Control[EB/OL]. (2017-12-30)[2020-03-25]. RFC 8283.

[12] Li Z, Dhody D, Chen H. Hierarchy of IP Controllers (HIC)[EB/OL]. (2020-03-08)[2020-03-25]. draft-li-teas-hierarchy-ip-controllers-04.
[13] Li Z, Zhuang Z. Use Cases and Framework of Service-Oriented MPLS Path Programming (MPP[EB/OL]. (2015-03-08)[2020-03-25]. draft-li-spring-mpls-path-programming-01.

Appendix A: IPv6 Basics

A.1 IPv6 OVERVIEW

Internet Protocol version 6 (IPv6), also called IP Next Generation (IPng), is a second-generation network layer protocol and an upgraded version of Internet Protocol version 4 (IPv4). It was designed by the IETF.

IPv4 is a widely used Internet protocol that developed rapidly in the early stages of the Internet due to its simplicity, ease of implementation, and interoperability. However, with the constant development of the Internet, some design defects have become more and more apparent. Namely, IPv4 has insufficient address space, its packet headers and options are complex to process, address maintenance is heavy, route aggregation is inefficient, and there is a lack of effective solutions for security, Quality of Service (QoS), and mobility.

IPv6 was developed to address these defects in the following ways:

1. Address space: IPv4 addresses are 32 bits long, theoretically providing 4.3 billion IPv4 addresses. IPv4 addresses are not only limited in number but also not fairly allocated. IPv4 addresses in the USA account for almost half of all addresses, leaving insufficient addresses for other areas such as Europe and Asia-Pacific. Furthermore, IPv4 addresses have already been exhausted with the development of mobile IPv4 and broadband technologies. While several solutions to IPv4 address exhaustion are currently in place, such as CIDR and NAT, they have some significant disadvantages, which have prompted the development of IPv6.

   IPv6 addresses are 128 bits long, allowing for an address space of 2^{128} possible addresses — that's enough for every grain of sand on earth, and as such unlikely to ever be exhausted.

2. Packet processing: An IPv4 header has an optional field (Options), including Security, Timestamp, and Record Route options. The variable length of the Options field results in an IPv4 header length range of 20–60 octets, and these often need to be forwarded by intermediate devices along with packets. The Options field occupies a large amount of resources and is therefore rarely used in practice.

   Unlike the IPv4 header, the IPv6 header does not carry IHL, Identification, Flags, Fragment Offset, Header Checksum, Options, or Padding fields, but instead carries the Flow Label field. This simplifies IPv6 packet processing and improves processing efficiency. The Extension Header information field is added to an IPv6 packet to support various options without changing the existing IPv6 packet format, improving extensibility.

3. Address maintenance: IPv4 addresses need to be reallocated often when networks are expanded or replanned. To make this easier, IPv4 depends on the Dynamic Host Configuration Protocol (DHCP) to automatically allocate and replan addresses.

   IPv6 takes this a step further and provides its own address autoconfiguration to allow hosts to automatically discover networks and obtain IPv6 addresses, improving network manageability.

4. Route aggregation: Many noncontiguous IPv4 addresses are allocated, making it impossible for routes to be aggregated effectively. This results in increasingly large routing tables which consume a lot of memory and affect forwarding efficiency. Manufacturers must continually upgrade devices to improve route addressing and forwarding performance.

   The enormous number of addresses provided by IPv6 allows us to design networks hierarchically, facilitating route aggregation and improving forwarding efficiency.

5. End-to-end security: IPv4 only supports retrofitted solutions for end-to-end security, as security was an afterthought during the initial design in the early 1980s. IPv6 supports IP security (IPsec) authentication and encryption at the network layer for end-to-end security.

6. QoS guarantee: IPv4 has no native support for QoS, which is particularly useful for real-time forwarding of voice, data, and video services such as network conferencing, telephones, and TVs. IPv6,

on the other hand, provides a Flow Label field, which can be used for QoS.

7. Mobility: Mobile IPv4 has experienced some significant issues such as triangular routing and source address filtering. To solve these issues, IPv6 provides native mobility support. Unlike mobile IPv4, mobile IPv6 uses the neighbor discovery function to discover a foreign network and obtain a care-of address without a foreign agent. In addition, the mobile and peer nodes can communicate using the Routing and Destination Options headers. Triangular routing and source address filtering are no longer an issue.

A.2 IPv6 ADDRESS

A.2.1 IPv6 Address Format

An IPv6 address is 128 bits long and is written as eight groups of four hexadecimal digits, each group separated by a colon (:). This is the preferred format of an IPv6 address. An IPv6 address example is FC00:0000:130F:0000:0000:09C0:876A:130B.

For convenience, IPv6 addresses can be written in a compressed format. For this example[1]:

- Leading zeros of each group can be omitted to form FC00:0:130F:0:0:9C0:876A:130B.
- Any number of consecutive groups of 0s can be replaced with two colons (::) to form FC00:0:130F::9C0:876A:130B.
- An IPv6 address can contain only one double-colon substitution. Multiple double-colon substitutions lead to ambiguity.

A.2.2 IPv6 Address Structure

An IPv6 address is composed of two parts:

- Network prefix: equivalent to the network ID of an IPv4 address. This prefix is variable in length.
- Interface ID: equivalent to the host ID of an IPv4 address. This ID is 128 bits minus the length of the prefix.

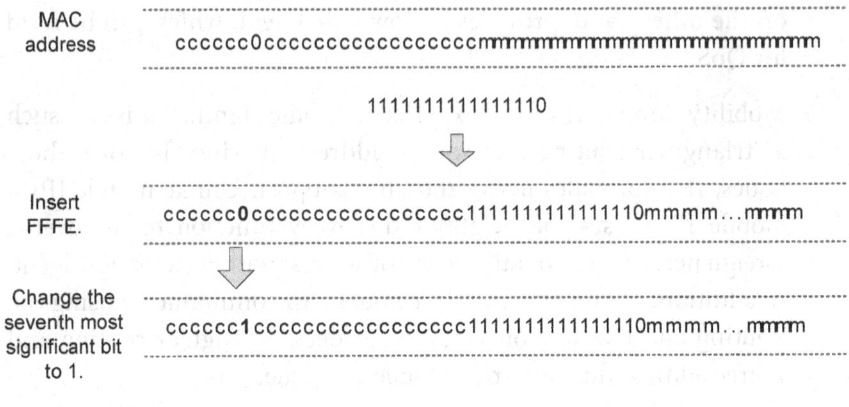

MAC address cccccc0cccccccccccccccccmmmmmmmmmmmmmmmmmmmmmmmm

1111111111111110

Insert FFFE. cccccc0ccccccccccccccccc1111111111111110mmmm...mmmm

Change the seventh most significant bit to 1. cccccc1ccccccccccccccccc1111111111111110mmmm...mmmm

FIGURE A.1 IEEE EUI-64 standard.

If the first 3 bits of an IPv6 unicast address are not 000, the interface ID must contain 64 bits. If the first 3 bits are 000, there is no such limitation.

An interface ID can be manually configured or automatically generated using software or the IEEE EUI-64 standard. Automatic generation using the IEEE EUI-64 standard is the most common practice.

The IEEE EUI-64 standard converts an interface MAC address into an IPv6 interface ID. Figure A.1 shows a 48-bit MAC address. When used as an interface ID, the first 24 bits (expressed as c below) are a vendor identifier, and the last 24 bits (expressed as m below) are an extension identifier. If the seventh most significant bit is 0, the MAC address is locally unique.

During conversion, EUI-64 inserts FFFE between the vendor and extension identifiers. The seventh most significant bit also changes from 0 to 1 to indicate that the interface ID is globally unique. This is because a globally unique IPv6 address needs to be generated.

For example, if the MAC address is 000E-0C82-C4D4, the interface ID is 020E:0CFF:FE82:C4D4.

Finally, we obtain an interface ID with 64 (48 + 16) bits, before which a 64-bit network prefix can be added to obtain a globally unique IPv6 address.

Converting MAC addresses into IPv6 interface IDs makes configuration easier, especially when you only need an IPv6 network prefix in stateless address autoconfiguration to form an IPv6 address. The disadvantage of this method is that IPv6 addresses can be deduced based on MAC addresses, which is insecure.

A.2.3 IPv6 Address Classification

IPv6 addresses are classified as unicast, multicast, or anycast addresses. Unlike IPv4, IPv6 does not implement broadcast addressing. Instead, multicast addresses are used to replace broadcast addresses. In addition, IPv6 supports anycast addresses.

Table A.1 lists the address segments allocated to various IPv6 addresses.[2]

A.2.3.1 IPv6 Unicast Address

An IPv6 unicast address identifies an interface. Since each interface belongs to a node, the IPv6 unicast address of any interface can identify its node. As you would expect, packets sent to an IPv6 unicast address are delivered to the interface identified by that address.

IPv6 defines multiple types of unicast addresses, including unspecified, loopback, global unicast, link-local, and unique local addresses.

- The IPv6 unspecified address is 0:0:0:0:0:0:0:0/128 or ::/128. It indicates that an interface or a node does not have an IP address. It can be used as the source IP address of some packets, such as Neighbor Solicitation (NS) messages, in duplicate address detection. Devices do not forward packets with an unspecified address as the source IP address.
- The IPv6 loopback address is 0:0:0:0:0:0:0:1/128 or ::1/128. Similar to the IPv4 loopback address 127.0.0.1, the IPv6 loopback address is used when a node needs to send IPv6 packets to itself. This IPv6 loopback address is usually used as the IP address of a virtual interface, such as a loopback interface. The loopback address cannot be used as the source or destination IP address of packets needing to be forwarded.

TABLE A.1 Address Segments Allocated to Various IPv6 Addresses

IPv6 Address Segment	Address Type
2000::/3	Global Unicast Address (GUA) and anycast address
FC00::/7	Unique Local Address (ULA)
FE80::/10	Link-Local Address (LLA)
FF00::/8	Multicast address
Others	Reserved

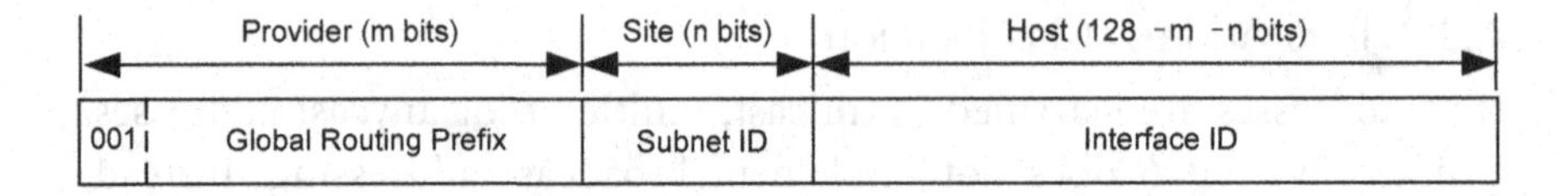

FIGURE A.2 Global unicast address format.

- An IPv6 global unicast address is an IPv6 address with a global unicast prefix and is similar to an IPv4 public address. IPv6 global unicast addresses support routing prefix aggregation, helping limit the number of global routing entries.

Figure A.2 shows a global unicast address consisting of a global routing prefix, subnet ID, and interface ID.

Table A.2 describes the related fields.

- Link-local addresses are used only in communication between nodes on a local link. A link-local address uses a link-local prefix of FE80::/10 as its 10 most significant bits (1111111010 in binary) and an interface ID as its 64 least significant bits.

  When IPv6 runs on a node, a link-local address that consists of a fixed prefix and an interface ID in EUI-64 format is automatically allocated to each interface of the node. This mechanism enables two IPv6 nodes on a link to communicate without any additional configuration. Therefore, IPv6 link-local addresses are widely used in scenarios, such as neighbor discovery and stateless address autoconfiguration.[4]

  Routing devices do not forward IPv6 packets to devices on other links when the source or destination addresses are link-local addresses. Figure A.3 shows the link-local address format.

TABLE A.2 Fields in a Global Unicast Address

Field	Description
Global Routing Prefix	A global routing prefix is allocated by a provider to an organization. It is usually comprised of at least 48 bits. Currently, the first 3 bits of every allocated global routing prefix is 001.[3]
Subnet ID	A subnet ID is used by an organization to construct a local network (site). Similar to an IPv4 subnet number, there are a maximum of 64 bits for both the global routing prefix and subnet ID.
Interface ID	An interface ID identifies a device (host).

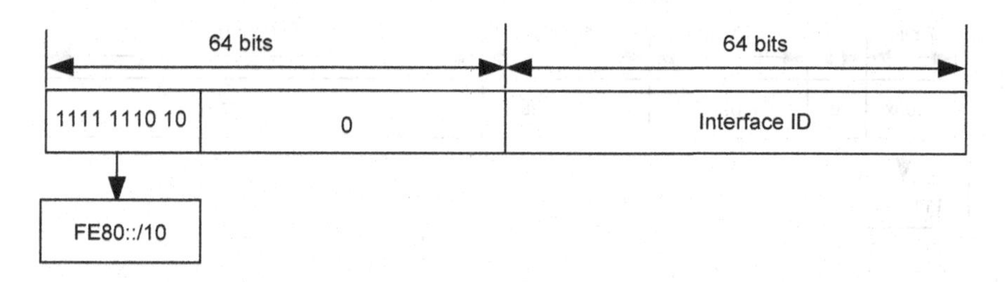

FIGURE A.3 Link-local address format.

- A ULA is also an address with a limited application scope. Its predecessor is a Site-Local Address (SLA), which has since been made obsolete due to its many problems.[5] To better understand a ULA, the following briefly describes an SLA.

  Similar to a private address in IPv4, an SLA is an address segment with the site-local prefix FEC0::/10 and can be used by a single domain. An SLA can be routed only within a single site and does not need to be obtained from an address allocation organization. Each site manages address segment division and address allocation.

  SLAs have similar problems to IPv4 addresses. For example, address conflicts may occur when networks are connected or converged. This tends to require replanning and reallocating network segments as well as addresses across multiple domains, complicating expansion, increasing workloads, and interrupting the network.

  ULAs are similar to private IPv4 addresses and can only be routed within a local network. However, they can be used without allocation from a registrar.

  Figure A.4 shows the ULA format.

  Table A.3 describes the related fields.

The unique local address block, FC00::/7, is planned separately from an IPv6 address segment and can be routed only within a domain. The prefix used by a single domain is Prefix + L + Global ID. The pseudo-randomness of a global ID ensures that addresses do not overlap over multiple domains that use a ULA.

The ULA is also a global unicast address. The only difference from a global unicast address is that the routing prefix of the ULA is not

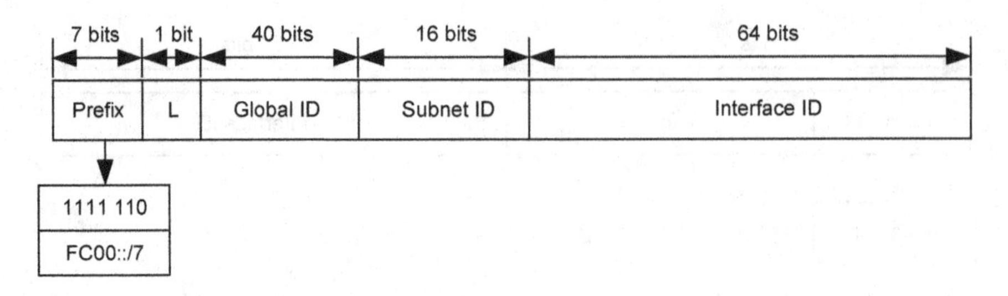

FIGURE A.4 ULA format.

TABLE A.3 Fields in a ULA

Field	Description
Prefix	The value is fixed at FC00::/7.
L	L flag. The value 1 indicates that the address is valid within a local network. The value 0 is reserved for future use.
Global ID	Global identifier, generated using a pseudo-random algorithm.
Subnet ID	Subnet identifier, used for subnetting.
Interface ID	Interface identifier.

advertised to the Internet. In addition, even if route leaking occurs, the original traffic on the Internet and the public network traffic in other domains are not affected. ULAs act as private addresses for traffic in a management domain and resolve the main problems of SLAs.

Specifically, a ULA has the following characteristics:

- Has a pseudo-randomly allocated prefix that is highly likely to be unique globally.
- Allows private connections to be established between sites without encountering address conflicts.
- Has a prefix of FC00::/7 that allows for easy route filtering at site boundaries.
- Does not conflict with any other addresses if it is leaked outside accidentally.
- Functions as a global unicast address for applications.
- Is not managed by ISPs.

A.2.3.2 IPv6 Multicast Address

Like IPv4 multicast addresses, IPv6 multicast addresses identify groups of interfaces, which usually belong to different nodes. A node may belong to any number of multicast groups. Packets sent to an IPv6 multicast address are delivered to all the interfaces identified by that address. For example, the multicast address FF02::1 indicates all nodes within the link-local scope, and FF02::2 indicates all routers within the link-local scope.

Figure A.5 shows the IPv6 multicast address format.[6]

An IPv6 multicast address is composed of Network Prefix, Flag, Scope, and Group ID fields.

- Network Prefix: is fixed at FF00::/8. The Plen field in Figure A.5 indicates the prefix length.
- Flag: is 4 bits long. The first 3 bits are reserved and must be set to 0s. The last bit set to 0 indicates a permanent multicast address allocated by the IANA, and the last bit set to 1 indicates a temporary multicast address.
- Scope: is 4 bits long. It limits the scope where multicast data flows are sent on the network. Figure A.5 shows the field values and meanings.
- Group ID: is 112 bits long. It identifies a multicast group. RFC 2373 does not define all the 112 bits as a group ID but recommends using

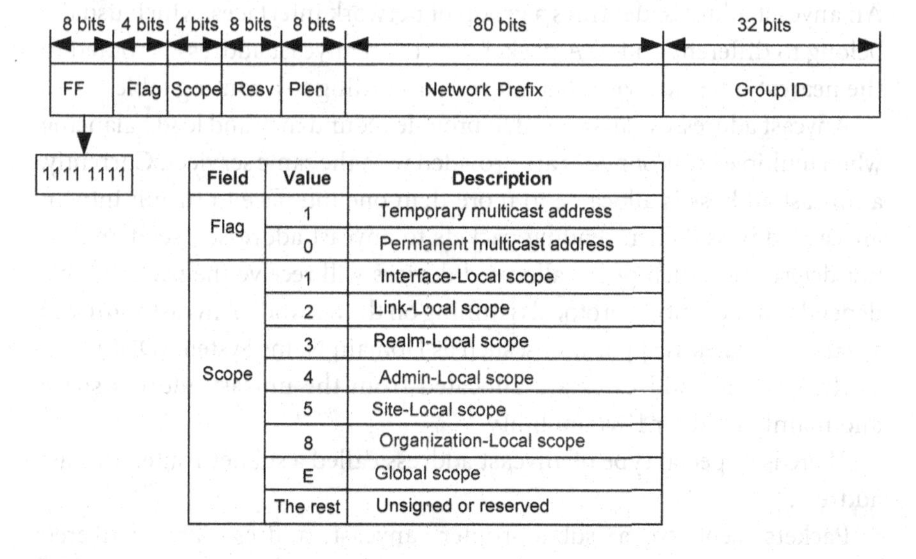

Field	Value	Description
Flag	1	Temporary multicast address
	0	Permanent multicast address
Scope	1	Interface-Local scope
	2	Link-Local scope
	3	Realm-Local scope
	4	Admin-Local scope
	5	Site-Local scope
	8	Organization-Local scope
	E	Global scope
	The rest	Unsigned or reserved

FIGURE A.5 IPv6 multicast address format.

the low-order 32 bits as the group ID and setting all of the remaining 80 bits to 0s.[7] In this case, each multicast group ID maps to a unique Ethernet multicast MAC address.[8]

There is a special type of multicast addresses called solicited-node multicast addresses.

A solicited-node multicast address is generated using an IPv6 unicast or anycast address of a node. If a node has an IPv6 unicast or anycast address, a solicited-node multicast address is generated for the node, and the node joins the corresponding multicast group. A unicast or anycast address corresponds to a solicited-node multicast address. This address is used for neighbor discovery and duplicate address detection.

IPv6 does not support broadcast addresses or the Address Resolution Protocol (ARP). In IPv6, NS messages are used to resolve IP addresses to MAC addresses. When a node needs to resolve an IPv6 address to a MAC address, it sends an NS message in which the destination IP address is the solicited-node multicast address corresponding to the IPv6 address. Only the node with this multicast address processes the NS message.

The solicited-node multicast address consists of the prefix FF02::1:FF00:0/104 and the last 24 bits of the corresponding unicast address.

A.2.3.3 IPv6 Anycast Address

An anycast address identifies a group of network interfaces, which usually belong to different nodes. A packet sent to an anycast address is routed to the nearest interface with that address, according to a routing table.

Anycast addresses can be used to provide redundancy and load balancing when multiple hosts or nodes are provided with the same services. Currently, a unicast address is allocated to more than one interface to turn it into an anycast address. When sending packets to anycast addresses, senders cannot determine which of the allocated devices will receive the packets. This depends on the routing protocols running on the network. Anycast addresses are used in stateless applications, such as Domain Name System (DNS).

IPv6 anycast addresses are allocated from the unicast address space and mainly used for IPv6 mobility.

There is a special type of anycast address called a subnet-router anycast address.

Packets sent to a subnet-router anycast address are delivered using routing protocols to the nearest device on the subnet identified by

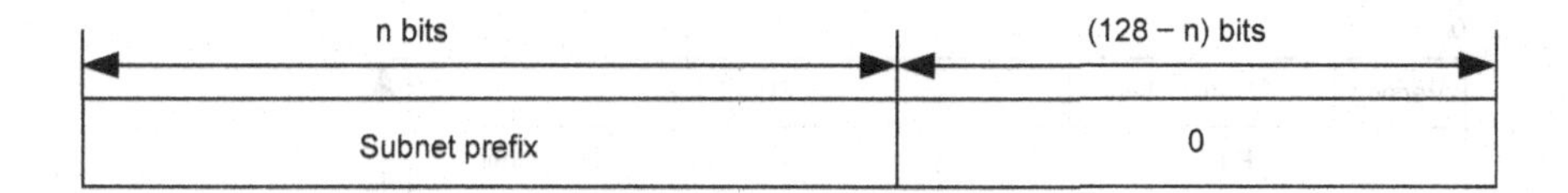

FIGURE A.6 Subnet-router anycast address format.

that address. A subnet-router anycast address is used when a node needs to communicate with any of the devices on the subnet identified by that address, for example, when a mobile node needs to communicate with one of the mobility agents on the home subnet. As a side note, all devices must support subnet-router anycast addresses.

In a subnet-router anycast address, the *n*-bit subnet prefix identifies a subnet, and the remaining bits are padded with 0s. Figure A.6 shows the subnet-router anycast address format.

A.3 IPv6 HEADER

An IPv6 packet consists of three parts: an IPv6 basic header, one or more IPv6 extension headers, and an upper-layer Protocol Data Unit (PDU).

An upper-layer PDU is usually composed of the upper-layer protocol header and its payload. The PDU can be an ICMPv6, TCP, or UDP packet.

A.3.1 IPv6 Basic Header

An IPv6 basic header has a fixed length of 40 octets with eight fields. An IPv6 basic header, required in every IPv6 packet, provides basic packet forwarding information, which all devices parse on the forwarding path. Figure A.7 shows an IPv6 basic header.

Table A.4 describes the main fields in an IPv6 basic header.

The IPv6 packet format is designed to simplify the basic header. In most cases, a device only needs to process the basic header to forward IP traffic. Unlike the IPv4 header, the IPv6 header does not carry the fields related to fragmentation, checksum, and options, but it carries the Flow Label field. This simplifies IPv6 packet processing and improves processing efficiency. To support various options without changing the existing packet format, the Extension Header information field is added to an IPv6 packet, improving flexibility.

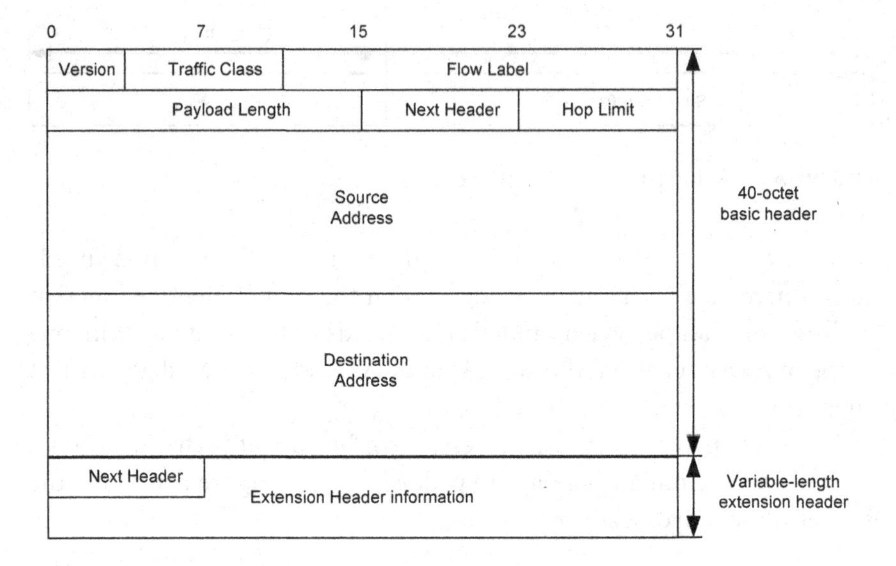

FIGURE A.7 IPv6 basic header.

TABLE A.4 Fields in an IPv6 Basic Header

Field	Length	Description
Version	4 bits	In IPv6, the value of this field is set to 6.
Traffic Class	8 bits	This field indicates the class or priority of an IPv6 packet. It is similar to the ToS field in an IPv4 packet and mainly used in QoS control.
Flow Label	20 bits	This field is added in IPv6 to differentiate real-time traffic. A flow label and source IPv6 address uniquely identify a data flow. Intermediate network devices can effectively differentiate data flows based on this field.
Payload Length	16 bits	The payload refers to the extension header and upper-layer PDU that follow the IPv6 basic header. If the Payload Length exceeds its maximum value of 65,535 bytes, the field is set to 0. The Jumbo Payload option in the Hop-by-Hop Options header is used to express the actual payload length.
Next Header	8 bits	This field identifies the type of the first extension header (if any) that follows the IPv6 basic header or the protocol type in the upper-layer PDU.
Hop Limit	8 bits	This field is similar to the TTL field in an IPv4 packet, defining the maximum number of hops that an IPv6 packet can pass through. The value is decremented by 1 each time it is forwarded by a device. If the field value is reduced to 0, the packet is discarded.
Source Address	128 bits	This field indicates the address of the packet originator.
Destination Address	128 bits	This field indicates the address of the packet recipient.

A.3.2 IPv6 Extension Header

An IPv4 header has an optional field (Options), which includes Security, Timestamp, and Record Route options. The variable length of the Options field results in an IPv4 header length range of 20–60 octets. Many resources are required to forward IPv4 packets with the Options field; therefore, such IPv4 packets are rarely used in practice.

To improve packet processing efficiency, IPv6 uses extension headers to replace the Options field in the IPv4 header. An IPv6 packet may or may not carry extension headers, and they are placed between the IPv6 basic header and upper-layer PDU. When the sender of a packet requests the destination device or other devices to perform special handling, the sender adds one or more extension headers to the packet. IPv6 has better extensibility than IPv4 with its variable-length extension headers, which are not limited to 40 octets. To improve extension header processing efficiency and transport protocol performance, IPv6 requires that the extension header length be an integral multiple of 8 octets.

When multiple extension headers are used, the Next Header field of an extension header indicates the type of the next header that follows. The Next Header field in the IPv6 basic header indicates the first extension header type, and the Next Header field in the first extension header indicates the next (second) extension header type, and so on. The Next Header field of the final extension header indicates the upper-layer protocol type. Figure A.8 shows an IPv6 packet with multiple extension headers.

Table A.5 describes the main fields in an IPv6 extension header.

A.3.3 Order of IPv6 Extension Headers

When more than one extension header is used in an IPv6 packet, those headers must appear in the following order[9]:

1. IPv6 header
2. Hop-by-Hop Options header
3. Destination Options header
4. Routing header
5. Fragment header
6. Authentication header

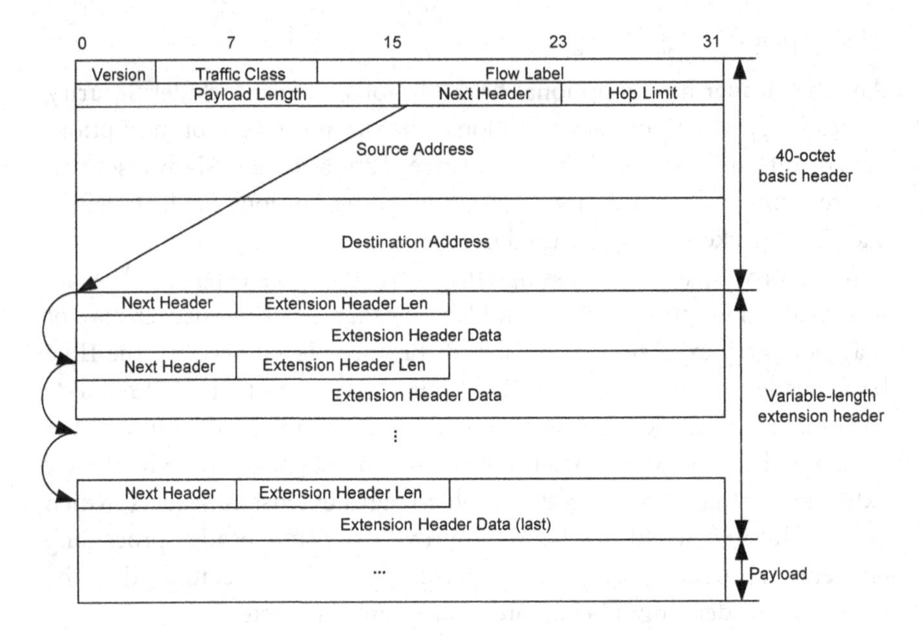

FIGURE A.8 IPv6 packet with multiple extension headers.

TABLE A.5 Fields in an IPv6 Extension Header

Field	Length	Description
Next Header	8 bits	This field is similar to the Next Header field in the IPv6 basic header, indicating the next extension header (if any) or upper-layer protocol type.
Extension Header Len	8 bits	This field indicates the extension header length excluding the first 8 octets, in 8-octet units. The shortest length of the extension header is 8 octets while this field is set to 0.
Extension Header Data	Variable	This field includes a series of options and the padding field.

7. Encapsulating Security Payload header
8. Destination Options header (options to be processed at the destination)
9. Upper-Layer header

Intermediate devices do not need to examine or process all extension headers, but only those based on the Next Header field value in the IPv6 basic header.

Each extension header can only occur once in an IPv6 packet, except for the Destination Options header which may occur twice (once before a Routing header and once before the Upper-Layer header).

The following briefly describes each extension header.

A.3.4 Hop-by-Hop Options Header

The Hop-by-Hop Options header is used to carry information that needs to be processed by each router on a forwarding path. It is identified by a Next Header value of 0. Figure A.9 shows the Hop-by-Hop Options header format.

The Value field of a Hop-by-Hop Options header consists of a series of options that allow this header to carry different types of information. Figure A.10 shows TLV-encoded options.

Table A.6 describes the fields in Options.

A.3.5 Destination Options Header

The Destination Options header is used to carry information that needs to be processed by the node of the destination address. This node can be the final destination of the packet or the endpoint node in a source routing scheme.

The Destination Options header is identified by a Next Header value of 60. The format and requirements of a Destination Options header are the same as those of a Hop-by-Hop Options header.

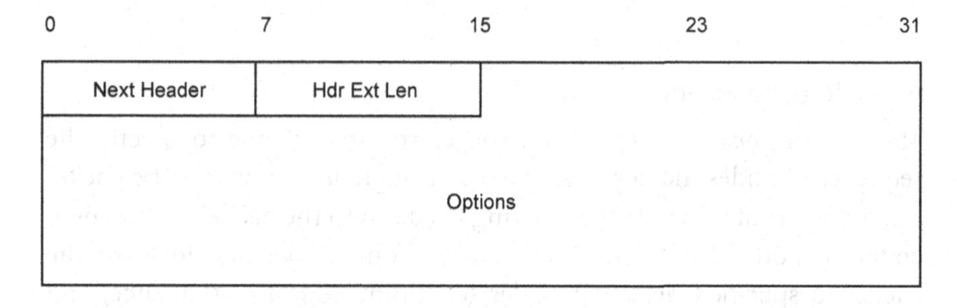

FIGURE A.9 Hop-by-Hop Options header format.

FIGURE A.10 Options.

TABLE A.6 Fields in Options

Field	Length	Description
Option Type	8 bits	Type of the current option. The data format in the Option Data field varies according to the option type. The requirements for using the Option Type field are as follows: 1. The first and second most significant bits specify the action to be taken when a node does not support the processing of this option. The values and their meanings are as follows: • 00: ignores this option and continues processing the next option. • 01: discards the packet. • 10: discards the packet and sends an ICMP Parameter Problem message to the source address. • 11: discards the packet and, only if the IPv6 destination address of the packet is not a multicast address, sends an ICMP Parameter Problem message to the source address. 2. The third most significant bit indicates whether the option can be modified during packet forwarding. The value 1 indicates that the option can be modified, and the value 0 indicates that the option cannot be modified. 3. The remaining 5 bits are reserved for future use. All 8 bits, together, are used to identify the type of an option.
Option Data Len	8 bits	Length of the Option Data field of the current option, in bytes.
Option Data	Variable	Data of the current option. The length of the Hop-by-Hop Options header must be an integral multiple of 8 octets. If the data length is insufficient, a padding option can be used.[9]

A.3.6 Routing Header

The Routing header is used in a source routing scheme to specify the sequence of nodes that a packet passes through on a network. The packet sender or a router inserts the Routing header into the packet. Subsequent endpoint nodes read node information from the header, forward the packet to specified next-hop nodes (endpoint nodes) in sequence, and finally forward the packet to the destination. The Routing header allows a packet to travel along a specified forwarding path instead of the default shortest path.

The Routing header is identified by a Next Header value of 43. Figure A.11 shows the Routing header format.

Table A.7 describes the main fields in the Routing header.

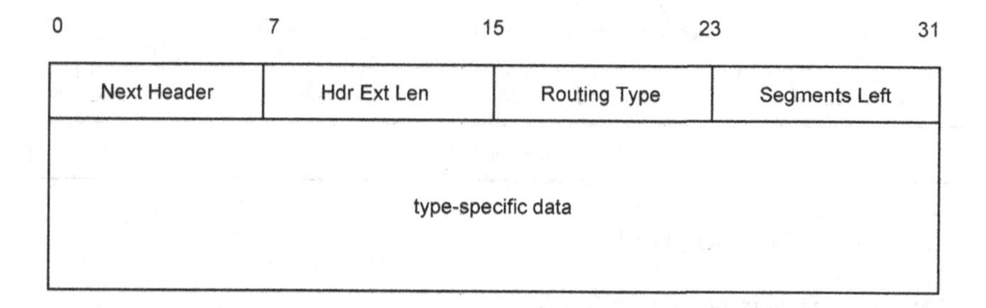

FIGURE A.11 Routing header format.

TABLE A.7 Main Fields in the Routing Header

Field	Length	Description
Routing Type	8 bits	Source routing scheme corresponding to the Routing header and data format of the routing data area.
Segments Left	8 bits	Number of remaining endpoint nodes, excluding the current endpoint node in the DA field.
type-specific data	Variable	Routing data of a specific routing scheme, which is usually the information about each endpoint node. The data format is defined in the routing scheme.

If a router does not support the source routing scheme specified by the Routing Type field in the Routing header of a packet, it processes the packet in the following ways:

- If Segments Left is 0, the router ignores the Routing header and proceeds to process the next header in the packet.
- If Segments Left is not 0, the router discards the packet and sends an ICMP Parameter Problem message to the source address.

A.3.7 Fragment Header

If the length of an application-layer packet exceeds the PMTU, the packet needs to be fragmented before transmission and reassembled after reception at the network layer. The Fragment header carries the identification information of each fragment, and it functions the same as corresponding fields in an IPv4 header. IPv6 allows only the packet sender to fragment packets.

The Fragment header is identified by a Next Header value of 44. Figure A.12 shows the Fragment header format.

Table A.8 describes the main fields in the Fragment header.

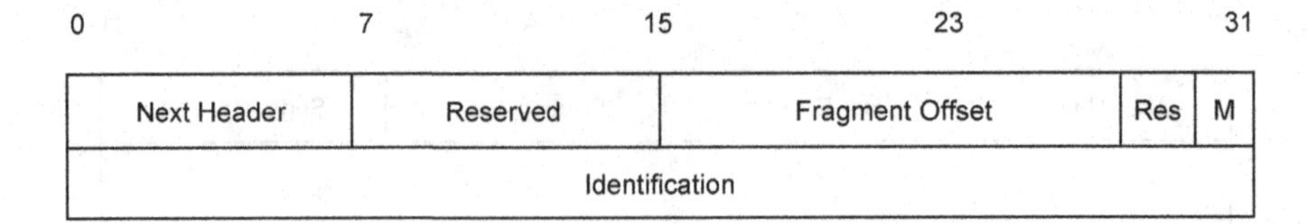

FIGURE A.12 Fragment header.

TABLE A.8 Main Fields in the Fragment Header

Field	Length	Description
Reserved	8 bits	Reserved field. This field should be the same as a length field in other extension headers. However, the length of the Fragment header is fixed and can be identified by a Next Header value of 44. Therefore, a length field does not need to be set. This field is reserved for future use and is currently set to 0.
Fragment Offset	13 bits	Offset of the fragment in the fragmentable part of the original packet, in 8 bytes. It can be used to calculate the length of a packet to be reassembled.
M (more)	1 bit	Indicates whether there are still more fragments to process. A value of 0 indicates that there are no more fragments to process. A value of 1 indicates that there are still fragments to process.
Identification	32 bits	Identifies the original packet to which the fragment belongs. All fragments of an original packet have the same Identification value. A receiver considers the fragments with the same source address, destination address, and Identification value to belong to the same original packet.

As shown in Figure A.13, an original packet is fragmented into three parts.

The IPv6 header and extension headers are divided into two parts. The first part is Per-Fragment Headers, which appear before the Fragment header. Each fragment carries the IPv6 header and Per-Fragment Headers. The second part is Extension & Upper-Layer Headers, which are carried after the Fragment header, and only in the first fragment.

The data following Extension & Upper-Layer Headers is called the fragmentable part. A packet needs to be fragmented for transmission

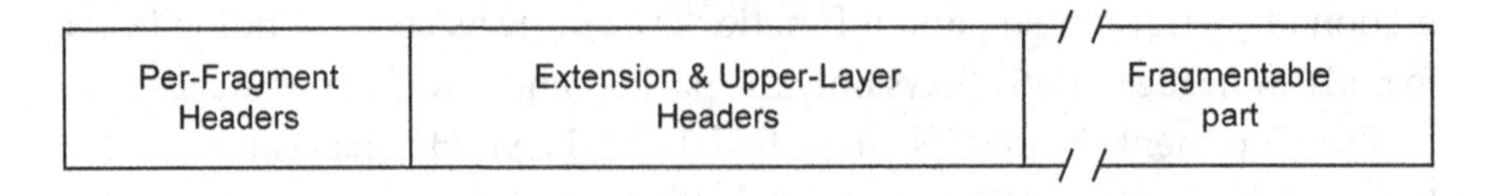

FIGURE A.13 Fragmentation of an original packet.

when its fragmentable part is too long. When calculating the Fragment Offset, the system calculates only the offset of the fragment data to the data of the fragmentable part. The lengths of all fragments, except the last fragment, are an integral multiple of 8 bytes.

When generating fragments, the system inserts a Fragment header after the first part. Figure A.14 shows the original and fragment packets.

During packet reassembly, the system obtains all fragments of an original packet, by looking for a matching source address, destination address, and Identification value. It uses the Fragment Offset field in the Fragment header to determine the sequence of a fragment in the original packet, and uses the Next Header value of the first part to identify the extension header of the second part according to the first fragment. The following formula is used to calculate the IPv6 payload length of the original packet:

PL.orig = (PL.first – FL.first – 8) + (8 × FO.last) + FL.last

PL.first – FL.first – 8: Payload length of the first fragment – Length of the first fragment data – Length of the Fragment header

8 × FO.last: 8 × Fragment offset of the last fragment, and we get the total length of other fragments except the last fragment.

FL.last: Length of the last fragment data

Original packet:

Per-Fragment Headers	Extension & Upper-Layer Headers	First fragment	Second fragment	...	Last fragment

Fragment packet:

Per-Fragment Headers	Fragment Header	Extension & Upper-Layer Headers	First fragment

Per-Fragment Headers	Fragment Header	Second fragment

⋮

Per-Fragment Headers	Fragment Header	Last fragment

FIGURE A.14 Original and fragment packets.

A.3.8 Authentication Header

The Authentication header is usually used in IPsec[10] to provide three security functions: connectionless integrity check, IP packet origin authentication, and anti-replay. For details about header processing, see RFC 4302.[11]

The Authentication header is identified by a Next Header value of 51. Figure A.15 shows the Authentication header format.

Table A.9 describes the main fields in the Authentication header.

0	7	15	23	31
Next Header	Payload Len	Reserved		
Security Parameters Index (SPI)				
Sequence Number Field				
Integrity Check Value-ICV (variable)				

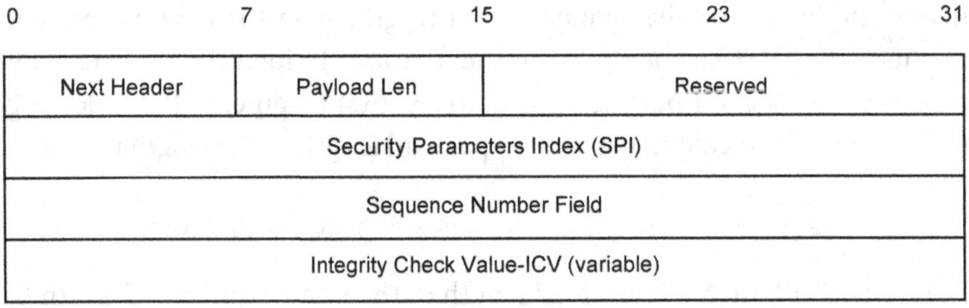

FIGURE A.15 Authentication header.

TABLE A.9 Main Fields in the Authentication Header

Field	Length	Description
Payload Len	8 bits	Specifies the length of the Authentication header. IPv6 requires that the length of the Authentication header be an integral multiple of 8 bytes; however, this field represents length as 4 bytes per unit, making it compatible with IPv4. In addition, the first 8 bytes are not included in the calculation.
Security Parameters Index	32 bits	Is used by a receiver to identify the Security Association (SA) to which the packet belongs.[11]
Sequence Number Field	32 bits	Specifies an ID that is incremented by one each time a packet is sent, and cannot decrease. It is bound to an SA to protect against replay attacks. The value is not allowed to increase back round to 0. That is, before the value increases to 2^{32}, the maximum value this field can hold, the two communication parties need to exchange a new key as well as establish and use a new SA.
Integrity Check Value	Variable, an integral multiple of 32 bits	Specifies an Integrity Check Value (ICV) for the related fields of the packet. The algorithm for generating the ICV is determined by an SA and can be a symmetric encryption algorithm or a hash algorithm.

A.3.9 Encapsulating Security Payload Header

The Encapsulating Security Payload header is usually used for IPsec[10] to provide security functions, such as connectionless integrity check, data origin authentication, anti-replay, and data encryption.[12]

The Encapsulating Security Payload header is identified by a Next Header value of 50. Figure A.16 shows the format of the Encapsulating Security Payload header.

Table A.10 describes the main fields in the Encapsulating Security Payload header.

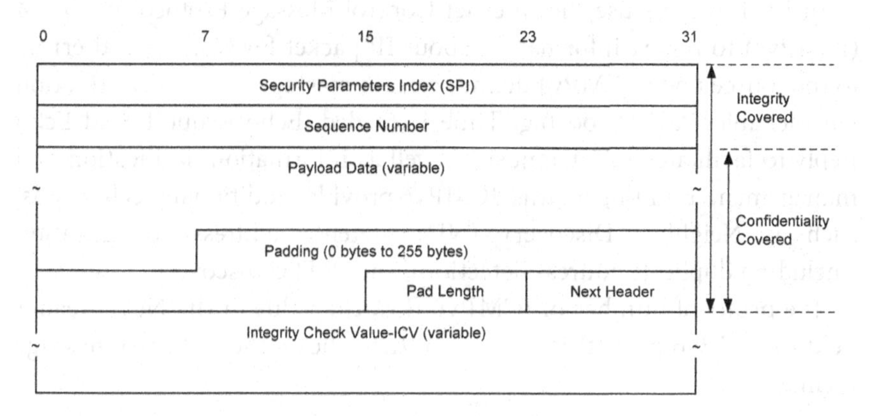

FIGURE A.16 Encapsulating Security Payload header.

TABLE A.10 Main Fields in the Encapsulating Security Payload Header

Field	Length	Description
Security Parameters Index	32 bits	Is used by a receiver to identify the SA to which the packet belongs.[12]
Sequence Number Field	32 bits	Specifies an ID that is incremented by one each time a packet is sent, and cannot decrease. It is bound to an SA to protect against replay attacks. The value is not allowed to increase back round to 0. That is, before the value increases to 2^{32}, the maximum value this field can hold, the two communication parties need to exchange a new key as well as establish and use a new SA.
Payload Data	Variable	Specifies the payload of the original IP packet.
Next Header	8 bits	Specifies the protocol type of the Payload Data.
Integrity Check Value	Variable, an integral multiple of 32 bits	Specifies an ICV. The algorithm for generating the ICV is determined by an SA.

In addition, if there is no data after the IPv6 header or extension headers, the value of the Next Header field needs to be set to 59. If the Payload Length field in the IPv6 header indicates that there is data following the extension header whose Next Header value is 59, the data must be transmitted transparently and cannot be changed during packet forwarding.

A.4 ICMPv6

Internet Control Message Protocol version 6 (ICMPv6) is a basic IPv6 protocol.

In IPv4, routers use the Internet Control Message Protocol version 4 (ICMPv4) to report information about IP packet forwarding and errors to the source node. ICMPv4 defines certain messages such as Destination Unreachable, Packet Too Big, Time Exceeded, Echo Request, and Echo Reply to facilitate fault diagnosis as well as information notification and management. Building on this, ICMPv6 provides additional mechanisms, such as Neighbor Discovery (ND), stateless address configuration (including duplicate address detection), and PMTU discovery.

The protocol number of ICMPv6 (i.e., the value of the Next Header field in an IPv6 packet) is 58. Figure A.17 shows the ICMPv6 message format.

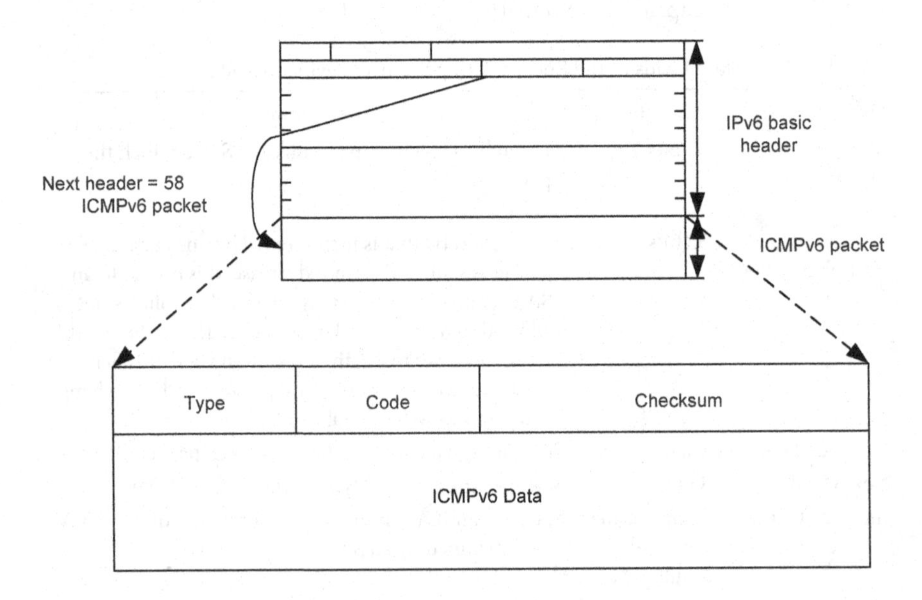

FIGURE A.17 ICMPv6 message format.

An ICMPv6 message contains the following fields:

- Type: message type. Values ranging from 0 to 127 indicate an error message, while those ranging from 128 to 255 indicate an informational message.
- Code: specific message type.
- Checksum: checksum of an ICMPv6 message.

A.4.1 ICMPv6 Error Messages

ICMPv6 error messages are used to report errors that occur during IPv6 packet forwarding, and these messages are classified into four types, as described in Table A.11.

TABLE A.11 ICMPv6 Error Messages

Name	Type Field Value	Function	Code
Destination Unreachable	1	When an IPv6 node forwards an IPv6 packet and discovers the packet's destination address is unreachable, it sends an ICMPv6 Destination Unreachable message to the source node of the packet. The value of the Code field, which is carried in the message, identifies the cause of the error.	Depending on the cause, the value of the Code field can be: • 0: no route to destination • 1: communication with destination administratively prohibited • 2: not assigned • 3: address unreachable • 4: port unreachable
Packet Too Big	2	When an IPv6 node forwards IPv6 packets and discovers the packet's size exceeds the outbound interface's PMTU, it sends an ICMPv6 Packet Too Big message, which carries the outbound interface's PMTU, to the source node of the packet. PMTU discovery is implemented based on Packet Too Big messages.	The value of the Code field is set to 0.

(*Continued*)

TABLE A.11 (*Continued*) ICMPv6 Error Messages

Name	Type Field Value	Function	Code
Time Exceeded	3	When a device receives a packet with a Hop Limit value of 0 or reduces this value to 0 during IPv6 packet transmission, it sends an ICMPv6 Time Exceeded message to the source node of the packet. An ICMPv6 Time Exceeded message is also generated if the fragment reassembly time exceeds the specified period.	Depending on the cause, the value of the Code field can be: • 0: hop limit exceeded in transit • 1: fragment reassembly time exceeded
Parameter Problem	4	When a destination node receives an IPv6 packet, it checks the validity of the packet. If it detects errors, it sends an ICMPv6 Parameter Problem message to the source node of the packet.	Depending on the cause, the value of the Code field can be: • 0: erroneous header field encountered • 1: unrecognized Next Header type encountered • 2: unrecognized IPv6 option encountered

A.4.2 ICMPv6 Informational Messages

ICMPv6 informational messages provide diagnosis and additional host functions such as Multicast Listener Discovery (MLD) and ND. Some of the common ICMPv6 informational messages include Echo Request and Echo Reply, which are both ping messages.

Echo Request message: A source node sends an Echo Request message to a destination node. The values of the Type and Code fields in this message are 128 and 0, respectively.

Echo Reply message: After receiving an Echo Request message, a destination node responds with an Echo Reply message. The values of the Type and Code fields in this message are 129 and 0, respectively.

A.5 PMTU

It is worth noting that in IPv4, oversized packets are fragmented. As such, when a transit node receives a packet exceeding the MTU of its outbound interface, the transit node fragments the packet before forwarding it.

In IPv6, this process is different in the sense that a source node fragments a packet to reduce the pressure on a transit node. To put it more precisely, when an interface on a transit node receives a packet that exceeds the MTU in size, the transit node discards the packet and sends an ICMPv6 Packet Too Big message (containing the MTU of the outbound interface) to the source node. The source node then fragments the packet based on the MTU and resends the fragment packets, which in turn increases traffic overhead. Path MTU Discovery (PMTUD) comes into play within this context as it can dynamically discover the MTU of each link on a transmission path, reducing additional traffic overhead.

PMTUD is implemented using ICMPv6 Packet Too Big messages. In this process, a source node first uses the MTU of its outbound interface as the PMTU and sends a probe packet. If a smaller PMTU exists on the transmission path, the transit node sends a Packet Too Big message containing the MTU of the outbound interface to the source node. After receiving the message, the source node changes the PMTU to the received MTU and sends packets based on the new MTU. This process repeats itself until packets are sent to the destination, after which the source node obtains the PMTU.

Figure A.18 is an example of PMTUD.

Breaking this process down, we can see that packets are transmitted through four links with MTUs of 1,500, 1,500, 1,400, and 1,300 bytes,

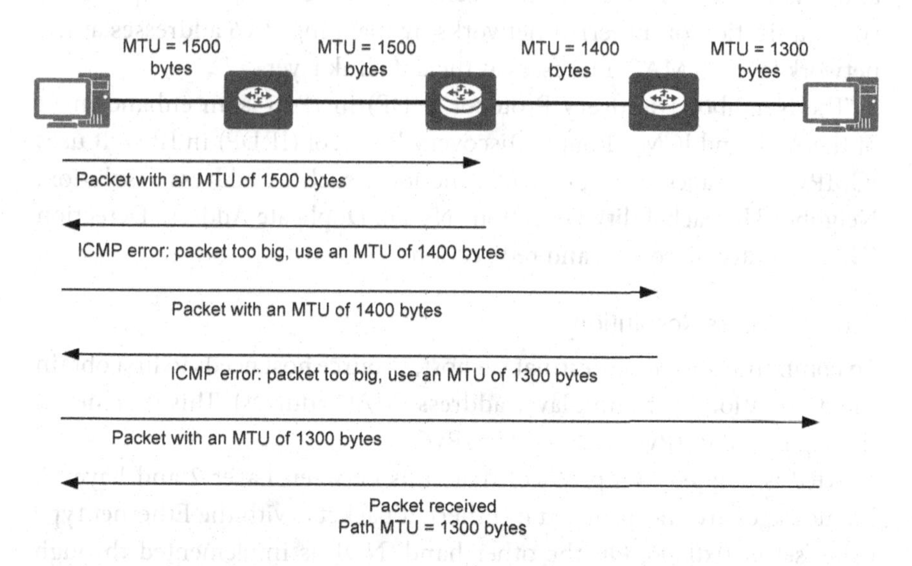

FIGURE A.18 PMTUD example.

respectively. Before sending a packet, the source node fragments it based on a PMTU of 1,500 bytes. Then, when a fragment is sent to the outbound interface with an MTU of 1,400 bytes, the corresponding node returns a Packet Too Big message carrying an MTU of this size. The source node then fragments the packet based on a PMTU of 1,400 bytes and resends a fragment. When the fragment is sent to the outbound interface with an MTU of 1,300 bytes, the corresponding node returns a Packet Too Big message carrying an MTU of this size. The source node then receives the message and fragments the packet based on a PMTU of 1,300 bytes. Through the process we just described, the source node sends the packet to the destination and discovers the PMTU of the path.

Do note that IPv6 requires a minimum MTU of 1,280 bytes at the link layer. In view of this, the PMTU must be greater than 1,280 bytes, and 1,500 bytes is recommended.

A.6 ND

If two hosts need to communicate, the sender must know the receiver's MAC address in addition to its network-layer IPv6 address. This is because IPv6 packets must be encapsulated with MAC addresses before they can be transmitted over the physical network. A mechanism is needed to learn mapping rules which map the receiver's IPv6 address to its MAC address, and this is where ND enters the scene. More precisely, ND implements communication on Ethernet networks by mapping IPv6 addresses at the network layer to MAC addresses at the data link layer.

The Neighbor Discovery Protocol (NDP) in IPv6 is an enhancement of the ARP and ICMP Router Discovery Protocol (IRDP) in IPv4. It uses ICMPv6 messages to implement functions, such as address resolution, Neighbor Unreachability Detection (NUD), Duplicate Address Detection (DAD), router discovery, and redirection.

A.6.1 Address Resolution

To communicate with a destination host, a source host needs to first obtain the destination host's link-layer address (MAC address). This is achieved through ARP in IPv4, and NDP in IPv6.

ARP is defined as a protocol that runs between Layer 2 and Layer 3. Its messages are encapsulated in Ethernet packets, with the Ethernet type value set to 0x0806. On the other hand, NDP is implemented through ICMPv6 messages, with the Ethernet type value set to 0x86DD. To signify that these are ICMPv6 messages, the Next Header value in the IPv6 header is set to 58. NDP is usually regarded as a Layer 3 protocol because it uses ICMPv6-encapsulated messages. Address resolution implemented at Layer 3 includes the following advantages:

- Different Layer 2 links can use the same address resolution protocol.
- Layer 3 security mechanisms can be used to prevent address resolution attacks.
- Request messages can be multicast, reducing loads on Layer 2 networks.

Address resolution uses ICMPv6 Neighbor Solicitation (NS) and Neighbor Advertisement (NA) messages.

- In NS messages, the Type and Code field values are 135 and 0, respectively. These messages are similar to IPv4 ARP Request messages.
- In NA messages, the Type and Code field values are 136 and 0, respectively. These messages are similar to IPv4 ARP Reply messages.

Figure A.19 shows the IPv6 address resolution process.

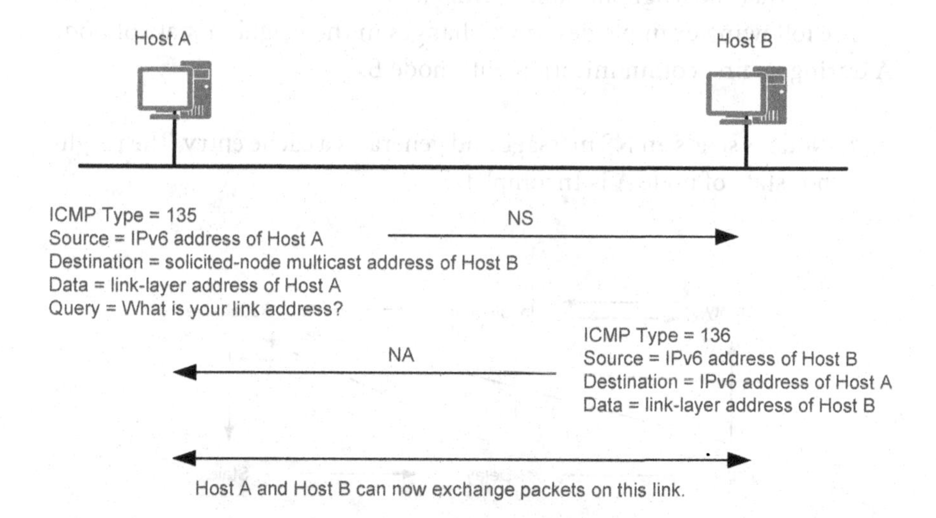

FIGURE A.19 IPv6 address resolution.

Before sending messages to Host B, Host A must obtain Host B's link-layer address. To accomplish this, Host A sends an NS message with its IPv6 address as the source address and the solicited-node multicast address of Host B as the destination address. In the NS message, the Options field carries Host A's link-layer address.

After receiving the NS message, Host B replies with an NA message. In the NA message, the source address is Host B's IPv6 address, the destination IPv6 address is Host A's IPv6 address, and the destination link-layer address is Host A's link-layer address, which is read from the NS message. Be aware that the Options field in the NA message carries Host B's link-layer address, and this address resolution process enables Host A to obtain Host B's link-layer address.

A.6.2 NUD

Hardware faults and the hot swapping of interface cards interrupt communication with neighboring devices. On top of that, communication cannot be restored if the destination of a neighboring device becomes invalid, but it can be restored if the path fails. So the device needs to maintain a neighbor table to monitor the state of each neighboring device.

Five neighbor states exist: Incomplete, Reachable, Stale, Delay, and Probe.

Figure A.20 shows the transition of neighbor states. The Empty state indicates that the neighbor table is empty.

The following example describes changes in the neighbor state of node A during its first communication with node B.

1. Node A sends an NS message and generates a cache entry. The neighbor state of node A is Incomplete.

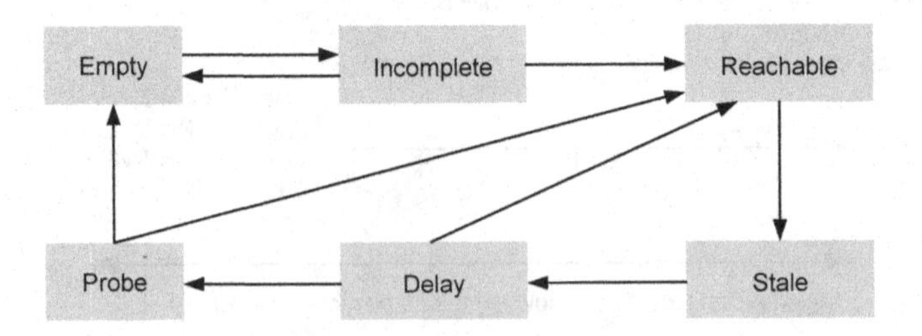

FIGURE A.20 Neighbor state transition.

2. If node B replies with an NA message, the neighbor state of node A changes to Reachable; otherwise, it changes to Empty after a specified period of time, and node A deletes this entry.

3. After the neighbor reachable time expires, the neighbor state changes from Reachable to Stale, indicating that it is not sure if node A can reach node B.

4. If node A receives an unsolicited NA message from node B in the Reachable state, and the link-layer address of node B carried in the message is different from the corresponding address learned by node A, the neighbor state of node A changes to Stale.

5. If node A needs to send data to node B in the Stale state, the neighbor state changes from Stale to Delay and node A sends an NS message.

6. After a specified period of time, the neighbor state changes from Delay to Probe. During this period, if node A receives an NA message, the neighbor state of node A changes to Reachable.

7. Node A sends a specified number of unicast NS messages at the configured interval while node B is in the Probe state. If node A receives an NA message, the neighbor state of node A changes from Probe to Reachable; otherwise, it changes to Empty and node A deletes the entry.

A.6.3 DAD

DAD checks whether an IPv6 unicast address is being used before the address is allocated to an interface, and is required if IPv6 addresses are configured automatically. An IPv6 unicast address that is allocated to an interface but not verified by DAD is called a tentative address. It is worth noting that an interface cannot use such an address for unicast communication but will still join two multicast groups: all-nodes multicast group and this address's solicited-node multicast group.

IPv6 DAD and IPv4 gratuitous ARP share some similarities. For instance, a node sends an NS message that requests the tentative address as the destination address to the solicited-node multicast group. If the node receives an NA message in response, then this means another node is already using the tentative address for communication, and the initial node therefore does not use it.

Figure A.21 shows a DAD example.

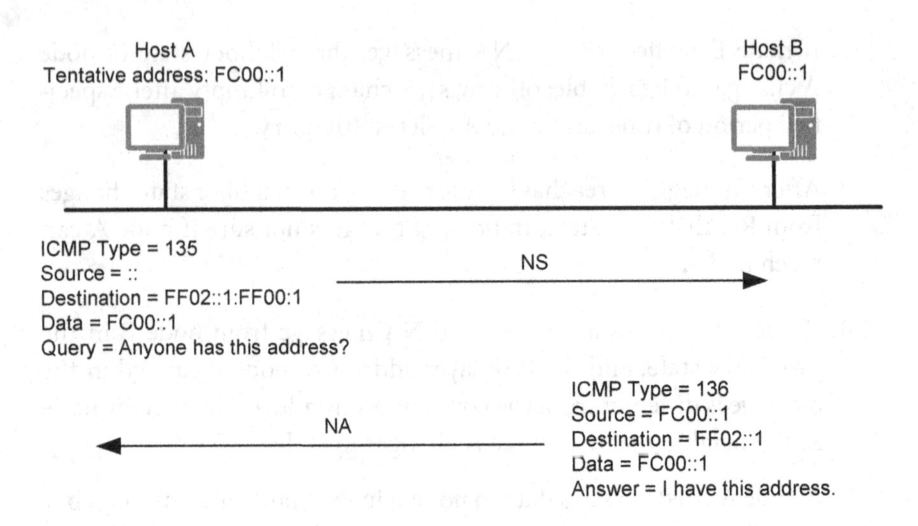

FIGURE A.21 DAD example.

First, FC00::1 is allocated to Host A as a tentative IPv6 address. To check the validity of this address, Host A sends an NS message containing the requested address FC00::1 to the solicited-node multicast group to which FC00::1 belongs. Because FC00::1 is not valid now, :: is used as the source address of the NS message. After receiving the NS message, Host B processes the message as follows:

If FC00::1 is a tentative address of Host B, Host B neither uses this address as an interface address nor sends an NA message.

If FC00::1 is being used on Host B, Host B sends an NA message carrying this address to FF02::1. The aim of this is to enable Host A to find and mark the duplicate tentative address after receiving the message, while also ensuring the address does not take effect on Host A.

A.6.4 Router Discovery

Router discovery is used to locate neighboring devices and learn their address prefixes as well as configuration parameters for address autoconfiguration. As IPv6 supports stateless address autoconfiguration, hosts can obtain network prefixes and automatically generate interface IDs. Router discovery is the basis of IPv6 address autoconfiguration and is implemented through the following two types of messages:

- Router Advertisement (RA) message: Each device periodically multicasts RA messages carrying network prefixes and flags to declare its existence to the hosts and devices on a Layer 2 network. This type of message has a Type field value of 134.
- Router Solicitation (RS) message: After connecting to a network, a host immediately sends an RS message to obtain network prefixes, and devices on the network reply with RA messages. An RS message has a Type field value of 133.

Figure A.22 shows a router discovery example.

A.6.4.1 Address Autoconfiguration

IPv4 uses DHCP to automatically configure IP addresses and default gateways, simplifying network management. That said, IPv6 addresses have an increased length of 128 bits, which means multiple terminal nodes have a more urgent requirement for the automatic configuration function. Fortunately, in addition to stateful address autoconfiguration, IPv6 supports stateless address autoconfiguration, which enables hosts to automatically generate link-local addresses. This process involves hosts automatically configuring global unicast addresses and obtaining additional information based on the prefixes in RA messages.

The stateless address autoconfiguration process is as follows:

1. A host automatically configures a link-local address based on the MAC address of an interface.
2. The host sends an NS message to detect duplicate addresses.

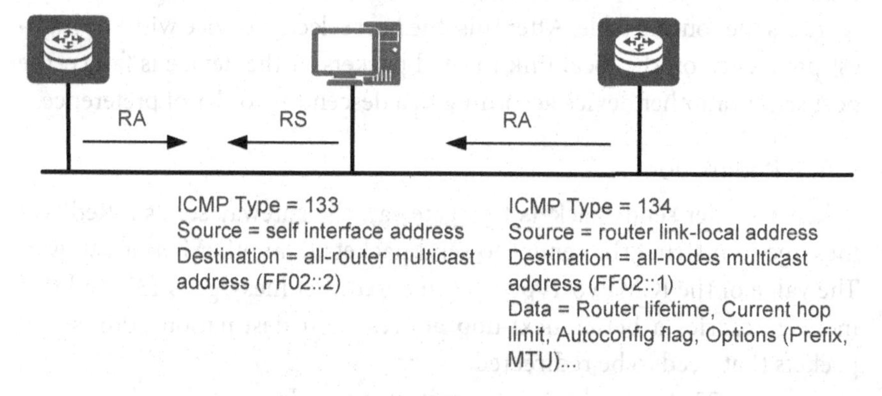

FIGURE A.22 Router discovery example.

3. If an address conflict occurs, the host stops address autoconfiguration, meaning addresses need to be configured manually.
4. If no address conflict occurs, the link-local address takes effect. The host can communicate with the nodes in this local link.
5. The host sends an RS message, and the requested device replies with an RA message (or the host receives the RA messages that devices periodically send).
6. The host obtains the IPv6 address based on the prefix carried in the RA message and the interface ID.

A.6.4.2 Default Router Preference and Route Information Discovery

If a host is connected to multiple devices, it must select a device to forward packets based on their destination addresses. In this case, the devices can advertise the default router preference and specified route information to the host, allowing the host to select a proper forwarding device based on the packets' destination addresses.

The Default Router Preference and Route Information fields are defined in an RA message to enable hosts to select an appropriate forwarding device.

After receiving an RA message carrying the route information, a host updates its routing table. Then, when sending packets to another device, the host searches the routing table and selects an appropriate router to send packets.

From a different perspective, if the RA message received by a host carries the default router preference, the host updates its routing table. Then, when sending packets to another device, if there is no route to be selected, the host searches the routing table. After this, the host selects a device with the highest preference on the local link to send packets. If the device is faulty, the host selects another device according to a descending order of preference.

A.6.5 Redirection

When a sender sends packets to a gateway, the gateway sends a Redirect message to instruct the sender to send packets to another better gateway. The value of the ICMPv6 Type field in a Redirect message is 137, and this message carries a better next-hop address and destination address for packets that need to be redirected.

Figure A.23 shows a packet redirection example.

Assume that Host1 needs to communicate with Host2, and the default gateway of Host1 is R1. When Host1 sends a packet to Host2, the packet is

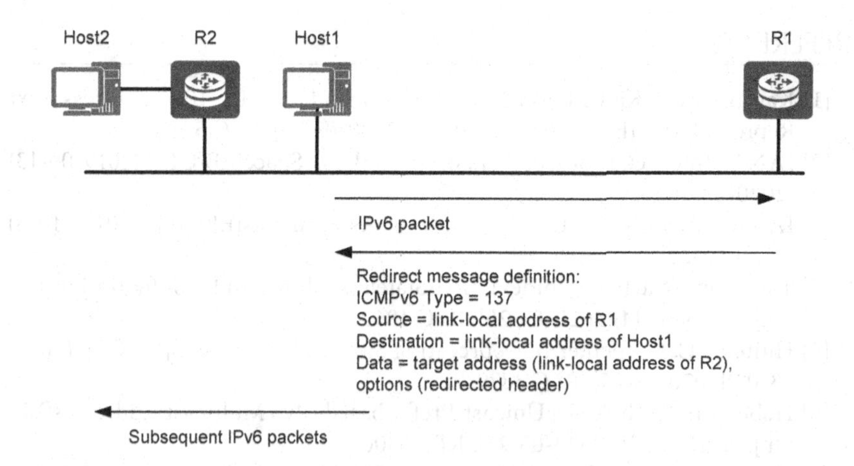

FIGURE A.23 Packet redirection example.

sent to R1. After receiving a packet from Host1, R1 discovers that it would be more fitting to directly send the packet to R2. Following this line of reasoning, R1 sends a Redirect message carrying the destination address of Host2 to Host1, notifying Host1 that R2 is a better next hop. After receiving the Redirect message, Host1 adds a host route to the default routing table, and the packets that are subsequently sent to Host2 will be sent directly to R2.

After receiving a packet, a device sends a Redirect message only in one of the following scenarios:

- The destination address of the packet is not a multicast address.
- The packet is not forwarded to the device through routing.
- After route computation, the outbound interface of the next hop is the interface that receives the packet.
- The device discovers that the optimal next-hop IP address of the packet is on the same network segment as its source IP address.
- After checking the source address of the packet, the device discovers that a neighboring device in the neighbor entries is using this address as a global unicast address or link-local address.

A final point is that, if the communication target is a host, the IPv6 address of the host is used as the destination address of a Redirect message. The link-layer address of the target host is included in an option that the message may contain.

REFERENCES

[1] Kawamura S, Kawashima M. A Recommendation for IPv6 Address Text Representation[EB/OL]. (2020-01-21)[2020-03-25]. RFC 5952.

[2] IANA. Internet Protocol Version 6 Address Space[EB/OL]. (2019-09-13)[2020-03-25].

[3] IANA. IPv6 Global Unicast Address Assignments[EB/OL]. (2019-11-06)[2020-03-25].

[4] Thomson S, Narten T, Jinmei T. IPv6 Stateless Address Autoconfiguration[EB/OL]. (2015-10-14)[2020-03-25]. RFC 4862.

[5] Huitema C, Carpenter B. Deprecating Site Local Addresses[EB/OL]. (2013-03-02)[2020-03-25]. RFC 3879.

[6] Haberman B, Thaler D. Unicast-Prefix-based IPv6 Multicast Addresses[EB/OL]. (2015-10-14)[2020-03-25]. RFC 3306.

[7] Haberman B, Thaler D. IP Version 6 Addressing Architecture[EB/OL]. (2020-01-21)[2020-03-25]. RFC 2373.

[8] Crawford M. Transmission of IPv6 Packets over Ethernet Networks[EB/OL]. (2020-01-21)[2020-03-25]. RFC 2464.

[9] Deering S, Hinden R. Internet Protocol Version 6 (IPv6) Specification[EB/OL]. (2020-02-04)[2020-03-25]. RFC 8200.

[10] Kent S, Seo K. Security Architecture for the Internet Protocol[EB/OL]. (2020-01-21)[2020-03-25]. RFC 4301.

[11] Kent S. IP Authentication Header[EB/OL]. (2020-01-21)[2020-03-25]. RFC 4302.

[12] Kent S. IP Encapsulating Security Payload[EB/OL]. (2020-01-21)[2020-03-25]. RFC 4303.

Appendix B: IS-IS TLVs

Table B.1 describes the mappings between TLVs and sub-TLVs used by IS-IS to generate routes. The SRv6 Locator TLV and SRv6 End SID sub-TLV have been newly defined to support SRv6.

The parent TLVs are described as follows:

- SRv6 Locator TLV: advertises SRv6 locators and the SIDs associated with them.
- Extended IP Reachability TLV: removes the restriction on the metrics carried in the traditional IP reachability TLVs, and carries

TABLE B.1 Mappings between TLVs and Sub-TLVs Used by IS-IS to Generate Routes

Sub-TLV/Parent TLV	27 (SRv6 Locator TLV)	135 (Extended IP Reachability TLV)	235 (MT IP. Reachability TLV)	236 (IPv6 IP. Reachability TLV)	237 (MT IPv6 IP. Reachability TLV)
1 (32-bit Administrative Tag Sub-TLV)	N	Y	Y	Y	Y
2 (64-bit Administrative Tag Sub-TLV)	N	Y	Y	Y	Y
3 (Prefix Segment Identifier)	N	Y	Y	Y	Y
4 (Prefix Attribute Flags)	Y	Y	Y	Y	Y
5 (SRv6 End SID sub-TLV)	Y	N	N	N	N
11 (IPv4 Source Router ID)	Y	Y	Y	Y	Y
12 (IPv6 Source Router ID)	Y	Y	Y	Y	Y

Note: Y indicates that a sub-TLV can be included in the corresponding parent TLV, and N indicates that it cannot be included.

a new flag to indicate that a route is leaked downwards in the IS-IS Level hierarchy.

- MT IP. Reachability TLV: Multi-Topology Reachable IPv4 Prefixes TLV, which extends TLV 135 with an additional 2 bytes to support multi-topology.
- IPv6 IP. Reachability TLV: contains the IPv6 reachability information.
- MT IPv6 IP. Reachability TLV: Multi-Topology Reachable IPv6 Prefixes TLV, which extends TLV 236 with an additional 2 bytes to support multi-topology.

Table B.2 describes the mappings between TLVs and sub-TLVs that are used to indicate adjacencies in IS-IS. The SRv6 End.X SID sub-TLV and SRv6 LAN End.X SID sub-TLV have been newly defined to support SRv6.

The parent TLVs are described as follows:

- Extended IS Reachability TLV: contains information about a series of IS-IS neighbors.
- IS Neighbor Attribute TLV: prevents IS-IS neighbor attribute information from being extended in the Extended IS Reachability TLV.
- L2 Bundle Member Attributes TLV: carries information about L2 bundle member interfaces.
- Inter-AS Reachability Information TLV: carries inter-AS link information.

TABLE B.2 Mappings between TLVs and Sub-TLVs Used to Indicate Adjacencies in IS-IS

Sub-TLV/ Parent TLV	22 (Extended IS Reachability TLV)	23 (IS Neighbor Attribute TLV)	25 (L2 Bundle Member Attributes TLV)	141 (Inter-AS Reachability Information TLV)	222 (MT IS Neighbor TLV)	223 (MT IS Neighbor Attribute TLV)
43 (SRv6 End.X SID sub-TLV)	Y	Y	Y	Y	Y	Y
44 (SRv6 LAN End.X SID sub-TLV)	Y	Y	Y	Y	Y	Y

Note: Y indicates that a sub-TLV can be included in the corresponding parent TLV, and N indicates that it cannot be included.

- MT IS Neighbor TLV: multi-topology IS-IS reachability TLV, which extends TLV 22 with an additional 2 bytes to support multi-topology.
- MT IS Neighbor Attribute TLV: multi-topology IS-IS neighbor attribute TLV, which is defined to prevent the neighbor attribute information from being extended in the MT IS Neighbor TLV.

Appendix C: OSPFv3 TLVs

Table C.1 describes the mappings between OSPFv3 LSAs and TLVs. Table C.2 describes the mappings between TLVs and sub-TLVs. Table C.3 describes the mappings between sub-TLVs and sub-sub-TLVs.

TABLE C.1 Mappings between OSPFv3 LSAs and TLVs

TLV/Corresponding LSA	Router Information LSA	SRv6 Locator LSA	E-Router-LSA
SRv6 Capabilities TLV	Y	N	N
SR Algorithm TLV	Y	N	N
Node MSD TLV	Y	N	N
SRv6 Locator TLV	N	Y	N
Router-Link TLV	N	N	Y

Note: Y indicates that a TLV can be included in the corresponding LSA, and N indicates that it cannot be included.

TABLE C.2 Mappings between TLVs and Sub-TLVs

Sub-TLV/Parent TLV	SRv6 Locator TLV	Router-Link TLV
SRv6 End SID sub-TLV	Y	N
SRv6 End.X SID sub-TLV	N	Y
SRv6 LAN End.X SID sub-TLV	N	Y
Link MSD sub-TLV	N	Y

Note: Y indicates that a sub-TLV can be included in the corresponding parent TLV, and N indicates that it cannot be included.

TABLE C.3 Mappings between Sub-TLVs and Sub-sub-TLVs

Sub-sub-TLV/ Corresponding Sub-TLV	SRv6 End SID Sub-TLV	SRv6 End.X SID Sub-TLV	SRv6 LAN End.X SID Sub-TLV
SRv6 SID Structure sub-sub-TLV	Y	Y	Y

Note: Y indicates that a sub-sub-TLV can be included in the corresponding sub-TLV, and N indicates that it cannot be included.

C.1 SRv6 END.X SID ADVERTISEMENT

OSPFv3 advertises link descriptions through Router-LSAs or Network-LSAs in order to describe adjacencies. Figure C.1 shows the format of the link description advertised in a Router-LSA.

OSPFv3 supports four types of links: P2P, P2MP, broadcast, and NBMA, which are described as follows:

- Each P2P link supports only one OSPFv3 adjacency. Table C.4 describes fields in a P2P link description.
- P2MP links support multiple adjacencies, and each link is described by a P2P link description.
- Broadcast or NBMA links support multiple adjacencies. As shown in Figure C.2, the following nodes of three different roles exist on the broadcast or NBMA links: the DR, BDR, and DR Others. We can also consider a Network-LSA to be a network node, and each adjacency consists of two segments of links. For example, the adjacency from the BDR/DR Other to the DR is described by both the link from the BDR/DR Other to the network node and the link from the network node to the BDR/DR Other.

0	7	15	23	31
Type	0	Metric		
Interface ID				
Neighbor Interface ID				
Neighbor Router ID				

FIGURE C.1 Format of the link description advertised in an OSPFv3 Router-LSA.

TABLE C.4 Fields in a P2P Link Description

Field	Length	Description
Type	8 bits	P2P type. The value is 1.
Metric	16 bits	Metric (or cost) of the link.
Interface ID	32 bits	ID of the local interface.
Neighbor Interface ID	32 bits	ID of the neighbor interface.
Neighbor Router ID	32 bits	Router ID of the neighbor.

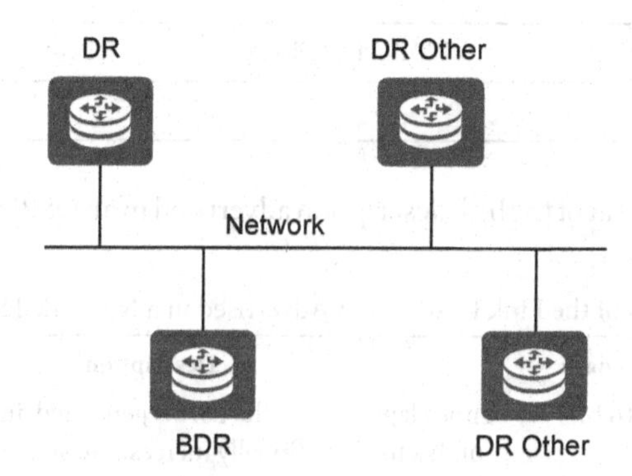

FIGURE C.2 Broadcast or NBMA links.

TABLE C.5 Fields in a Broadcast or NBMA Link Description

Field	Length	Description
Type	8 bits	Transit type. The value is 2.
Metric	16 bits	Metric (or cost) of the link.
Interface ID	32 bits	ID of the local interface.
Neighbor Interface ID	32 bits	ID of the DR interface.
Neighbor Router ID	32 bits	Router ID of the DR.

Table C.5 describes fields in the description of the link from the DR/BDR/DR Other to the network node.

The link from the network node to the DR/BDR/DR Other is described by the multiple Attached Router fields carried in a Network-LSA. Figure C.3 shows the format of the link description advertised in a Network-LSA.

Table C.6 describes fields of the link description advertised in a Network-LSA.

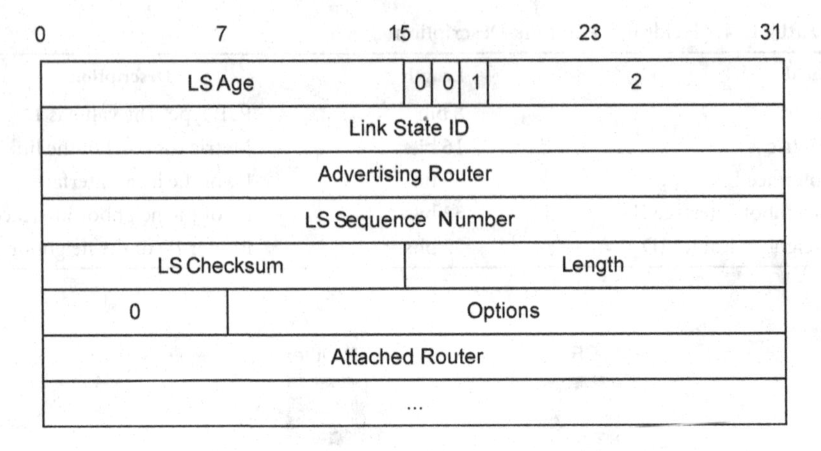

FIGURE C.3 Format of the link description advertised in an OSPFv3 Network-LSA.

TABLE C.6 Fields of the Link Description Advertised in a Network-LSA

Field	Length	Description
LS Age	16 bits	Time elapsed since the LSA is generated, in seconds. The value of this field continually increases regardless of whether the LSA is transmitted over a link or saved in an LSDB.
Link State ID	32 bits	This field, together with the LS Type field, uniquely identifies the LSA in an OSPFv3 area.
Advertising Router	32 bits	Router ID of the device that generates the LSA.
LS Sequence Number	32 bits	Sequence number of the LSA. Neighbors can use this field to identify the latest LSA.
LS Checksum	16 bits	Checksum of all fields in the LSA except the LS Age field.
Length	16 bits	Total length of the LSA, including the LSA header.
Options	24 bits	Optional capabilities supported by the device.
Attached Router	32 bits	Router ID of each device attached to the broadcast or NBMA network, including the router ID of the DR.

C.2 OSPFv3 ROUTER-LINK TLV

The OSPFv3 Router-Link TLV is the parent TLV in an OSPFv3 E-Router-LSA and is used to advertise extended OSPFv3 link information. Figure C.4 shows the OSPFv3 Router-Link TLV format.

Descriptions of the fields in the Router-Link TLV and descriptions of the fields in the link description carried in an OSPFv3 Router-LSA are the same, and the Router-Link TLV can carry the sub-TLVs related to links or adjacencies. For P2P or P2MP links, a link description contains

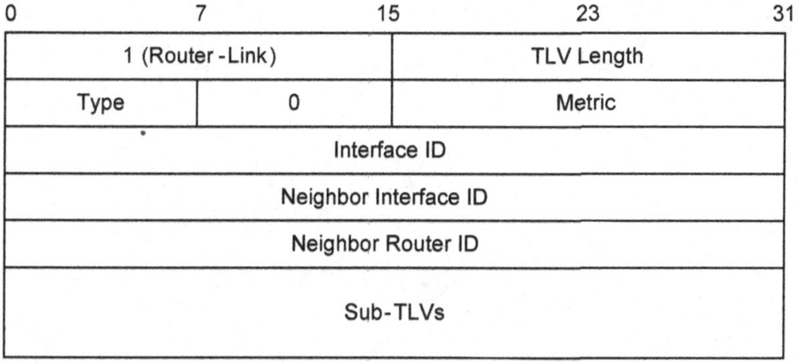

0	7	15	23 31
1 (Router-Link)		TLV Length	
Type	0	Metric	
Interface ID			
Neighbor Interface ID			
Neighbor Router ID			
Sub-TLVs			

FIGURE C.4 OSPFv3 Router-Link TLV.

information about only one adjacency, and the Router-Link TLV can directly carry the sub-TLVs related to the adjacency.

For broadcast or NBMA links, if a link points to the BDR or a DR Other, the Router-Link TLV can carry a sub-TLV associated with an adjacency pointing to the DR, for example, the SRv6 End.X SID sub-TLV; if a link points to the DR, the Router-Link TLV can carry multiple sub-TLVs, such as SRv6 LAN End.X SID sub-TLVs, to describe adjacencies pointing to the BDR and different DR Others.

Appendix D: Acronyms and Abbreviations

Acronym and Abbreviation	Full Name
3G	Third Generation/3rd Generation
3GPP	3rd Generation Partnership Project
4G	Fourth Generation/4th Generation
5G	Fifth Generation/5th Generation
ABR	Area Border Router
AC	Attach Circuit
ACC	Access
ACL	Access Control List
AD	Auto-Discovery
AFI	Address Family Identifier
AGG	Aggregation
AH	Authentication Header
AI	Artificial Intelligence
AMBR	Aggregate Maximum Bit Rate
AMF	Access and Mobility Management Function
API	Application Programming Interface
APN6	Application-aware IPv6 Networking
AR	Augmented Reality
ARP	Address Resolution Protocol
ARPANET	Advanced Research Projects Agency Network
AS	Autonomous System
ASBR	Autonomous System Boundary Router
ASG	Aggregation Site Gateway
ASM	Any-Source Multicast

ATM	Asynchronous Transfer Mode
AUSF	Authentication Server Function
BAC	Bandwidth Admission Control
BBF	Broadband Forum
BBU	Baseband Unit
BD	Bridge Domain
BE	Best Effort
BESS	BGP Enabled ServiceS
BFD	Bidirectional Forwarding Detection
BFER	Bit-Forwarding Egress Router
BFIR	Bit-Forwarding Ingress Router
BGP	Border Gateway Protocol
BGPCC	Central Controlled Border Gateway Protocol
BGP-LS	Border Gateway Protocol - Link State
BIER	Bit Index Explicit Replication
BIERv6	Bit Index Explicit Replication IPv6 Encapsulation
BIFT	Bit Index Forwarding Table
BNG	Broadband Network Gateway
BSL	BitString Length
BSR	BootStrap Router
BTV	Broadcast TV
BUM	Broadcast & Unknown-unicast & Multicast
CAE	Chinese Academy of Engineering
CAICT	China Academy of Information and Communications Technology
CAPEX	Capital Expenditure
CC	Continuity Check
CCN	Content-Centric Network
CDC	Central Data Center
CE	Customer Edge
CGN	Carrier-Grade NAT
CIDR	Classless Inter-Domain Routing
CL	C-SID Left
CLI	Command-Line Interface
CNF	Cloud-native Network Function
COC	Continuation of Compression
CPE	Customer-Premises Equipment
CPU	Central Processing Unit

C-RAN	Cloud RAN
CRH	Compressed Routing Header
CR-LDP	Constraint-based Routing Label Distribution Protocol
CR-LSP	Constraint-based Routed Label Switched Path
CSG	Cell Site Gateway
C-SID	Compressed SID
CSLB	Cloud Service Load Balancer
CSPF	Constrained Shortest Path First
CSQF	Cycle Specified Queuing and Forwarding
CU	Central Unit
CV	Connectivity Verification
DA	Destination Address
DAD	Duplicate Address Detection
DC	Data Center
DCI	Data Center Interconnect
DCN	Data Center Network
DEX	Directly EXport
DF	Designated Forwarder
DHCP	Dynamic Host Configuration Protocol
DIS	Designated Intermediate System
DM	Delay Measurement
DMM	Distributed Mobility Management
DN	Data Network
DNS	Domain Name System
DPI	Deep Packet Inspection
DR	Designated Router
D-RAN	Distributed Radio Access Network
DSCP	Differentiated Services Code Point
DU	Distributed Unit
DVMRP	Distance Vector Multicast Routing Protocol
EAM	Enhanced Alternate Marking
EANTC	European Advanced Networking Test Center
ECMP	Equal-Cost Multiple Path
EDC	Edge Data Center
E-LAN	Ethernet LAN
ENLP	Explicit NULL Label Policy
EOC	End-Of-Carrier
EPC	Evolved Packet Core

EPE	Egress Peer Engineering
ERO	Explicit Route Object
ES	Ethernet Segment
ESI	Ethernet Segment Identifier
ESP	Encapsulating Security Payload
EVI	EVPN Instance
EVPL	Ethernet Virtual Private Line
EVPN	Ethernet Virtual Private Network
FBM	Forwarding Bit Mask
FEC	Forwarding Equivalence Class
FHR	First Hop Router
FM	Fault Management
FMC	Fixed Mobile Convergence
FR	Frame Relay
FRR	Fast Reroute
GNE	Gateway Network Element
GPB	Google Protocol Buffer
G-SID	Generalized SID
G-SRH	Generalized SRH
GTP-U	GPRS Tunneling Protocol-User Plane
GUI	Graphical User Interface
HD	High-Definition
HSB	Hot Standby
IAB	Internet Architecture Board
IANA	Internet Assigned Numbers Authority
IBGP	Internal Border Gateway Protocol
ICMP	Internet Control Message Protocol
ICMPv4	Internet Control Message Protocol version 4
ICMPv6	Internet Control Message Protocol version 6
ICV	Integrity Check Value
IDC	Internet Data Center
IDR	Inter-Domain Routing
IEEE	Institute of Electrical and Electronics Engineers
IESG	Internet Engineering Steering Group
IETF	Internet Engineering Task Force
IFIT	In-situ Flow Information Telemetry
IGMP	Internet Group Management Protocol
IGP	Interior Gateway Protocol

IGPCC	Central Controlled Interior Gateway Protocol
IGW	Internet Gateway
IKE	Internet Key Exchange
IMS	IP Multimedia Subsystem
INT	In-band Network Telemetry
IOAM	In-situ Operations, Administration, and Maintenance
IPFIX	Internet Protocol Flow Information Export
IP FPM	IP Flow Performance Measurement
I-PMSI	Inclusive-Provider Multicast Service Interface
IPS	Intrusion Prevention System
IRB	Integrated Routing and Bridging
IRDP	ICMP Router Discovery Protocol
IS-IS	Intermediate System to Intermediate System
ISP	Internet Service Provider
ITU-T	International Telecommunication Union-Telecommunication Standardization Sector
L2VPN	Layer 2 Virtual Private Network
L3VPN	Layer 3 Virtual Private Network
LANE	Local Area Network Emulation
LDP	Label Distribution Protocol
LFA	Loop-Free Alternate
LFIB	Label Forwarding Information Base
LHR	Last Hop Router
LIR	Leaf Information Required
LM	Loss Measurement
LPM	Longest Prefix Match
LSA	Link State Advertisement
LSDB	Link-State Database
LSM	Label Switched Multicast
LSP	Label Switched Path
LSR	Link State Routing
MAC	Media Access Control
MC	Metro Core
MDT	Multicast Distribution Tree
MEC	Mobile Edge Computing
MLD	Multicast Listener Discovery
MME	Mobility Management Entity
MOSPF	Multicast Open Shortest Path First

MP2MP	Multipoint-to-Multipoint
MP-BGP	Multiprotocol Extensions for BGP
MPLS-TP	Multiprotocol Label Switching-Transport Profile
MRT	Maximally Redundant Trees
MSD	Maximum SID Depth
MT	Multi-Topology
MTP	Motion-to-Photons
MTU	Maximum Transmission Unit
MVPN	Multicast VPN
NA	Neighbor Advertisement
NAAF	Network as a Fabric
NAT	Network Address Translation
ND	Neighbor Discovery
NDN	Named Data Network
NDP	Neighbor Discovery Protocol
NE	Network Element
NETCONF	Network Configuration Protocol
NFV	Network Functions Virtualization
NG MVPN	Next-Generation MVPN
NH	Next Header
NLRI	Network Layer Reachability Information
NS	Neighbor Solicitation
NSH	Network Service Header
NSMF	Network Slice Management Function
NSSA	Not-So-Stubby Area
NUD	Neighbor Unreachability Detection
NVO3	Network Virtualization over Layer 3
O&M	Operations and Maintenance
OAM	Operations, Administration, and Maintenance
OLT	Optical Line Terminal
ONUG	Open Network User Group
OPEX	Operating Expense
OS	Operating System
OTT	Over the Top
P2MP	Point-to-Multipoint
PBR	Policy-Based Routing
PBT	Postcard-Based Telemetry
PBT-I	Postcard-Based Telemetry with Instruction Header

PC	Program Counter
PCC	Path Computation Client
PCE	Path Computation Element
PCECC	Path Computation Element Central Controller
PCEP	Path Computation Element Communication Protocol
PCF	Policy Control Function
PDU	Packet Data Unit
PDU	Protocol Data Unit
PE	Provider Edge
PFL	Pruned-Flood-List
PGW	Provisioning Gateway
PIC	Prefix Independent Convergence
PIM	Protocol Independent Multicast
PLR	Point of Local Repair
PM	Performance Measurement
PMSI	Provider Multicast Service Interface
PMTU	Path Maximum Transmission Unit
PMTUD	Path MTU Discovery
POF	Protocol Oblivious Forwarding
PPSI	Per-Path Service Instruction
PSP	Penultimate Segment Pop of the SRH
PSSI	Per-Segment Service Instruction
PW	Pseudo Wire
QFI	QoS Flow Identifier
R&D	Research and Development
RA	Router Advertisement
RD	Route Distinguisher
RDC	Regional Data Center
RFC	Requirement For Comment
RH0	Routing Type 0
RIB	Routing Information Base
RIP	Routing Information Protocol
RLFA	Remote Loop-Free Alternate
ROI	Return on Investment
RP	Rendezvous Point
RPF	Reverse Path Forwarding
RPT	Rendezvous Point Tree
RQI	Reflective QoS Indicator

RR	Route Reflector
RRO	Reported Route Object
RRU	Remote Radio Unit
RS	Router Solicitation
RSG	Radio Network Controller Site Gateway
RSVP-TE	Resource Reservation Protocol-Traffic Engineering
RT	Route Target
SA	Secure Association
SAFI	Subsequent Address Family Identifier
SBA	Service-Based Architecture
SBFD	Seamless Bidirectional Forwarding Detection
SDH	Synchronous Digital Hierarchy
SDN	Software Defined Networking
SF	Service Function
SFC	Service Function Chain
SFF	Service Function Forwarder
SFP	Service Function Path
SGW	Serving Gateway
SI	Service Index
SID	Segment Identifier
SIGCOMM	Special Interest Group on Data Communication
SLA	Service Level Agreement
SLA	Site-Local Address
SMF	Session Management Function
SP	Service Provider
SPBM	Shortest Path Bridging for MAC
SPF	Shortest Path First
SPI	Service Path Identifier
SPRING	Source Packet Routing in Networking
SPT	Shortest Path Tree
SRGB	Segment Routing Global Block
SRH	Segment Routing Header
SRLG	Shared Risk Link Group
SR-MPLS	Segment Routing over MPLS
SRMS	Segment Routing Mapping Server
SRP	Stateful PCE Request Parameters
SR-TP	Segment Routing-Transport Profile
SRv6	Segment Routing over IPv6

SSM	Source Specific Multicast
STB	Set-Top Box
TAS	Time Aware Shaping
TCAM	Ternary Content Addressable Memory
TDM	Time Division Multiplexing
TE	Traffic Engineering
TEAS	Traffic Engineering Architecture and Signaling
TEDB	Traffic Engineering Database
TIH	Telemetry Information Header
TI-LFA	Topology Independent Loop-Free Alternate
TLV	Type Length Value
T-MPLS	Transfer MPLS
TOR	Top of Rack
TRILL	Transparent Interconnection of Lots of Links
TSN	Time-Sensitive Networking
TTL	Time to Live
TWAMP	Two-Way Active Measurement Protocol
UCMP	Unequal-Cost Multiple Path
UE	User Equipment
UPF	User Plane Function
USD	Ultimate Segment Decapsulation
USP	Ultimate Segment Pop of the SRH
VAS	Value-Added Service
vBNG	Virtual Broadband Network Gateway
VCI	Virtual Channel Identifier
vEPC	Virtualized Evolved Packet Core
VLAN	Virtual Local Area Network
VM	Virtual Machine
VNF	Virtual Network Function
VPC	Virtual Private Cloud
VPI	Virtual Path Identifier
VPLS	Virtual Private LAN Service
VPN	Virtual Private Network
VPWS	Virtual Private Wire Service
VR	Virtual Reality
VRF	Virtual Routing and Forwarding
VTN	Virtual Transport Network
VXLAN	Virtual eXtensible Local Area Network

WAF	Web Application Firewall
XML	Extensible Markup Language
YANG	Yet Another Next Generation
ZTP	Zero Touch Provisioning